PRACTICAL GUIDELINES FOR THE ANALYSIS OF SEAWATER

PRACTICAL GUIDELINES FOR THE ANALYSIS OF SEAWATER

Edited by
Oliver Wurl

Institute of Ocean Sciences
Sidney, British Columbia, Canada

CRC Press
Taylor & Francis Group
Boca Raton London New York

CRC Press is an imprint of the
Taylor & Francis Group, an **informa** business

CRC Press
Taylor & Francis Group
6000 Broken Sound Parkway NW, Suite 300
Boca Raton, FL 33487-2742

First issued in paperback 2019

ISBN-13: 978-1-4200-7306-5 (hbk)
ISBN-13: 978-0-367-38557-6 (pbk)

Library of Congress Cataloging-in-Publication Data

Practical guidelines for the analysis of seawater / editor, Oliver Wurl.
    p. cm.
  Includes bibliographical references and index.
  ISBN 978-1-4200-7306-5 (alk. paper)
  1. Seawater--Analysis. I. Wurl, Oliver, Dr. II. Title.

GC101.2.P73 2009
551.46'6--dc22

2008048755

Visit the Taylor & Francis Web site at
http://www.taylorandfrancis.com

and the CRC Press Web site at
http://www.crcpress.com

Contents

Chapter 10
Karen Helen Wiltshire

Chapter 11
Jacqueline Stefels

Chapter 12
Andrew R. Bowie and Maeve C. Lohan

Chapter 13
Mark Baskaran, Gi-Hoon Hong, and Peter H. Santschi

Chapter 14
Sylvia G. Sander, Keith Hunter, and Russell Frew

Chapter 15
Oliver Wurl

Chapter 16
John L. Zhou and Zulin Zhang

Preface

The ocean is the largest water body on our planet and interacts with the atmosphere and land masses through complex cycles of biogeochemical and hydrological processes. It regulates the climate by the adsorption and transportation of an enormous amount of energy and material, plays a critical role in the hydrological cycle, sustains a beautiful portion of the earth's biodiversity, supplies essential food and mineral sources, and its shorelines offer attractive places for living and recreation. Understanding the chemical composition and processes of the ocean becomes more and more important, because of the major function played by the ocean in regulating changes in the global environment. The science community moves toward a greater awareness and understanding of the ocean's role in global changes such as climate change, invasion of CO_2, eutrophication and decrease of fish stocks. However, to understand oceanic processes a wide range of measurements are required in the vast ocean, from the sea surface to deep-ocean trenches, as well from the tropics to the poles.

Analytical chemistry is a very active and fast-moving field in the science of chemistry today due to advances in microelectronics, computer, and sensor technologies. Despite the development of innovative new analytical techniques for chemical trace element research and greater awareness of quality assurance, today's marine chemists face formidable obstacles to obtain reliable data at ultratrace levels. The aim of the book is to provide a common analytical basis for generating quality-assured and reliable data on chemical parameters in the ocean. It is not attempted to describe the latest innovation of analytical chemistry and its application in the analysis of seawater, but methodologies proved to be reliable and to consistently yield reproducible data in routine work.

The book serves as a source of practical guidelines and know-how in the analysis of seawater, including sampling and storage, description of analytical technique, procedural guidelines, and quality assurance schemes. The book presents the analytical methodologies in a logical manner with step-by-step guidelines that will help the practitioner to implement these methods successfully into his or her laboratory and to apply them quickly and reliably.

After an introductory chapter of a general description of sampling of seawater and its treatments (e.g., filtration and preservation), Chapters 2–6 are dedicated to describe methodologies for the analysis of carbon in seawater, from dissolved organic carbon to complex chromophoric dissolved organic matter. For methodologies of carbon dioxide measurements, the reader is referred to Dickson et al.'s *Guide to Best Practices for Ocean CO₂ Measurements* (PICES, 2007). Chapter 7 describes the analysis of marine gel particles, a relatively new field in chemical oceanography, but it is well known that such particles hold an important function in biogeochemical cycles. The segmented flow analysis of nutrients in seawater has been used for more than four decades and is the subject of Chapter 8, whereas the analytical procedure for organic nitrogen and phosphorous is described in Chapter 9. Many studies in chemical oceanography include the analysis of photo pigments (Chapter 10) due to the impact of primary productivity in many oceanic processes. Chapter 11 deals with analysis of dimethylsulfide produced by phytoplankton communities and well known to impact the climate, being the initial stage in the production of sulfate-containing aerosols. The role of iron in the formation of phytoplankton blooms has been under investigation since the 1990s, and rapid developments in analytical techniques have led to standard procedures, described in Chapter 12. Chapter 13 describes the analytical procedure for radionuclides used as tracer material, an essential tool in studying the dynamic of oceanic processes. Marine chemists have been interested in the distribution of heavy metals for several decades because at elevated levels they cause a wide range of ecotoxicologal effects, but at trace levels some heavy metals take over important biogeochemical functions. The analysis of heavy metals as well their specifications is detailed in

Chapter 14. Finally, Chapters 15 and 16 are the subject of the analysis of various man-made organic contaminants, often present at elevated levels in coastal waters accumulating in marine food webs. Chapter 16 presents suggestions and first steps in the standardization of procedures for the analysis of pharmaceutical compounds in seawater, as concern over such compounds in the marine environment has risen more recently and procedures for routine analysis have not been established yet.

I thank the authors for their enthusiastic cooperation in the preparation of the book. It was a pleasure to work with all of them. The chapters were reviewed by other scientists, whose efforts and time are very much appreciated. I thank CRC Press for giving me the opportunity to publish this book and for guidance at various stages in the process. My work on the book was accomplished while I was a postdoctoral scholar at the Institute of Ocean Sciences, Sidney (Canada); I am most grateful for that scholarship provided by the Deutsche Forschungsgemeinschaft (German Research Foundation). I thank my loving wife, Ching Fen, for her understanding and encouragement at critical stages during the preparation and publication process of the book.

Finally, I hope the book will contribute much in future studies of oceanography and will go some way toward removing some of the mysteries that the ocean still holds for us.

Editor

Oliver Wurl received his BA with a diploma from the Hamburg University of Applied Sciences in 1998. After a 1-year scholarship at the GKSS Research Centre and 2 years' working experience as an application chemist for Continuous Flow Analyzer with Bran+Luebbe GmbH, he began studying the fate and transport mechanisms of organic pollutants in the marine environment of Asia. He received his PhD from the National University of Singapore in 2006. His current research field includes the formation and chemical composition of the sea-surface microlayer and its impact on air-sea gas exchange. Dr. Wurl is currently affiliated with the Institute of Ocean Sciences, British Columbia, Canada, as a postdoctoral researcher through a scholarship provided by the Deutsche Forschungsgemeinschaft (DFG).

Contributors

Alain Aminot
Institut Français de Recherche pour
  L'Exploitation de la Mer
Plouzané, France

Mark Baskaran
Department of Geology
Wayne State University
Detroit, Michigan

Andrew R. Bowie
Antarctic Climate and Ecosystems
  Cooperative Research Centre
University of Tasmania
Tasmania, Australia

Jennifer Cherrier
Florida Agricultural and Mechanical
  University
Environmental Sciences Institute
Tallahassee, Florida

Paula G. Coble
College of Marine Sciences
University of South Florida
St. Petersburg, Florida

Stephen C. Coverly
SEAL Analytical GmbH
Norderstedt, Germany

Thorsten Dittmar
Department of Oceanography
Florida State University
Tallahassee, Florida

Anja Engel
Alfred Wegener Institute for Polar
  and Marine Research
Bremerhaven, Germany

Russell Frew
Department of Chemistry
University of Otago
Dunedin, New Zealand

Laodong Guo
Department of Marine Science
University of Southern Mississippi
Stennis Space Center, Mississippi

Gi-Hoon Hong
Korea Oceanographic Research
  and Development Institute
Ansan, South Korea

Keith Hunter
Department of Chemistry
University of Otago
Dunedin, New Zealand

Gerhard Kattner
Alfred Wegener Institute for Polar
  and Marine Research
Ecological Chemistry
Bremerhaven, Germany

Roger Kérouel
Institut Français de Recherche pour
  L'Exploitation de la Mer
Plouzané, France

Maeve C. Lohan
School of Earth Ocean
  and Environmental Science
University of Plymouth
Devon, United Kingdom

Kai-Uwe Ludwichowski
Alfred Wegener Institute for Polar
  and Marine Research
Bremerhaven, Germany

Norman B. Nelson
Institute for Computational Earth
  System Science
University of California
Santa Barbara, California

Christos Panagiotopoulos
Géochimie et Ecologie Marines (LMGEM)
Université de la Méditerranée
Centre d'Océanologie de Marseille
Marseille, France

Sylvia G. Sander
Department of Chemistry
University of Otago
Dunedin, New Zealand

Peter H. Santschi
Department of Marine Sciences
  and Oceanography
Texas A&M University
Galveston, Texas

Tsai Min Sin
Tropical Marine Science Institute
National University of Singapore
Singapore

Jacqueline Stefels
Laboratory of Plant Physiology
University of Groningen
Haren, The Netherlands

Ming-Yi Sun
Department of Marine Sciences
University of Georgia
Athens, Georgia

Karen Helen Wiltshire
Biologische Anstalt Helgoland
Alfred Wegener Institute for Polar
  and Marine Research
Helgoland, Germany

Oliver Wurl
Centre for Ocean Climate Chemistry
Institute of Ocean Sciences
Sidney, British Columbia, Canada

Zulin Zhang
The Macaulay Institute
Craigiebuckler, Aberdeen, United Kingdom

John L. Zhou
Department of Biology
  and Environmental Science
University of Sussex
Falmer, Brighton, United Kingdom

1 Sampling and Sample Treatments

Oliver Wurl

CONTENTS

1.1 INTRODUCTION

One of the most remarkable achievements in chemical oceanography in recent decades has been the clarification of the distribution of trace levels of biogeochemically active elements, metals, and organic pollutants. This success is attributed not only to the development of sophisticated analytical techniques, but also to the continuous and strenuous efforts of marine chemists to develop clean sampling and noncontaminating treatment techniques for seawater.

The use of inappropriate material or erroneous handling of sampling equipment and treatment leads to an enormous risk of sample contamination and consequently to incorrect data. These errors cannot be corrected afterwards, and sampling, treatment, and storage of samples are very critical steps in the analysis of seawater.

Developments made during the last two decades include the availability of clean sampling devices and laboratory facilities on research vessels (Gustafsson et al., 2005; Helmers, 1994), analytical techniques for shipboard measurements (Achterberg, 2000; Croot and Laan, 2002), and increased awareness of contamination sources associated with sampling and sample treatment by scientists (Hillebrand and Nolting, 1987). However, contamination lurks everywhere, often originating from ship operations and materials in contact with the sample, such as closure mechanisms, sealing, and containers to collect and store samples. Sample handling requires considerable attention from marine analytical chemists through rigorous following of protocols and constant awareness of contamination sources.

The distribution of trace constituents is being affected by the dynamic of oceanic processes, which can greatly disturb the representativeness of samples collected. Physical processes include turbulences, diffusion, advection, and convection of water masses. Chemical reactions can rapidly change concentrations of biogeochemical elements and micropollutants, in particular at boundary layers such as particle surfaces, water-sediment interfaces, and the sea-surface microlayer. Vast communities of microorganisms in the ocean, including phytoplankton, bacteria, protists, and zooplankton, influence the distribution of organic matter, nutrients, and trace metals through uptake and remineralization processes. The dynamic of such processes needs to be addressed in the sampling strategy, and requires a reasonable understanding of oceanography from the marine analytical chemist conducting the sampling.

Overall, the responsibility of the marine analytical chemists conducting the sampling is to ensure that (1) the sample represents the properties of the study area, that is, two samples collected from the same water mass are not discriminable from each other (representativeness), and (2) the sample keeps the properties of interests from the point of collection to the final analytical measurement (stability).

The chapter is divided into five sections, beginning with a discussion on sampling strategies. The second section provides an overview of errors typically occurring during sampling and sample treatments. This is followed by the three main sections, in which selections of sampling, filtration, and sample preservation techniques are discussed. Different techniques for various analytes are briefly described, and more details on sampling and sample handling for individual analytes are provided in Chapters 2–16.

1.2 SAMPLING STRATEGY

A sampling strategy is defined as a procedure for the selection, collection, preservation, transportation, and storage of samples. It also includes the assessment of quality assurance (QA) data, for example, to ensure representativeness of collected samples, to meet required levels of confidence, and to estimate sampling errors (Figure 1.1).

The sampling strategy depends on the study area and the objectives of the investigation. It defines the locations and numbers of stations, vertical resolution, depths and frequency of sampling, and suitable sampling techniques. Even though the sample strategy depends on the objectives of the study, some general rules can be applied for the density of stations and frequency of sampling, as

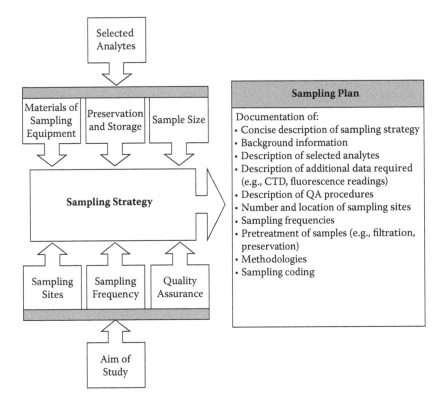

FIGURE 1.1 Elements of a sampling strategy.

pointed out by Capodaglio (1997). For example, in bays and harbors, a high density of stations and frequent sampling are required to account for effects by local inputs and tidal changes. Coastal waters are considerably affected by human activities and experience seasonal changes, and the sampling strategy depends much on hydrological conditions and their variability. Oceanic waters present high horizontal homogeneity, but require sampling with a higher vertical resolution due to the presence of stratified water layers with different properties.

In situ measurement of salinity and temperature gives important and readily accessible information of homogeneity of the water masses within the study area, whereas fluorescence in situ as a proxy for Chl-a provides data about depth and zones of maximum biomass of phytoplankton communities (Capodaglio, 1997).

Standard depths are commonly used to collect oceanographic parameters for global databases, such as World Ocean Database (NOAA) and Joint Global Ocean Flux Study (JGOFS). However, the standard depths are clearly not applicable for studies addressing specific objectives, such as at boundary layers and stratified water masses.

Oceanography is a broad and multidisciplinary field of science, and biologists, chemists, geologists, and physicists often participate together in cruises. The selection of sampling sites and depth resolution depends on several requirements, and compromises need to be made. Consequently, the chemists often share water samples with other scientists onboard, and the sampler device should be checked prior to the cruise to ensure it fulfills the requirements of the trace constituents to be analyzed. Special requests on depth, sampler device, and required data have to be sent well in advance to the chief scientist of the cruise for arrangements and preparations.

The methodologies for sampling, preservation, storage, and analysis that are required in the field should be described as step-by-step procedures and included in the sampling strategy. They should include information on method performance and validation, and requirements for quality assurance.

1.3 SAMPLING AND ANALYTICAL ERRORS

Analytical chemists distinguish between random and systematic errors. Random errors are statistical fluctuations in both directions in the measured data due to the limitations of the analytical instrument. Such errors are also caused by variations in the handling of samples and interferences from the chemistry of the analytical methods itself to the instrumental output. Random errors caused by instrumental noise can be minimized by providing optimal laboratory conditions, including constant temperature and stabilized power supply. Adequate estimates of random errors caused by personnel handling and methodologies can be achieved by participating in intercomparison studies, which include independent analyses of various laboratories using different analytical methods (for example, Bowie et al., 2006). Random errors affect the precision, or reproducibility, of a measurement. Random errors are always existent, but can usually be estimated and minimized through statistical analysis of repeated measurements (see Section 1.4.1.4).

Systematic errors are more serious, not only because they affect the accuracy, that is, the proximity to the true value, but for their detection the true value needs to be known—a most unlikely case in oceanography and other scientific disciplines. Systematic errors may arise during sampling, either through the improper determination of a property of the collected water mass (depth, salinity, temperature) or the use of inappropriate sampler devices causing changes in the analytes' concentrations during their operation. Sampling implies the deployment of alien material to the depth of sample collection, which includes hydrographic wires, container, messengers, sealing, and weights. Such material can cause contamination or adsorptive losses to the analytes. For example, metallic weights and hydrographic wires can cause severe contamination to samples subjected to trace metal analysis. Certain plastic material adsorbs metals and organic compounds, such as pesticides. Hydrographic wires, messengers, weights, and containers are nowadays commercially available made from various materials or are Teflon coated for noncontaminating sampling. Proper selection of sampling equipment and its maintenance can minimize undetectable systematic errors. The research vessel itself can be a source of systematic errors through physical mixing of surface waters to be collected, and continuous contamination of the surrounding waters and sampling devices. Discharge of waste and cooling waters, corrosion processes, leakages, and depositions from exhaust emissions are of most concern. Another type of systematic error occurs with false assumptions made about the accuracy of analytical instruments. In particular, in the computer era, an inviting description in manuals, such as "self-calibrating" or "self-adjusting," lowers the skills required of operators, although the operation of sophisticated instruments still requires well-trained technicians. A simple example of a systematic error is the gravimetric measurements of suspended particulate matter on a self-calibrating but improperly tared microbalance. Systematic errors are often hidden and difficult to detect. However, precautions taken before each step of sampling, and analytical procedures can greatly reduce the risk of the appearance of systematic errors. A conscientious marine analytical chemist carefully considers analytical procedures to be adopted, instruments to be used, and analytical steps to be performed. Systematic errors of instruments can be detected using reference materials with known value (see Section 1.4.1.5). If the known value lies outside the confidence level of repetitive measurements, it is likely that a systematic error occurred. Minor revisions in the design of the experiment can avoid the occurrence of systematic errors. For example, weighing differences removes such errors in gravimetric measurements as described above. Forethoughts of this kind are very valuable.

1.4 METHOD VALIDATION AND STATISTICAL TESTS ON QUALITY ASSURANCE

The concept of method validation and quality assurance (QA) is an inherent element of analytical protocols and has been a concern in laboratory management for a few decades (Keith et al., 1983). Analytical chemists use QA programs to identify unreliable values from data sets and to show attainment of a defined level of statistical control of the analytical methods. The process of method validation directly affects the quality of future data sets, and therefore is a key element in analytical chemistry.

The text presented here gives a brief overview of statistical tests that are most appropriate to problems a marine analytical chemist is faced with in his or her daily work. The statistical tests are described in a way to emphasize their practical aspect rather than to present details of theoretical background. Numerous references on statistics in analytical chemistry exist in the literature, covering the subject in a comprehensive and more mathematical approach (Anderson, 1987; Brereton, 2007).

1.4.1 Method Validation

Method validation in analytical chemistry ensures that the analytes of interests are determined accurately and precisely within acceptable and specified uncertainty to the true value. As the true value of the analyte in collected samples is unknown to the marine analytical chemist, certified reference material (CRM) and statistical tests are used to estimate accuracy and precision. A further objective of method validation is to ensure that the methodology is robust and provides consistent quality-assured data in daily routine work.

1.4.1.1 Selectivity and Specificity

Selectivity is defined as the extent to which an analytical method can quantify a certain analyte in the presence of interferences without diverting from the defined performance criteria. Specificity is the instrumental output for the pure analyte in an aqueous standard solution, that is, specificity = 100% selectivity (Taverniers et al., 2004).

In practice, the instrumental output of the analyte in the presence of all potential sample components is compared to the output of a standard solution containing only the analyte. As seawater has typically a complex matrix, it is often acceptable to consider selected matrix components, which are expected to cause interferences to the highest degree (Thompson et al., 2002), that is, to test pure standard solution against solutions of different salinities, and dissolved and particulate organic carbon contents. Knowledge of the chemistry of the analyte is requisite for the selection of the potential interferences.

1.4.1.2 Linearity and Calibration

The test on the linearity of analytical techniques confirms the concentration range where the analytical output response is linearly proportional to the concentration of the analyte. The test is performed with standard solutions at concentration levels representing 50% to 150% of the expected analyte concentration in real samples. As some analytes can vary widely in their concentration in seawater, the range of linearity test should be extended to ensure that analyte concentrations lie within the tested range. At least five concentration levels are required to detect any diversion from a linear response. A typical approach to estimate the acceptability of a linear range is the examination of the correlation coefficient and y-intercept (a) of the linear regression line for the instrumental output versus analyte concentration. The correlation coefficient r^2 should be typically greater than 0.9990, whereas a coefficient of 1.0000 represents the perfect fit of the data points to the regression line. The intercept a of the regression line with the y-axis should not exceed 5% of the expected analyte concentration in the samples. In Figure 1.2 a procedure for the evaluation of a calibration is outlined to illustrate that examination of correlation coefficient alone can be misleading. The correlation coefficient of 0.9886 of calibration set 2 is close to 0.9990, but the error in the determination of an unknown is X_0 is unacceptably high compared to the good calibration set 1. It can be seen that the relative error and confidence level of slope b and y-intercept a are higher for the second calibration set, and the negative y-intercept causes further reduction in the quality of the calibration. For linearity, the y-residuals should be randomly scattered when plotted against X, that is, analyte concentration. Systematic trends in the scatter plots indicate nonlinearity. Theories of the calculations outlined in Figure 1.2 are given in detail in the literature (Anderson, 1987).

Procedural Steps for Calibration Evaluation

Procedural Step	Calibration Set 1		Calibration Set 2	
	X_1	Y_1	X_2	Y_2
1. Obtain calibration data.	0.25	0.723	12	0.187
$X_{1,2}$ = Concentration of standard solution	0.50	1.323	40	0.723
$Y_{1,2}$ = Output of analytical instrument	0.75	1.954	65	1.275
	1.00	2.613	100	3.201
	1.50	3.921	150	5.289
	2.00	5.100	250	8.453

2. Calculate standard deviations (s_x, s_y) of the differences of the observed pairs.

$$s_x = \sqrt{\frac{\sum X^2 - \left(\sum X^2\right)/n}{n-1}}$$

         0.952   9.318

$$s_y = \sqrt{\frac{\sum Y^2 - \left(\sum X\right)^2/n}{n-1}}$$

         1.647   1.568

3. Calculate the correlation coefficient r^2.

$$r^2 = \left\{ \frac{\sum\left[(X-\bar{X})(Y-\bar{Y})\right]}{\sqrt{\sum(X-\bar{X})^2 \sum(Y-\bar{Y})}} \right\}$$

         0.9997   0.9886

4. Estimate the calibration equation $Y = a + bX$.

$$b = \frac{\sum XY - \left(\sum X \sum Y\right)/n}{\sum X^2 - \left(\sum X\right)^2/n}$$

         2.526   0.037

$$a = \bar{Y} - b\bar{X}$$

         0.079   −0.577

5. Calculate residual Y_R for all Y-values.

		0.013	0.325
$Y_R = Y - \hat{Y}$		−0.019	−0.164
		−0.019	−0.527
		0.008	0.118
		0.053	0.376
		−0.031	−0.120

6. Calculate standard deviation of Y on X ($s_{y/x}$).

$$S_{y/x} = \sqrt{\frac{\sum Y_R^2}{n-2}}$$

         0.034   0.381

FIGURE 1.2 Procedural steps for the statistical evaluation of calibration data.

Procedural Steps for Calibration Evaluation

Procedural Step	Calibration Set 1	Calibration Set 2
7. Calculate error and confidence level in b.		
$S_b = \dfrac{S_{y/x}}{\sqrt{\sum (X - \bar{X})^2}}$	0.023	0.00197
Confidence level: $b \pm ts_b$ with $t = 2.78$ based on 95% confidence level and freedom of $4 = n - 2$	2.526 ± 0.065	0.0366 ± 0.0055
8. Calculate error and confidence level in a.		
$S_a = Sy/x \sqrt{\dfrac{\sum X^2}{n \sum (X - \bar{X})^2}}$	0.027	0.2550
Confidence level: $a \pm ts_a$ with $t = 2.78$ based on 95% confidence level and freedom of $4 = n - 2$	0.079 ± 0.075	-0.5770 ± 0.7089

9. Plot regression line using calibration equation $Y = a + bX$.

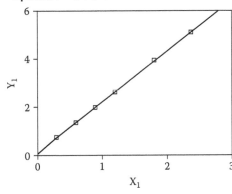

A: Regression line of calibration set 1

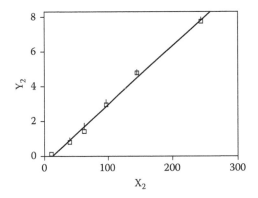

B: Regression line of calibration set 2

10. Plot residual Y (Y_R) against X.

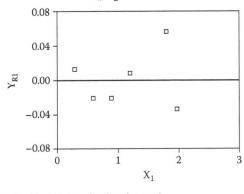

C: Residual Y-plot of calibration set 1

D: Residual Y-plot of calibration set 2

FIGURE 1.2 Procedural steps for the statistical evaluation of calibration data. (Continued)

Procedural steps for calibration evaluation

Procedural Step	Calibration Set 1	Calibration Set 2
11. Calculate error of X_0 estimated from calibration equation.		

X_0 can be calculated from a measured Y_0 using calibration equation (step 4). Error of X_0 is estimated through s_{X_0}.

$$S_{X_0} = \frac{S_{y/x}}{b} \sqrt{1 + \frac{1}{n} + \frac{(Y_0 - \bar{Y})^2}{b^2 \sum (X - \bar{X})^2}}$$

	Calibration Set 1	Calibration Set 2
	$Y_0 = 0.722$	$Y_0 = 0.593$
	$X_0 = 0.255$	$X_0 = 31.967$
	$s_{X_0} = \pm 0.016$	$s_{X_0} = \pm 11.864$

$\bar{x}, \bar{y}$ averaged values of X and Y

^{n} number of calibration points ($n = 6$)

^{b} slope of regression line

^{a} intercept of regression line with y-axis (Fig. A)

^{YR} Y-residuals (Fig. B)

^{Y} Y values on the calculated regression line $Y = a + bX$ (Fig. B)

FIGURE 1.2 Procedural steps for the statistical evaluation of calibration data. (Continued)

Certain analytical techniques have a typical nonlinear response to the analyte concentration, and the calibration of such techniques requires a more careful evaluation, for example, for the electrochemical determination of surfactants in seawater (Ćosović and Vojvodić, 1998).

1.4.1.3 Limit of Detection

The limit of detection (LOD) is defined as the lowest concentration level that is statistically different from a blank within a specified confidence level. The LOD is often the ultimate criterion for the performance of analytical methods presenting the upper limit at which the instrument can differentiate between a signal due to noise and a signal due to low analyte concentration in the sample. The technology is so advanced that the sensitivity of analytical instruments is often not the limiting factor for low LODs, but the level of the analyte and its variability in the blanks. The relationship between instrumental noise and analyte signal is expressed by the average value of the blank ($\bar{c}_{Blank}$) plus three times its standard deviation (SD_{Blank}) (Equation 1.1).

$$LOD = \bar{c}_{Blank} + 3 * SD_{Blank} \tag{1.1}$$

In trace metal analysis, it is acceptable to define the LOD as three times the standard deviation only (Bowie et al., 2006; see Chapter 12). If the blank is relatively high but not very variable, the subtraction of such defined LOD still yields good estimates of the concentration.

To ensure that the concentration of the analyte is significantly higher than that found in blanks, its concentration must be equal to or higher than the LOD. In practice, ten independently prepared blanks are analyzed in the same way as the samples (see Section 1.4.2). The analytical results are averaged and the SD calculated according to Equation 1.2:

$$SD = \sqrt{\frac{\sum (c - \bar{c})^2}{(n - 1)}} \tag{1.2}$$

where c is the concentration of the analyte (i.e., in the blanks), $\bar{c}$ is the averaged value of c, and n is the number of measurements.

However, in the trace analysis of seawater the determination of a representative SD_{Blank} is challenging and can be erroneous due to the presence of sampling errors, as outlined in the discussion above. An alternative approach is to collect low-concentration samples frequently (e.g., deep water) during a cruise, which is substituted for the blank samples.

1.4.1.4 Precision

Analysts differentiate between two types of precision: (1) instrument precision and (2) inter- or intralaboratory precision, also called ruggedness.

Instrument precision is a measure of the scatter in the analytical results obtained from a homogeneous sample. Instrument precision represents the reproducibility of the analytical methodology under the same conditions and environment and is expressed as relative standard deviation (RSD%). In practice, ten aliquots are taken from a homogeneous sample and prepared strictly according to the method description. It is recommended to collect a homogeneous water mass by filling several samplers on a rosette at the same depth. It is important to shake each sampler just prior to withdrawing the aliquots to ensure its homogeneity, as particulate matter can rapidly settle at the bottom. All containers for the storage of aliquots have to be cleaned exactly in the same manner. Furthermore, it is important that the aliquots are prepared on a single day, in the same laboratory, by a single analyst, and analyzed in a single run on the instrument. The SD is calculated according to Equation 1.2 and related to the average concentration of the ten aliquots. RSD% tends to be higher at low concentration ranges (i.e., analyte ratio of 1.00^{-09} or lower), as typically encountered by marine analytical chemists, and a RSD% of 30% has been reported to be acceptable for trace analysis (Table 1.1).

Intralaboratory precision is defined as the reproducibility of an analytical methodology performed by multiple analysts, using diverse instruments, on different days, but in one laboratory. It is important that each analyst prepares new batches of reagents required for the analysis, including calibration standards and QA samples. Interprecision is assessed through participation in community-wide intercomparison studies (Bowie et al., 2006; Landing et al., 1995; Sharp et al., 1995), in which a homogeneous sample is distributed among the participants. All sample treatments, analytical procedures, and results are rigorously documented, and statistically evaluated by the scientific leadership of the intercomparison study. If participation in an intercomparison study for analytes of interest is available, it should be the ultimate aim of a marine analytical chemist to participate and validate his or her analytical methodology.

TABLE 1.1
Acceptable %RSD according to Horwitz Function and AOAC, and Acceptable Recovery Percentage from CRM and Spiked Samples as Function of Analyte Concentrations[*]

Analyte Concentration	Precision		Accuracy
	Horwitz %RSD	AOAC %RSD	Acceptable Recovery Range
100 mg L^{-1}	8	5.3	90–107
10 mg L^{-1}	11.3	7.3	80–110
1 mg L^{-1}	16	11	80–110
100 mg L^{-1}	22.6	15	80–110
10 mg L^{-1}	32	21	60–115
1 mg L^{-1}	45.3	30	40–120

[*] *CRM*, certified reference material.

1.4.1.5 Accuracy

Accuracy describes the correctness of the data, that is, closeness of the measured value to the true value. As the true value is unknown, accuracy is more difficult to assess than precision, but is the most informative and important QA criteria. Certified reference materials (CRMs) are well-characterized standard samples with certified values representing true values. CRMs are analyzed according to the analytical protocol to be validated. Accuracy is expressed as recovery of the certified value. Acceptable recovery ranges are a function of analyte concentration (Table 1.1). CRMs are commercially available for a wide range of analytes in different sample matrices (e.g., seawater, freshwater, sediments, marine organisms). Table 1.2 contains commercially available CRMs for seawater, including some noncertified reference values. Matrix composition of seawater can vary widely in its chemical and biological properties. This means that good results obtained from a

TABLE 1.2

Typical Certified Reference Materials (CRMs) for Various Analytes in Seawater

Code	Parameter	Source	Remark
Dissolved Organic Carbon			
Batch 8/9	DOC	Dr. Hansell, University of Miami, Rosenstiel School of Marine and Atmospheric Science	Website: www.rsmas.miami.edu/groups/biogeochem/CRM.html Oceanic seawater
Dissolved Inorganic Carbon and Alkalinity			
Batch 84		Dr. Dickson, Scripps Institute of Oceanography, San Diego	Website: http://andrew.ucsd.edu/co2qc/index.html Oceanic seawater
Nutrients			
MOOS-1	PO_4, Si, NO_2, NO_3	National Research Council Canada (NRCC)	Oceanic seawater
Batch 8/9	Total nitrogen	University of Miami, Rosenstiel School of Marine and Atmospheric Science	Website: www.rsmas.miami.ed/groups/biogeochem/CRM.html Oceanic seawater
RMNS	PO_4, Si, NO_2, NO_3	Geochemical Research Department, Meteorological Research Institute, Japan	Website: www.mri-jma.go.jp/Dep/ge/RMNScomp.html Oceanic seawater
Trace Metals			
NASS-5	As, Cd, Cr, Co, Cu, Fe, Pb, Mn, Mo, Ni, Se*, U*, V*, Zn	NRCC	Oceanic water
CASS-4	As, Cd, Cr, Co, Cu, Fe, Pb, Mn, Mo, Ni, U*, V, Zn	NRCC	Near-shore seawater
SLEW-3	Ag*, As, Cd, Cr, Co, Cu, Fe, Pb, Mn, Mo, Ni, U*, V, Zn	NRCC	Estuarine water
CRM-403	Hg	Institute for Reference Materials and Measurements (IRMM)	Oceanic seawater
CRM-505	Cd, Cu, Ni, Zn	IRMM	Estuarine water
LGC-6016	Cd, Cu, Mn, Ni, Pb, Zn*	LGC standards	Estuarine water
Radioisotopes[a]			

[a]See Table 14.3.

matrix-matching CRM may not necessarily guarantee trueness of analytical results from unknown samples. However, CRMs represent an efficient tool for the verification of trueness and to assess the performance of a laboratory at any time. Currently available CRMs in seawater are limited, and an alternative approach to test accuracy is to determine the recovery of a known amount of the analyte spiked into a matrix-matching solution. As a matrix-matching solution artificial seawater can be prepared according to Kester et al. (1967). However, artificial seawater is a very simplified matrix for seawater, and a better approach is to use the technique of standard addition. A series of aliquots are taken from homogeneous seawater samples. The aliquots are spiked with varying amounts of the analyte and analyzed in the same way as the bulk sample containing no additional spike. The analytical results minus the analyte concentration of the unspiked bulk sample represent the recovered concentrations. Although with this approach a close match of the matrix is achieved, it should be noted that the behavior of spikes may be different from that of the analyte in the sample. Recovery or spiking studies should be performed for different types of matrices, several examples of each matrix type, and each matrix type at different levels of analyte concentration. CRM or spiked control samples should be analyzed systematically with each batch of samples, preferably at the start, middle, and end of the batch.

Reanalysis of samples using a different analytical technique or laboratories may provide further indication of how close the analytical results are to the true value. It should be noted that using different analytical techniques and interlaboratory comparison increases the confidence for the correctness of the results, but never can replace CRMs as a tool to assess accuracy, as the results obtained from different techniques and laboratories may be similarly biased. Naturally, a conscientious scientist examines thoroughly the plausibility of his or her data in context of the objectives of the study, which provides further confidence in the correctness of the data.

1.4.1.6 Stability and Robustness

Routine measurements with high sample throughput require a stability study of the analytical technique. During stability studies the analyst gains information on the stability of reagents, standards, and sample solutions. Stable solutions allow for delays during instrument breakdowns and overnight analysis using an autosampler. Samples and reagents should be tested over at least 48 hours using aliquots from a homogeneous bulk sample, CRM, or standard solutions. The onset of instability is indicated by an increasing or decreasing trend in the analytical output compared to the output measured at the beginning of the testing period.

Robustness of an analytical technique is defined as its ability to remain unaffected by minor changes in its operation. Such changes include small variations in reagent concentrations and pH values, ambient temperature, and vibrations. A marine analyst should not neglect a study on robustness, and in particular, temperature changes and vibrations on research vessels can cause interferences to analytical detectors used for shipboard measurements. Shipboard preparations of reagents requiring weighing and volumetric equipment can be difficult under rougher sea conditions, leading to variations compared to laboratory-based preparations. An example to test on robustness is to prepare required buffer solutions with adjusted pH values of ±0.2 units compared to the original pH value. The analytical outputs of a single sample using the different buffers should be within the interlaboratory precision to define it as stable. As several parameters can affect the robustness of an analytical technique, factorial design analysis (Brereton, 2007) is a useful tool to investigate which changes in the methodology cause critical influences on the robustness leading to questionable results.

1.4.2 BLANKS

A procedural blank is a sample that is presumed to be free of the analyte, but subjected to the entire analytical procedure in the same way as real samples. An instrument blank represents the analytical signal for the injection of purified and analyte-free water into the instrument. Routine measurements on blanks are an essential part of a QA program to detect contamination sources throughout the analytical procedure. Typical contamination sources are reagents, catalysts, ambient air, labware, and parts of the instrument in contact with the sample. Low blank values are important to achieve LODs for ultratrace analysis, and therefore typically of concern to any marine analyst. The marine analyst is often faced with the challenge to produce and store analyte-free water to be used as a blank sample for reliable determination of LODs (see Section 1.4.1.3). Procedures for the preparation and determination of blanks are given in the following chapters in detail.

1.4.3 DOCUMENTATION OF QA DATA

It is very important for any analytical measurements to be under statistical control to ensure that the results are reliable and within the specified certainty. The most common way is to prepare so-called control charts, which gives the analyst quick access to information about the statistical control of the analytical procedure (Mullins, 1994).

Control charts are plots of multiple data points from QA samples (y-axis) versus time (x-axis) (Figure 1.3). Control charts are often used to visualize and monitor the following data:

- %RSD of repetitive measurements (precision)
- Percentage recovery of CRMs or spiked samples (accuracy)
- Procedural and instrumental blanks (detection limits)
- Calibration data, such as slope (sensitivity)

At the beginning of the control chart, several replicate measurements are carried out in order to determine mean value $\bar{Y}$ and standard deviation (SD). The number of the initial measurements should be $n > 30$ to obtain representative values (Mullins, 1994), preferably under different conditions as defined in the validation process (different batches of reagents, analyzed by multiple

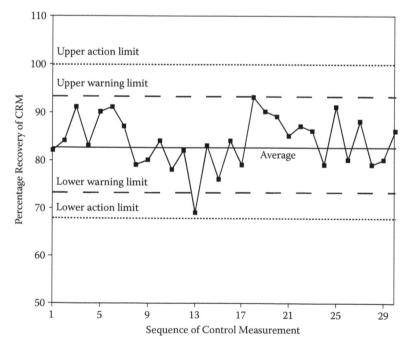

FIGURE 1.3 Control chart for the recovery from an analyte of certified reference material.

analysts). The warning and action limits are defined as $\bar{Y} \pm 2*SD$ and $\bar{Y} \pm 3*SD$, respectively, and are indicated as horizontal lines in the control charts (Figure 1.3). Any measurement or analysis result that falls between the warning limits and the action limits should signal the analyst regarding the need for a careful observation of future QA data. Any measurement that falls outside the action limits requires action to identify and eliminate sources for the inconsistency in the trend of the QA data. Detailed rules of out-of-control values are described by Mullins (1994). Possible sources of such inconsistencies include but are not limited to: contamination in the ultrapure water supply, incomplete digestion, drift in analytical output caused by electronic problems, sudden change in analytical conditions, and new analyst doing the job. As $\pm 2*SD$ and $\pm 3*SD$ are theoretically associated with 95.5% and 99.7% certainty, respectively, an analytical procedure is statistically under control if only 4.5% of the data points fall outside the $\pm 2*SD$ limit and 0.3% fall outside the $\pm 3*SD$ limit; that is, about 1 of 20 and 1 of 1,000 samples, respectively.

1.5 SAMPLING DEVICES

1.5.1 STANDARD WATER SAMPLER

Water samplers should fulfill various requirements. To collect representative samples, it is important that the bottle of the sampler is filled rapidly or its content exchanged at desired depths completely, that is, to have good flushing characteristics. The closing mechanism, triggered by a "messenger" or electronically through a remote control from the surface, has to work reliably under harsh conditions. The bottle has to be sealed completely, triggering the closing mechanism to avoid internal water exchange on the way to the surface. The material the sampler (bottle and closing mechanism) is made of has to be chemically inert not to contaminate the surrounding waters or the collected water during retrieval.

The Nansen sampler was the standard water sampler on research vessels for many decades, but has been replaced with modern water samplers on most research vessels. The original Nansen sampler was fabricated from brass for robustness and safe handling at all water depths, but was not suitable for trace metal analysis for obvious reasons. In an oceanographic cast, several Nansen samplers with reversing thermometers have been attached at certain intervals on a wire and lowered into the water. When the samplers reached the desired depth and were conditioned for several minutes, a messenger was dropped down the wire to trigger the closing mechanism of the uppermost bottle by turning around at 180°. As it turned, the mercury column inside the thermometer was fixed for readings onboard. The same mechanism released a new messenger from the bottle; that messenger now traveled down the wire to close the second bottle, and so on until the last bottle was reached. Nowadays, such oceanographic casts are often replaced with a rosette sampler (see Section 1.5.3). The Knudsen sampler is very similar in design to the Nansen sampler, but has small spring-operated lids sealing the bottle. Its major disadvantage is the poor flushing characteristics through the design of the closing mechanism. Therefore, original Nansen and Knudsen samplers are not recommended for modern oceanographic work; even so, they have been proven to be reliable instruments in the early years of oceanography. The Niskin and Go-Flo sampler are nowadays commonly found on research vessels. The Niskin sampler is based on a principle similar to that of the Nansen sampler, but the major modification is that the top and bottom valves are held open by strings and closed by an internal elastic band or a coated stainless steel spring. The design significantly improved the flushing characteristic of the bottle. Later, an external closing mechanism for the Niskin bottle was developed (Niskin et al., 1973). As no turning over is required to close the Niskin bottle, large-volume Niskin samplers are easy to handle and operate (available up to a volume of 30 L), whereas Nansen bottles had a limited volume of up to 1.7 L. Niskin bottles are usually made of PVC and suitable for most oceanographic work, including the analysis of trace metals after cleaning and proper flushing. Typical Niskin bottles have glued-PVC mounts to hold the closing and sealing mechanism, which can become brittle and prone to breakage in cold water. Advanced designs substitute a rugged Ti base and Delrin mount blocks for the glued-PVC components (Model 110, SeaBird) (Figure 1.4a). The Go-Flo bottle is tripped in the same way as the Niskin bottle, but the main difference in its operation is that the sampler is sealed by two large PVC ball valves (GeneralOceanics) (Figure 1.4b). These valves are set in a way that the sampler passes through the air-water interface sealed, opens automatically at a depth of 10 m, and is finally closed at the desired depth through the messenger. The design avoids serious contamination when the sampler passes through surface films, that is, sea-surface microlayer, which are often enriched in organic matter, trace metal, and particulates. The so-called Free Flow Water Sampler (Hydrobios) (Figure 1.4c) is similar in design to the Niskin bottle, but no cone or ball valve hinders the flow through the bottle, therefore offering optimal flushing characteristics. The LIMNOS sampler (Figure 1.4d) is a surface sampler (down to depths of 100 m) consisting of two to four 500 mL glass bottles, which are opened at the desired depth. The advantage is that the bottles can be used as storage bottles and therefore avoid possible contamination during transfer.

Specially designed seawater intake systems are reported to pump water directly in clean rooms located on deck (Gustafsson et al., 2005). Such intake systems are often located at the prow of the ship to collect surface waters from depths of several meters. Therefore, such a sampling system is suitable for the study of horizontal distribution of constituents in surface waters, but not to investigate vertical profiles. A more recent development is the Lamont Pumping SeaSoar (LPS) (Figure 1.5), a combination measurement and sampling platform towed by a research ship at speeds of 6–7 knots (Hales and Takahashi, 2002). The system allows not only measurement of a suite of oceanographic parameters with in situ sensors, but also collection of seawater from a depth down to 200 m through a 750 m tube (5/16 in. inner diameter) to a shipboard laboratory for chemical analyses. The LPS has been successfully tested during the Joint Global Ocean Flux Study (JGOFS) in the Ross Sea.

1.5.2 WATER SAMPLER FOR TRACE CONSTITUENTS

In oceanographic work, the water sampler often has to be modified to meet special requirements for the analysis of trace constituents. Comparison studies of modified and unmodified Niskin and Go-Flo bottles showed that the modifications were necessary to collect uncontaminated samples for trace metal analysis (Berman et al., 1983; Bewers and Windom, 1982). Such modifications include the use of easily cleaned Teflon-coated bottles, the replacement of all O-rings and seals by equivalents made of silicone, the replacement of internal stainless steel springs with silicone tubing or Teflon-coated springs (Niskin bottle), and the replacement of drain cocks by ones made of Teflon. The Go-Flo sampler is the preferred sampling device for trace metals, as it avoids contamination with metals often enriched in the sea-surface microlayer. It has been reported that the interior of standard PVC Go-Flo bottles was sprayed with a Teflon coating to minimize contamination and adsorption effects (De Baar et al., 2008). The WATES sampler (Warnemünder Teflon Schöpfer) (Brügmann et al., 1987) passes the sea-surface microlayer in a closed position, and enclosed water is only in contact with Teflon. Mercury is easily lost in standard PVC bottles through adsorption processes on the wall, and therefore sampling requires such devices made of Teflon. For the determination of trace levels of nutrients, standard Niskin and Go-Flo bottles are usually employed for water collection. As for metal analysis, the Niskin and Go-Flo bottles have to be scrubbed with an acid solution and thoroughly rinsed with ultrapure water prior to water collection.

For trace organic contaminants, the sampling is challenging, as high volumes of seawater are needed (10–400 L) to preconcentrate and detect the contaminants. Teflon-membrane pumps driven by compressed air were used for the collection of surface waters onboard ships (IOC, 1993). For deeper waters, a large stainless steel sampler with a capacity of 400 L has been deployed (IOC, 1993), but it is difficult to clean such devices in an appropriate way, that is, solvent rinsed. The KISP pump (Figure 1.6) has been specially designed to filter and extract contaminants in situ down to depths of 6,000 m (Petrick et al., 1996). The device consists of pump, filter holder, adsorbent cartridges, battery, and electronic unit to control the operation. The filtered water flows through

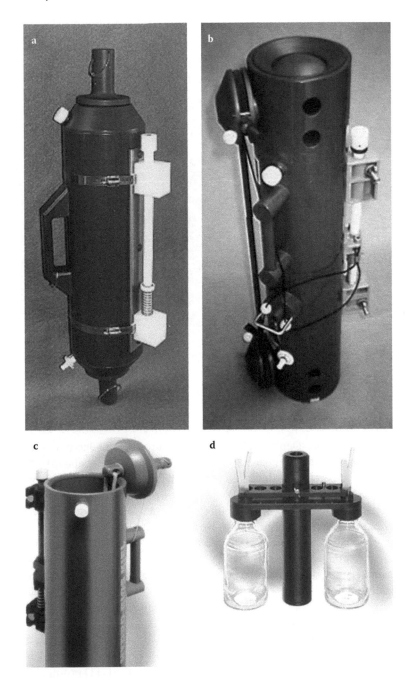

FIGURE 1.4 (a) Niskin bottle (Model SBE 110, Sea-Bird Electronics, Inc., Bellevue, Washington). (b) Nono-metallic Go-Flo bottle (Model 1080, General Oceanics, Inc., Miami, Florida). (c) Free Flow bottle without cone or ball valve for free flushing (Hydro-Bios GmbH, Kiel, Germany). (d) LIMNOS surface water sampler (Hydro-Bios GmbH).

adsorbent cartridges to trap the contaminants from the dissolved phase of the sample. The cartridges can be easily kept at low temperature for storage until the contaminants can be eluted for further processing in the laboratory. Kelly et al. (1993) describe a similar pumping device to filter and extract organochlorine compounds from coastal seawater onboard.

FIGURE 1.5 The towed instrument SeaSoar modified to an online sampling platform. (Photo kindly provided by Dr. Burke Hales, Oregon State University.)

1.5.3 CTD PROFILERS AND ROSETTE SYSTEMS

The Conductivity-Temperature-Depth (CTD) sensor measures the conductivity and temperature at the depth where the instrument is situated. The data are sent directly onboard via computer. The density of the water at a certain depth can be calculated from its salinity (i.e., conductivity), temperature, and the pressure (i.e., depth). A CTD can produce a continuous reading of temperature and conductivity as functions of depth at a rate of up to thirty samples per second, a vast improvement over the twelve data points produced by the twelve Nansen or Niskin bottles that could be used on a single hydrographic cast. However, reverse thermometers are still used together with CTD for comparison and calibration purposes.

The rosette sampler consists of a circular frame holding six to thirty-six Niskin or Go-Flo bottles with a CTD usually mounted underneath or in the center (Figure 1.7). Bottle sizes may vary from 1.2 to 30 L capacity. The deck command unit of the rosette sampler allows the control of the closing mechanism of the bottles electronically with a remote. This means that the sample depths do not have to be set before the bottles are lowered, as for the classical hydrographic cast. As the rosette is lowered and data are received from the CTD, the operator can look for stratified water layers of particular interest and take water samples at any depth based on the CTD profile. Rosettes with aluminum and titanium frames can be deployed down to depths of 6,800 and 10,500 m, respectively. A rosette for ultraclean sampling has been successfully deployed recently to collect samples for trace metals and isotopes (De Baar et al., 2008). The maximum vertical resolution for a typical rosette is 5 m. A recent development is the PUMP-CTD (Strady et al., 2008), which allows water sampling for trace metal with a vertical resolution of 1 m. With the PUMP-CTD, the water is pumped directly onboard into the clean laboratory through a nylon hose, which limits the application to a depth down to 350 m.

A standard deployment of a rosette/CTD, depending on water depth, requires 2 to 5 hours of station time, which is when a ship remains stationary at a given location. The order of sample collection from the Niskin/Go-Flo bottles is important to minimize artifacts in changes of concentrations

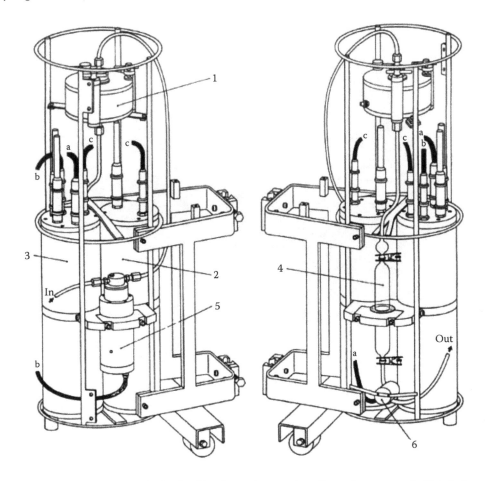

FIGURE 1.6 KISP in situ pump for in situ filtration and extraction of organic compounds (Scholz Ingenieur Büro, Fockbek, Germany). Filter holder (1), battery (2), electronic unit (3), resin cartridge (4), pump (5), and flowmeter (6).

during sampling. Typically, samples for gases are collected first, followed by samples subjected to nutrients, trace metal, and organic carbon analysis, and salinity samples are often collected last.

1.5.4 SEA-SURFACE MICROLAYER (SML) SAMPLER

The SML (uppermost 30–1,000 μm layer) forms the boundary layer interface between the atmosphere and ocean. The collection of water from this very thin surface layer is very difficult and challenging, and therefore sampling procedures are outlined here in more detail.

The SML is enriched in naturally occurring surface-active organic compounds, including carbohydrates, proteins, and lipids, giving it a distinct chemical composition. The SML has been of much interest to oceanographers, earth scientists, and meteorologists, as it represents a critical interface for biological, chemical, and physical processes between the ocean and atmosphere. Recent studies using in situ microelectrodes suggested that the actual thickness of the SML (γ_{SML}) is about 50–60 μm based on the sudden change of physicochemical properties of the water (Zhang et al., 2003).

The first attempts to collect such thin layers of water from the ocean's surface dated back to the 1960s and 1970s (Garrett, 1965; Harvey, 1966; Harvey and Burzell, 1972). In the past 40 years, several techniques for sampling the SML have been deployed, and the advantages and disadvantages

FIGURE 1.7 Rosette samplers with (a) twelve Niskin bottles (Sea-Bird Electronics, Inc., Bellevue, Washington) and (b) six Free Flow bottles (Hydro-Bios GmbH, Kiel, Germany).

of twenty-one different SML sampling techniques were reviewed by Hühnerfuss (1981). The metal screen (Garrett, 1965), glass plate sampler (Harvey and Burzell, 1972), and rotating drum sampler (Harvey, 1966) are the most widely used SML sampling techniques and are described in the following sections. However, collection of SML samples of an acceptable integrity and reproducibility remains a challenge. Reasons for this are:

1. The period of sampling may be excessive in order to collect a sufficient volume of SML sample for analysis at trace levels.
2. The SML is physically, chemically, and biologically very heterogeneous; that is, the characteristics and concentrations of materials in the SML may change rapidly during sampling.
3. Operation of sample techniques requires experience to collect samples with high reproducibility. For example, a relative standard deviation (RSD) of 12% in the collected volume of SML has been reported for a group of ten scientists using a screen sampling technique (Knap et al., 1986).

1.5.4.1 General Remarks on SML Sampling

SML sampling should be performed from a smaller boat or a fixed platform. The boat should move slowly forward while collecting the samples from the bow. A bulbous bow may negatively affect the sampling by creating turbulence and consequently altering the characteristic of the SML. For the same reason, sampling should never be performed from the stern of the boat, unless anchored. Sampling from a platform should be performed from the upcurrent corner or edge, so that ambient water is constantly flowing toward the sampling site and interferences (i.e., contamination, restricted water flow, etc.) from the platform are minimized. Samples collected from the bulk water underneath the SML (typically at a depth of 1 m) can be easily collected through a 12 V DC pump and suitable tubing material. It is not recommended to collect bulk water samples by opening a sample container at the desired depth by hand, as pushing the bottle through the SML can pull down dissolved and particulate fractions from the SML into the bulk water.

Wind speed and direction can greatly affect the presence and distribution of the SML and should be frequently recorded during the sampling period. Furthermore, wind speed and water temperature affect the thickness of collected SML using the glass plate/drum technique (Carlson, 1982). Rainfall events can cause a quick accumulation of material, such as particulates and pollutants (Wurl and Obbard, 2005; Lim et al., 2007), and it is recommended to record such events and its occurring intensity several hours before and during sampling.

Natural slicks, that is, a visible sea-surface pattern in which capillary ripples are absent, are SMLs with an intensive enrichment of surface-active material enabling the dampening of ripples through high surface tensions. The coverage of such slicks of the water surface should be recorded and estimated through photographs and visual observations. The slick is usually lighter in appearance than the rippled water, but may be seen as a darker zone when viewed toward the sun. In the presence of slicks, sampling should be performed either outside or within slicks with an appropriate record.

Sampling parameters, such as collection area of sampler, sample volume, sampling time, number of dips (screen and glass plate sampler), and rotating speed of drum (drum sampler) need to be recorded for the calculation of the thickness of the collected SML (see Sections 1.5.4.2 to 1.5.4.4).

1.5.4.2   Screen Sampler

1.5.4.2.1   Design and Characteristics

A metal screen dipped repeatedly in the water and slowly withdrawn collects SML samples of a thickness of typical 150–400 µm. Garrett (1965) determined experimentally that a SML of a thickness of 150 µm can be collected with a 16-mesh stainless steel screen and 60% open space made from 0.14 mm diameter wire, and is therefore recommended, as it is the thinnest layer to be collected with a screen. A smaller mesh size of 4 caused the collected water film to rupture after withdrawal. The thickness γ_{SML} depends primarily on the wire diameter and opening space (Carlson, 1982). Carlson (1982) used screens with a wire diameter of 0.4 mm (80% open space) and 0.55 mm (52% open space) to collect SML samples with a thickness of 222 ± 8 µm and 465 ± 4 µm, respectively. Fine-mesh sizes, that is, large wire diameter and small open spacing, decrease the sampling efficiency of the screen and are not recommended. For fine meshes, a larger wire area is available for adsorption processes, while less open spacing exists for the entrapment of water. Plastic screens can be used for study on trace metals and inorganic constituents in the SML. The screen (typically 50 cm × 65 cm) is held in a frame with a bent handle made from noncontaminating material (Figure 1.8).

The thickness γ_{SML} collected by a screen is independent of the wind speed (i.e., wave state), water temperature, and salinity (Carlson, 1982). However, Falkowska (1999) reported that the γ_{SML} is influenced by the wind speed using screen samplers. It was supposed that the increased advective transport of organic matter and its accumulation in the SML caused the increase of the γ_{SML} with the wind speed. In all cases, it is recommended to record the prevailing wind speeds during collection of SML samples. It should be noted that in the literature the screen sampler is often recommended for the advantage of the higher sampling rate, that is, higher volume collected per dip compared to a glass plate sampler (Falkowska, 1999; Momzikoff et al., 2004; Garcia-Flor, 2005). However, the screen sampler collects a substantial amount of subsurface waters based on the actual thickness of about 50–60 µm (Zhang et al., 2003), and therefore dilutes the collected SML samples, leading to increased sampling rates.

1.5.4.2.2   Procedure and Handling

In use, the precleaned screen is dipped vertically and slowly through the water surface and positioned parallel to the water surface (Figure 1.8). The sampler is held in the position for about

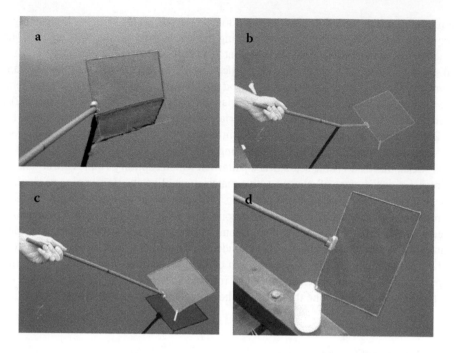

FIGURE 1.8 Sea-surface microlayer sampling with a screen sampler: (a) immersion into water, (b) parallel orientation to water surface, (c) withdrawal, and (d) dripping sample in container.

5 seconds, slowly withdrawn, and replaced in the water. This step should be repeated several times prior to each sampling to condition the screen wires. As the initial adsorption of surface-active material deactivates the sampling efficiency of the screen by 25%, the conditional step is important for accurate and effective collection of the SML. The sampling proceeds as described above, but after withdrawal the sampler is slanted toward one of the rear corners to allow the drainage of the entrapped water into a clean sample container. The screen drains typically within 45 to 60 seconds. The procedure is repeated until the desired volume has been collected. However, it is recommended to limit the volume to 1 L to avoid long sampling periods.

The following equation is used to calculate the average thickness of the collected SML:

$$\gamma_{SML} = \frac{V_{Sample} * 10^4}{N * A_{Screen} \cdot 2}$$

where γ_{SML} is the average thickness of collected SML (µm), V_{Sample} the volume of sample (mL), N the number of dips, and A_{Screen} the surface area of the screen (cm²).

1.5.4.3 Glass Plate Sampler

1.5.4.3.1 Design and Characteristics

The glass plate sampler (Harvey and Burzell, 1972) collects SML samples with thicknesses of about 50–60 µm by repeatedly dipping and withdrawing, in a manner similar to that for the screen sampler. The glass plate is designed for easy and safe handling, typically 30 cm × 30 cm in size, 4 mm thick, and with an attached handle made of rigid plastic material. The adhering water is

FIGURE 1.9 Sea-surface microlayer sampling with a glass plate. (a) Immersion and withdrawal of glass plate. (b) Collecting sample into container.

scraped off with a noncontaminating wiper into a clean sample container (Figure 1.9). A typical characteristic of a glass plate sampler is that the thickness γ_{SML} is associated with the withdrawal rates. Harvey and Burzell (1972) used a withdrawal rate of 20 cm s^{-1}, and the layer collected can be thicker than 100 µm (Carlson, 1982). Carlson (1982) suggested withdrawing the glass plate at a rate of about 5 cm s^{-1} to collect a thinner layer of about 50 µm from clean, nonslicked surfaces, which is consistent with the in situ measured SML thickness (Zhang et al., 2003). However, the thickness γ_{SML} increased with the wind speed from 33 µm (wind speed: <1 m s^{-1}) to 62 µm (wind speed: 8 m s^{-1}) (Carlson, 1982). Furthermore, Carlson (1982) reported that the thicknesses γ_{SML} decreased with increasing water temperature, presumably due to decreasing viscosity. Therefore, a careful record of wind speeds and water temperature is required, in particular when seasonal processes in the SML are studied.

1.5.4.3.2 Procedure and Handling

The precleaned glass plate is dipped vertically into the water and withdrawn at a rate of 5 cm s^{-1} without collecting the adhering water (Figure 1.9). This step should be repeated several times to condition the glass plate. For collection, the glass plate is dipped and withdrawn in the same way. Then the adhering water is removed from both sides of the plate with the wiper and drained into the sample container. The collection is easier using a wide-neck bottle or a small funnel.

The collected thickness γ_{SML} in micrometers is determined with the following equation:

$$\gamma_{SML} = \frac{V_{Sample} * 10^4}{N * A_{Screen} . 2}$$

where V_{Sample} is the volume of sample (mL), N the number of dips, and A_{Screen} the surface area of the glass plate (cm^2). The glass plate should be consistently dipped into the water to a certain depth to ensure accurate determination of the thickness γ_{SML}. For example, a mark just underneath the attachment of the handle to the glass plate can be used as an upper limit.

1.5.4.4   Rotating Drum Sampler

1.5.4.4.1   Design and Characteristics

The screen and glass plate samplers are suitable to collect small-volume samples of up to about 500 to 1,000 mL, as many repeated dips are required to collect large volumes. A rotating drum sampler using ceramic (Harvey, 1966), Teflon-coated (Hardy et al., 1988), or glass drums (Carlson et al., 1988) has been designed to collect large volumes of SML of up to 20 L h^{-1}. The drum is rotated through a DC motor at rotation speeds of about 8–10 rpm. The drum (typically 30–40 cm in diameter and 50 cm in length) is partly immersed in the water, and the water is drawn on to the ascending side of the rotating drum and then scraped off by an inclined wiper located on the downward side for collection. The drum sampler has been designed to operate by remote control via navigation systems, like the SESAMO (Caccia et al., 2005) (Figure 1.10a). Alternatively, a drum sampler can be attached to smaller research boats through a beam and suspension springs (Figure 1.10b). The tension of suspension is adjustable through the length of attachment to stabilize the sampler during operation. It is advantageous to use a hollow glass cylinder and to orient the rotation axis parallel to the travel direction, rather than perpendicular (Carlson et al., 1988). With this design, accumulation of material in front of the drum is avoided through the small forward cross section. Drum samplers are often equipped with in situ sensors, such as fluorescence and UV detectors, for continuous measurements of SML properties. Special construction may be required for the analysis of trace constituents. For example, the drum sampler OceanSwab (Wurl and Obbard, 2005) (Figure 1.10b) was designed for analysis of trace organic pollutants, where only stainless steel, anodized aluminum, glass, and Teflon materials were used for construction. An alternative technique is to replace the drum by multiple rotating disks as developed by Magnus Eek (SOLAS, 2005). The main advantage of the rotating disk sampler is less disturbance of the water surface as the sampling vessel moves up and down with the waves, and such a sampler has been deployed for collecting oceanic SML samples in rougher sea conditions (Figure 1.10c).

Overall, the drum sampler can be considered as the state-of-the-art SML sampling device, but questions remain for all sampling techniques as to whether water adhering to the drum dilutes the SML sample and whether there are compromises to sample integrity from device fabrication materials.

1.5.4.4.2   Procedure and Handling

Drum samplers are often self-designed and fabricated equipment. The operation and handling may differ among the designs. In general, the drum sampler is launched into the water according to the available equipment onboard the vessel. The launching of a drum sampler was practical from a small landing craft workboat used as a research boat in the coastal waters of Singapore (Figure 1.10b). After launching and getting the operation ready to start, the rotation speed is set to about 8–10 rpm to be consistent with the withdrawal rates of glass plate sampling. The drum should run for about 10 minutes to condition all material in contact with the sample prior to starting the actual collection. Sample containers are loaded on the drum sampler, and an automatic forwarding tray holding a number of containers is useful for the continuous collection of discrete water samples. Binoculars are helpful for observing the operation of the sampler and controlling navigation. During the operation, the water surface should be observed for the appearance of slicks and bands of accumulated debris, waste, and oil, in particular in coastal waters. Crossing slicks and bands may result in poor representative samples.

FIGURE 1.10 (a) Remote-controlled glass drum sampler SESAMO (photo kindly provided by Massimo Caccia, CNR-ISSIA U.O. Genova.) (b) Glass drum sampler. OceanSwab attached to a small research vessel. (c) Remote-controlled glass disk sampler SKIMMER (photo kindly provided by Svein Vagle, Institute of Ocean Sciences, Sidney, British Columbia.)

1.6 FILTRATION OF SEAWATER

Filtration is often required to discriminate between constituents in the dissolved and particulate phases in seawater. Filtration of oceanic waters may not be necessary, as the concentration of suspended particulate material (SPM) is generally very low except in surface waters at times of high production, that is, phytoplankton blooms. Coastal and estuarine waters, where the concentration of suspended material can be several orders higher, are invariably subjected to filtration processes. Traditionally, the dissolved phase has been defined as the fraction of seawater that passes through a filter with a pore size of 0.45 μm, for operational reasons in the earlier years of oceanography. Nowadays, it is known that the boundary between dissolved and particulate fractions is a continuum, and the traditionally defined term *dissolved* contains a large fraction of colloidal material (Koike et al., 1990; Wells and Goldberg, 1993). The colloidal material has a typical size range of 1 to 200 nm, and therefore easily passes through a 0.45 μm filter. Correctly, the term *dissolved* is defined as the sum of "truly" dissolved and colloidal fractions. Colloidal material can make up a large fraction of the dissolved material, and given its tremendous surface area and the importance of surface reactions in marine systems, it is indubitable that colloids play a very important role in the cycling of trace elements and contaminants. In the mid-1990s, efforts were made in intercomparison studies on cross-filtration techniques to separate colloids from the truly dissolved phase (special issue 55 in *Marine Chemistry*, Vols. 1–2, 1996), but it is still very challenging to separate these two fractions as outlined below.

Nevertheless, the separation of dissolved and particulate fractions through filters with defined pore sizes is still the most commonly used technique in oceanography for practical reasons. There are several reasons to filter seawater samples as soon after collection as possible: (1) to obtain important information about the fractionation of trace elements and contaminants, (2) to remove solids from solution that can interfere with various detectors in analytical systems, and (3) to avoid leaching of analytes from particles into solution during sample preservation and storage (Kremling and Brügmann, 1999). However, the marine analyst should be aware of potential contamination of the samples during the filtration process. Exposure to air, filter material, glass apparatus, and pumps can irreversibly cause contamination of the sample and therefore lead to diversions from the true value of the sample. Recommendations are given in the following sections for suitable techniques, materials, and equipment to be used to minimize the risk of contamination, but it is strongly advised to check blank samples regularly during filtration.

1.6.1 PRESSURE FILTRATION

Filtration is carried out under a pressure of pure air or inert gas. To avoid changes in the redox state of the sample, an inert gas such as nitrogen has to be used to apply the pressure. This is of particular importance for the collection samples subjected to trace metal analysis to prevent precipitation of metal hydroxides (Kremling and Brügmann, 1999). The marine chemist has to select the material and size of the filter holder according to the analytes and required sample volume. Filter holders are available in various plastic materials (Teflon, polypropylene, polycarbonate) and stainless steel with silicone O-rings in sizes between 25 and 293 mm. The largest sizes are required to filter high volumes (5–100 L) of water, depending on the concentration of SPM. Silicone O-rings can be replaced with Teflon or Viton O-rings, but reduce the maximum pressure to be applied. Several types of filter holders have an air vent valve to prevent pressure increases due to filter clogging.

It is an advantage if the water sampler itself can be pressurized and the filter holder directly connected. The sample can then be forced into the receiving bottle while passing through the filter, thus minimizing the exposure of the sample to the air, and therefore potential contaminations. The off-line filtration system consisted of cylinders (typically 250–1,000 mL) screwed directly onto the top of the filter supports. The cylinder is filled with the sample, a pressure adapter is screwed onto the top, and the sample is filtered into the bottle for sample storage by applying pressure. Several such

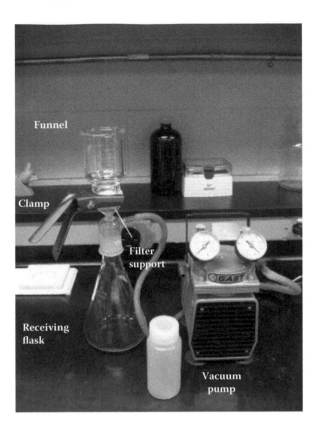

FIGURE 1.11 Typical setup for vacuum filtration of seawater.

units can be placed in a frame to process replicate samples. With both approaches, three fractions of 50 mL of sample are used to rinse the filtration unit, filter, and receiving sample bottle prior to collecting the filtered sample.

1.6.2 Vacuum Filtration

Filtration is carried out by suction with a vacuum pump. A vacuum filtration unit consists of a receiving flask with a hose connector (typically 500–2,000 mL), filter support with glass frit, funnel (typically 250–1,000 mL), and a clamp to attach the funnel on the filter support (Figure 1.11). As for pressure filtration, the vacuum pressure should be minimal, with a reasonable flow rate, to avoid cell rupture and leaching of dissolved material on the filter. Typical vacuum pressure is below $26*10^3$ Pa (200 mmHg). The first fractions of filtered sample are used for repetitive rinsing of the filtration unit.

1.6.2.1 Filter Types

Ideal filters have to fulfill a number of requirements, described in Kremling and Brügmann (1999), which include:

- Uniform and reproducible pore size
- High filtration rate without clogging easily
- Quick equilibration with the surrounding atmosphere for reliable gravimetric measurements

TABLE 1.3
Various Filter Materials and Their Characteristics

Filter Material	Manufacturer[a]	Typical Pore Size [µm][b]	Typical diameter	Characteristics
Glass fiber	MI, WH	0.7, 1.0, 1.2, 1.6, 2.7	25, 47, 90, 142, 293	Free of organic binder
Polycarbonate	MI, ST, SA,	0.01[ST], 0.05[ST], 0.22, 0.04	25, 47	Absolute pore size and density, superior strength
Polypropylene	ST, WH	0.1, 0.2, 0.45	25, 47	Broad chemical compatibility
Polyethersulfone	MI, ST, SA	0.1[SA], 0.22, 0.45	25, 47	Low extractables, superior strength
Nylon	ST, WH	0.1, 0.2, 0.45	25, 47	Low extractables, superior strength
Mixed cellulose esters	MI, ST, WH	0.1[SA], 0.2, 0.45	25, 47	Uniform pore structure
Cellulose acetate	ST, SA, WH	0.2, 0.45	25, 47	Uniform pore structure
Cellulose nitrate	SA, WH	0.1[WH], 0.2[WH], 0.45	25, 47	Superior strength
Teflon, hydrophilic	MI	0.1, 0.2, 0.45	25, 47	Chemically inert
Teflon, hydrophobic	MI, ST, SA	0.22, 0.45	25, 47	Hydrophobic, chemically inert
Silver	MI, ST	0.2[ST], 0.45	25	Bacteriostatic, nonadsorptive
Alumina	WH	0.02, 0.1, 0.2	25, 47	No lateral crossovers between pores, i.e., precise cutoffs

[a] MI, Millipore; ST, SterilTech; SA, Sartorius; WH, Whatman. When superscript, the sizes are manufactured by only the indicated company.
[b] For glass fiber filter, particle retention size.

- Retaining of particles on the filter surface to support microscopic and spectroscopic observations
- Not adsorbing dissolved constituents to be measured in the filtrate
- Free of significant amounts of constituents to be measured
- Providing reasonable mechanical strength and integrity

Kremling and Brügmann (1999) emphasized that none of the existing filter material fulfilled all of the requirements, and this has not changed during the last decade. The marine chemist needs to make compromises, and the highest priority is often given to potential contamination risk and adsorption losses through the filter material. A reasonable filtration rate is desirable without quick clogging of the filter, which can occur by processing high productive waters, that is, water samples collected during phytoplankton blooms or from coastal and estuarine areas.

Various commercial filter materials are listed in Table 1.3. It is always practical if the filter material can be easily precleaned according to the sample protocol. Chemical compatibilities of filter materials with various chemical solutions are provided in Appendix B (Table B.2). Glass fiber filters (GFFs) are the best choice for the determination of dissolved and particulate organic carbon (DOC/POC), nitrogen (DON/PON), and phosphorous (DOP/POP). GFFs have a poor uniform pore size, but they can be easily cleaned by baking at high temperatures (typically 450°C) for several hours, to produce low blanks for these elements. For the same reason, and because they provide good flow rates for high-volume samples, GFFs are commonly used to separate dissolved and particulate organic contaminants. GFFs are the classical filter material for the determination of chlorophyll pigments, although recent studies showed that polycarbonate filters are a good alternative (Knefelkamp et al., 2007). GFFs are also suitable for the filtration of nutrient samples except for silica, in which polycarbonate filters are mostly used. Due to their low metal concentration and easiness to clean with acid solutions, polycarbonate filters are widely applied for trace metal analysis. However, these membranes are not suitable for Hg analysis due to adsorption and contamination problems. Membranes made of Teflon-like materials minimize such problems in the trace analysis of Hg.

1.6.3 CROSS-FLOW FILTRATION (CFF)

In pressure or vacuum filtration, fluid flow is perpendicular to the filter surface and the filter rapidly becomes clogged with particles. In cross-flow filtration (CFF), called ultrafiltration (see Chapter 14), fluid flows parallel to the filter surface and particles become more concentrated as filtrate leaves through the filter's pores. CFF is the only practical technique that allows the separation of collodial from the truly dissolved phase by filtration of 10–100 L of seawater. The CFF technique originates from the industrial application with the objective to recover certain products from a concentrated stream. In contrast, the marine chemist's objective is to achieve an accurate and precise size separation in very diluted streams with commercial CFF systems designed for industrial applications. Laboratory-scaled systems are now available (e.g., Osmonics HEPA, Millipore PELLICON) for easier handling. During an intercomparison study on CFF systems, it was shown that significant differences in the quantity and quality of collected collodial material occurred among the participants (Buesseler, 1996), which has been well documented in several papers in special issue 55 of *Marine Chemistry*. The differences have been attributed to the operator protocols and different designs and operational factors of the CFF systems used, such as flow rate, transmembrane pressure, and membrane characteristics. For example, scanning electron microscopy (SEM) combined with atomic force microscopy (AFM) revealed colloids smaller than 50 nm in all fractions, indicating that CFF fractionation is not fully quantitative (Doucet et al., 2005). Furthermore, permeate concentration of DOC increased with the concentration factor, that is, the ratio of initial sample volume to final retentate volume (Guo et al., 2000; Larsson et al., 2002). Further studies on the calibration of CFF systems are needed for a better understanding and standardization of them and their operation. More details on CFFs are given in Chapters 6 and 13.

1.7 SAMPLE PRESERVATION AND STORAGE

Bacteria and micro- and nanoplanktons remain in filtered samples and continue to alter the concentration of bioreactive elements and organic matter through digestion and excretion. Ideally samples should be measured in situ, or at least shipboard immediately after collection. However, required instrumentation is often not available on research vessels through technical and logistical constraints. Shipboard measurements in a routine manner have been reported for nutrients (Raimbault et al., 1999), dissolved organic carbon (DOC) (Alvarez-Salgado and Miller, 1998), and trace metals (Achterberg, 2000; Wurl et al., 2001). Nevertheless, the marine analytical chemist needs to have a protocol in hand to preserve samples in case of circumstances leading to a delay in analysis. For example, shortage of personnel and laboratory space, breakdown of instruments, and rough weather are common situations at sea preventing the prompt processing of collected samples. An experienced analytical chemist keeps part of a sample as a backup in case a reanalysis is required at a later time, which makes preservation necessary. Another reason for sample preservation is the absence of clean laboratories on research vessels, as sample processing and analysis under normal conditions means higher risk of contamination. Preservation protocols for various analytes have been reported in the literature, but unfortunately not always with consistent results and conclusions. Furthermore, each class of constituents (or even individual analytes) with their own chemistry require storage under different conditions. The complex interaction of different analytes with various materials needs to be carefully considered for long-term storage (>2 weeks). For example, nutrients are best stored in polypropylene (PP) bottles, whereas such containers might be contaminating to samples subjected to the analysis of DOC. Overall, no single and universal preservation protocol can be recommended. Here general preservation techniques are reviewed and detailed protocols can be found in the following chapters.

1.7.1 NUTRIENTS

Nutrients are subject to rapid changes in their concentration within a few hours in unpreserved samples. Traditionally, $HgCl_2$ has been used for decades to stop biological activity, and therefore to preserve nutrients in seawater samples (Kirkwood, 1992). Poisoning with 105 mg L^{-1} of $HgCl_2$ preserved nutrients (except ammonia) in filtered seawater samples for up to 2 years (Kattner, 1999). Kirkwood (1992) used a five times lower concentration of $HgCl_2$ and reported effective preservation at room temperature for 1 year. Concern has been reported that $HgCl_2$ at a concentration of >59 mg L^{-1} causes interferences in the measurement of nitrate using copperized cadmium reductors (Kirkwood, 1992). Furthermore, $HgCl_2$ has lost its status as a standard preservation technique over the years due to the environmental impacts of mercury. Quick freezing in liquid nitrogen and storage at –20°C has been reported to preserve nutrients in filtered samples for several months (Gardolinski et al., 2001; Macdonald and McLaughlin, 1982; Venrick and Hayward, 1985). It is essential that the process of freezing and thawing is quick to prevent losses. Silicate tends to polymerize in low-salinity samples during the freezing process, and storage of separate subsamples for silicate analysis is necessary under acidic conditions (pH = 2.5) (Venrick and Hayward, 1985). However, freezing under acidic conditions or acidification without freezing was unsatisfactory for nitrate and phosphate in the same study. Good agreement of measurements with those of autoclaved seawater samples has been reported after a storage time of 4 months (Aminot and Kérouel, 1995). The same authors have tested successfully heat treatment and pasteurization for the preservation of nutrients (Aminot and Kérouel, 1997). However, an airtight storage container is of utmost importance to avoid losses through evaporation, in particular for ammonia. Unpreserved coastal and oceanic seawater samples subjected to nutrients analysis can be stored in the dark at 4°C for up to 2 and 8 hours, respectively.

1.7.2 TRACE METALS

The levels of trace metals are extremely low in seawater, and even minor losses or sample contamination may have a great influence on the result, making "clean" sample preservation critical. It was recognized in the 1960s and 1970s that losses of trace metals from seawater samples are caused by adsorption on container walls, and acidifying is an effective approach to minimize such losses for long-term storage (1–2 years). Acidifying with HNO_3 to pH < 1.5 has been adopted as a standard preservation method (Subramanian et al., 1978; Berman et al. 1983). However, proper selection of the type of container is requisite for the storage of seawater subjected to trace metal analysis, and Nalgene (fluorinated ethylene-polypropylen) (Cuculiv and Branica, 1996) and polyethylene bottles (Berman et al., 1983) have been reported to be the best material. Exceptional samples for the analysis of mercury need to be stored in Teflon or quartz glass bottles. To avoid contamination of samples, a clean laboratory (Helmers, 1994) and ultrapure HNO_3 are required, for example, ULTREX II (J.T. Baker) and ULTRAPUR (Merck).

Chelating resins have been used to separate trace metals from seawater shipboard with subsequent elution in onshore laboratories (e.g., Chelex 100, XAD 4, NTA Superflow). The resin cartridges can be easily stored onboard at <4°C. However, disadvantages include the necessary adjustment of the pH of the sample for the uptake of metals on resin material. The optimal pH range varies among the metals of common interest and between different resin materials (Lohan et al., 2005). Furthermore, incomplete recovery for some metals during elution can lead to underestimations of the results (Hirata et al., 2003; Ramesh et al., 2002).

1.7.3 ORGANIC MATTER

When immediate analysis is not possible the samples are acidified with H_3PO_4 shortly after collection. High-purity concentrated H_3PO_4 is used to adjust the pH value of the samples to pH = 2–3 (about 4 µl concentrated H_3PO_4/mL sample). This pH range is sufficient to eliminate biological activity and convenient, as this pH adjustment is required for removal of inorganic carbon from the sample prior to analysis. Deep ocean samples can be stored after acidification in sealed glass ampoules (precombusted) at 4°C for 1 year (Hansell, 2005), but in general should be analyzed as soon as possible, in particular samples from surface waters. A pH of <2 is not recommended, as strong acidification may lead to the enhanced production of volatile products from organic matter and precipitation of macromolecules. Storage at –20°C in sealed glass ampoules has been reported (Pakulski and Benner, 1994), but since analytical procedures require acidification, it is advisable to preserve the samples with the addition of H_3PO_4.

Carbohydrates in seawater samples have been preserved with $HgCl_2$ (Ahel et al., 2005; Ittekot et al., 1981) or stored in precombusted and sealed glass vials at –20°C (Pakulski and Benner, 1994). However, Hiroshi (1978) reported that carbohydrates in seawater could only be preserved in unused PP containers at –20°C for 23 days, but not in used PP or glass bottles because of adsorption processes on the container walls. Samples subjected to the analysis of amino acids were also stored frozen without defined duration of storage (Horiuchi et al., 2004; Kuznetsova et al., 2005) or treated with $HgCl_2$ (Svensson et al., 2004). Samples for the analysis of individual carbohydrates and amino acid enantiomers cannot be acidified, as hydrolysis causes changes in these molecules.

Most bioaccumulative organic pollutants occur at trace concentrations in seawater, and their analysis often requires the processing of several hundreds of liter of seawater, including in situ filtration and adsorption on an adsorbent. Adsorbed on such material is often the only way to store and analyze such a great volume of oceanic water for longer periods. Dachs and Bayona (1997) compared C_{18} extraction disks and Amberlite XAD-2 as adsorbents for the sorption and recovery of dissolved n-alkanes, polyaromatic hydrocarbons (PAHs), and polychlorinated biphenyls (PCBs). Polyurethane foam (PUF) plugs have also been used to extract PCBs from seawater (Gustafsson et al., 2005). Preparation of analyte-free adsorption material, its breakthrough volume, and complete recovery of analyte are of concern (see Chapter 15), but it often provides the only way to store organic pollutants extracted from seawater for a longer period (typically at < 4°C), that is, to conduct the elution and analysis in a clean environment in the home-based laboratory. Petrick et al. (1996) developed an in situ sampler for the filtration and extraction of organic pollutants on XAD-2 cartridges at the depth of collection (maximum depth, 6,000 m), and it can be considered state-of-the-art sampling of organic pollutants from seawater (see Figure 1.6).

REFERENCES

Achterberg, E. P. 2000. Automated techniques for real-time shipboard determination of dissolved trace metals in marine surface waters. *International Journal of Environment and Pollution* 13:249–61.

Ahel, M., N. Tepic, and S. Terzic. 2005. Spatial and temporal variability of carbohydrates in the northern Adriatic—A possible link to mucilage events. *Science of the Total Environment* 353:139–50.

Alvarez-Salgado, X. A., and A. E. J. Miller. 1998. Simultaneous determination of dissolved organic carbon and total dissolved nitrogen in seawater by high temperature catalytic oxidation: Conditions for precise shipboard measurements. *Marine Chemistry* 62:325–33.

Aminot, A., and R. Kérouel. 1995. Reference material for nutrients in seawater: Stability of nitrate, nitrite, ammonia and phosphate in autoclaved samples. *Marine Chemistry* 49:221–32.

Aminot, A., and R. Kérouel. 1997. Assessment of heat treatment for nutrient preservation in seawater samples. *Analytica Chimica Acta* 351:299–309.

Anderson, R. L. 1987. *Practical statistics for analytical chemistry*. New York: Van Nostrand Reinhold Company.

Berman, S. S., R. E. Sturgeon, J. A. H. Desaulniers, and A. P. Mykytiuk. 1983. Preparation of the sea water reference material for trace metals, NASS-1. *Marine Pollution Bulletin* 14:69–73.

Bewers, J. M., and H. L. Windom. 1982. Comparison of sampling devices for trace metal determinations in seawater. *Marine Chemistry* 11:71–86.

Bowie, A. R., E. P. Achterberg, P. L. Croot, H. J. de Baar, P. Laan, J. W. Moffett, et al. 2006. A community-wide intercomparison exercise for the determination of dissolved iron in seawater. *Marine Chemistry* 98:81–99.

Brereton, R. G. 2007. *Applied chemometrics for scientists*. Chichester, UK: John Wiley & Sons.

Brügmann, L., E. Geyer, and R. Kay. 1987. A new Teflon sampler for trace metal studies in seawater—Wates. *Marine Chemistry* 21:91–99.

Buesseler, K. 1996. Introduction to "Use of cross-flow filtration (CFF) for the isolation of marine colloids." *Marine Chemistry* 55:vii–viii.

Caccia, M., R. Bono, G. Bruzzone, E. Spirandelli, G. Veruggio, A. M. Stortini, and G. Capodaglio. 2005. Sampling sea surfaces with SESAMO: An autonomous craft for the study of sea-air interactions. *Robotics & Automation Magazine* 12:95–105.

Capodaglio, G. 1997. Sampling techniques for sea water and sediments. In *Marine Chemistry*, ed. A. Gianguzza, E. Pelizzetti, and S. Sammartano. Dordrecht, The Netherlands: Kluwer Academic.

Carlson, D. J. 1982. A field evaluation of plate and screen microlayer sampling techniques. *Marine Chemistry* 11:189–208.

Carlson, D. J., J. L. Cantey, and J. J. Cullen. 1988. Description of and results from a new surface microlayer sampling device. *Deep Sea Research: Oceanographic Research Papers* 35A:1205–13.

Ćosović, B., and V. Vojvodić. 1998. Voltammetric analysis of surface active substances in natural seawater. *Electroanalysis* 10:429–34.

Croot, P. L., and P. Laan. 2002. Continuous shipboard determination of Fe(II) in polar waters using flow injection analysis with chemiluminescence detection. *Analytica Chimica Acta* 466:261–73.

Cuculiv, V., and M. Branica. 1996. Adsorption of trace metals from sea-water onto solid surfaces: Analysis by anodic stripping voltammetry. *Analyst* 121:1127–31.

Dachs, J., and J. M. Bayona. 1997. Large volume preconcentration of dissolved hydrocarbons and polychlorinated biphenyls from seawater. Intercomparison between C18 disks and XAD-2 column. *Chemosphere* 35:1669–79.

De Baar, H., K. Timmermans, P. Laan, H. De Porto, S. Ober, J. Blom, M. C. Bakker, J. Schilling, G. Sarthou, M. G. Smit, and M. Klunder. 2008. Titan: A new facility for ultraclean sampling of trace elements and isotopes in the deep oceans in the international Geotraces program. *Marine Chemistry*, 111:4–21.

Doucet, F. J., L. Maguire, and J. R. Lead. 2005. Assessment of cross-flow filtration for the size fractionation of freshwater colloids and particles. *Talanta* 67:144–54.

Falkowska, L. 1999. Sea surface microlayer: A field evaluation of Teflon plate, glass plate and screen sampling techniques. Part 1. Thickness of microlayer samples and relation to wind speed. *Oceanologia* 41:211–21.

Garcia-Flor, N., C. Guitart, L. Bodineau, J. Dachs, J. M. Bayona, and J. Albaiges. 2005. Comparison of sampling devices for the determination of polychlorinated biphenyls in the sea surface microlayer. *Marine Environmental Research* 59:255–75.

Gardolinski, P. C. F. C., G. Hanrahan, E. P. Achterberg, M. Gledhill, A. D. Tappin, W. A. House, and P. J. Worsfold. 2001. Comparison of sample storage protocols for the determination of nutrients in natural waters. *Water Research* 35:3670–78.

Garrett, W. D. 1965. Collection of slick-forming material from the sea surface. *Limnology and Oceanography* 10:602–5.

Guo, L., L. Wen, D. Tang, and P. H. Santschi. 2000. Re-examination of cross-flow ultrafiltration for sampling aquatic colloids: Evidence from molecular probes. *Marine Chemistry* 69:75–90.

Gustafsson, O., P. Andersson, J. Axelman, T. Bucheli, P. Komp, M. McLachlan, A. Sobek, and J.-O. Thörngren. 2005. Observations of the PCB distribution within and in-between ice, snow, ice-rafted debris, ice-interstitial water, and seawater in the Barents Sea marginal ice zone and the North Pole area. *Science of the Total Environment* 342:261–79.

Hales, B., and T. Takahashi. 2002. The Pumping SeaSoar: A high-resolution seawater sampling platform. *Journal of Atmospheric and Oceanic Technology* 19:1096–1104.

Hansell, D. A. 2005. Dissolved organic carbon reference material program. *Eos* 86:318–19.

Hardy, J. T., J. A. Coley, L. D. Antrim, and S. L. Kiesser. 1988. A hydrophobic large-volume sampler for collecting aquatic surface microlayers: Characterization and comparison with the glass plate method. *Canadian Journal of Fisheries and Aquatic Sciences* 45:822–26.

Harvey, G. W. 1966. Microlayer collection from the sea surface: A new method and initial results. *Limnology and Oceanography* 11:608–14.

Harvey, G. W., and L. A. Burzell. 1972. A simple microlayer method for small samples. *Limnology and Oceanography* 17:156–57.

Helmers, E. 1994. Sampling of sea and fresh water for the analysis of trace elements. In *Sampling and sampling preparation*, ed. M. Stoeppler. Berlin: Springer-Verlag.

Hillebrand, M. T. J., and R. F. Nolting. 1987. Sampling procedures for organochlorines and trace metals in open ocean waters. *TrAC Trends in Analytical Chemistry* 6:74–77.

Hirata, S., T. Kajiya, N. Takano, M. Aihara, K. Honda, O. Shikino, and E. Nakayama. 2003. Determination of trace metals in seawater by on-line column preconcentration inductively coupled plasma mass spectrometry using metal alkoxide glass immobilized 8-quinolinol. *Analytica Chimica Acta* 499:157–65.

Hiroshi, H. 1978. Preservation of sea water samples for determination of carbohydrates. *Japan Analyst* 27:252–55.

Horiuchi, T., Y. Takano, J. Ishibashi, K. Marumo, T. Urabe, and K. Kobayashi. 2004. Amino acids in water samples from deep sea hydrothermal vents at Suiyo Seamount, Izu-Bonin Arc, Pacific Ocean. *Organic Geochemistry* 35:1121–28.

Hühnerfuss, H. 1981. On the problem of sea surface film sampling: A comparison of 21 microlayer-, 2 multilayer-, and 4 selected subsurface samplers. Part 1. *Sonderdruck aus Meerestechnik* 12:136–42.

Intergovernmental Oceanograhic Commission (IOC). 1993. *Chlorinated biphenyls in open ocean waters: Sampling, extraction, clean-up and instrumental determination*. IOC Manuals and Guides 27, UNESCO, Paris.

Ittekot, V., U. Brockmann, W. Michaelis, and E. T. Degens. 1981. Dissolved free and combined carbohydrates during a phytoplankton bloom in the northern North Sea. *Marine Ecology Progress Series* 4:299–305.

Kattner, G. 1999. Storage of dissolved inorganic nutrients in seawater: Poisoning with mercuric chloride. *Marine Chemistry* 67:61–66.

Keith, L. H., W. Crummett, J. Deegan, R. A. Libby, J. K. Taylor, and G. Wentler. 1983. Principles of environmental analysis. *Analytical Chemistry* 55:2210–18.

Kelly, A. G., I. Cruz, and D. E. Wells. 1993. Polychlorobiphenyls and persistent organochlorine pesticides in sea water at the pg 1-1 level. Sampling apparatus and analytical methodology. *Analytica Chimica Acta* 276:3–13.

Kester, D. R., I. W. Duedall, D. N. Connore, and R. M. Pytkowicz. 1967. Preparation of artificial seawater. *Limnology and Oceanography* 12:176–79.

Kirkwood, D. S. 1992. Stability of solutions of nutrient salts during storage. *Marine Chemistry* 38:151–64.

Knap, A. H., K. A. Burns, R. Dawson, M. Ehrhardt, and K. H. Palmork. 1986. Dissolved/dispersed hydrocarbons, tarballs and the surface microlayer: Experiences from an IOC/UNEP workshop in Bermuda, December, 1984. *Marine Pollution Bulletin* 17:313–19.

Knefelkamp, B., K. Carstens, and K. H. Wiltshire. 2007. Comparison of different filter types on chlorophyll-a retention and nutrient measurements. *Journal of Experimental Marine Biology and Ecology* 345:61–70.

Koike, I., S. Hara, K. Terauchi, and K. Kogure. 1990. Role of sub-micrometer particles in the ocean. *Nature* 345:242–43.

Kremling, K., and L. Brügmann. 1999. Filtration and storage. In *Methods of seawater analysis*, ed. K. Grasshoff, K. Kremling, and M. Ehrhardt. Weinheim, Germany: Wiley-VCH.

Kuznetsova, M., C. Lee, and J. Aller. 2005. Characterization of the proteinaceous matter in marine aerosols. *Marine Chemistry* 96:359–77.

Landing, W. M., G. A. Cutter, J. A. Dalziel, A. R. Flegal, R. T. Powell, D. Schmidt, A. Shiller, P. Stratham, S. Westerlund, and J. Resing. 1995. Analytical intercomparison results from the 1990 Intergovernmental Oceanographic Commission open-ocean baseline survey for trace metals: Atlantic Ocean. *Marine Chemistry* 49:253–65.

Larsson, J., Ö. Gustafsson, and J. Ingri. 2002. Evaluation and optimization of two complementary cross-flow ultrafiltration systems toward isolation of coastal surface water colloids. *Environmental Science and Technology* 36:2236–41.

Lim, L., O. Wurl, S. Karuppiah, and J. P. Obbard. 2007. Atmospheric wet deposition of PAHs to the sea-surface microlayer. *Marine Pollution Bulletin* 54:1212–19.

Lohan, M. C., A. M. Aguilar-Islas, R. P. Franks, and K. W. Bruland. 2005. Determination of iron and copper in seawater at pH 1. 7 with a new commercially available chelating resin, NTA Superflow. *Analytica Chimica Acta* 530:121–29.

Macdonald, R. W., and F. A. McLaughlin. 1982. The effect of storage by freezing on dissolved inorganic phosphate, nitrate and reactive silicate for samples from coastal and estuarine waters. *Water Research* 16:95–104.

Momzikoff, A., A. Brinis, S. Dallot, G. Gondry, A. Saliot, and P. Lebaron. 2004. Field study of the chemical characterization of the upper ocean surface using various samplers. *Limnology and Oceanography: Methods* 2:374–86.

Mullins, E. 1994. Introduction to control charts in the analytical laboratory. *Analyst* 119:369–75.

Niskin, S., D. Segar, and P. Betzer. 1973. New Niskin sampling bottles without internal closures and their use for collecting near bottom samples for trace metal analysis. *Transactions of the American Geophysics Union* 54:1110.

Pakulski, J. D., and R. Benner. 1994. Abundance and distribution of carbohydrates in the ocean. *Limnology and Oceanography* 39:930–40.

Petrick, G., D. E. Schulz-Bull, V. Martens, K. Scholz, and J. C. Duinker. 1996. An in-situ filtration/extraction system for the recovery of trace organics in solution and on particles tested in deep ocean water. *Marine Chemistry* 54:97–105.

Raimbault, P., W. Pouvesle, F. Diaz, N. Garcia, and R. Sempere. 1999. Wet-oxidation and automated colorimetry for simultaneous determination of organic carbon, nitrogen and phosphorus dissolved in seawater. *Marine Chemistry* 66:161–69.

Ramesh, A., K. Rama Mohan, and K. Seshaiah. 2002. Preconcentration of trace metals on Amberlite XAD-4 resin coated with dithiocarbamates and determination by inductively coupled plasma-atomic emission spectrometry in saline matrices. *Talanta* 57:243–52.

Sharp, J. H., R. Benner, L. Bennett, C. A. Carlson, S. E. Fitzwater, E. T. Peltzer, and L. M. Tupas. 1995. Analyses of dissolved organic carbon in seawater: The JGOFS EqPac methods comparison. *Marine Chemistry* 48:91–108.

Surface Ocean Lower Atmosphere Study (SOLAS). 2005. Microlayer: μystery and μagic! *SOLAS News* 1:11.

Strady, E., C. Pohl, E. V. Yakushev, S. Kruger, and U. Hennings. 2008. PUMP-CTD-system for trace metal sampling with a high vertical resolution. A test in the Gotland Basin, Baltic Sea. *Chemosphere* 70:1309–19.

Subramanian, K. S., C. L. Chakrabarti, J. E. Sueiras, and I. S. Maines. 1978. Preservation of some trace metals in samples of natural waters. *Analytical Chemistry* 50:444–48.

Svensson, E., A. Skoog, and J. P. Amend. 2004. Concentration and distribution of dissolved amino acids in a shallow hydrothermal system, Vulcano Island (Italy). *Organic Geochemistry* 35:1001–14.

Taverniers, I., M. De Loose, and E. Van Bockstaele. 2004. Trends in quality in the analytical laboratory. II. Analytical method validation and quality assurance. *TrAC Trends in Analytical Chemistry* 23:535–52.

Thompson, M., S. L. R. Ellison, and R. Wood. 2002. Harmonized guidelines for single-laboratory validation of methods of analysis. *Pure and Applied Chemistry* 74:835–55.

Venrick, E. L., and T. L. Hayward. 1985. Evaluation of some techniques for preserving nutrients in stored seawater samples. *CalCOFI* XXVI:160–68.

Wells, M. L., and E. D. Goldberg. 1993. Colloid aggregation in seawater. *Marine Chemistry* 41:353–58.

Wurl, O., O. Elsholz, and R. Ebinghaus. 2001. On-line determination of total mercury in the Baltic Sea. *Analytica Chimica Acta* 438:245–49.

Wurl, O., and J. P. Obbard. 2005. Chlorinated pesticides and PCBs in the sea-surface microlayer and seawater samples of Singapore. *Marine Pollution Bulletin* 50:1233–43.

Zhang, Z., L. Liu, C. Liu, and W. Cai. 2003. Studies on the sea surface microlayer. II. The layer of sudden change of physical and chemical properties. *Journal of Colloid and Interface Science* 264:148–59.

2 Analysis of Dissolved and Particulate Organic Carbon with the HTCO Technique

Oliver Wurl and Tsai Min Sin

CONTENTS

2.1 INTRODUCTION

The dissolved organic matter (DOM) pool in the ocean is equivalent to that of the atmospheric CO_2 and is measured at ~10^{18} g (McCarthy et al., 1996). Net oxidation of only 1% of the oceanic DOM pool within 1 year would be sufficient to generate a CO_2 flux larger than that produced annually by fossil fuel combustion (Hedges, 2002). The great size and dynamics of the DOM pool have brought it within focus of global proportions. The DOM pool provides a large reservoir of substrates for life, a source for nutrient regeneration, ion exchange capacity, binding capacity of contaminants, light, and heat absorption. The carbon pool drives the microbial loop in the ocean, and therefore the marine food web. Overall, accurate quantification of DOM pools, fluxes, and their controls is critical to understanding oceanic carbon cycling and how the oceans will respond to increasing concentrations of atmospheric CO_2 and climate change. However, knowledge of the marine carbon cycle, including production, recycling, and burial of organic matter, is still limited, and so is the precise estimation of its reservoirs.

DOM consists of three subsets: (1) dissolved organic carbon (DOC), (2) dissolved organic nitrogen (DON), and (3) dissolved organic phosphorous (DOP). The determination of DON and DOP and their particulate forms is described in Chapter 9, whereas this chapter addresses the measurements of DOC and particulate organic carbon (POC). POC is defined as the fraction typically retained on a filter with a pore size of 0.7 µm, whereas DOC is the fraction that passes through it. In comparison to DOC and dissolved inorganic carbon (DIC), the POC pool comprises a rather small amount of carbon in oligotrophic waters, that is, oceanic waters. However, it represents an important fraction of the total organic carbon (TOC) pool by forming sinking aggregates (marine snow) together with gel particles (see Chapter 7) and being a source of food for organisms in the deep ocean.

The measurement of DOC and POC in seawater has been difficult and controversial (Dafner and Wangersky, 2002a; Spyres et al., 2000), which has resulted in limited and questionable information on the distribution of organic carbon in the global ocean as reported for the international Joint Global Ocean Flux Study (JGOFS) (Hansell and Carlson, 2001). All techniques to measure DOC are based on the conversion of all organic carbon to CO_2 through a chemical, photo-, or high-temperature catalytic oxidation. Prior to the oxidative step, the inorganic carbon is removed through acidification (to convert inorganic carbon to CO_2) and purging with a purified gas (acid-purging step). Two general approaches in the estimation of DOC have been used since the 1950s: (1) wet chemical oxidation (WCO) with persulfate or UV light (Menzel and Vaccaro, 1964) and (2) high-temperature combustion (HTC) on dried samples (Skopintsev, 1969). The resultant oxidation product CO_2 has been detected using colorimetric techniques or through the loss of mass after combustion. Nowadays, a nondispersive infrared (IR) analyzer is commonly used as a detector. The WCO method (Menzel and Vaccaro, 1964) was widely used among marine chemists in the 1960s and 1970s to estimate DOC in seawater. It is based on the chemical oxidation of organic compounds by persulfate, but the limitations were low sensitivity and high blanks. In the 1970s, automated photooxidation in the presence of oxidants was developed on the Technicon AutoAnalyzer II with a coupled IR detector (Collins et al., 1977). Dillido (1976) designed the Technicon AutoAnalyzer II with photooxidation, in which the generated CO_2 is dialyzed through a silicone rubber membrane and reacted with a buffered phenolphthalein indicator. The decrease in color of the indicator was recorded with a spectrophotometer at a wavelength of 550 nm. The limitation of these techniques was the unreliable estimation of blanks, which is a critical step in the accurate and precise measurement of DOC. HTC methods were believed to be superior to WCO because of a more efficient oxidation via a combustion tube streamed with oxygen. However, the HTC method became popular in the late 1980s as the technique was modified by Sugimura and Suzuki (1988). The modification included a Pt-catalyzed combustion unit (high-temperature catalytic oxidation [HTCO]) for efficient oxidation. Furthermore, the HTCO technique is faster, easier, and more readily automated than WCO techniques. The unusually elevated levels of DOC measured by Sugimura and Suzuki with their newly developed HTCO led to excitement and new interest in the study of DOC profiles and distributions. However, it was reported by Benner and Strom (1993) and widely accepted that Sugimura and Suzuki's analysis suffered from high instrument blanks. For this reason, Suzuki (1993) retracted the results that had caused so much excitement. However, because of the excitement generated by their original paper, many researchers began to think in new ways about DOC research and its analysis. Community-wide intercomparison of different methods and between laboratories has been conducted and summarized by Sharp (2002), and the greater awareness of the importance of low instrument blanks and control of oxidation efficiency (Dafner and Wangersky, 2002a; Spyres et al., 2000) has increased the accuracy and precision of the HTCO method over the last two decades. The HTCO technique is currently the preferred analytical technique for the measurement of organic carbon in aqueous media. Modern HTCO analyzers can be upgraded with units to measure POC (this chapter) and total nitrogen (Chapter 9). Production and storage of low-carbon water (LCW) still needs attention by the DOC analysts. Numerous sources of contamination during processing of samples (i.e., sampling, filtration, preservation, and analysis) shipboard and in laboratories exist through the high abundance of organic compounds. Overall, the measurement of DOC and POC in seawater remains an analytical challenge due to low levels in oceanic waters and potential contamination sources during sample processing and analysis, difficulties in maintaining instrument blanks at a consistent and low level, and a lack of the use of certified reference materials (CRMs) to validate methods within and among laboratories. For procedures of the measurement of DIC and other CO_2 parameters, the reader is referred to the PICES special publication *Guide to Best Practices for Ocean CO_2 Measurements* (Dickson et al., 2007), which is available for free on the Internet (www.pices.int).

2.2 SAMPLING PROCEDURE

Sampling, like other trace analysis, is a very critical step in the analysis of organic carbon. Inadvertent contamination can occur at any time during sample collection and treatment due to the high abundance of carbonaceous compounds in the environment. Contamination sources include sampling devices (i.e., bottles, conductivity-temperature-depth (CTD) cable, hydrowire, and storage container), vessel operations (i.e., ship's waste disposal and emissions, winches, dust dispersion), and the analysts (i.e., fingerprints). Further problems include sorption onto bottle walls and changes in concentration through biological and physical processes during filtration and storage.

2.2.1 PREPARATION OF STORAGE CONTAINER

The cleaning procedure and storage of a cleaned sample container before and after collection are of utmost importance. A sample container should be made from a noncontaminating material, low in the adsorption of organics to its walls, and economical. Glass bottles are the classical sample container for DOC analysis, but their cleaning is more time-consuming and needs more attention due to the bigger volume. Glass ampoules are a good choice, although they require some experience and practice for sealing with a portable gas burner. Typical pore-like openings formed during glass sealing were smallest using the method of draw sealing (Greiff et al., 1975), as illustrated in Figure 2.1. Dafner and Wangersky (2002a) reported that no contamination was introduced using 10 mL Wheaton glass ampoules, precombusted (i.e., 500°C for several hours) and prescored, but ampoules of smaller volumes (i.e., 2 mL) can cause contamination during flame sealing due to low headspace volume. Glass bottles or ampoules should be soaked for at least 12 hours (2–3 days is better) in 10% HCl and then rinsed thoroughly with ultrapure water. Prior to combustion, the glassware is covered with aluminum foil, heated to 500°C for >12 hours, and uncovered only before drawing the sample from the sampling bottles. Alternatively, plastic containers made of high-density polyethylene (HDPE) and polypropylene bottles are suitable for storage of DOC samples, provided they have been soaked in 10% HCl for at least 1 week (i.e., removing leachable components) and

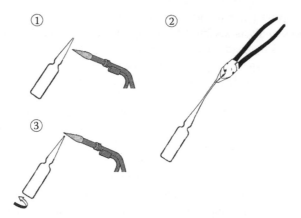

FIGURE 2.1 Draw sealing according to Greiff et al. (1975). ① Neck of ampoule is carefully heated and ② removed from the flame and drawn out for about 2.5 cm. ③ After cooling the ampoule is sealed by holding the restricted tip in the flame for a few seconds while turning the ampoule.

rinsed thoroughly with ultrapure water. It is recommended to store a precleaned sample container filled with deep-sea water (<2,000 m) and in Ziploc bags until use. The losses of DOC through sorption on HDPE or polypropylene (PP) container walls are negligible, and no significant changes in DOC concentration have been observed in bulk samples stored in glass ampoules, PP and HDPE bottles for 5 months (Tupas et al., 1994).

2.2.2 PREPARATION AND OPERATION OF SAMPLING DEVICE

Specially modified Niskin and Go-Flo bottles are most suitable for the collection of DOC/TOC samples. The Go-Flo bottle has the advantage of passing through the sea-surface microlayer (SML) (see Section 1.5) closed, and thereby avoids coating its inside with surface film rich in organic carbon. It should be noted that the larger the volume of the sampler, the less impact any inadvertent contamination will have. Typical modifications include an interior Teflon coating of the bottle, replacement of the standard latex tubing spring closure with Teflon-coated springs, and replacement of standard sealing with heavy-walled silicone tubing and O-rings. The bottle should be equipped with monofilament nylon or Kevlar lanyards. In general, plastic material should be avoided, although properly cleaned nylon, polypropylene (PP), and high-density polyethylene (HDPE) materials are considered to be free of contamination. Soft plastic material (i.e., LDPE, latex, rubber) and materials with distinctive smells are likely to be contaminating. For example, no Tygon®- or phthalate-containing material should be used as part of the sampling device. Niskin or Go-Flo bottles have to be carefully washed and rinsed just before deployment using a solution of 10% HCl and ultrapure water to remove traces of contamination. Organic solvents were used in combination with 10% HCl (Loh and Bauer, 2000) or as a single washing solution (Thomas et al., 1995) to clean sampling bottles or parts of it, but this is not recommended here. After cleaning the sampling bottles need to be covered with a clean plastic cover (acid rinsed) until deployment. It is recommended to estimate blanks from the sampling and filtration device, as suggested in Section 2.6. While the sampling bottles pass through the SML, organics enriched in this layer can be carried down by adsorption to the surfaces of the bottles. This is of particular concern when Niskin bottles are deployed because they pass through the SML while open. Contamination can be minimized by lowering the rosette (or single bottle on hydrocast) as deep as possible to rinse it with clean deep-sea water before activating the collection at the desired depths. After retrieval of the rosette sampler, the sampling bottles have to be shaken prior to drawing the subsamples to ensure even distribution of particulates. The samples should be drawn directly from the sampling port using nylon, PP, or a Teflon stopcock without any transfer tubing, and phthalate-containing tubing used for the foregoing collection was suspected to be contaminating to DOC samples (Peltzer and Brewer, 1993). The sample should be allowed to flow freely from the Niskin/Go-Flo bottle for a few seconds to clean the port. The receiving sample container should not be in contact with the port. Transferring the sample from the Niskin/Go-Flo bottle into the sample container should be done as quickly as possible to minimize the contact time with the air. The sample container and caps are rinsed three times with a small volume of sample prior to filling. Overflow of the sample container should be avoided in all cases. Water can be drawn back into the bottle after the contact with the dirty exterior of the container, for example, the lip of the bottle (Cauwet, 1999).

2.3 SAMPLE PROCESSING

2.3.1 FILTRATION

2.3.1.1 Filtration Procedure for DOC

The decision to filter the sample prior to storage (or analysis) or not has to be made on a case-by-case basis. Estuarine and coastal waters require the filtration of samples prior to storage during all seasons. However, POC in oligotrophic waters (i.e., oceanic waters) often contributes only 1%–2% to the TOC, and filtration is not necessary and even represents a potential source or loss of DOC in oceanic samples with typical low DOC concentrations (e.g., 40–60 μM). However, it cannot be used as an explict guideline because phytoplankton blooms in certain oceanic regions cause the POC concentration to rise substantially and filtration becomes necessary.

Filtration through gravity directly from the sampling bottle (see Figure 8.3) is recommended using a precleaned (soaked in 10% HCl and rinsed with ultrapure water) 25 or 47 mm filtration holder made of PP, PE, or Teflon. The connection tube between stop cock and filter holder has to be made of PP, PE, or Teflon, and be kept as short as possible. The filtered sample should be collected directly into the sample container. Glass fiber filters type F (GF/F) with a pore size of 0.7 μm are the classical filter material due to the easiness of cleaning and the resulting low blanks. GF/Fs are wrapped in aluminum foil, precombusted at 500°C for at least 12 hours, and stored in an airtight container. They are unwrapped from the aluminum foil just before loading into the filtration unit. If only DOC data are of interest, we found that polycarbonate filters (soaked in 10% HCl for several hours and rinsed thoroughly with ultrapure water) are noncontaminating (Table 2.1). Compared to GF/Fs, their advantage is a well-defined pore size (see Table 1.3) and less adsorption of water, that is, of DOC. Other tested filters (precleaned in 10% HCl) made from organic materials were contaminating (Table 2.1).

Alternatively, vacuum filtration can be used (see Section 1.6) with a low vacuum pressure (<150 mmHg) to avoid cell lysis. The pump should be a nonoil system, and it is advisable to

---

TABLE 2.1
Concentration of DOC [μM C] in Filtered Ultrapure Water through Different Types of Filter Material ($n = 3$)

Filter Material	Concentration [μM C]
Unfiltered	4.0 ± 1.9
Glass fiber GF/F	5.3 ± 2.0
Polycarbonate	4.3 ± 0.9
Cellulose acetate	22.2 ± 4.5
HVLP (polyvinylidene fluoride)	17.1 ± 7.3

disconnect the vacuum from the receiving flask before turning off the pump to avoid back vapor contamination. Rinsing the precombusted filtration unit five times with ultrapure water and twice with the sample results in very low DOC blank values (see Section 2.4). Vacuum filtration requires a clean laboratory workspace, preferably clean room container laboratories with an ISO class 5 HEPA filtered air system. A laminar airflow work bench can be useful as an alternative measurement to avoid contamination through ambient air (see Section 2.4).

2.3.1.2   Filtration Procedure for POC

Vacuum filtration is required for the filtration of higher volumes (~1 L for coastal and up to 10 L for oceanic waters) of samples to collect sufficient material for POC analysis. Precombusted and preweighted GF/Fs (on a microbalance) are used to calculate the dry weight of filtered material after drying. After reduction of the water column to a few milliliters, the filtration funnel is rinsed with a few milliliters of prefiltered seawater. Some air is drawn through the filter to remove much of the water in the filter. It is recommended to use GF/Fs with a diameter of 25 mm to accommodate for the limited space of the combustion chamber of typical TOC analyzers. We usually store filters for POC analysis in hinged-lid and airtight PP containers with an inner diameter slightly larger than the filter diameter. Alternative sampling methods are described in Section 6.5.

2.3.2   Preservation

2.3.2.1   Preservation of DOC Samples

The samples are transferred into 10 mL glass ampoules (recommended for storage of up to 1 year) with precombusted pasteur pipettes and acidified with H_3PO_4 or HCl under clean working conditions (i.e., clean room or laminar airflow workbench) shortly after collection. Samples can be stored in glass or plastic bottles (up to 5 months) used for the collection of the subsamples from the Niskin/Go-Flo bottles, and this avoids the transfer step into the glass ampoules. However, the tightness of the closure of the bottles needs to be careful checked. Alternatively, samples can be directly collected and preserved in autosampler vials (see Section 1.4.2) provided they can be tightly closed. High-purity concentrated H_3PO_4 or HCl (e.g., ULTREX II, J.T. Baker; SUPRAPUR, Merck) is used to adjust the pH value of the samples to pH = 2–3 (about 4 µl H_3PO_4 or 12 µl HCl per mL sample). This pH range is sufficient to eliminate biological activity and preserve organic molecules. A pH of <2 is not recommended, as strong acidification may lead to the enhanced production of volatile products from organic matter and precipitation of macromolecules. Glass ampules are flame-sealed after acidification, whereas bottles are tightly closed and wrapped in clean Ziploc bags. The samples are stored at 4°C in a refrigerator until the analysis. Generally, the storage time should be as short as possible, although storage for several months without significant changes in DOC concentration has been reported (Wiebinga and de Baar, 1998). The acidification requires clean working conditions as outlined above. If such conditions are not available onboard, it is recommended to collect subsamples in HDPE bottles (filled to three-quarters), immerse in liquid nitrogen for quick freezing, and store at –20°C. Samples are then thawed at room temperature prior to analysis. It was shown that this alternative preservation technique allows storage for at least 5 months (Tupas et al., 1994). Storage at –20°C without quick freezing is not recommended.

2.3.2.2   Preservation of POC Samples

Filters in small PP storage containers are dried at 60°C in a clean oven. Blank filters should be dried with each batch for control of potential contamination during drying. The dried samples are weighed on a microbalance in a room with low humidity. Samples and blanks are stored at –20°C in the airtight PP container.

2.3.3 Removing Particulate Inorganic Carbon

The automated removal of DIC is well established in the HTCO analyzer (see Section 2.5). Particulate inorganic carbon (PIC) is often negligible in its concentration compared to POC. However, PIC needs to be removed in some cases, and the best-known case is the presence of coccolithophorid blooms forming calcite shells. Materials from sediments and sediment traps can also contain higher amounts of $CaCO_3$. Removal of PIC within HTCO systems involves the addition of acid and heat, but such an approach can lead to significant POC losses, in particular at low concentration ranges. Alternatively, the vapor acidification method (Hedges and Sterm, 1984) is described here for suspended particulates in seawater samples, although it is more time-consuming. Samples are placed in ceramic boats (supplied with the solid combustion unit of the HTCO analyzer). The boats are kept in a desiccator containing a small beaker with concentrated HCl for 48 hours. Samples rich in coccolithophores should be analyzed visually under a microscope after fuming to ensure the calcite shells have been disappeared. If not, the fuming needs to be repeated for another 24–48 hours. The PIC-free samples are dried for 1 hour at 50°C to remove residual HCl and water.

2.4 PREPARATION OF HTCO ANALYSIS

The preparation of the analysis includes (1) preparation of calibration standards, (2) cleaning of autosampler vials, and (3) production of low-carbon water (LCW) for assessment of instrument blanks. The preparation should be done at a clean workspace separated from other laboratory activities, in particular from the use of solvents. It is recommended to prepare LCW and calibration standards in a clean room or under a laminar airflow workbench, as it offers protection against dust. We have intentionally exposed autosampler vials containing ultrapure water to ambient air in the laboratory and filtered air under a laminar airflow workbench. The results are summarized in Table 2.2. As the extent of contamination is not related to the exposure time, it is suggested that contamination originates from airborne particles randomly and not from continuous adsorption of volatile organic compounds from the air. No contamination was detected in the samples protected under the laminar airflow workbench.

TABLE 2.2
Concentration of DOC [μM C] in Duplicate Samples of Ultrapure Water Exposed to the Ambient Air and Protected under a Laminar Flow Bench

Exposure Time (min)	Sample Set 1	Sample Set 2
	Unprotected	
0.5	35.4	28.4
2	6.2	4.5
10	13.7	3.4
30	2.6	2.8
	Laminar Flow Bench	
0.5	2.4	4.3
2	7.2	5.8
10	5.0	3.1
30	3.5	6.3

2.4.1 Preparation of Calibration and Oxidation Efficiency (OE) Standards

The volumetric flasks and other labware used for the preparation of the standards are rinsed with ultrapure water, soaked in 10% HCl for at least 24 hours, rinsed with ultrapure water, and dried at <100°C while wrapped in aluminum foil. Volumetric glassware is not combusted at high temperatures, as the accuracy of the volumetric graduation will suffer under such treatment. Two OE standards prepared from EDTA and sulfathiazole are recommended to check the efficiency of oxidation of the catalyst. The known concentration of these standards should be in the same range as expected for the unknown samples.

2.4.1.1 Organic Carbon Stock Solution

Dry about 2.5 g of potassium biphthalate in a precleaned beaker at 100°C for 4 hours in a clean oven. Weigh 2.1254 g of the dried potassium biphthalate and dissolve it in 1,000 mL of LCW (see Section 2.4.3). One milliliter of the stock solution is equivalent to 1.00 mg C. Prepare calibration solution by stock dilution in LCW. Adjust the pH value of the stock and calibration solution to 2–3 (about 4 µl H_3PO_4/mL solution).

2.4.1.2 Inorganic Carbon Stock Solution

Dry about 4.7 g of sodium carbonate and 3.8 g of sodium bicarbonate in precleaned beakers at 100°C for 4 hours in a clean oven. Weigh 4.4122 g and 3.4970 g of the dried sodium carbonate and sodium bicarbonate, respectively, and dissolve it in 1,000 mL of LCW. One milliliter of the stock solution is equivalent to 1.00 mg C. Prepare calibration solution by stock dilution in LCW. Adjust the pH value of the stock and calibration solution to 2–3 (about 4 µl H_3PO_4/mL solution).

2.4.1.3 OE Standards

Dry about 2.8 g of sulfathiazole and 3.3 g of EDTA in precleaned beakers at 100°C for 4 hours in a clean oven. Weigh 2.3641 g of the dried sulfathiazole and dissolve it in 1,000 mL of LCW (OE1). Weigh 3.0992 g of EDTA and dissolve it in 1,000 mL of LCW (OE2). One milliliter of each OE stock solution is equivalent to 1.00 mg C. Prepare OE standards by stock solution in LCW. Adjust the pH value of the stock and OE standard solution to 2–3 (about 4 µl H_3PO_4/mL solution).

2.4.2 Cleaning of Autosampler Vials

The autosampler vials are rinsed with a copious amount of ultrapure water and then soaked in 10% HCl for at least 24 hours. After rinsing with ultrapure water, the vials are combusted at 500°C for at least 5 hours. The septum caps are rinsed several times with ultrapure water, soaked in 10% HCl, rinsed with ultrapure water, wrapped in aluminum foil, and dried at 100°C. The vials are immediately closed with the caps after combustion and stored in an airtight plastic container in a dry and clean place. We observed an increase of blank samples by a factor of 3 in vials stored open and with exposure to the ambient air compared to blanks analyzed in properly stored vials, as described above.

2.4.3 Production of Low-Carbon Water (LCW)

It is a very challenging task to produce and store low-carbon water (LCW), which is required for the assessment of the instrument blank and preparation of field blanks and standards. Erroneously high instrument blanks lead to an overestimation of DOC in the samples, and the blanks therefore need to be accurately measured. The instrument blank represents the blank of any HTCO analyzer (system blank) plus the carbon content of the water used for the blank estimation. The catalyst is

TABLE 2.3
Concentrations of DOC in Different Types of Purified Waters

Type of Ultrapure Water	Concentration ± SD [μM C]
Distilled water[a]	40.5 ± 7.0 ($n = 10$)
Non-UV-irradiated ultrapure water[b]	12.4 ± 5.2 ($n = 10$)
UV-irradiated ultrapure water (old UV lamp)[c]	6.1 ± 2.5 ($n = 5$)
UV-irradiated ultrapure water (new UV lamp)[d]	4.0 ± 2.0 ($n = 11$)
Blanks estimated during routine analysis of samples:	
UV-irradiated ultrapure water (new UV lamp)[d]	6.0 ± 2.1 ($n = 15$)
UV-irradiated ultrapure water with additional UV exposure in autosampler vials	3.4 ± 1.8 ($n = 8$)

[a] Fistream Cyclon, Sanyo.
[b] PureLab, Elga.
[c] Gradient, millipore (1-year-old UV lamp).
[d] Gradient, millipore (6-week-old UV lamp).

considered to be a major source of carbon contamination in the system, but can vary between different types of catalysts (Benner and Strom, 1993; Cauwet, 1999).

We investigated the concentrations of carbon in distilled (Fistreem Cyclon, Sanyo), non-UV-irradiated (PureLab, Elga), and UV-irradiated (Gradient, Millipore) water samples (Table 2.3). As expected, distilled water is not suitable, as it contains as much carbon as seawater from the deep sea. The non-UV-irradiated water contained levels of carbon lower by a factor of 3 than those in the distilled water. Aliquots of UV-irradiated water were analyzed before and after the replacement of the UV lamp in the Milli-Q system. The new UV lamp was more efficient in removing remaining carbon in the ultrapure water compared to the old UV lamp (in use for 1 year). It indicates that a well-maintained water purification system is essential to obtain LCW with lowest carbon content.

Typical blanks obtained with the Gradient water purification system in routine work were in the range of 3 to 10 μM C (mean = 6.0 ± 2.1 μM C, $n = 15$). A mean instrument blank of 6 μM C for the analysis of DOC and TOC in oceanic waters is in the typical range, as reported by other studies (Doval and Hansell, 2000; Fransson et al., 2001; Misic et al., 2005; Wiebinga and de Baar, 1998). A simple setup has been used to keep levels of carbon in ultrapure water low by using autosampler vials custom-made from quartz glass. UV-purified water in the presence of H_2O_2 is further irradiated under UV light in the custom-made autosampler vials. Twenty microliters of 30% H_2O_2 is added per 20 mL ultrapure water and the quartz glass vial tightly capped. The vial is then placed under a UV-light cabinet (wavelength = 180–400 nm). After 8–10 hours another 20 μL of 30% H_2O_2 is added and placed under the UV light. This process is repeated three or four times. The vial is placed directly on the autosampler tray to estimate the instrument blank without reopening and transfer of the freshly produced LCW. The carbon content observed in water treated with this procedure was lower (3.4 ± 1.8 μM C, $n = 8$) in routine work than those obtained from the Gradient purification system. However, well-maintained water purification systems with integrated photo-oxidation (UV irradiation) can produce LCW suitable for the assessment of instrument blanks and preparation of field blanks.

2.5 PERFORMANCE CHECK AND ANALYSIS ON HTCO ANALYZER

The principle of the HTCO technique is illustrated and described in Figure 2.2.

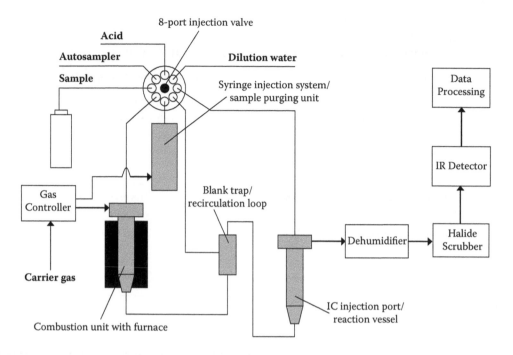

FIGURE 2.2 Simplified principle of HTCO technique (e.g., Shimadzu TOC-V$_{CPN}$). Acidified sample is drawn into a syringe, purged inside the syringe, and released CO_2 carried through the nondispersive IR detector, bypassing the combustion unit (inorganic carbon [IC] measurement). Then the syringe injects the sample into the catalyst-filled combustion unit, heated to 680°C and flushed with oxygen, to oxidize and decompose organic carbon into CO_2. CO_2 streams are dehumidified and cleaned in a scrubber before passing the flow cell of the nondispersive IR detector.

2.5.1 PERFORMANCE CHECK

Prior to the analysis of samples, the analyst needs to ensure that the instrument is in good condition to achieve highest accuracy and precision of the analysis. The described performance check and analysis are routinely used on Shimadzu HTCO analyzers, but applicable to most other commercial HTCO analyzers (e.g., Teledyne Tekmar, Ankersmid, etc.), although minor changes in the procedure may be required. Typical HTCO instrument settings are summarized in Table 2.4. Autosampler functions usually include various options and should be set as in Table 2.4. If samples have been stored frozen without addition of acid, automatic acid addition is set with an adjusted volume of H_3PO_4 (85%) to lower the pH of samples to 2–3.

Various types of catalysts have been used in the past (Sharp et al., 1995), but a more recent study suggested that a catalyst of Al_2O_3 impregnated with 0.5% Pt achieves highest oxidation efficiency, although a catalyst filled with quartz beads had a lower blank due to the amphoteric property of Al_2O_3 (Benner and Strom, 1993; Watanabe et al., 2007). However, we recommend using catalysts of Al_2O_3 impregnated with 0.5% Pt (Shimadzu type), as poor oxidation recoveries of alternative catalytic material are unacceptable, in particular for catalysts based on quartz beads (Watanabe et al., 2007). In case a new or reconditioned catalyst (for cleaning a catalyst, refer to Section 9.4) has been installed prior to the analysis, the catalyst has to be conditioned for 24 hours at a typical furnace temperature and repeatedly injected with LCW until the blank values have been stabilized and fall in a typical range for the instrument. A large number of injections (up to 100) may be required to bring the blank down to an acceptable level (Skoog et al., 1997). Skoog et al. (1997) also reported that no deterioration in catalyst (Al_2O_3 impregnated with 0.5% Pt) performance occurred up to a

TABLE 2.4
Instrument Conditions and Typical Analytical Performance Data for the Shimadzu TOC Analyzers Instrument Conditions/Settings

Instrument Conditions/Settings

Carrier gas	Oxygen (ultrapure, >99.999%)
Flow rate	150 mL min^{-1}
Purge flow rate	75 mL min^{-1}
Injection volume	80 mL
Catalyst	0.5% Pt-coated aluminum oxide
Furnace temperature	680°C
Oxidation product	CO_2
Detection	Infrared detector
Recording	Peak area

Autosampler Settings

No. of needle rinses	2
No. of flow line rinses	2
Auto addition of acid	Off
Acid volume	0
Rinse after addition	Off

Analytical Performance

Linearity	20 to >1,000 µM; R^2 > 0.9998
Limit of detection (mean blank + 3*σ)	DOC and TOC: 8 µM
Precision as coefficient of variation (CV)	DOC: <3.6% at 60 µM; TOC: <4.0% at 60 µM
Sample throughput	2 samples/h, including 2 washing cycles and 3–5 repeat injections for each sample

total injection volume of 250 mL of seawater, that is, >2,500 injections. However, salt abrasion in injection mechanisms may occur earlier (Spyres et al., 2000), and thorough cleaning to remove salt deposits is necessary (see below).

After switching on the analyzer, the combustion unit needs to heat up (typical to 680°C) for about 1 hour. If POC analysis is performed, switch on the module for solid sample (e.g., Shimadzu SSM-5000A) and open the valve for the purified oxygen supply; wait until the combustion unit stabilizes at the set temperature (typically 900°C). Meanwhile, the flow rate of CO_2-free gas (purified nitrogen, typically 150 mL min^{-1}) and reservoirs for water, acid, and waste are checked. The purge gas flow is checked and adjusted to typical values of 50–75 mL min^{-1}. The injection port, syringe, and combustion column have to be free of any salt residues, which are easily deposited after a large number of seawater sample injections. Software-controlled cleaning procedures are very helpful (e.g., Shimadzu) to flush and keep injection ports, syringes, valves, and combustion columns in good condition for prolonged lifetimes. Repeated runs of cleaning procedures are performed after the furnace reaches the desired temperature and stabilizes, or LCW is injected repeatedly if such cleaning is not controlled by the software.

As a performance check, it is recommended to run four warm-up seawater samples, four LCW blanks, a calibration set, OE standards (1 + 2), and two LCW blanks. This is a typical run prior to the analysis of any unknown samples. The warm-up sample, calibration standards, LCW blanks, and OE standards have to be acidified in the same manner as the unknown samples. The warm-up seawater samples are run to minimize and stabilize the instrument background/blank. The same sample will be injected repeatedly so it will be possible to see the stabilization of the instrument.

If the instrument is still drifting after four warm-up samples, repeat injection until a reproducible signal is obtained. The warm-up sample is ideally a deep-sea water sample or one with a low DOC content. The recoveries of the OE standards should be better than 90% (see Section 2.6.4). Lower values are an indication of the inefficiency of the catalyst to oxidize organic matter and have to be replaced. With modern HTCO analyzers (e.g., Shimadzu) it is possible to condition the catalyst and effectively measure the instrument blank with an internal recirculation of injected LCW. An LCW sample is injected onto the catalyst, where it is combusted and collected downstream as pyrolyzed water (e.g., theoretically carbon-free water). The latter is reinjected to determine the instrument blank.

2.5.2 CALIBRATION AND ANALYSIS OF DOC SAMPLES

The calibration sets should include six standards (each for DOC and DIC), including a zero standard (i.e., LCW), and cover the expected concentration range of unknown samples. For the calibration of the detector, the oxidation step can be bypassed using a standard of CO_2 in air (e.g., NIST). The instrument response (i.e., slope of the calibration line) should be identical with both calibration modes.

After a satisfactory performance check and full calibration, a typical run of unknown samples consists of two LCW blanks, a series of samples with LCW blanks interdispersed, an LCW blank, and two calibration standards identical to the initial set and LCW blank. By repeating the two calibration points, it is possible to identify any drift in the instrument response.

2.5.3 CALIBRATION AND ANALYSIS OF POC SAMPLES

If removed through the fuming method (see Section 2.3.3), calibration with dextrose is sufficient for the POC determination. Otherwise, an additional calibration set of Na_2CO_3 (to determine PIC) is required. Three ceramic boats are loaded with three different amounts of dextrose on a microbalance according to the concentration range expected from the samples (typically 20 to 200 µg L^{-1}). The last sample boat is analyzed empty (blank). Each standard should be analyzed in triplicate. Filter samples are loaded in ceramic boats with forceps and analyzed by pushing the sample boats into a combustion unit (for Shimadzu SSM-5000A). Alternatively, samples (free of PIC) can be analyzed with a CHN analyzer (see Chapter 9), which separates H, CO_2, and N_2 through a gas chromatographic column, and subsequent detection with a thermal conductivity detector.

2.6 QUALITY ASSURANCE (QA)

A rigorous QA is essential for obtaining accurate and precise analytical data. For DOC analysis, QA samples often contribute 20% to the total number of samples. For procedures to estimate accuracy and precision of analytical methodologies, refer to Chapter 1.

2.6.1 FIELD BLANKS

Field blanks should be collected regularly (i.e., for each batch of samples collected/processed) to monitor potential contamination during sample processing. Field blanks include blanks for (1) storage, (2) filtration, and (3) the sampler device. Ideally, the various blanks are insignificantly different from each other.

2.6.1.1 Storage Field Blanks

On the day of sample collection, at least two precleaned sample containers (same type as used for sample storage) are carefully filled with LCW in a clean room or laminar airflow workbench, preserved as the samples (4°C with acidification or –20°C without acidification), and tightly closed. The bottles are stored in the laboratory in the refrigerator or freezer, dedicated for the sample storage for the same duration as the samples. These bottles represent the field blanks for storage.

2.6.1.2 Filtration Field Blanks

A volume of freshly produced LCW equivalent to the sample volume is filtered through the precleaned filtration unit just before the filtration of the samples. The filtered LCW is treated (preserved, stored, and analyzed) in the same way as the sample and other field blanks. The storage field blank is subtracted from this blank to obtain the filtration field blank.

2.6.1.3 Sampling Field Blanks

Freshly produced LCW is poured into a precleaned and rinsed Niskin bottle, preferably in a clean room container. If a sufficient number of sampling bottles are available, this Niskin bottle is closed, mounted on the rosette, and deployed, but not used for the actual collection. After sampling, the LCW is filtered, stored, and analyzed in the same way as the samples. Alternatively, the whole procedure can be done prior to sampling by keeping the LCW inside the Niskin bottle for a time equivalent to a typical sampling procedure without using a Niskin bottle for the actual sampling. The filtration field blank is subtracted from this blank to obtain the sampling field blank.

2.6.2 BLANK CORRECTION

All results need to be blank-corrected, typically with the instrument blank. The area units of the instrument blanks (LCW) are plotted with the run number. If no trend is evident, the average of blanks obtained during the run is calculated and used for correction. Otherwise, a linear interpolation is applied between two successive blanks. The difference between these two blanks is divided by the number of samples between them plus 1. The quotient is the stepwise blank for successive samples to be corrected. If major contamination occurred during sampling and treatments, a correction using field blanks may be adequate with an appropriate remark in the final data report.

2.6.3 CERTIFIED REFERENCE MATERIAL

Certified reference materials (see Chapter 1) are essential for checking the accuracy of the analysis. For DOC analysis, there is currently only one CRM available, coordinated by Dr. Dennis Hansell's laboratory at Rosenstiel School of Marine and Atmospheric Science (see Table 1.2). The CRM consists of two samples, a low-carbon water (2 μM C) and a deep-sea sample from the Atlantic (about 45 μM C, depending on the batch). The CRM is already acidified with H_3PO_4 and stored in a 10 mL ampoule. It can be stored for 1 year at room temperature. Laboratories involved in the analysis of DOC in seawater should obtain this CRM material, and they are encouraged to report QA data from CRM in scientific publications.

2.6.4 OXIDATION EFFICIENCY (OE STANDARDS)

To determine the completeness of oxidation the recovery of a known concentration of a recalcitrant organic compound is measured during the performance check of the instrument (see Section 2.5). Suitable test compounds are alinic acid, caffeine, EDTA, soluble starch, urea, oxalic acid, and sulfathiazole. The use of EDTA and sulfathiazole is suggested here, and the preparation of OE standards as described in Section 2.4. The recovery should be greater than 90% to ensure good oxidation efficiency. Recoveries below 90% are indicative of the need to replace the catalyst. Sulfur-containing compounds (e.g., sulfathiazole) are known to be more recalcitrant.

The recovery is calculated according to

$$R = C_{measured}/C_{OE-Std} \times 100\%$$ (2.1)

with the measured concentration of DOC ($C_{measured}$) and the known concentration of OE standard (C_{OE-Std}).

2.7   CONCLUSION AND FUTURE TRENDS

Carbon is a key element in oceanic biogeochemical cycles, and its accurate and precise measurement has been the focus of marine chemists for many years. The HTCO technique is nowadays the preferred analytical technique to estimate DOC in seawater. The technique has been developed over the last two decades to a well-accepted standard method (e.g., see special issue 41 in *Marine Chemistry*, 1993; Spyres et al., 2000; Skoog et al., 1997; Watanabe et al., 2007) since its appearance in the literature (Sugimura and Suzuki, 1988). The HTCO technique has the ability to measure nitrogen with an additional module, and simultaneous measurement of carbon and nitrogen is often advantageous in biogeochemical studies. However, further distribution of available CRM and international intercalibration exercises are crucial for marine analytical chemists to test and validate their DOC analysis, in particular in the context of complete oxidation of organic material, reproducibility of data, and robustness of the analyzer. Technical developments in the HTCO techniques will occur in the future (Dafner and Wangersky, 2002b; Peterson et al., 2003). Furthermore, analytical problems such as control of contamination and absolute stop of bacterial degradation in stored DOC samples still exist and require awareness and skill from the marine analytical chemist. For example, it is still not resolved how to measure DOC adsorption on filters used for POC analysis, therefore leading to an overestimation (Gardner et al., 2003). Many studies nowadays require smaller spatial and temporal data resolution than in the past, and such a trend is likely to continue in the future. DOC measurement onboard research vessels has been successfully reported, but the sample rate on a typical HTCO analyzer is about 45 minutes per sample (triplicate injection), that is, 8 to 10 hours for a depth profile at twelve depths without QA samples. In practice, samples are often accumulated in the fridge for storage. Seagoing DOC measurement avoids physical and chemical treatments of samples, and the analysis of fresh and unaffected DOC samples is of great benefit to understand oceanic carbon cycling better (Dafner and Wangersky, 2002b). As sparging time to remove DIC is the most time-consuming step with the HTCO analysis (typically 10 minutes for each analytical cycle), technical developments for a quicker but complete removal of DIC will lead to an improved sampling rate.

REFERENCES

Benner, R., and M. Strom. 1993. A critical evaluation of the analytical blank associated with DOC measurements by high-temperature catalytic oxidation. *Marine Chemistry* 41:153–60.

Cauwet, G. 1999. Determination of dissolved organic carbon and nitrogen by high temperature combustion. In *Methods of seawater analysis*, ed. K. Grasshoff, K. Kremling, and M. Ehrhardt. Weinheim, Germany: Wiley-VCH.

Collins, K. J., P. J. Le, and B. Williams. 1977. An automated photochemical method for the determination of dissolved organic carbon in sea and estuarine waters. *Marine Chemistry* 5:123–41.

Dafner, E. V., and P. J. Wangersky. 2002a. A brief overview of modern directions in marine DOC studies. I. Methodological aspects. *Journal of Environmental Monitoring* 4:48–54.

Dafner, E. V., and P. J. Wangersky. 2002b. A brief overview of modern directions in marine DOC studies. II. Recent progress in marine DOC studies. *Journal of Environmental Monitoring* 4:55–69.

Dickson, A. G., C. L. Sabine, and J. R. Christian. 2007. *Guide to best practices for ocean CO₂ measurements*. PICES Special Publication 3, Sidney, British Columbia.

Dillido, J. 1976. On-line monitoring for organic carbon detection. *Advances in Automated Analysis* 2:22–26.

Doval, M. D., and D. A. Hansell. 2000. Organic carbon and apparent oxygen utilization in the western South Pacific and the central Indian Oceans. *Marine Chemistry* 68:249–64.

Fransson, A., M. Chierici, L. G. Anderson, I. Bussmann, G. Kattner, E. Peter Jones, and J. H. Swift. 2001. The importance of shelf processes for the modification of chemical constituents in the waters of the Eurasian Arctic Ocean: Implication for carbon fluxes. *Continental Shelf Research* 21:225–42.

Gardner, W. D., M. J. Richardson, C. A. Carlson, D. Hansell, and A. V. Mishonov. 2003. Determining true particulate organic carbon: Bottles, pumps and methodologies. *Deep Sea Research Part II* 50:655–74.

Greiff, D., H. Melton, and T. W. G. Rowe. 1975. On the sealing of gas-filled glass ampoules. *Cryobiology* 12:1–14.

Hansell, D. A., and C. A. Carlson. 2001. Marine dissolved organic matter and the carbon cycle. *Oceanography* 14:41–49.

Hedges, J. I., and J. H. Sterm. 1984. Carbon and nitrogen determinations of carbonate-containing solids. *Limnology and Oceanography* 29:657–63.

Hedges, J. I. 2002. Why dissolved organics matter. In *Biogeochemistry of marine dissolved organic matter*, ed. D. A. Hansell and C. A. Carlson. San Diego: Academic Press.

Loh, A. N., and J. E. Bauer. 2000. Distribution, partitioning and fluxes of dissolved and particulate organic C, N and P in the eastern North Pacific and Southern Oceans. *Deep Sea Research Part I* 47:2287–2316.

McCarthy, M., J. Hedges, and R. Benner. 1996. Major biochemical composition of dissolved high molecular weight organic matter in seawater. *Marine Chemistry* 55:281–97.

Menzel, D. W., and R. F. Vaccaro. 1964. The measurement of dissolved organic and particulate carbon in seawater. *Limnology and Oceanography* 9:138–42.

Misic, C., M. Giani, P. Povero, L. Polimene, and M. Fabiano. 2005. Relationships between organic carbon and microbial components in a Tyrrhenian area (Isola del Giglio) affected by mucilages. *Science of the Total Environment* 353:350–59.

Peltzer, E. T., and P. G. Brewer. 1993. Some practical aspects of measuring DOC—Sampling artifacts and analytical problems with marine samples. *Marine Chemistry* 41:243–52.

Peterson, M. L., S. Q. Lang, A. K. Aufdenkampe, and J. I. Hedges. 2003. Dissolved organic carbon measurement using a modified high-temperature combustion analyzer. *Marine Chemistry* 81:89–104.

Sharp, J. H. 2002. Analytical methods for total DOM pools. In *Biogeochemistry of marine dissolved organic matter*, ed. D. A. Hansell and C. A. Carlson. San Diego: Academic Press.

Sharp, J. H., R. Benner, L. Bennett, C. A. Carlson, S. E. Fitzwater, E. T. Peltzer, and L. M. Tupas. 1995. Analyses of dissolved organic carbon in seawater: The JGOFS EqPac methods comparison. *Marine Chemistry* 48:91–108.

Skoog, A., D. Thomas, R. Lara, and K. Richter. 1997. Methodological investigations on DOC determinations by the HTCO method. *Marine Chemistry* 56:39–44.

Skopintsev, B. A. 1969. Organic carbon in the equatorial and southern Atlantic and Mediterranean. *Okeanologiya* 6:201–10.

Spyres, G., M. Nimmo, P. J. Worsfold, E. P. Achterberg, and A. E. J. Miller. 2000. Determination of dissolved organic carbon in seawater using high temperature catalytic oxidation techniques. *TrAC Trends in Analytical Chemistry* 19:498–506.

Sugimura, Y., and Y. Suzuki. 1988. A high-temperature catalytic oxidation method for the determination of non-volatile dissolved organic carbon in seawater by direct injection of a liquid sample. *Marine Chemistry* 24:105–31.

Suzuki, Y. 1993. On the measurement of DOC and DON in seawater. *Marine Chemistry* 41:287–88.

Thomas, C., G. Cauwet, and J. Minster. 1995. Dissolved organic carbon in the equatorial Atlantic Ocean. *Marine Chemistry* 49:155–69.

Tupas, L. M., B. N. Popp, and D. M. Karl. 1994. Dissolved organic carbon in oligotrophic waters: Experiments on sample preservation, storage and analysis. *Marine Chemistry* 45:207–16.

Watanabe, K., E. Badr, X. Pan, and E. P. Achterberg. 2007. Conversion efficiency of the high-temperature combustion technique for dissolved organic carbon and total dissolved nitrogen analysis. *International Journal of Environmental Analytical Chemistry* 87:387–99.

Wiebinga, C. J., and H. J. W. de Baar. 1998. Determination of the distribution of dissolved organic carbon in the Indian sector of the Southern Ocean. *Marine Chemistry* 61:185–201.

3 Spectrophotometric and Chromatographic Analysis of Carbohydrates in Marine Samples

Christos Panagiotopoulos and Oliver Wurl

CONTENTS

3.1 INTRODUCTION

Carbohydrates are ubiquitous in the marine environment and comprise about 15% to 35% of the dissolved organic carbon (DOC) (Burney et al., 1982; Romankevich, 1984; Pakulski and Benner, 1993; Myklestad and Børsheim, 2007). Carbohydrates are versatile compounds of the dissolved organic matter (DOM) serving numerous functions in cell metabolism, for example, as energy, storage, and structural components.

In most phytoplanktons, structural carbohydrates are water-insoluble polysaccharides (typically cellulose, β-1,4-linked polymer of β-D glucose) that are mainly present as fibrillar cell wall constituents forming the skeleton of the wall, and thus providing structural support and protection (Lee, 1999). The amorphous component of the wall forms a more complex matrix of other carbohydrates, in which the fibrillar component is embedded. Bacterial cell walls are made of peptidoglycan (also called murein), which is made from polysaccharide chains cross-linked by unusual peptides containing D-amino acids.

Phytoplankton-derived carbohydrates are the major form of storage of chemical energy and provide, in turn, nonphotosynthesizing organisms with energy through glycolysis and respiration. Storage carbohydrates are produced in the light and serve then as an internal energy and carbon reserve. In low light conditions, or at night, this reserve can be used for maintaining cell metabolism and protein synthesis (Lancelot and Mathot, 1985; Granum and Myklestad, 2001). Photosynthesis under high light conditions and nutrient limitation can lead to excess production of storage carbohydrates (Myklestad and Haug, 1972; Myklestad, 1988) and their accumulation within the cells. Such conditions enhance the extracellular release of carbohydrates (Staats et al., 2000), typically released as large heteropolymers. While amino acids (see Chapter 4) are the major cellular component in rapidly growing cells, carbohydrates are the dominating compounds among extracelluar release products (Myklestad and Haug, 1972; Penna et al., 1999; Granum et al., 2002).

Besides the extracellular release of carbohydrates by phytoplankton as a significant part of the DOC in seawater, sloppy feeding and egestion by zooplankton (Strom et al., 1997) and cell lysis (Fuhrmann, 1999; Suttle, 2005) are further major sources of carbohydrates in seawater cycling through the marine carbon system. As dissolved fractions, carbohydrates can rapidly grow to complex polymers (Chin et al., 1998), which aggregate further to gel-like particles. Such particles (particulate organic matter [POM]) play a significant role in biogeochemical cycles (see Chapter 7), including carbon sequestration to the deep ocean through the formation of marine snow. Overall, understanding the sources, transformation, and fate of carbohydrates in the marine environment can provide insight into the overall cycling of photosynthetically produced organic carbon.

The analysis of carbohydrates in seawater is challenging for various reasons. First, typical concentrations of carbohydrates are low in seawater (nanomolar concentration range), and for molecular separation techniques desalting of the sample is necessary, which can lead to potential losses of analytes (Borch and Kirchman, 1997). Second, their detection is difficult as carbohydrates lack light-absorbing chromophores and chemical derivatization is required prior to analysis. Last, but not least, carbohydrates exhibit multiple charge states at a typical pH value for seawater, for example, neutral sugars, amino sugars (positively charged), and uronic acids (negatively charged), challenging the molecular isolation of these compounds.

A review of various analytical techniques for carbohydrate analysis in seawater is given by Panagiotopoulos and Sempéré (2005a). Bulk analysis of carbohydrates using spectrophotometric methods includes diverse classes of carbohydrates with the advantage of simplicity of instrumental technique and short analysis time, but lack of molecular identification. The most common spectrophotometric methods are the phenol-sulfuric acid (PSA) method (Dubois et al., 1956), the 3-methyl-2-benzo thiazoline hydrazone hydrochloride (MBTH) method (Burney and Sieburth, 1977) and the 2,4,6-tripyridyl-s-triazine (TPTZ) method (Myklestad et al., 1997). The TPTZ method combines the low detection limits and good precision of the MBTH method with the rapidity and simplicity of the PSA method. For these reasons, the TPTZ method is most suitable for shipboard analysis.

In order for polysaccharides to be detected, they must first be hydrolyzed in acid to convert nonreducing sugars to reducing sugars. Various hydrolysis protocols are reported in the literature using different acids and concentrations as well as varying hydrolysis times and temperatures (Borch and Kirchman, 1997; Panagiotopoulos and Sempéré, 2005a).

The advent of high-performance anion exchange chromatography (HPAEC) in the early 1980s revolutionized sugar analysis on a molecular level compared to classical separation techniques using borate complexes (Mopper, 1978; Mopper et al., 1980). The introduction of pulsed amperometric detection (PAD) permitted carbohydrate detection at high sensitivity (down to 10 pmol) without pre- or postcolumn derivatization (Rocklin and Pohl, 1983; Mopper et al., 1992; Jørgensen and Jensen, 1994; Borch and Kirchman, 1997; Skoog and Benner, 1997). However, the technique requires considerable effort for sample treatments, for example, removal of salts and metals by resins (Wicks et al., 1991; Mopper et al., 1992). Various compounds were employed to form volatile derivatives for the gas chromatographic (GC) detection of carbohydrates. Detection was most frequently performed using flame ionization detection (FID). In contrast to liquid chromatographic techniques, GC can be more easily coupled to mass spectrometry (MS), providing important structural information (Klok et al., 1984; Sigleo, 1996). However, the multiplicity of sugar peaks in the chromatograms and the potentially complex chemical manipulations require a demand of careful laboratory techniques and interpretation.

It is difficult to compare reported concentration of carbohydrates in seawater, given the wide range of hydrolysis protocols used by investigators and the lack of intercomparison studies between analytical procedures. This chapter provides standard procedures for spectrophotometric and chromatographic techniques.

3.2 SAMPLING, FILTRATION, AND STORAGE

Carbohydrates are a major fraction of the DOC in seawater, and therefore similar precautions in sampling as for DOC need to be taken to avoid contamination of samples (see Chapter 2). Contamination sources include sampling devices (e.g., bottles, conductivity-temperature-depth [CTD] cable, hydrowire, and storage container), vessel operations (e.g., ship's waste disposal and emissions, winches, dust dispersion), and the analysts (e.g., fingerprints). Further problems include adsorption onto bottle walls and changes in concentration through biological and physical processes during filtration and storage.

3.2.1 PREPARATION OF STORAGE CONTAINER

The sample container should be made from a noncontaminating material, low in the adsorption of organics to its walls, and economical. As the required sample volume is low (typically 4 mL), 10 mL glass ampoules are well suited as storage containers. The sealing of the glass ampoules requires some experience, and a procedure is outlined in Chapter 2 (Section 2.2.1). The glass ampoules are soaked in 10% HCl for at least 2 days, rinsed with ultrapure water, and precombusted at 450°C for >4 hours. The cleaned glass ampoules are stored in an airtight container. For the collection of subsamples from Niskin or Go-Flo bottles, 60 mL screw-capped bottles made of glass or HDPE can be used after thorough cleaning in 10% HCl and rinses with ultrapure water. Glass bottles can be combusted at 450°C for >4 hours.

3.2.2 SAMPLING PROCEDURE

The procedure for DOC described in Section 2.2.2 should be followed. Briefly, precleaned (brushed with 10% HCl and rinsed with ultrapure water) Niskin and Go-Flo bottles are most suitable for the collection of seawater for the analysis of carbohydrates. The sampling bottle should be flushed at a deeper depth before retrieval to the desired depth and collection of water. Although all research personnel involved in taking subsamples should wear disposable gloves, any touching of the opening and port of the sampling bottle should be avoided. Before taking subsamples, the port is rinsed with the sample by free flow for a few seconds. The sample is then collected in a 60 mL glass or HDPE bottle, either online filtered or unfiltered (see Section 3.2.3).

3.2.3 Filtration

Filtration may contribute not only to sample contamination, but also to other systematic errors, such as adsorption of analytes on filter material. Collection of oceanic waters may not require filtration considering that particulate-associated carbohydrates are negligible compared to the dissolved fraction (at most a few percent), and filtration may cause higher errors. However, unfiltered samples need to be processed shortly after collection, as organisms in the sample (although low in abundance) may alter concentration and composition of carbohydrates. Preservation of samples is difficult and involves unpredictable risks of alterations of sample integrity (Section 3.2.4) due to the lack of validated preservation techniques. Samples collected in estuarine, coastal, and oceanic waters with high primary production need to be filtered as soon as possible after collection, preferably directly online from the sampling bottle by gravity or in a clean laboratory by vacuum or syringe filtration as soon as possible after collection (<1 hour). Polycarbonate membranes (pore size of 0.22 or 0.45 µm) are suitable for filtration, as potential contamination can be easily removed by soaking filters in 10% HCl for >12 hours, and the cutoff size is well defined. All other filtration materials (filter holder, syringe, vacuum filtration unit) are also soaked in 10% HCl for several hours, rinsed with ultrapure water, and glass material combusted at 450°C for >4 hours.

The preferred filtration technique is online filtration directly from the sampling bottle through gravity, as described and illustrated in Chapter 8 (Section 8.3.1 and Figure 8.3). Vacuum filtration should be performed at low and constant vacuum pressure of <150 mmHg to ensure that polysaccharide-rich gel particles (see Chapter 7) are retained on the filter and do not contribute to the fraction of dissolved carbohydrates. High vacuum or pressure during filtration forces those flexible particles through filters, although their size is larger than the pore size. Low vacuum pressures also prevent cell lysis on the filter. For these reasons, a low and constant pressure during syringe filtration is required, but this technique is often preferable for shipboard filtration compared to the relative extensive setup of vacuum filtration and difficulties to clean glass apparatus (see Section 3.4.2).

To estimate the content carbohydrates from particulate matter, particles are collected from 250 to 2,000 mL of sample, and a low vacuum pressure is commonly used to complete filtration of the larger volume in a reasonable time.

3.2.4 Storage and Preservation of Samples

Storage of samples may be subject to alteration during storage even when deep frozen (Liebezeit and Behrends, 1999), but not within 2 weeks (Mopper et al., 1992). Any addition of preservative agents to eliminate metabolic activity is likely to change the composition of carbohydrates, although Mopper et al. (1992) reported no significant changes within 2 weeks of addition of acetonitrile (10%, v/v). Based on Mopper's report, it can be concluded that the storage in the freezer without any addition of preservatives seems to be adequate, provided the samples are analyzed within 2 weeks. For the measurement of bulk concentrations, samples can be stored frozen (quick freezing in liquid nitrogen) for a longer period of up to 5 months in sealed glass ampoules (similar to DOC, as carbohydrates are a major component of DOC; see Chapter 2). Filters for the analysis of particulate carbohydrates are stored at −80°C in screw-capped glass tubes.

3.3 SAMPLE PREPARATION FOR SUGAR ANALYSIS: EXTRACTION PROTOCOLS, HYDROLYSIS, AND DESALINATION PROCEDURES

The efficiency of extraction of sugars from marine samples is of critical importance in order to maximize their amount for subsequent spectrophotometric or chromatographic analyses. Sugars can be extracted from POM with boiling water for 1 hour (Handa, 1967; Handa and Tominaga, 1969), while alkali can also be used when extracting sugars from sediments (Miyajima et al., 2001). This procedure generally targets the water-soluble carbohydrates, while the residue contains the structural (non-water-soluble) polysaccharides. Once water-soluble carbohydrates are released to the water phase, they can be precipitated using hydrophilic organic solvents such as methanol or ethanol (Handa and Tominaga, 1969; Underwood et al., 1995) and further analyzed.

The most common extraction procedure that releases both structural and storage carbohydrates from a marine sample (POM, sediments, DOM, etc.) is acid hydrolysis (Mopper et al., 1977, 1992; Panagiotopoulos and Sempéré, 2005a, and references therein). This procedure requires the use of a strong acid (HCl, H_2SO_4, trifluoroacetic acid [TFA]) at high temperatures (100–120°C) for several hours (1–22 hours) and yields a pool of monomers (monosaccharides) (Table 3.1). Several factors can influence the hydrolysis yield and depend on the type of acid used, acid strength (pH), the duration and temperature of the hydrolysis, and finally the nature of the sample (DOM, POM, sediment, high molecular weight dissolved organic matter (HMWDOM), riverine POM sample, etc.).

Both strong and mild hydrolysis have been widely employed by marine biogeochemists (Table 3.1). Strong hydrolysis is usually performed in two steps: samples are treated with concentrated H_2SO_4 (72 wt% ~12 M H_2SO_4) for 2 hours at ambient temperature (pretreatment), and then the solution is diluted to 1.86–1.2 M H_2SO_4 and heated at 100°C for 3–4 hours (Mopper, 1977; Cowie and Hedges, 1984; Pakulski and Benner, 1992). Although the two-step hydrolysis may induce losses of some monosaccharides (notably of rhamnose, fucose, and xylose; Mopper, 1977), several investigators reported that concentrated H_2SO_4 gave higher total aldose yields than classical 1.86–1.2 M H_2SO_4 hydrolysis (Cowie and Hedges, 1984; Pakulski and Benner, 1992). By contrast, mild hydrolysis (0.09–2 M) is usually performed without pretreatment using various acids (H_2SO_4, HCl, TFA, etc.). This type of hydrolysis has been applied to all kinds of marine samples, including sinking or suspended POM, HMWDOM, DOM, and sediments. Earlier investigations indicated that H_2SO_4 (0.85 M, 100°C, 4 hours, or 0.85 M, 100°C, 24 hours) gave the same or higher concentrations for dissolved carbohydrates in natural samples than hydrolysis in dilute HCl (0.09 M, 100°C, 20 hours, or 0.25 M, 100°C, 24 hours; Mopper, 1977; Hanisch et al., 1996; Borch and Kirchman, 1997). Borch and Kirchman (1997) suggested that 0.85 M HCl hydrolysis (100°C, 24 hours) of DOM resulted in yields similar to those of the pretreatment method (Pakulski and Benner, 1992). Trifluoroacetic acid (TFA) at 121°C for 1 to 2 hours has also been used by marine biogeochemists for mild hydrolysis of environmental samples (Wicks et al., 1991; Aluwihare et al., 1997; Repeta et al., 2002). A suite of seven neutral sugars (fucose, rhamnose, arabinose, galactose, glucose, mannose, and xylose) have been measured in marine samples, while fructose and ribose concentrations are only occasionally reported because they are partially or completely destroyed after acid hydrolysis.

Recoveries of monosaccharide standards using mild hydrolysis conditions (2 M TFA at 121°C for 1 hour; 0.1 M HCl at 100°C for 20 hours; 0.85 M H_2SO_4 at 100°C for 24 hours) fall into the range of 70%–95% (Wicks et al., 1991; Borch and Kirchman, 1997). It is interesting to note that whether mild or strong hydrolysis is used, sugar loss will occur and should always be taken into account when interpreting the results. The most common sugars used as spiked internal standards for correction of these losses are adonitol and 2-deoxyribose (Cowie and Hedges, 1984; Borch and Kirchman, 1997). In a recent study Panagiotopoulos and Sempéré (2005a) indicated that mild and strong hydrolysis give comparable results for open ocean samples, including POM, DOM, and HMWDOM (except for sediments), and therefore other parameters should be considered before using HCl, H_2SO_4, or TFA acids.

TABLE 3.1

Common Hydrolysis Conditions for Carbohydrate Extraction for POM, DOM, Sediment, and HMWDOM (>1 kDa)

	Mild Hydrolysis				Strong Hydrolysis			
	Molarity (M)	Duration (h)	Temperature (°C)	Type of Sample	Molarity (M)	Duration (h)	Temperature (°C)	Type of Sample
HCl	0.1	20	100	Sink. POM, DOM	12[a]	12	Ambient	Sink. POM
	0.1	1	150	DOM	1.2[b]	3	100	
	0.24	12	100	DOM				
	0.5	1	100	Susp. POM				
	1	20	100	DOM				
	1.5	4	100	DOM				
	1.8	3.5	100	Susp. POM, DOM				
	2	3.5	100	Susp. and sink. POM				
	3	5	100	Susp. POM, HMWDOM				
	3	1	100	DOM				
	4	3	110	Susp. POM, HMWDOM				
H_2SO_4	0.5	4	100	Susp. and sink. POM	12[a]	2	Ambient	Susp. and sink. POM, HMWDOM, sediment
	0.25	18	100	Sediment	1.2[b]	3	100	
	0.85	24	100	Sediment				
	1	4	90	DOM				
	1.2	3	100	Susp. and sink. POM, sediment				
TFA	0.5	2	135	Sink. POM, sediment				
	2	2	120	HMWDOM				

Strong hydrolysis is performed in two steps:

[a] Treatment with concentrated HCl or H_2SO_4 at ambient temperature

[b] Hydrolysis with 1.86–1.2 M H_2SO_4/HCl at 100°C for 3–4 hours.

For example, H_2SO_4 is not volatile compared to HCl or TFA acids. In this context, hydrolysis employing sulfuric acid requires additional neutralization steps (use of $Ba(OH)_2$ or precombusted $CaCO_3$ powder). These steps involve precipitate formation of $BaSO_4$ or $CaSO_4$ salts that may provide a substrate for absorption and possible sugar loss. In general, marine biogeochemists prefer $CaCO_3$ over $Ba(OH)_2$ because the pH remains lower with $CaCO_3$ (pH ~ 6), and hydrolyzed sugars do not undergo rearrangements occurring at pH > 7 that would lower the recovery (Borch and Kirchman, 1997; Skoog and Benner, 1997).

HCl acid is a very good alternative over H_2SO_4 because it does not require neutralization steps. In our laboratory, we used HCl acid for hydrolysis of our marine samples, which results in very low sugar losses (Panagiotopoulos and Sempéré, 2005b, 2007). Briefly, samples are hydrolyzed under a

N_2 atmosphere with 0.1 M HCl at 100°C for 20 hours (Burney and Sieburth, 1977) in precombusted Pyrex vials (450°C for 6 hours). After cooling, the acid solution is evaporated in a rotary evaporator at ~30°C, and the residue is washed with a small volume of ultrapure water (200 μL), which is further removed by a second evaporation (Panagiotopoulos and Sempéré, 2005b; Cheng and Kaplan, 2001). Repetition of the evaporation procedure two to three times by adding 200 μL of ultrapure water until the pH is ~7 does not produce detectable sugar losses.

The desalination of DOM samples consists of salt removal (mostly NaCl) by ion exchange prior to HPAEC-PAD or GC-FID/MS analysis (Mopper et al., 1992; Skoog and Benner, 1997; Sempéré et al., 2008). This step is of great importance in order to minimize the salt content of samples, which may harm the column performance and the detector of the chromatographic systems (e.g., shift of the retention times because of the absorption of Cl^- on the HPAEC ion exchange column, detection of other ionic species by PAD, possible variations of the flow rate during analysis, and salt deposit to the detector).

AG2-X8 resin in the carbonate form and AG50W-X8 resin in the hydrogen form are generally used for desalting DOM samples. Before use both resins are Soxhlet extracted with CH_3CN and ultrapure water to minimize possible organic contamination. Then, resins are successively washed with 1 M NaOH, ultrapure water, 1 M HCl, and ultrapure water. The anion exchange resin AG2-X8 in the Cl^- form is transformed to AG2-X8 in the HCO_3^- form after addition of 1 M $NaHCO_3$. One to two drops of 10 mM $AgNO_3$ are added to the filtrate to check if any Cl^- is left in the resin (precipitation of AgCl). Two milliliters of each resin bed is mixed and washed abundantly with ultrapure water. This amount of resins is sufficient to desalt 4 mL of seawater sample.

Samples are desalted immediately after hydrolysis samples according to Mopper et al. (1992). The reactions between resins and acidified samples (pH 1) favored the elimination of carbonates (which were released into the sample by anionic exchange with Cl^-) into CO_2, allowing a partial neutralization of sugar samples (pH 3–4.5 after desalting) (Kaiser and Benner, 2000; Sempéré et al., 2008). Briefly, two or three aliquots (0.5 mL) of the seawater sample are used to rinse the resin bed. Then 2 mL of the sample is applied to the resin for 5 minutes or until effervescence ceases. When most of the bubbles are gone, the column is drained with a N_2 stream and the sample is recovered into a clean vial for further chromatographic analysis. Although some investigators (Rich et al., 1996; Borch and Kirchman, 1997) reported sugar losses of 40%–50% during deionization, we found lower sugar losses using the above protocol. The recoveries of desalted sugar NaCl solutions (sugar standards 20–100 nM prepared in NaCl) were 85% ± 9% (n = 4) for fucose, rhamnose, arabinose, and xylose (deoxysugars and pentoses), and 86% ± 12% (n = 4) for galactose, mannose, and glucose (hexoses), respectively. Procedural blanks run on the HPAEC-PAD with desalted sodium chloride solutions showed only a small peak of glucose (~5 nM), even though a systematic peak induced by desalting was coeluted with fructose, avoiding its quantification (Sempéré et al., 2008). Therefore, we strongly recommend to run constantly blanks with ultrapure water passing through the resin to check if any ionic species released from the column interfere in the sugar analysis.

The desalination procedure described here works for neutral sugars, while charged sugars (amino sugars and uronic acids) are stuck to the column. Analysis of amino sugars is feasible using other types of resins, like AG11 18-Biorad for neutralization and AG50-X8 in the Na form for desalting. The amino sugars are subsequently eluted from the column with 20 mM NaOH (Kaiser and Benner, 2000). Desalination of samples containing uronic acids is performed with AG50W-X8 resin in the hydrogen form after conditioning with 1 M HCl solution. Uronic acids are then eluted with ultrapure water (Hung and Santschi, 2001).

3.4   SPECTROPHOTOMETRIC METHODS: ANALYTICAL PROCEDURES

All labware used in the following procedures is cleaned by soaking in 10% HCl for at least 2 days, rinsed with a copious amount of ultrapure water, and dried in a clean oven at 60°C. In addition, all glassware is combusted (except volumetric glassware) for at least 4 hours at 450°C. All labware is covered with aluminum foil and stored in a dry place. Ultrapure water for rinsing and preparation of reagents should be obtained from a water purification system equipped with a multiwavelength UV lamp and organic acid polishing cartridges (i.e., Millipore TOC plus).

3.4.1 Procedure for PSA Method

3.4.1.1 Preparation of Reagents

> **Reagent A:** Phenol solution 5% (m/v). 5 mg of phenol is dissolved in 100 mL of ultrapure water. The solution can be stored for 2 months at 4°C in an amber bottle.
> **Reagent B:** Concentrated H_2SO_4 (95%).

3.4.1.2 Procedure and Recommendations

This method is based on the dehydration of sugars in the presence of concentrated H_2SO_4 at high temperature, producing furfurals (from pentoses) and hydroxyfurfurals (from hexoses). The latter compounds are condensed with phenol, which produces orange-yellow substances that absorb at 480–490 nm. The color produced is proportional to the amount of sugar originally present (Dubois et al., 1956; Gerchakov and Hatcher, 1972).

The analytical procedure for sample preparation is as follows: In a 20 mL glass vial, we add 1 mL of sample and slowly (exothermic reaction) 5 mL of concentrated H_2SO_4. The mixture is well vortexed and placed in a water bath at 100°C for half an hour. After cooling for 2–3 minutes at ambient temperature, the vials are placed into an ice bath for 5 minutes. Then 1 mL of phenol 5% (m/v) is added and the mixture is again well vortexed. The absorbance is measured at 485 nm no later than 1 hour in a 20 or 50 mm cuvette. We found that this procedure gave higher absorption values for standard glucose (up to 10%–40% higher signal at the 50–1,000 mg L^{-1} level, $n = 5$) than those originally produced by Dubois et al. (1956) or other investigators (Liu et al., 1973; Underwood et al., 1995, Miyajima et al., 2001), where all reagents with the sample were mixed and heated at 100°C.

We also recommend to run each sample in triplicate and to perform the whole procedure in the dark by covering the reagent tubes with aluminum foil. By doing this, we found that precision of the technique is improved to about 10% ($n = 5$).

The detection limit of the method is between 25 and 50 μM (in glucose equivalents), with a precision of <20% ($n = 3$) at the 50 μmol level. At the 130 μmol level the precision is better than 10% ($n = 3$). We also investigated possible interferences from other compounds, such as amino acids, as well as salt effects in sugar determination. Our results showed that several amino acids (leucine, phenylalanine, serine, glutamic acid, and glycine) including BSA do not give a positive reaction with the PSA reagents. Low signals (two times the detection limit of the method, ~45 μM) have been found only for aspartic acid and alanine (at a final concentration of each compound of 100 mg L^{-1}). The effect of salts on the absorbance of standard glucose solution in ultrapure water and salt water (10, 20, and 40 g L^{-1}) is shown in Figure 3.1. The results show that high salinity (40 g L^{-1}) decreases the absorbance of glucose, and thus the sensitivity of the method by 14%–38%. Therefore, we recommended using this technique for samples with low salt content, such as sediments, POM, and DOM from rivers.

3.4.2 Procedure for TPTZ Method

3.4.2.1 Preparation of Reagents

> **Reagent A:** Potassium ferricyanide (0.7 mM). 400 mg NaOH, 20 g Na_2CO_3 and 230 mg $K_3(Fe(CN)_6)$ are dissolved in 1,000 mL of ultrapure water. The solution can be stored for several months at 4°C in an amber bottle.

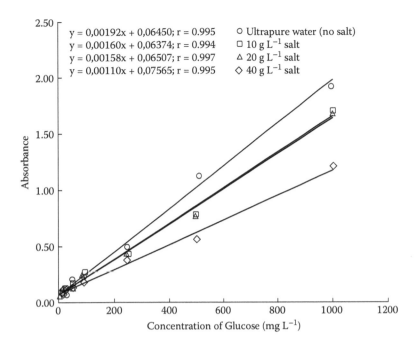

FIGURE 3.1 Influence of salts on standard glucose solution (5 mg L^{-1}, 10 mg L^{-1}, 25 mg L^{-1}, 50 mg L^{-1}, 100 mg L^{-1}, 250 mg L^{-1}, 500 mg L^{-1}, and 1000 mg L^{-1}) made in ultrapure water (no salt), 10 g L^{-1}, 20 g L^{-1}, and 40 g L^{-1} of seawater salts. Each point corresponds to a triplicate measurement.

Reagent B: Ferric chloride (2 mM). 164 g sodium acetate (CH$_3$COONa, anhydrous) and 42 g citric acid (C$_6$H$_8$O$_7$) are dissolved in 650 mL of ultrapure water. Three hundred grams of acetic acid is added, mixed well, and the volume is made up to 1,000 mL with ultrapure water. The solution can be stored for several months at 4°C. On the day of analysis, 32.4 mg of FeCl$_3$ (anhydrous) is dissolved in 100 mL of this solution to prepare the final reagent B. After adding FeCl$_3$, the solution is stable for 2 days at 4°C stored in an amber bottle.

Reagent C: TPTZ (2,4,6-tripyridyl-s-triazine, 2.5 mM). 78 mg of TPTZ is dissolved in 3 M acetic acid. Reagent C is stable for 1 week and stored at 4°C at darkness.

3.4.2.2 Procedure

As the reactions are very photosensitive, light exposure to the samples and reagents is minimized by conducting the analytical procedures in dimmed light, and test tubes are wrapped in aluminum foil (Panagiotopoulos and Sempéré, 2005a). One milliliter of seawater (for analysis of monosaccharides), hydolysate (for analysis of mono- + polysaccharides), or standard solution of D-glucose is mixed with reagent A in a test tube (16 mL screw-capped test tube). The solution is heated for 10 minutes in a boiling water bath. Then 1 mL of reagent B and 2 mL of reagent C are added immediately and mixed well. For adding the reagents, Eppendorf pipettes with acid-cleaned tips are most practical. The solution is allowed to stand for 30 minutes to develop color complex before the absorbance is read at 595 nm in a 20 or 50 mm (microvolume) cuvette. The spectrophotometer is zeroed with ultrapure water and calibrated with D-glucose standards before analysis of samples. A standard curve is obtained by the analysis of five standards with concentrations of 0.2, 0.25, 0.5, 1.0, and 2.0 mg L^{-1}. The standard solutions are prepared by dilution from a stock solution, which is made from D-glucose dried in a desiccator for 24 hours under vacuum. No differences in absorbance were reported for standard solutions prepared in ultrapure water and seawater (Myklestad et al., 1997). Correlation coefficient R^2 should be better than >0.999. For the quantification of monosaccharides, blank values (obtained from ultrapure water) are subtracted from the sample absorbance and are typically in a range of 0.03 to 0.05 absorbance unit with a 20 mm cuvette. Witter and Luther (2002) suggested using an aliquot of each seawater sample as a blank prior to reagent addition to correct for additional absorbance from light-absorbing chromophores that may be present in the samples prior to derivatization with TPTZ. Blanks (ultrapure water) are hydrolyzed for the quantification of polysaccharides, which is calculated as the difference between the total monosaccharide concentrations measured before and after acid hydrolysis. Hydrolyzed blanks are typically higher and range between 0.036 and 0.06 absorbance unit with a 20 mm cuvette. The concentration of total monosaccharide after hydrolysis needs to be corrected with a dilution factor according to the volume of acid and neutralization agent added during the hydrolysis step. All samples should be run in triplicate, and precision is typically better than 10%. The limit of detection is 0.4 μM of glucose (i.e., 2.4 μM C).

3.5   CHROMATOGRAPHIC METHODS: ANALYTICAL PROCEDURES

All labware used for sample preparation for chromatographic techniques is cleaned as described above, while syringes, inserts, and injection vials are cleaned successively with CH_3CN, MeOH, and ultrapure water and dried in a clean oven at 60°C. Ultrapure water was used for solvent (i.e., NaOH solution for HPLC) and sugar standards preparation.

3.5.1   HIGH-PERFORMANCE ANION EXCHANGE CHROMATOGRAPHY WITH PULSED AMPEROMETRIC DETECTION (HPAEC-PAD)

This method is based on anion exchange mechanisms between sugars and the stationary phase, and the order of elution of sugars depends mainly on their dissociation constants (pKa = 12–13). Thus, using a strong base such as NaOH (range = 12–25 mM) as an eluent, sugars are either partially or completely ionized, permitting their separation (Rocklin and Pohl, 1983). Detection is performed without derivatization by a pulsed amperometric detector (PAD) that applies a triple sequence of potential to a gold electrode and allows the determination of sugars at nanomolar levels (Johnson and LaCourse, 1990; Mopper et al., 1992).

3.5.1.1   Solvent Preparation

For sugar analysis four mobile phases are used (1) 20 mM NaOH (working solution), (2) 1 M NaOH (cleaning solution), (3) ultrapure water for dilution of the NaOH solutions, and (4) 1 M NaOH (post-column solution). The NaOH solutions are kept in 1 L polycarbonate bottles. The NaOH mobile phases are prepared by diluting low-carbonate NaOH (J. T. Baker, analytical concentrate) into ultrapure water, which has been sparged with an ultrapure inert gas (He or N_2) for at least 10 minutes at a flow rate of 100 mL min^{-1}. To avoid absorption of carbonates by NaOH, all solutions are constantly purged with He at a flow rate of 4 mL min^{-1}. In contrast to other HPLC mobile phases (e.g., CH_3CN, MeOH), the NaOH solution does not require any filtration prior to use, which may induce contamination. Sugars are separated with 19 mM NaOH solution, which consists of a 95/5 (v/v) mixture of the working solution (20 mM NaOH) and ultrapure water (Panagiotopoulos et al., 2001). Assuming four injections per day that include HPAEC-PAD runtime (35–40 minutes), column regeneration (35–40 minutes), and equilibration of the system with 19 mM NaOH (30–35 minutes), the working solution (20 mM NaOH) can be used for only 4–5 days. If the 20 mM NaOH solution is not consumed within a week, replace it with a fresh one.

3.5.1.2   HPAEC-PAD System Description

The most common HPAEC-PAD system used for sugar analysis in environmental samples is from Dionex Corp., although other systems have also been used (Mopper et al., 1992; Jørgensen and

Jensen, 1994; Panagiotopoulos et al., 2001). The later systems employed different modules (pumps, detectors) from various manufacturers; however, all of them used Dionex CarboPac™ PA-1 analytical columns, which are the best for monosaccharide separation. The analytical column is generally placed into an oven in order to control its temperature (see below).

The electrochemical detector operates in the pulsed mode (PAD) and consists of a gold electrode (working electrode) and an Ag/AgCl electrode (reference electrode). Although several electrochemical detectors (Dionex, DECADE, and EG&G) exist in the market with similar detection limits (5–10 nM), the operating parameters/settings of one detector may not be well adapted for another. Therefore, the optimization of the applied PAD potentials lies in the hands of the scientist to get the highest response of the detector. All other components of the chromatographic system, including pumps, tubing, and solvent bottles, must be made by inert materials such as polyether ether ketone (PEEK), polypropylene, or Teflon because NaOH is highly corrosive.

3.5.1.3 Operating Procedure and Recommendations

The analysis of monosaccharides (fucose, rhamnose, arabinose, galactose, glucose, mannose, xylose, fructose, and ribose) is made by isocratic elution with the use of a mobile phase set at 19 mM NaOH and a flow rate of 0.7 mL min^{-1}, while the column is placed into an oven at 17°C (Panagiotopoulos et al., 2001). The gradient of temperature inside the oven is about 1°C, and 30 minutes is needed to reach equilibration inside the oven. Although several investigators proposed different elution conditions (12–24 mM NaOH) at different temperatures, we found that the above conditions were the optimal in order to achieve the best separation of closely eluting pairs of sugars, such as rhamnose/arabinose and xylose/mannose (Rs$_{rha/ara}$ = 1.02; Rs $_{man/xyl}$ = 0.70). In addition, small changes in temperature (±5°C) that can be found in uncontrolled laboratory conditions may significantly influence the retention times of sugars, resulting in coelutions or poor reproducibility, and therefore the control of temperature is fundamental to obtain consistent results (Panagiotopoulos et al., 2001).

Before detection (i.e., between the column and the detection cell), a strong base (1 M NaOH, post-column solution) at a flow rate of 0.2 mL min^{-1} may be added to the eluent stream to enhance PAD sensitivity and minimize baseline drifting (Dionex, 1989). We recommend using this postcolumn addition when working at <20 mM NaOH; however, it should be kept in mind that this addition induces a dilution of the sample before detection.

As indicated above, HPAEC-PAD eluents are continuously purged; nevertheless, this can be insufficient, and traces of carbonate may be present in the chromatographic system. This problem is most pronounced with eluents <20 mM NaOH. Although previous research (Cataldi et al., 1998) indicated that the addition of Ba(OH)$_2$ into the eluents reduces significantly (by production of BaCO$_3$) the amount of carbonate, we found that the use of Ba(OH)$_2$ also decreased the resolution factor of closely eluted sugars (mannose/xylose). As such, a carbonate retarder column ATC-1 (Dionex Chrom Ion PAC) is installed between the injector and the pump. The carbonate retarder column is regenerated once every 2 months using a Na$_2$B$_4$O$_7$ solution (final concentration = 70 mM) at a flow rate of 1 mL min^{-1}. Additionally, all samples are sparged for 10–15 seconds with He/N$_2$ prior to injection to remove any dissolved carbonate.

Finally, flushing the column between two sample injections using 1 M NaOH (cleaning solution) for 30–40 minutes not only regenerates the column (elimination of carbonate traces), but also reestablishes its performance by removing other organic or inorganic contaminants associated with sugar samples.

Precision of the method falls in the 5%–10% range (coefficient of variation, $n = 5$; based on peak areas) at the 50 nM level for sugar standards, with a detection limit of 5–10 nM (S/N = 3, loop 200 µL; Panagiotopoulos et al. 2001; Mopper et al., 1992). Analytical errors determined from duplicate analysis of environmental samples are <8% for all sugars except ribose (15%–17%).

The procedure described here is optimized for the separation and detection of neutral sugars. For charged sugars (amino sugars and uronic acids) other elution conditions, including gradient with

CH$_3$COONa/NaOH, should be used to achieve a good separation of these compounds (Kaiser and Benner, 2000; Colombini et al., 2002).

3.5.2   Gas Chromatography with Flame Ionization Detection (GC-FID) or Mass Spectrometry (GC-MS)

As indicated above, sugars are not volatile, and therefore, they must be derivatized prior to the GC analysis. Several derivatization procedures can be found in the literature, including trimethylsilyl ester derivatives (Modzeleski et al., 1971; Cowie and Hedges, 1984; Hernes et al., 1996), trifluoacetate esters (Eklund et al., 1977), and alditol acetate derivatives (Klok et al., 1982; Tanoue and Handa, 1987; Aluwihare et al., 2002). All these techniques have advantages and drawbacks. Here we will present the alditol acetate procedure because it reproduces easily interpretable chromatograms (no multiple peaks for each sugar). Briefly, this derivatization procedure is based on the reduction of the sugar carbonyl group using NaBH$_4$-producing sugar alcohols (alditols). The latter compounds are acetylated, resulting in volatile alditol acetate derivatives ready for GC analysis.

3.5.2.1   Procedure for Alditol Acetate Derivatives

3.5.2.1.1   Preparation of Reagents

Reagent A: 1 M NH$_4$OH/NaBH$_4$ solution. 100 mg of NaBH$_4$ is dissolved to 10 mL of 1 M NH$_4$OH. The 1 M NH$_4$OH solution is prepared by dilution of stock solution ammonium hydroxide, ~13.2 M (25% w/w). The 1 M NH$_4$OH/NaBH$_4$ solution is not stored in the refrigerator and should be fresh.

Reagent B: MeOH/CH$_3$COOH solution at 9:1 (v/v).

Reagent C: Glacial CH$_3$COOH.

Reagent D: CH$_2$Cl$_2$.

Reagent E: MeOH.

Reagent F: Acetic anhydride (kept in the dessicator).

Reagent G: 1-Methyl imidazole (kept in the dessicator).

Reagent H: Anhydrous sodium sulfate (precombusted at 450°C for 4 hours and kept in a dessicator).

All solutions can be stored in the refrigerator, but solutions should be made fresh every 3 months.

3.5.2.1.2   Alditol Acetate Procedure

The procedure described here is based on the preparation of alditol acetate derivatives by York et al. (1985) and Aluwihare et al. (2002). In a 4–5 mL vial sugar samples/standards are dissolved in 0.25 mL of 1 M NH$_4$OH/NaBH$_4$. The solution is flushed with N$_2$, and kept at room temperature for a minimum of 1 hour. The reaction is quenched with a few drops of glacial acetic acid until bubbling is ceased. Glacial acetic acid is further removed by adding 0.5 mL methanol/acetic acid of 9:1 (v/v), and the solution is evaporated under N$_2$. The last step (addition of methanol/acetic acid and subsequent evaporation) is repeated three more times.

Then 0.5 mL of 100% methanol is added to the vial and the sample is evaporated to complete dryness. The last step is repeated two or three times.

The dry alditols are acetylated with 100 µL of acetic anhydride and 20 µL of 1-methyl imidazole. The sample is mixed thoroughly and allowed to stand for 15 minutes. The excess of acetic anhydride is quenched with 0.5 mL of ultrapure water. After 10 minutes at room temperature, 0.5 mL of dichloromethane is added and the solution is vortex mixed. The organic layer (lower) containing the alditol acetates is transferred to a 4 mL clean vial. Trace amounts of water are further removed by adding a pinch of sodium sulfate. The solution is then filtered through quartz wool (precombusted at 450°C for 4 hour) and the organic layer is transferred to another 4 mL vial. The dichloromethane is removed under a stream of N_2. The alditol acetates are then dissolved in methanol and analyzed by GC-FID or GC-MS. All sugar standards and samples are run in duplicate. The above procedure can be broken to leave samples standing only at a dry step, and then samples must kept refrigerated. Alditol acetate derivatives are stable for 3–4 months when kept at 4°C in the dark.

3.5.2.2 GC-FID/MS System Description

Gas chromatography of the alditol acetates is performed using a DB-5 (5% phenyl-methylpolysiloxane) fused silica capillary column (0.25 mm ID, 0.20 mm film, 30 m length), installed in a Carlo Erba, Agilent, or HP gas chromatograph equipped with a flame ionization or mass spectrometer detector. Helium is generally used as carrier gas at 0.5 or 1 mL min^{-1}. Both the FID and the port injector (split/splitless) are maintained at a constant temperature of 300°C. The injector generally operates in the splitless mode (valve reopened 1 minute after injection). When mass spectrometers are employed as the detection mode they are commonly operated at 70 eV (electron impact), while the source temperature is set to 250°C.

3.5.2.3 Operating Procedure and Recommendations

Alditol acetate derivatives are injected at 90°C column temperature. After 8 minutes the oven temperature is rapidly raised by 10°C min^{-1} to 150°C and then to 230°C by 2°C min^{-1}. The oven is maintained at this temperature for about 47 minutes. Alditol acetates are eluted within 20–45 minutes (Figure 3.2). Peaks are identified (through retention times or their mass spectra) using a standard mixture of seven neutral monosaccharides: arabinose, rhamnose, fucose, xylose, mannose, galactose, and glucose (Figure 3.2). Myo-inositol or inositol is generally employed as an internal standard for quantification.

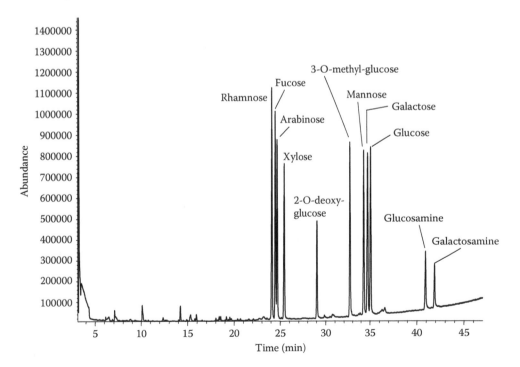

FIGURE 3.2 Typical GC-MS chromatogram of a standard mixture of sugars (16.4 ng μL^{-1} e.a.) after derivatization to their alditol acetate derivatives. Analytical conditions as described in the text (see Sections 3.5.2.2 and Sections 3.5.2.3).

Precision of both GC-FID/MS techniques fall in the <20% range at the 150–200 nM level, and the detection limit in the low ppb level ($\mu g\ L^{-1}$) (Panagiotopoulos and Sempéré, 2005a). GC-MS is generally more sensitive and more expensive than the GC-FID. Nevertheless, sugar compounds can be identified not only by their retention times but also by their specific mass spectral fragmentation pattern, which provides additional verification about the identity of the sugar. In addition, unknown sugars present in a sample may be identified by MS. From this perspective it is preferable to choose an MS detector than an FID.

3.6 CONCLUSION AND FUTURE TRENDS

Sugars are important organic components of the DOM; however, literature results cannot be compared to each other because of the various extraction protocols from marine samples as well as the different analytical techniques used by marine biogeochemists. Here we present the most employed procedures (extraction protocols and analytical conditions) for spectrophotometric and chromatographic sugar analysis. The two most common spectrophotometric methods are the PSA and TPTZ, while for individual sugar determination HPAEC-PAD and GC-FID/MS have been used.

The HPAEC-PAD technique is unique in that detection is performed without derivatization and has a very low detection limit compared to GC-MS. On the other hand, although the GC-MS technique is tedious (preparation of volatile derivatives; see above), it includes a much broader spectra of sugars in the analysis (not only neutral sugars), including amino sugars, methylated, methyldeoxy, 2-deoxy sugars, and 6-deoxy sugars, for which authentic standards are not always available (Figure 3.2; Moers et al., 1989; Klok et al., 1984; Panagiotopoulos et al., 2007). Future analytical developments consist of developing an extraction procedure and analytical techniques for the analysis of acidic sugars (uronic acids, sulfated and phosphorylated sugars), for which little biogeochemical information is available. In this regard, the combination of HPAEC and MS may provide crucial structural information for these compounds. However, interfacing anion exchange chromatography with MS detection is a technological challenge. Typical alkali acetate and hydroxide eluents are not compatible with atmospheric pressure ionization due to their nonvolatility and high conductance, and therefore, a desalting device needs to be installed between the column and the MS.

Recently, the Dionex Corp. made this coupling feasible (Cataldi et al., 2000; Bruggink et al., 2005), and the first applications were made in DNA-glucosylated compounds (Bidmon et al., 2007), glucosylflavones from flowering plants (Kite et al., 2006), and intracellular metabolites (Van Dam et al., 2002). To the best of our knowledge there are no environmental applications for sugar determination using HPAEC-MS/MS, and therefore it will be very interesting to perform such sugar analysis with this new technique.

ACKNOWLEDGMENTS

We are grateful to Dr. J.-F. Rontani for valuable advice and assistance on the GC-MS analysis, and Dr. D. Repeta, who kindly provided the alditol acetate derivatization protocol. We also express our gratitude for their interest in this work and fruitful discussions to Drs. R. Sempéré and B. Charrière. Funding was provided by the region Provence Alpes Côte d'Azur (PACA) and the Conseil General Bouches du Rhône (UV PACA project).

REFERENCES

Aluwihare, L. I., D. J. Repeta, and R. F. Chen. 1997. A major biopolymeric component to dissolved organic carbon in surface water. *Nature* 387:166–69.
Aluwihare, L. I., D. J. Repeta, and R. F. Chen. 2002. Chemical composition and cycling of dissolved organic matter in the Mid-Atlantic Bight. *Deep-Sea Research II* 49:4432–37.

Bidmon, C., M. Frischmann, and M. Pischetsrieder. 2007. Analysis of DNA-bound advanced glycation end-products by LC and mass spectrometry. *Journal of Chromatography* 855B:51–58.

Borch, N. H., and D. L. Kirchman. 1997. Concentration and composition of dissolved combined neutral sugars (polysaccharides) in seawater determined by HPLC-PAD. *Marine Chemistry* 57:85–95.

Bruggink, C., R. Maurer, H. Herrmann, S. Cavalli, and F. Hoefler. 2005. Analysis of carbohydrates by anion exchange chromatography and mass spectrometry. *Journal of Chromatography* 1085A:104–9.

Burney, C. M., P. G. Davis, K. M. Johnson, and J. McN. Sieburth. 1982. Diel relationships between microbial tropics groups and in-situ dissolved carbohydrate dynamics in the Carribean Sea. *Marine Biology* 67:311–22.

Burney, C. M., and J. McN. Sieburth. 1977. Dissolved carbohydrates in seawater. II. A spectrophotometric procedure for total carbohydrate analysis and polysaccharide estimation. *Marine Chemistry* 5:15–28.

Cataldi, T. R. I., C. Campa, and G. E. de Benedetto. 2000. Carbohydrate analysis by high-performance anion-exchange chromatography with pulsed amperometric detection: The potential is still growing. *Fresenius Journal of Analytical Chemistry* 368:739–58.

Cataldi, T. R. I., C. Campa, G. Margiotta, and S. A. Bufo. 1998. Role of barium ions in the anion-exchange chromatographic separation of carbohydrates and pulsed amperometric detection. *Anal. Chem.* 70:3940–3945.

Cheng, X., and L. A. Kaplan. 2001. Improved analysis of dissolved carbohydrates in stream water with HPLC-PAD. *Analytical Chemistry* 73:458–61.

Chin, W. C., M. V. Orellana, and P. Verdugo. 1998. Spontaneous assembly of marine dissolved organic matter into polymer gels. *Nature* 391:568–72.

Colombini, M. P., A. Ceccarini, and A. Carmignani. 2002. Ion chromatography characterization of polysaccharides in ancient wall paintings. *Journal of Chromatography* 968A:79–88.

Cowie, G. L., and J. I. Hedges. 1984. Determination of neutral sugars in plankton, sediments, and wood by capillary gas chromatography of equilibrated isomeric mixtures. *Analytical Chemistry* 56:497–504.

Dionex Corporation Sunnyvale. 1989. Dionex Technical Note 21 TN21.

Dubois, M., K. A. Gilles, J. K. Hamilton, P. A. Rebers, and F. Smith. 1956. Colorimetric method for determination of sugar and related substances. *Analytical Chemistry* 28:350–56.

Eklund, G., B. Josefsson, and C. Roos. 1977. Gas-liquid chromatography of monosaccharides at the pictogram level using glass capillary columns, trifluoacetyl derivatization and electron-capture detection. *Journal of Chromatography* 142:575–85.

Fuhrman, J. A. 1999. Marine viruses and their biogeochemical and ecological effects. *Nature* 399:541–48.

Gerchakov, S. M., and P. G. Hatcher. 1972. Improved technique for analysis of carbohydrates in sediments. *Limnology and Oceanography* 17:938–43.

Granum, E., S. Kirkvold, and S. M. Myklestad. 2002. Cellular and extracellular production of carbohydrates and amino acids by the marine diatom *Skeletonema costatum*: Diel variations and effects of N depletion. *Marine Ecology Progress Series* 242:83–94.

Granum, E., and S. M. Myklestad. 2001. Mobilization of beta-1,3-glucan and biosynthesis of amino acids induced by NH_4^+ addition to N-limited cells of the marine diatom *Skeletonema costatum* (Bacillariophyceae). *Journal of Phycology* 37:772–82.

Handa, N. 1967. Identification of carbohydrates in marine particulate matters and their vertical distribution. *Records of Oceanographic Works in Japan* 9:65–73.

Handa, N., and H. Tominaga. 1969. A detailed analysis of carbohydrates in marine particulate matter. *Marine Biology* 2:228–35.

Hanisch, K., B. Scweitzer, and M. Simon. 1996. Use of dissolved carbohydrates by planktonic bacteria in a mesotrophic lake. *Microbial Ecology* 31:41–55.

Hernes, P. J., J. I. Hedges, M. L. Peterson, S. G. Wakeham, and C. Lee. 1996. Neutral carbohydrate geochemistry of particulate material in the central equatorial Pacific. *Deep-Sea Research II* 43:1181–1204.

Hung, C. C., and P. H. Santschi. 2001. Spectrophotometric determination of total uronic acids in seawater using cation-exchange separation and pre-concentration lyophilization. *Analytical Chimica Acta* 427:111–17.

Johnson, D. C., and W. R. LaCourse. 1990. Liquid chromatography with pulsed amperometric detection at gold and platinum electrodes. *Analytical Chemistry* 62:589–97.

Jørgensen, N. O. G., and R. E. Jensen. 1994. Microbial fluxes of free monosaccharides and total carbohydrates in freshwater determined by PAD-HPLC. *FEMS Microbial Ecology* 14:79–94.

Kaiser, K., and R. Benner. 2000. Determination of amino sugars in environmental samples with high salt content by high-performance anion-exchange chromatography and pulsed amperometric detection. *Analytical Chemistry* 72:2566–72.

Kite, G. C., E. A. Porter, F. C. Denison, R. J. Grayer, N. C. Veitch, I. Butler, and M. S. J. Simmonds. 2006. Data-directed scan sequence for the general assignment of C-glycosylflavone O-glycosides in plant extracts by liquid chromatography-ion trap mass spectrometry. *Journal of Chromatography* 1104A:123–31.

Klok, J., H. C. Cox, M. Baas, J. W. de Leeuw, and P. A. Schenck. 1982. Analysis of complex mixtures of partially methylated alditol acetates by capillary gas chromatography, gas chromatography-electron impact mass spectrometry and gas chromatography-chemical ionization mass spectrometry. *Journal of Chromatography* 253A:53–64.

Klok, J., H. C. Cox, M. Baas, P. J. W. Schuyl, J. W. de Leeuw, and P. A. Schenck. 1984. Carbohydrates in recent marine sediments. I. Origin and significance of deoxy- and O-methyl-monosaccharides. *Organic Geochemistry* 7:73–84.

Lancelot, C., and S. Mathot. 1985. Biochemical fractionation of primary production by phytoplankton in Belgian coastal waters during short- and long-term incubations with ^{14}C-bicarbonate. II. *Phaeocystis pouchetii* colonial population. *Marine Biology* 86:227–32.

Lee, R. E. 1999. *Phycology.* Cambridge, UK: Cambridge University Press.

Liebezeit, G., and B. Behrends. 1999. Determination of amino acids and carbohydrates. In *Methods of seawater analysis*, ed. K. Grasshoff, K. Kremling, and M. Ehrhardt. Weinheim, Germany: Wiley-VCH.

Liu, D., P. T. S. Wong, and B. J. Dutka. 1973. Determination of carbohydrates in the lake sediment by a modified phenol-sulfuric acid method. *Water Research* 7:741–46.

Miyajima, T., H. Ogawa, and I. Koike. 2001. Alkali-extractable polysaccharides in marine sediments: Abundance, molecular size distribution, and monosaccharide composition. *Geochimica et Cosmochimica Acta* 65:1455–66.

Modzeleski, J. E., W. A. Laurie, and B. Nagy. 1971. Carbohydrates from Santa Barbara Basin sediments: Gas chromatography-mass spectrometric analysis of trimethylsilyl derivatives. *Geochimica et Cosmochimica Acta* 35:825–38.

Moers, M. E. C., J. J. Boon, J. W. de Leeuw, M. Baas, and P. A. Schenck. 1989. Carbohydrate speciation and Py-MS mapping of peat samples from a subtropical open marsh environment. *Geochimica et Cosmochimica Acta* 53:2011–21.

Mopper, K. 1977. Sugars and uronic acids in sediment and water from the Black Sea and North Sea with emphasis on analytical techniques. *Marine Chemistry* 5:585–603.

Mopper K. 1978. Improved chromatographic separations on anion-exchange resins. III. Sugars in borate medium. *Analytical Biochemistry* 87:162–68.

Mopper K., R. Dawson, G. Liebezeit, and H. P. Hansen. 1980. Borate complex ion exchange chromatography with fluorimetric detection for determination of saccharides. *Analytical Chemistry* 52:2018–22.

Mopper K., C. Schultz, L. Chevolot, C. Germain, R. Revuelta, and R. Dawson. 1992. Determination of sugars in unconcentrated seawater and other natural waters by liquid chromatography. *Environmental Science and Technology* 26:133–37.

Myklestad, S. M. 1988. Production, chemical structure, metabolism, and biological function of the (1,3)-linked, beta-D-glucans in diatoms. *Biological Oceanography* 6:313–26.

Myklestad, S. M., and K. Y. Børsheim. 2007. Dynamics of carbohydrates in the Norwegian Sea inferred from monthly profiles collected during 3 years at 66°N, 2°E. *Marine Chemistry* 107:475–85.

Myklestad, S. M., and A. Haug. 1972. Production of carbohydrates by the marine diatom *Chaetoceros affinis* var. Willei (Gran) Hustedt. II. Peliminary investigation of the extracellular polysaccharide. *Journal of Experimental Marine Biology and Ecology* 9:137–44.

Myklestad, S. V., E. Skånøy, and S. Hestmann. 1997. A sensitive method for analysis of dissolved mono- and polysaccharides in seawater. *Marine Chemistry* 56:279–86.

Pakulski, J., and R. Benner. 1992. An improved method for the hydrolysis and MBTH analysis of dissolved and particulate carbohydrates in seawater. *Marine Chemistry* 40:143–60.

Panagiotopoulos, C., R. Sempéré, R. Lafont, and P. Kervevé. 2001. Sub-ambient temperature effects on separation of monosaccharides by HPAEC-PAD. Application to marine chemistry. *Journal of Chromatography* 920A:13–22.

Panagiotopoulos, C., D. J. Repeta, and C. G. Johnson. 2007. Identification of methyl and deoxy sugars in marine high molecular weight dissolved organic matter (HMWDOM). *Organic Geochemistry* 38:884–96.

Panagiotopoulos, C., and R. Sempéré. 2005a. Analytical methods for the determination of sugars in marine samples: A historical perspective and future directions. *Limnology and Oceanography: Methods* 3:419–54.

Panagiotopoulos, C., and R. Sempéré. 2005b. The molecular distribution of combined aldoses in sinking particles in various oceanic conditions. *Marine Chemistry* 95:31–49.

Panagiotopoulos, C., and R. Sempéré. 2007. Sugar dynamics in large particles during in vitro degradation experiments. *Marine Ecology Progress Series* 330:67–74.

Penna, A., S. Berluti, N. Penna, and M. Magnani. 1999. Influence of nutrient ratios on the in vitro extracellular polysaccharide production by marine diatoms from the Adriatic Sea. *Journal of Plankton Research* 21:1681–90.

Repeta, D. J., T. M. Quan, L. I. Aluwihare, and A. Accardi. 2002. Chemical characterization of high molecular weight dissolved organic matter in fresh and marine waters. *Geochimica et Cosmochimica Acta* 66:955–62.

Rich, J. H., H. W. Ducklow, and D. L. Kirchman. 1996. Concentrations and uptake of neutral monosaccharides along 140°W in the equatorial Pacific: Contribution of glucose to heterotrophic bacterial activity and the DOM flux. *Limnology and Oceanography* 41:595–604.

Rocklin, R. D., and C. A. Pohl. 1983. Determination of carbohydrates by anion exchange chromatography with pulsed amperometric detection. *Journal of Liquid Chromatography* 6:1577–90.

Romankevich, E. A. 1984. *Geochemistry of organic matter in the ocean.* Berlin: Springer-Verlag.

Sempéré, R., M. Tedetti, C. Panagiotopoulos, B. Charrière, and F. Van Wambeke. 2008. Distribution and bacterial availability of dissolved neutral sugars in the South East Pacific. *Biogeosciences*, 5:1165–1173.

Sigleo, A. C. 1996. Biochemical components in suspended particles and colloids: Carbohydrates in the Potomac and Patuxent estuaries. *Organic Geochemistry* 24:83–93.

Skoog, A., and R. Benner. 1997. Aldoses in various size fractions of marine organic matter: Implications for carbon cycling. *Limnology and Oceanography* 42:1803–13.

Staats, N., L. J. Stal, and L. R. Mur. 2000. Exopolysaccharide production by the epipelic diatom *Cylindrotheca closterium*: Effects of nutrient conditions. *Journal of Experimental Marine Biology and Ecology* 249:13–27.

Strom, S. L., R. Benner, S. Ziegler, and M. J. Dagg. 1997. Planktonic grazers are a potentially important source of marine dissolved organic carbon. *Limnology and Oceanography* 42:1364–74.

Suttle, C. A. 2005. Viruses in the sea. *Nature* 437:356–61.

Tanoue, E., and N. Handa. 1987. Monosaccharide composition of marine particles and sediments from the Bering Sea and northern North Pacific. *Oceanologica Acta* 10:91–99.

Underwood, G. J. C., D. M. Paterson, and R. J. Parkes. 1995. The measurement of microbial carbohydrate exopolymers from intertidal sediments. *Limnology and Oceanography* 40:1243–53.

Van Dam, J. C., M. R. Eman, J. Frank, H. C. Lange, G. W. K. van Dedem, and S. J. Heijnen. 2002. Analysis of glycolytic intermediates in *Saccharomyces cerevisiae* using anion exchange chromatography and electrospray ionization with tandem mass spectrometric detection. *Analytica Chimica Acta* 460:209–18.

Wicks, R. J., M. A. Moran, L. J. Pittman, and R. E. Hodson. 1991. Carbohydrate signatures of aquatic macrophytes and their dissolved degradation products as determined by a sensitive high-performance ion chromatography method. *Applied and Environmental Microbiology* 57:3135–43.

Witter, A. E., and G. W. Luther III. 2002. Spectrophotometric measurement of seawater carbohydrate concentrations in neritic and oceanic waters from the U.S. Middle Atlantic Bight and the Delaware estuary. *Marine Chemistry* 77:143–56.

York, W. S., A. G. Darvill, M. McNeil, T. T. Stevenson, and P. Albersheim. 1985. Isolation and characterization of plant cell walls and cell wall components. *Methods in Enzymology* 118:3–40.

4 The Analysis of Amino Acids in Seawater

Thorsten Dittmar, Jennifer Cherrier,
and Kai-Uwe Ludwichowski

CONTENTS

4.1 INTRODUCTION

Amino acids are essential building blocks of life. A number of mechanisms are responsible for the presence of amino acids in seawater, including direct release by phytoplankton via passive diffusion or active exudation, zooplankton excretion, and bacterial exoenzyme production (Nagata, 2000; Bronk, 2002). Indirect mechanisms include viral lysis, heterotrophic grazing, fecal pellet dissolution, and death and decay of both autotrophic and heterotrophic organisms (Jumars et al., 1989; Carlson, 2002).

In the course of bacterial degradation the amino acid content of organic matter strongly decreases, and amino acids turn from a major component of living biomass into a minor component of the nonliving organic matter in seawater and sediments (Cowie and Hedges, 1992; Dittmar et al., 2001; Bronk, 2002). The residence times of amino acid compounds in a water column are a function of their molecular structure as well as the physiologic status and taxonomic makeup of the bacterial communities where the compounds are released and metabolized (del Giorgio and Cole, 2000; Giovanni and Rappé, 2000; Cherrier and Bauer, 2004). Due to the tight coupling between production and utilization processes of free amino acids (Billen and Fontigny, 1987; Fuhrman, 1987; Suttle et al., 1991; Keil and Kirchman, 1999) their concentrations are generally very low and range from undetectable concentrations (<1 nM) to 70 nM in surface waters and 4 nM in the deep ocean (Lee and Bada, 1977; Suttle et al., 1991; Keil and Kirchman, 1999; Bronk, 2002; Cherrier and Bauer, 2004). Dissolved combined amino acid concentrations have been observed to be 5 to 20 times higher than those of free amino acids and range from 150 to 4,200 nM in surface waters and 150 to 550 nM in deeper waters (Keil and Kirchman, 1991a, 1999; Dittmar et al., 2001; Cherrier and Bauer, 2004). Particulate amino acid concentrations vary strongly and can be >600 nM at the sea surface and <20 nM in the deep sea (Dittmar et al., 2001).

Compared to free amino acids and proteins, amino acids in structural polymers are less accessible to bacterial attack and become enriched over the course of degradation. Because of the preferential decay of free amino acids and proteins, and the relative enrichment of structural polymers, the amino acid composition changes characteristically over the course of degradation (Wakeham et al., 1997; Dauwe et al., 1999). Most amino acids have two (D and L) stereoisomers. D-amino acids are an important component of bacterial cell walls. Nonbacterial phytoplankton, however, does not contain D-amino acids. The relative ratio of D-amino acids increases during bacterial degradation. In the deep sea more than 40% of alanine is D-alanine, indicating that dissolved organic matter has been extensively reworked by bacteria (McCarthy et al., 1998; Dittmar et al., 2001). Amino acids abiotically racemize, but at ambient seawater temperatures the process is slow and does not contribute significantly to the natural abundance of D-amino acids (Dittmar et al., 2001). Because of these characteristic degradation patterns, amino acids are excellent molecular indicators for the degradation history of dissolved organic matter.

Modern analytical techniques are sensitive enough to quantify amino acids in seawater in natural abundance, and desalting is not required. Dissolved free amino acids (DFAAs) are analyzed directly in seawater. Bound or dissolved combined amino acids (DCAAs) must be hydrolized (Section 4.3) prior to amino acid analysis and are reported as the difference between total hydrolizable amino acids (THAAs) and DFAA.

Amino acids in seawater can be determined on the molecular level or as total amines. The analysis of total amines in seawater has the advantage of easy operation that allows its routine application on a research vessel. However, the reagents used for amino acid analysis (see below) are not specific to the amino acid functional group, $-(NH_2)COOH$, but to the amine group alone, $-NH_2$. Therefore, in contrast to chromatographic methods, which separate amino acids from each other and from the matrix, bulk methods include amines other than amino acids (North, 1975; Aminot and Kérouel, 2006). A current review and an optimized procedure for the determination of total dissolved free amines in seawater can be found in Aminot and Kérouel (2006). After hydrolysis (see below), the same method can be used for the determination of total combined amines.

For the molecular-level determination of amino acids, samples are analyzed after derivatization via high-performance liquid chromatography (HPLC) with fluorescence detection (Section 4.4.2). If chiral derivatization reagents are used, D- and L-amino acids are also separated, which provides an additional level of information (Section 4.4.1). Alternatively to HPLC, amino acid enantiomers can also be analyzed via gas chromatography (GC; McCarthy et al., 1998). If the GC is coupled to an isotope-ratio mass spectrometer, stable carbon and nitrogen isotopes ($\delta^{13}C$, $\delta^{15}N$) can be analyzed on individual amino acids, which provides further detail on the origin and cycling of the amino acids in seawater (McCarthy et al., 2004, 2007) (see also Chapter 6). The GC method is less sensitive than the HPLC method, and preconcentration of the seawater samples, for example, via ultrafiltration or solid-phase extraction, is required. In the following sections, practical guidelines are given for sampling, acid hydrolysis, and the molecular-level determination of amino acids and their stereoisomers in seawater via HPLC.

4.2 SAMPLING AND AVOIDANCE OF SAMPLE CONTAMINATION

Amino acids are abundant in the human environment but present in seawater at low concentrations. A human fingerprint can contain more amino acids than a liter of seawater. Extreme care must therefore be taken to avoid sample contamination at all analytical steps, including sampling. A careful determination of the procedural blank is mandatory for accurate and precise determination of amino acids in seawater. Prior to use, all nonvolumetric glassware should be rinsed with ultrapure water, covered with aluminum foil, and muffled at 500°C for at least 4 hours. Volumetric glassware and plastics should be rinsed with 1 M NaOH, soaked in 1 M HCL, thoroughly rinsed with ultrapure water, and dried in a dust-free environment. Material can be stored in muffled aluminum foil or cleaned glass or plastic containers. All reagents should be of the highest available quality. Reagents can easily contaminate after opening, especially through dust. Handling of all materials and open containers in a laminar flow hood with filtered air supply can reduce the risk of contamination.

Thorough blank tests using all materials and reagents should precede sampling and analysis. Many reagents and even UV-treated ultrapure water can contain considerable amounts of amino acids. The use of reagents from a different manufacturer or the use of bottled HPLC-quality water may be required in this case. Nonpowdered nitrile gloves (rinsed with ultrapure water) should be worn at all times during sampling and analysis, and contact with contaminated surfaces (e.g., door knobs, skin, bench tops) has to be avoided. If samples are drawn from a shared sampling system (e.g., rosette of Niskin bottles), other persons collecting subsamples from the sampling system have to be made aware of these standard precautions that must be maintained. Common sources of contamination on research vessels are lubricants on the hydro wire or conductivity-temperature-depth (CTD) cable, smoke, and the use of grease or sealants on the rosette system.

To separate particulate from dissolved amino acids, water samples have to be filtered immediately after sampling. For dissolved amino acid analysis, the collection of 10 mL water is recommended. For the recovery of sufficient particulate amino acids, the filtration of 100 mL is sufficient, even for samples collected from the deep sea. However, more representative particulate matter samples are obtained if larger volumes, usually 1 L, are filtered. Filters should be stored frozen at −20°C or below for subsequent analysis. Whatman GF/F glass fiber filters (nominal pore size = 0.7 μm) are most commonly used for this purpose. These filters are binder-free and can be muffled at 450°C. Higher combustion temperatures can negatively impact the performance of the filters. Excessive pressure or vacuum (exceeding 0.5 bar of pressure gradient) during filtration can break cells and has to be avoided. Gravity filtration directly from the Niskin bottles is preferable. For this purpose, a filtration cartridge containing a precombusted GF/F filter is attached to the spigot of the Niskin bottle with a Teflon tube (see Figure 8.3). The first 20 mL of permeate should be discarded before collecting the filtered sample. The filter can be recovered for particulate amino acid analysis. If only dissolved amino acids are of interest, disposable syringe filter cartridges can be used. If rinsed with 20 mL sample, these cartridges generally do not contaminate the samples and are easier to use than reusable cartridge filter holders. However, as a precaution, blanks for the disposable filter cartridges should always be collected. Some protocols require that samples be filtered through a smaller pore size filter than GF/F. In this case, samples can be cleanly filtered through thoroughly rinsed (>20 mL of sample) polycarbonate filters (0.2 μm). Depending on the manufacturer, soaking of the polycarbonate filters with 0.1 M HCl for several hours might be required to obtain acceptable blanks.

If gravity filtration is not practical, samples can be filtered via either vacuum filtration or syringe filtration. Vacuum filtration entails a relatively extensive setup. Because of the additional procedural steps, particular care must be taken to avoid sample contamination. All glassware must be precombusted and filtration should be carried out under a laminar flow hood, if available. Alternatively, syringe filtration is simpler and, as such, may be a preferred method if the filtration is to be done in the field. Simple polyethylene syringes (cleaned with NaOH and HCl as described above), without rubber stoppers or lubricants, are most suitable for filtration. When syringe filtering, the sample has to be pushed very slowly through the filter cartridge to avoid excess pressure and breakage of cells. Filtered samples can then be collected into either precombusted glass ampoules (20 mL) or precombusted glass vials (20 mL) equipped with acid-washed Teflon-lined closures. If the former, ampoules are fire-sealed immediately after sampling. Handheld butane burners are now widely available for this purpose (see Figure 2.1). After filtration, samples should be immediately frozen at −20°C or below. To avoid breakage, glass ampoules or vials should be filled half-full only (10 mL) and be frozen in an angular position. If freezers or dry ice are not available on site, samples must be preserved chemically. Acidification to pH 2 with HCl is the method of choice (10 M HCl; 10 μL per 10 mL sample) if free amino acids are not of major interest. Weak hydrolysis may occur at this pH, which may impact the distribution of free amino acids. Addition of sodium azide (sample concentration 1.5 mM) keeps samples stable at room temperature for over a month (Kaufman and Manley, 1998). The use of mercury chloride as a preservative may impact the amino acid distribution in the samples because of its strong complexation properties and affinity to organic sulfur. If possible, the

TABLE 4.1
Materials and Reagents for General Preparation and Sampling

Equipment and Materials	Reagents
Whatman binder-free GF/F glass fiber filters (25 or 47 mm)	Hydrochloric acid (10 M)
Where applicable, polycarbonate filters (0.2 μm)	Acid bath (HCl, 1 M)
Glass ampoules (20 mL) and butane gas burner or glass vials (20 mL) with Teflon-lined enclosures	Hydroxide bath (NaOH, 1 M)
Micropipette	Ultrapure water
Forceps	
Teflon tubing and fittings	
Nitrile gloves	
Furnace (for combustion of glassware and filters)	
Additional equipment for gravity filtration:	
GF/F filter housing (47 or 25 mm)	
Additional equipment for syringe filtration:	
Syringe filter cartridge (47 or 25 mm)	
Polyethylene syringes	
Additional equipment for vacuum filtration:	
Vacuum filtration unit (precombusted glass or acid-washed polycarbonate), including vacuum pump	
Laminar flow hood	

use of azide and mercury should be avoided in the field, because of their environmental and human toxicity. A list of materials and reagents needed for sampling is presented in Table 4.1.

4.3   HYDROLYSIS

Dissolved free amino acids (DFAAs) are analyzed directly in the filtered seawater samples, without the need for additional procedural steps. For the analysis of hydrolyzable amino acids (THAAs, i.e., free plus combined), samples must first be hydrolyzed prior to analysis. Acidic hydrolysis can be performed in liquid or vapor phase using HCl. Liquid-phase hydrolysis induces considerably less racemization than vapor-phase hydrolysis (Kaiser and Benner, 2005). For enantiomeric amino acid analysis liquid-phase hydrolysis is therefore preferable. However, it has been reported that vapor-phase hydrolysis is more efficient for the hydrolysis of nonprotein amino acids, and it yields up to 300% higher amino acid concentrations in seawater than traditional liquid-phase hydrolysis (Keil and Kirchman, 1991b). Details on vapor-phase hydrolysis can be found in Tsugita et al. (1987) and Keil and Kirchman (1991b). A recent update and detailed discussion of the hydrolysis-induced racemization of vapor-phase hydrolysis is in Kaiser and Benner (2005).

During acidic hydrolysis glutamine (Gln) and asparagine (Asn) react into glutamic acid (Glu) and aspartic acid (Asp), respectively. Consequently, these amino acids should be reported as the sum of Gln + Glu (Glx) and Asp + Asn (Asx). Methionine and tryptophan are usually omitted because they are recovered inefficiently from acid hydrolysis (Kaiser and Benner, 2005). As a result, these amino acids are rarely determined in marine samples. If tryptophan is of interest, alkaline hydrolysis as described in Wu and Tanoue (2001) can be used.

The acidic liquid-phase hydrolysis method described in the following is a streamlined version of the approach described in Fitznar et al. (1999). For hydrolysis of water samples, 0.5 mL of the sample is transferred into a 2 mL glass ampoule, and 0.5 mL HCl (12 M) and 5 μL ascorbic acid (11 mM) are added. Ascorbic acid is used to prevent sample oxidation during hydrolysis (Robertson et al., 1987). For the analysis of particulate amino acids, a 25 mm GF/F filter or a quarter of a 47 mm

---

TABLE 4.2
Materials and Reagents for Hydrolysis

Equipment and Materials	Reagents
Heating and N_2 evaporation manifold, or freeze drier with HCl trap	Hydrochloric acid (12 M for water samples, or 6 M for filter samples)
Furnace	Sodium hydroxide solution (0.1 M)
pH meter	Ascorbic acid solution (11 mM, 2 g L^{-1})
Butane gas burner	Compressed nitrogen (ultrapure)
Micropipette	
Pasteur pipettes	
Glass ampoules (2 mL)	
Autosampler vials (1 mL)	
Forceps	

GF/F filter is folded and transferred into the 2 mL ampoule, and 1 mL HCl (6 M) and 5 µL ascorbic acid (11 mM) are added. The headspace of the ampoules is then flushed with ultrapure N_2 for about 1 minute, using Pasteur glass pipettes. Immediately thereafter, the ampoules are fire-sealed (see Figure 2.1). The ampoules are then placed into a furnace, and the temperature is kept stable (within ±2°C) at 110°C for 24 hours. After hydrolysis, aliquots of the samples (0.5 mL) are transferred into HPLC autosampler vials. The supernatant of the particulate samples has to be removed very carefully to avoid particles in the samples. Centrifugation of the supernatant helps to remove particles from the supernatant but is usually not required. The samples are then evaporated to complete dryness in a heated manifold (80°C) under a N_2 stream or lyophilized in a freeze drier, equipped with an HCl trap. The samples are rehydrated with 100 µL ultrapure water whose pH is carefully adjusted to 10.0 with NaOH (0.1 M). Losses of amino acids do not occur during this procedure. Nevertheless, L-glutamic acid-methyl-ester may be added as an internal standard after hydrolysis to monitor potential losses during drying and redissolution. The sample hydrolysates should be stored at 4°C and analyzed within a day of hydrolysis. If immediate analysis is not possible, the hydrolysates can be stored frozen (–20°C) or preserved with sodium azide (1.5 mM) for more than a month (Kaufman and Manley, 1998). A summary of materials and reagents needed for hydrolysis is presented in Table 4.2.

Hydrolysis-induced racemization should convert less than 5% of free amino acids from L- into D-enantiomers. If within this limit, a numeric correction of the sample results is usually not performed. However, hydrolysis-induced racemization should be frequently monitored by exposing an amino acid standard mix to the same hydrolytic procedure as for samples.

4.4 AMINO ACID ANALYSIS

4.4.1 ENANTIOMERIC AMINO ACID ANALYSIS

The enantiomers of the individual amino acids, which are released by hydrolysis or are freely dissolved in seawater, can be analyzed via HPLC and fluorescence detection after derivatization with a chiral reagent (Kaufman and Manley, 1998; Fitznar et al., 1999). Derivatization is performed with *o*-phthaldialdehyde (OPA) and *N*-isobutyryl-L/D-cysteine (IBLC or IBDC, respectively) before injection onto the HPLC column. The amino acid enantiomers (D-AA, L-AA) react into diastereomers that can be separated on a conventional reversed-phase C_{18} column:

$$D\text{-AA} + IBLC + OPA \neq L\text{-AA} + IBLC + OPA$$

and

$$D\text{-AA} + IBDC + OPA \neq L\text{-AA} + IBDC + OPA$$

However, the reaction products of IBLC/OPA and a D-amino acid is the same as the products from the reaction of IBDC/OPA with the respective L-amino acid:

$$D\text{-AA} + IBLC + OPA = L\text{-AA} + IBDC + OPA$$

and

$$D\text{-AA} + IBDC + OPA = L\text{-AA} + IBLC + OPA$$

This fact can be used for unambiguous peak identification and to identify coelution. Each sample is analyzed twice, first with IBLC + OPA and then with IBDC + OPA. The retention times of D- and L-amino acids are inverted in the two runs, but nonstereoisomeric compounds always elute at the same time (Figure 4.1). Only the amino acids that are unambiguously identified in both runs should be reported.

Depending on the manufacturer, some of the reagents used for derviatization contain significant amounts of amino acids; for example, the IBC reagents can contain detectable amounts of racemic alanine. Blank correction of the results is important. For blank correction of hydrolyzed samples, hydrolysates of ultrapure water are derivatized and analyzed using the same procedure as for samples. For the determination of free amino acids, ultrapure water is directly analyzed without preceding hydrolysis. A blank is unacceptable if it exceeds 20% of the individual amino acid concentrations in the sample. In this case, the use of new reagents and thorough cleaning of all analytical material might be required prior to the analysis of seawater samples.

A consistent sample pH is essential for reproducible derivatization and retention time. Therefore, OPA is dissolved in a borate buffer solution, whose pH is carefully adjusted to 9.5. However, even traces of hydrochloride acid from the hydrolysis can considerably reduce the efficiency of the derivatization and change retention times. For the preparation of the derivatization solutions, first a borate buffer solution (0.5 M, 31.8 g L^{-1} of boric acid) is prepared. The pH is adjusted to 9.5 by adding successively sodium hydroxide solution (12 M) and continuously monitoring the pH with a pH meter. OPA is dissolved in the buffer at a concentration of 5 g L^{-1}. The IBLC and IBDC reagents (5 g L^{-1}) are dissolved first in methanol, and then water is added at a ratio of 4:6 (methanol:water). The OPA, IBLC, and IBDC solutions are unstable at room temperature. The solutions are stable for a few days at 4°C. For storage, the derivatization solutions can be divided into autosampler vials and stored frozen (–20°C) for more than 1 year. The reagent solutions are thawed at room temperature immediately prior to analysis. A summary of materials and reagents needed for derivatization is presented in Table 4.3.

Derivatization is performed immediately before injection with an intelligent autosampler either in a sample vial or within the sample loop. Two or three mixing cycles are usually performed. Manual derivatization is also possible, but has to be performed immediately before injection onto the HPLC column. In order to keep the blank signal at a minimum, the use of the smallest possible amount of derivatization reagents is recommended. For derivatization, 100 μL of sample is thoroughly mixed with 2 μL OPA and 2 μL IBC (IBLC or IBDC) reagents. The reagent:sample ratio and the reaction time have considerable influence on the reaction efficiency (Kaufman and Manley, 1998). For reproducible results, the observation of precise reaction times and mixing rates is essential. The autosampler should be kept at 4°C to prevent sample and reagent decay.

The chromatographic separation of the amino acid enantiomers is performed on a conventional reversed-phase C$_{18}$ column, and a linear binary solvent gradient between an aqueous buffer solution (pH 6.0) and acetonitrile. The buffer solution consists of 25 mM sodium acetate hydrate

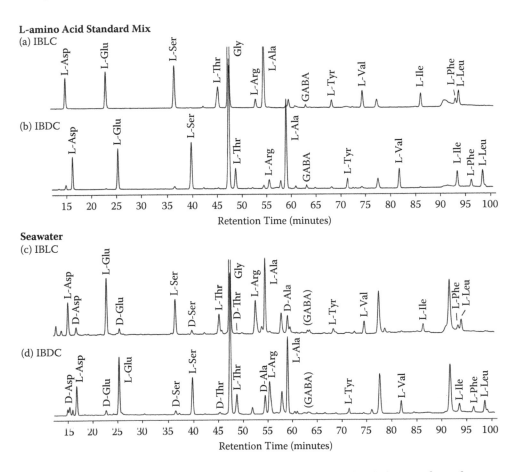

FIGURE 4.1 Chromatograms of an L-amino acid standard mix (a, b) and hydrolysates of a surface seawater sample from the central East Sea, offshore Korea (c, d). Duplicate runs are shown, using IBLC (a, c) and IBDC (b, d) as derivatization reagents. The composition of the standard mix was chosen to reflect approximately the concentration of the individual amino acids in seawater. In an IBLC run, the L-amino acids elute at the same retention time as the respective D-amino acids in an IBDC run. The samples were run on a Shimadzu HPLC system (10ADvp), equipped with a high-pressure binary solvent delivery system, degasser, intelligent autosampler, sample cooler, column heater, and fluorescence detector. Samples and derivatization reagents were kept at 4°C in the autosampler. For derivatization 90 μL sample was mixed with 2 μL OPA and 2 μL IBC reagents at 4°C in the autosampler. An Alltech Alltima HP C18 column (5 μm; 4.6 × 150 mm) with an Alltech Alltima HP C18 guard column (5 μm; 4.6 × 7.5 mm) was used for separation. The column temperature was regulated at 25°C. Sample injection volume was 10 μL, and the flow rate of the eluent was 1.1 mL min⁻¹. Fluorescence was detected at 330/445 nm. A gradient program with acetate buffer (Table 4.3) and acetonitrile was used for separation: 0–5 minutes 2% acetonitrile, 5–105 minutes linear gradient 2%–24% acetonitrile, 105–110 minutes linear gradient 24%–80% acetonitrile, 110–115 minutes hold at 80%, 115–120 minutes linear gradient 80%–2% acetonitrile, hold at 2% for 5 minutes before next injection.

(3.4 g L⁻¹), whose pH is carefully adjusted to 6.0 with hydrochloric acid (0.1 M). Careful pH adjustment of the mobile phase is essential, because even small variations of pH can cause considerable changes in retention time. Injection volume, gradient program, flow rates, and column temperature depend on the column used. These parameters should be optimized for baseline separation of all amino acids. To protect the main chromatographic column, a guard column should be used and exchanged when fluctuations in retention time or pressure are observed. An example of chromatographic parameters is given in Figure 4.1. After each run, the column is rinsed with

TABLE 4.3
Materials and Reagents for Enantiomeric Amino Acid Analysis

Equipment and Materials	Reagents
Binary solvent gradient HPLC system	OPA reagent: o-phthaldialdehyde (5 g L^{-1}) in boric acid (0.5 M,
Intelligent autosampler for sample	31.8 g L^{-1}), adjusted to pH 9.5 with sodium hydroxide (12 M)
derivatization and sample cooling (4°C)	IBC reagent: N-isobutyryl-D-cysteine (5 g L^{-1}) or N-isobutyryl-L-
HPLC column heate	cysteine (5 g L^{-1}) in methanol (400 mL L^{-1}) and water
HPLC fluorescence detector r	(600 mL L^{-1})
HPLC reversed-phase C$_{18}$ column	Mobile phase A: sodium acetate hydrate 25 mM (3.4 g L^{-1}),
pH meter	adjusted to pH 6.0 with hydrochloric acid (0.1 M)
Microbalance	Mobile phase B: acetonitrile
Micropipettes	
Volumetric flasks	

100% acetonitrile. Fluorescence is detected at an excitation wavelength of 330 nm and an emission wavelength of 445 nm. Sensitivity can be increased approximately fivefold at an excitation of 230 nm. However, more interfering peaks are usually observed at this wavelength (Fitznar et al., 1999). If a dual-wavelength detector is available, simultaneous detection at 330/445 nm and 230/445 nm is recommended. For quantification, the strongest and best-resolved peak can then be chosen from the two detector signals.

The elution time of the individual amino acids depends on the chromatographic conditions. For initial peak identification, a solution of each individual amino acid should be injected. For external calibration, a standard mix containing at least glycine (Gly), γ-amino-butyric acid (GABA), and all L-amino acids can be used. The proportion of each amino acid in the standard solution should reflect approximately the expected proportions in the sample. Lysine is known to form unstable OPA-IBC derivates and is therefore not considered. An example of stock solution concentration that reflects deep sea concentrations is given in Table 4.4. The amino acids used for the preparation of the stock solution can be obtained as powder, which should be kept cool and dry, ideally in a desiccator at 4°C. For the quantification of D-amino acids in an IBLC run, the IBDC run of the L-amino acids standard mix can be used, and vice versa. A stock solution of the standard mix can be stored frozen (–20°C) in 1 mL glass ampoules for years. Immediately before use, standard solutions are prepared from the stock. The remaining stock solution from an opened ampoule is discarded. For calculating the sample concentration, the procedural blank has to be subtracted and an enrichment factor from evaporating and redissolving (2.5 for dissolved amino acid analysis) has to be taken into account.

The detection limit for each individual amino acid should be 1 nM or less. The coefficient of variation between multiple runs should be <5% for each individual amino acid. One duplicate run (with IBLC and IBDC) takes approximately 4 hours.

4.4.2 SIMPLIFIED AMINO ACID ANALYSIS

If the analysis time of the above-described method is unacceptably long, and if the advantages of determining enantiomer ratios and obtaining unambiguous peak identification can be compromised, the HPLC approach described by Lindroth and Mopper (1979) and Cowie and Hedges (1992) can be used. A recent update of the method can be found in Maie et al. (2006). The main difference between this and the enantiomeric approach is that no chiral derivatization reagent is used. The IBC reagent is substituted by mercaptoethanol. The number of peaks is therefore smaller, and chromatography can be performed faster, employing a steeper gradient program.

TABLE 4.4

Suggested Composition of the Amino Acid Stock Solution, Reflecting the Approximate Concentrations of the Individual Amino Acids in Hydrolyzed Deep-Sea Water (Dittmar et al., 2001)

Amino Acid	Abbreviation	Stock Solution (μM)	Stock Solution (mg L^{-1})
Aspartic acid	Asp	2.0	0.27
Glutamic acid	Glu	1.5	0.22
Serine	Ser	1.5	0.16
Threonine	Thr	1.0	0.12
Glycine	Gly	5.0	0.38
Arginine	Arg	1.0	0.14
Alanine	Ala	5.0	0.45
γ-Amino butyric acid	GABA	0.5	0.05
Tyrosine	Tyr	0.5	0.09
Valine	Val	1.0	0.12
Isoleucine	Ile	0.5	0.07
Phenylalanine	Phe	0.5	0.08
Leucine	Leu	1.5	0.20

Note: The stock solution is prepared in ultrapure water. For the preparation of calibration standard solutions the following volumes (μL) of stock solution are diluted to 1 mL with ultrapure water: 5, 10, 15, 20, 25.

Source: Adapted from Dittmar et al. (2001).

OPA reagent is prepared in a borate buffer solution as described above, and 100 mL L^{-1} mercaptoethanol is added, as the second derivatization reagent. The derivatization reagents are also commercially available readily mixed. The same chromatographic conditions and detection parameters as described above are used, but a faster gradient between acetate buffer and acetonitrile can be used to save analysis time. Depending on the column used, acetonitrile may be substituted for methanol. A typical linear gradient starts with 25% methanol to 45% at 20 minutes and further increases to 80% methanol at 35 minutes at a flow rate of 0.9 mL min^{-1}. Then the system is returned to initial conditions and allowed to equilibrate for 3 minutes before the next injection. A standard buffering system and elution gradient is outlined in Hill et al. (1979), but as described above, the exact chromatographic conditions will have to be modified according to the specific column used to obtain baseline separation of the amino acids listed in Table 4.4. Peak identification is done by retention time only, and coelutions may not always be identifiable. Detection limits, precision, and reproducibility are similar to those in the enantiomeric analytical method.

4.5 PERSPECTIVES

Marine dissolved organic matter is one of the largest organic carbon pools on earth, but also one of the most ill-defined components in global biogeochemical models. The history and source of dissolved organic matter is imprinted in its molecular composition, and major advances in this field of research are closely related to our ability to characterize natural organic matter at the molecular level. Amino acids are among the very few groups of organic components that can be analyzed at the molecular level at ambient seawater concentrations. The recent advent of advanced analytical techniques, in particular ultra-high-resolution MS (Fourier transform ion cyclotron resonance MS) and high-resolution multidimensional nuclear magnetic resonance spectroscopy (NMR), has revealed new insights into the molecular composition of marine dissolved organic matter in unsurpassed detail (Koch et al., 2005; Dittmar and Koch, 2006; Hertkorn et al., 2006). However, preconcentration and desalting steps required for these novel analytical techniques inevitably introduce artifacts. Major efforts are being undertaken for a more efficient isolation of dissolved organic matter from the saline seawater matrix (Vetter et al., 2007; Dittmar et al., 2008), but complete recovery has not been achieved yet. Another limitation of the novel analytical techniques is the enormous instrumental requirements and costs involved. The analysis of both DFAA and THAA, on the other hand, is comparatively simple and can be performed in most environmental analytical laboratories. The analytical techniques are well established, and will remain as standard analytical tools in marine chemistry. Reduced analysis time may be achieved in the future through the use of novel specialty HPLC columns, or the adoption of the newest ultra-high-performance liquid chromatography (UPLC) techniques. The data generated by amino acid analysis have probably not been explored to their full potential. In particular, new multivariate statistical approaches for the interpretation of amino acid data will probably lead to new insights into the biogeochemical cycling of organic matter in the ocean (Dauwe et al., 1999; Dittmar, 2004).

REFERENCES

Aminot, A., and R. Kérouel. 2006. The determination of total dissolved free primary amines in seawater: Critical factors, optimized procedure and artifact correction. *Marine Chemistry* 98:223–40.

Billen, G., and A. Fontigny. 1987. Dynamics of a *Phaeosystis*-dominated spring bloom in Belgian coastal waters. II. Bacterioplankton dynamics. *Marine Ecology Progress Series* 37:249–57.

Bronk, D. A. 2002. Dynamics of DON. In *Biogeochemistry of marine dissolved organic matter*, ed. D. A. Hansell and C. Carlson. San Diego: Academic Press.

Carlson, D. A. 2002. Production, consumption and cycling of DOM. In *Biogeochemistry of marine dissolved organic matter*, ed. D. A. Hansell and C. Carlson. San Diego: Academic Press.

Cherrier, J., and J. E. Bauer. 2004. Bacterial utilization of transient plankton-derived dissolved organic carbon and nitrogen inputs in surface ocean waters. *Aquatic Microbial and Ecology* 35:229–41.

Cowie, G. L., and J. I. Hedges. 1992. Sources and reactivities of amino acids in coastal marine environment. *Limnology and Oceanography* 37:703–24.

Dauwe, B., J. J. Middelburg, P. M. J. Herman, and C. H. R. Heip. 1999. Linking diagenetic alteration of amino acids and bulk organic matter reactivity. *Limnology and Oceanography* 44:1809–14.

del Giorgio, P. A., and J. J. Cole. 2000. Bacterial energetics and growth efficiency. In *Microbial ecology of the oceans*, ed. D. L. Kirchman. New York: Wiley-Liss.

Dittmar, T. 2004. Evidence for terrigenous dissolved organic nitrogen in the Arctic deep-sea. *Limnology and Oceanography* 49:148–56.

Dittmar, T., H. P. Fitznar, and G. Kattner. 2001. Origin and biogeochemical cycling of organic nitrogen in the eastern Arctic Ocean as evident from D- and L-amino acids. *Geochimica et Cosmochimica Acta* 65:4103–14.

Dittmar, T., and B. P. Koch. 2006. Thermogenic organic matter dissolved in the abyssal ocean. *Marine Chemistry* 102:208–17.

Dittmar, T., B. P. Koch, N. Hertkon, and G. Kattner. 2008. A simple and efficient method for the solid-phase extraction of dissolved organic matter (SPE-DOM) from seawater. *Limnology and Oceanography: Methods* 6:230–35.

Fitznar, H. P., J. M. Lobbes, and G. Kattner. 1999. Determination of enantiomeric amino acids with high-performance liquid chromatography and pre-column derivatisation with *o*-phthaldialdehyde and *N*-isobutyrylcysteine in seawater and fossil samples (mullusks). *Journal of Chromatography* 832A:123–32.

Fuhrman, J. 1987. Close coupling between release and uptake of dissolved free amino acids in seawater studied by an isotope dilution approach. *Marine Ecology Progress Series* 37:45–52.

Giovanni, S., and M. Rappé. 2000. Evolution, diversity, and molecular ecology of marine prokaryotes. In *Microbial ecology of the oceans*, ed. D. L. Kirchman. New York: Wiley-Liss.

Hertkorn, N., R. Benner, M. Frommberger, P. Schmitt-Kopplin, M. Witt, K. Kaiser, A. Kettrup, and J. I. Hedges. 2006. Characterization of a major refractory component of marine dissolved organic matter. *Geochimica et Cosmochimica Acta* 70:2990–3010.

Hill, D. W., F. H. Walters, T. D. Wilson, and J. D. Stuart. 1979. High performance liquid chromatographic determination of amino acids in the picomole range. *Analytical Chemistry* 51:1338–41.

Jumars, P. A., D. L. Penry, J. A. Baross, M. J. Perry, and B. W. Frost. 1989. Closing the microbial loop: Dissolved carbon pathway to heterotrophic bacteria from incomplete ingestion, digestion and absorption in animals. *Deep Sea Research* 36:483–95.

Kaiser, K., and R. Benner. 2005. Hydrolysis-induced racemization of amino acids. *Limnology and Oceanography: Methods* 3:318–25.

Kaufman, D. S., and W. F. Manley. 1998. A new procedure for determining DL amino acid ratios in fossils using reversed phase liquid chromatography. *Quaternary Geochronology* 17:987–1000.

Keil, R. G., and D. L. Kirchman. 1991a. Contribution of dissolved free amino acids and ammonium to the nitrogen requirements of heterotrophic bacterioplankton. *Marine Ecology Progress Series* 73:1–10.

Keil, R. G., and D. L. Kirchman. 1991b. Dissolved combined amino-acids in marine waters as determined by a vapor phase hydrolysis method. *Marine Chemistry* 33:243–59.

Keil, R. G., and D. L. Kirchman. 1999. Utilization of dissolved protein and amino acids in the northern Sargasso Sea. *Aquatic Microbial and Ecology* 18:293–300.

Koch, B. P., M. Witt, R. Engbrodt, T. Dittmar, and G. Kattner. 2005. Molecular formulae of marine and terrigenous dissolved organic matter detected by electrospray ionisation Fourier transform ion cyclotron resonance mass spectrometry. *Geochimica et Cosmochimica Acta* 69:3299–3308.

Lee, C., and J. L. Bada. 1977. Dissolved amino acids in the equatorial Pacific, the Sargasso Sea, and Biscayne Bay. *Limnology and Oceanography* 22:502–10.

Lindroth, P., and K. Mopper. 1979. High performance liquid chromatographic determination of subpicomole amounts of amino acids by pre-column fluorescence derivatization with *o*-phthaldialdehyde. *Analytical Chemistry* 51:1667–74.

Maie, N., K. J. Parish, A. Watanabe, H. Knicker, R. Benner, T. Abe, K. Kaiser, and R. Jaffé. 2006. Chemical characteristics of dissolved organic nitrogen in an oligotrophic subtropical coastal ecosystem. *Geochimica et Cosmochimica Acta* 70:4491–4506.

McCarthy, M. D., R. Benner, C. Lee, and M. L. Fogel. 2007. Amino acid nitrogen isotopic fractionation patterns as indicators of heterotrophy in plankton, particulate, and dissolved organic matter. *Geochimica et Cosmochimica Acta* 71:4727–44.

McCarthy, M. D., R. Benner, C. Lee, J. I. Hedges, and M. L. Fogel. 2004. Amino acid carbon isotopic fractionation patterns in oceanic dissolved organic matter: An unaltered photoautotrophic source for dissolved organic nitrogen in the ocean? *Marine Chemistry* 92:123–34.

McCarthy, M. D., J. I. Hedges, and R. Benner. 1998. Major bacterial contribution to marine dissolved organic nitrogen. *Science* 281:231–34.

Nagata, T. 2000. Production mechanisms of dissolved organic matter. In *Microbial ecology of the oceans*, ed. D. L. Kirchman. New York: Wiley-Liss.

North, B. B. 1975. Primary amines in California coastal waters: Utilisation by phytoplankton. *Limnology and Oceanography* 20:20–27.

Robertson, K., P. M. Williams, and J. L. Bada. 1987. Acid hydrolysis of dissolved amino acids in seawater: A precautionary note. *Limnology and Oceanography* 32:996–97.

Suttle, C. A., A. M. Chan, and J. A. Fuhrman. 1991. Dissolved free amino acids in the Sargasso Sea: Uptake and respiration rates, turnover times, and concentrations. *Marine Ecology Progress Series* 70:189–99.

Tsugita, A., T. Uchida, H. W. Mewes, and T. Atake. 1987. A rapid vapor-phase (hydrochloric and trifluoroacetic acid) hydrolysis of peptide and protein. *Journal of Biochemistry* 102:1593–97.

Vetter, T. A., E. M. Perdue, E. Ingall, J.-F. Koprivnjak, and P. H. Pfromm. 2007. Combining reverse osmosis and electrodialysis for more complete recovery of dissolved organic matter from seawater. *Separation and Purification Technology* 56:383–87.

Wakeham, S. G., C. Lee, J. I. Hedges, P. J. Hernes, and M. L. Peterson. 1997. Molecular indicators of diagenetic status in marine organic matter. *Geochimica et Cosmochimica Acta* 61:5363–69.

Wu, F., and E. Tanoue. 2001. Sensitive determination of dissolved tryptophan in freshwater by alkaline hydrolysis and HPLC. *Analytical Science* 17:1063–66.

5 Optical Analysis of Chromophoric Dissolved Organic Matter

Norman B. Nelson and Paula G. Coble

CONTENTS

5.1 INTRODUCTION

Chromophoric dissolved organic matter (CDOM) is the term that is most commonly used to describe compounds dissolved in natural waters that absorb ultraviolet or short-wavelength visible light energy. Other terms in past and present use include gelbstoff (Kalle, 1938), gilvin (Kirk, 1994), and yellow substance. These terms refer to the fact that natural waters with high CDOM concentration are yellow to brown in color due to absorption of blue and green light. For the purposes of optical analysis of CDOM, *dissolved* is defined operationally as the material that will pass a submicron filter (see Section 5.2), so CDOM may include certain colloids and very small particles in a typical analysis.

There is a rich history of research on CDOM in natural waters, so the following is an abbreviated overview of current topics. CDOM is important in natural waters as a major factor controlling UV and visible light penetration (Kirk, 1994). Hence, CDOM is an important factor influencing photobiology as well as ocean color-based remote sensing of aquatic environments (Carder et al., 1989; Arrigo and Brown, 1996; Nelson and Siegel, 2002; Mueller et al., 2003; Siegel et al., 2005). CDOM also sensitizes photochemical reactions in the aquatic environment, including production or photolysis of climatically relevant trace compounds such as carbon monoxide, carbonyl sulfide, and dimethyl sulfide (Mopper and Kieber, 2002). Furthermore, CDOM appears to be present in all natural waters and at all ocean depths (Nelson et al., 2007). Abundance and distribution of CDOM in the global ocean appears to be controlled by in situ production, photochemical bleaching, terrestrial input, and deep ocean circulation (Blough et al., 1993; Nelson et al., 1998, 2007; Blough and Del Vecchio, 2002; Siegel et al., 2002; Spencer et al., 2007a). There is therefore considerable interest in accurate and reproducible quantification and characterization of CDOM, and this chapter is intended as a practical guide.

At present, little is known about the chromophoric compounds that make up CDOM, but it is speculated that they include humic and fulvic acids derived from terrestrial plant matter, as well as compounds derived from decomposing particulate material or excreted by organisms in situ (Blough and Del Vecchio, 2002; Rochelle-Newall and Fisher, 2002; Nelson et al., 2004; Del Vecchio and Blough, 2004; Steinberg et al., 2004). Recent chemical characterizations of open-ocean and lake samples suggest that CDOM contains polychlorinated biphenyl carboxylic acids thought to be derived from in situ biological activity (Repeta et al., 2004), and also contains quinone residues, which may indicate redox potential and functional group composition (Cory and McKnight, 2005). Fluorescence spectra (Coble, 1996) have suggested that aromatic amino acids and humic materials of terrestrial and marine origin are important contributors to CDOM and can be used to trace its origin (Coble, 1996; Conmy et al., 2004).

Since the exact chemical composition of CDOM is undefined and potentially variable, it is not possible currently to quantify CDOM in terms of mass or carbon equivalents separately from the total dissolved organic matter pool. Hence, CDOM is operationally quantified and characterized by spectroscopic methods, in particular UV-visible absorption spectroscopy (see Section 5.3) and fluorescence spectroscopy (see Section 5.4). CDOM absorbance or fluorescence as a function of dissolved organic carbon (DOC) concentration is variable, so there is no single expression relating CDOM optical properties to DOC (Ferrari, 2000; Ferrari et al., 1996; Blough and Del Vecchio, 2002). In fact, open-ocean abundances of CDOM (as absorption coefficient) are often not correlated or are very weakly negatively correlated with DOC concentration (Nelson and Siegel, 2002). Changes in chemical composition due to source material and solar bleaching also influence the relationship between CDOM optical activity and DOC concentration, so any such relationships tend to be localized in space and time.

A fraction of CDOM is fluorescent and is referred to as FDOM. Differences between absorption and excitation spectra of CDOM in natural waters indicate the fact that not all CDOM is fluorescent. The variability in absorption to fluorescence ratios globally shows a threefold variation (Blough and Del Vecchio, 2002); however, variability within a given geographical area is smaller, or even constant (Conmy et al., 2004). Fluorescence spectroscopy is more sensitive than absorption spectroscopy, and both excitation and emission spectra show greater detail and provide more information as to chemical composition than do absorbance spectra.

Collection of hyperspectral (high-wavelength-resolution) fluorescence data has been shown to provide an enormous benefit over collection of individual excitation/emission pairs, and two techniques have been employed. Excitation-emission matrix spectroscopy (EEMS) involves collection of multiple emission spectra at a range of excitations that are concatenated into a matrix

(Coble et al., 1990). Synchronous scanning (SS) involves increasing both excitation and emission wavelengths simultaneously to produce a single scan (Cabaniss and Shuman, 1987). Although less time-consuming, SS provides less information, as it diagonally samples an EEM spectrum.

The use of EEMS permits discrimination of CDOM sources based on which fluorophores are present and their relative concentrations. Eight general types of fluorescence peaks have been identified in natural waters (Coble et al., 1998). These groups include humic-like, protein-like, and pigment-like fluorescence (Coble et al., 1993, 1998; Coble, 1996). The reader is referred to a recent review (Coble, 2007) for additional information about composition and distribution of CDOM in the ocean.

5.2 SAMPLE COLLECTION AND PREPARATION

5.2.1 OVERVIEW

There are two main objectives with regard to sample collection and preparation for CDOM analysis: avoid contamination and separate particles from the solution. Initial collection of water samples should be carried out using clean equipment that has been well rinsed with the sample water in situ (i.e., on the upcast after a CTD profile). Niskin bottles or similar samplers should have silicone seals and Teflon-coated springs, as is customary for trace metal or trace gas analysis. Go-Flo bottles (General Oceanics, Miami) are ideal, as they open beneath the surface and are not exposed to the air-water interface, which can be contaminated with hydrocarbons from the vessel, which can be highly UV absorbing or fluorescent. If Niskin bottles are used to collect water near the surface, care should be taken to ensure that the bottles do not touch the air-water interface before closing. In cases where prefiltration is necessary, samples should be filtered directly into storage bottles whenever possible. All sample bottles should be rinsed three times with sample prior to sample collection.

Samples should be collected such that contamination is minimized. Polypropylene or nitrile gloves should be worn during all collection and sample handling procedures to prevent contamination from oils or cosmetics on hands. Latex gloves should be avoided, as they can leach absorbing compounds rapidly into seawater. Airborne contamination from tobacco smoke, cleaning solutions, or internal combustion engines should also be avoided. Many types of plastics are highly fluorescent and should be tested before use. Preferred materials for sample collection and storage are Teflon, stainless steel, glass, and nylon. All materials that come into direct contact with the sample should be precleaned as described below.

The purpose of the cleaning procedure is to remove as much organic material as possible, hence the widespread practice of washing all glassware. After cleaning, materials should be kept dry and covered, to prevent microbial growth or collection of airborne particles. The following has been found to be an acceptable procedure. Sample storage bottles are first cleaned with soapy water, and then rinsed six times with ultrapure water. An acceptable alternative is to wash bottles in a mild (10%) acid solution and rinse thoroughly with clean water. Bottles are capped with foil and dried in a lab oven at 60°C. Glass bottles are then baked at 450°C for 4–12 hours in a muffle furnace. Glassware that cannot be baked (such as volumetrics, beakers, pipettes, etc.), should be rinsed with methanol, rinsed again with ultrapure water, and allowed to dry before use. Teflon-lined caps are also precleaned separately by first soaking in soapy water, rinsing six times in ultrapure water, then wiping with a methanol-saturated laboratory wipe. Care should be taken to avoid contact between the Teflon surface and the black Bakelite cap material. Caps are dried in the oven at 60°C. Caps are screwed on to bottles over the piece of foil used to close bottles in the muffle furnace and stored without ever exposing the inside to air until the time of sampling. All cleaning procedures are performed using polypropylene gloves (never latex). In cases of severe organic contamination, a rinse with 1 N NaOH can be used on glassware (followed by rinsing six times in ultrapure water). Washing procedures should be checked by filling a cleaned bottle with ultrapure water and analyzing for fluorescence or absorbance.

5.2.2 SAMPLE FILTRATION

Particles can interfere with measurements and can also contain live organisms that produce or metabolize CDOM, thereby altering concentrations and compositions during sample storage. It is recommended that water samples, even those with very low particle abundance, be filtered prior to analysis. Blanks should be run regularly to establish overall effects of filtration, storage, and other sample handling artifacts, as CDOM from different environments can exhibit varying susceptibilities to contamination and alteration.

The choice of filter type generally includes nylon, polysulfone, and polycarbonate filters with pore sizes between 0.2 and 0.45 μm, although glass fiber (GF/F) filters with a nominal pore size of <0.7 μm are also used due to their lack of contamination, ease of cleaning (bake at 450°C for 12 hours), and high flow rates. Glass fiber filters are also available with finer effective pore sizes (e.g., Advantech GF-75, 0.3 μm), but at present GF/F filters are most commonly used.

Nylon filters should be solvent rinsed with methanol and ultrapure water, then rinsed thoroughly with the sample water. Disadvantages of this procedure are the problem of solvent disposal and the possible introduction of contamination. When large numbers of samples will be filtered over a short period of time, it is helpful to reserve a filtration rig specifically for precleaning filters. This saves time and also avoids contamination of the sample with methanol. This is particularly important if dissolved organic carbon (DOC) will also be measured on the filtered water. Although methanol is not fluorescent, it will contaminate DOC analyses. Other types of filters should be preconditioned with ultrapure water and then rinsed thoroughly with the sample water.

Filtration pressure should be the lowest necessary to pass the filter in a reasonable time period, relative to potential microbial growth or photodestruction of CDOM during filtration. If at all possible, gravity filtration directly into sample bottles should be used.

Filter holders must also be precleaned by solvent rinsing or baking. A stainless steel apparatus is preferred, but if this is not possible, be sure to test materials for fluorescent contamination. In the field, it is not possible to bake before each use. All surfaces that come into contact with the sample should be methanol rinsed, then very thoroughly rinsed with sample before a sample is collected, including sample bottles.

5.2.3 SAMPLE STORAGE AND STABILITY

The two most important causes of sample instability are microbial growth and exposure to light. Samples should be refrigerated for short-term storage in the dark, and analyzed as soon as possible after collection. There is no consensus on longer-term (weeks or longer) storage of samples, so the following is a brief discussion of current practices and their possible advantages and drawbacks. We recommend that researchers document their procedures and examine the effects of the storage method used for each study. Results may vary with type and origin of sample (e.g., high vs. low CDOM, freshwater vs. seawater) and filtration technique.

Although there have been reports of changing fluorescence intensity and peak position following freeze and thaw in a wide range of freshwater samples (Spencer et al., 2007c), others have reported no effect after filtration of pore water samples (Otero et al., 2007). It is always advised that any and all treatment effects are investigated prior to beginning a new study. Long-term (up to 1 year) frozen storage has been used routinely for CDOM fluorescence, absorbance, and DOC determinations without observable changes (Coble, unpublished). In some samples, DOM may precipitate out of solution on freezing and cause scatter interference in absorption measurements. Samples previously frozen that have precipitated should be discarded. Samples from water containing high CDOM concentrations are especially susceptible to loss of fluorescence from freezing, and if freezing is necessary, samples should be diluted prior to freezing. Refrigerated samples should be analyzed in 1–2 weeks; however, this is very dependent on water type, and storage artifacts should be determined for each study. Open-ocean 0.2 μm filtered seawater samples from the Equatorial Pacific have been stored at 4°C with no measurable change in absorbance for over a year at UCSB (C. Swan, UCSB, unpublished data), so freezing is not recommended for open-ocean samples if prompt analysis is possible. Refrigerators and freezers used for CDOM sample storage should not be used to store other biological samples, as some volatile organic compounds can contaminate. Preservatives, including acidification, are not recommended. Solvents such as acetone and methanol should be stored separately.

Care should be taken that caps are screwed on tightly to prevent water leakage and evaporation, since if water can leak out, contaminants can also leak in. One solution is to wrap caps with Teflon tape after filling. Bottles for freezing must be thick walled and filled to at most an inch below the shoulder so as not to cause breakage. Also, when freezing, avoid bottle breakage by not overfilling with sample, and if possible, first cool the sample with refrigeration, then freeze in a household-type freezer, and then move to a freezer at –20°C. Going directly from room temperature to –20°C often results in bottle breakage. During shipping, samples need to remain frozen, or have at least some ice remaining when they are unpacked. All sample handling should take place in low light; bottles should be amber colored or protected by foil. Storage blanks should be collected before and after storage to determine effects of the protocol used.

5.2.4 ENVIRONMENTAL INFLUENCES

Environmental factors influencing the optical properties of CDOM should be recorded in the field and controlled in the laboratory. Temperature (Baker, 2005), pH (Spencer et al., 2007c), salinity (Zepp et al., 2004), and interactions of DOM with metal ions have been shown to cause shifts in excitation and emission wavelength maxima as well as intensities. Refractive index of the sample water relative to ultrapure water blanks is also potentially an issue in spectrophotometry (see Section 5.3). Since salinity is an important determinant of refractive index, salinity of samples should be determined as well.

5.3 ANALYSIS BY SPECTROPHOTOMETRIC METHODS

5.3.1 PRINCIPLES OF OPERATION: DEFINITIONS

Absorption of light in a liquid medium is commonly defined in the context of a collimated beam of light passing through a sample of known thickness, with loss of the incident light beam being attributable only to absorption. Absorbance of a substance in solution, also known as optical density (abbreviated as A or OD), is defined as the base 10 logarithm of the quotient of the intensity of the light passing through a sample and the intensity of the light passing through a blank (Kirk, 1994). This dimensionless term is commonly used in colorimetric analysis, and is usually the data returned by commercial spectrophotometers. In environmental optics the dimensional quantity *absorption coefficient* (*a*, units of inverse length) is preferred. Absorption coefficient is defined in terms of the natural logarithm, so it is related to absorbance (optical density) as follows:

$$a(\lambda) = 2.303\, A(\lambda)/d \qquad (5.1)$$

where *d* is the distance over which the beam travels through the sample (or *pathlength*), λ is the wavelength, and the scalar 2.303 converts the base 10 logarithm to the natural logarithm. It is important to carefully specify which quantities are reported, as ambiguity in the use of the terms *absorbance* and *absorption* in the literature has led to confusion (Hu et al., 2002).

The absorption spectrum of CDOM is largely featureless, declining with wavelength between 250 and 700 nm. Peaks in the spectrum may be present at shorter wavelengths, but in natural waters ions such as nitrate are present and absorb in the 200–250 nm wavelength range, so these peaks do not necessarily represent organic compounds. Over short discrete wavelength intervals the absorption spectrum of CDOM may be accurately parameterized by a single exponential equation with a negative exponent (Bricaud et al., 1981).

5.3.2 Sample and Blank Preparation

Sample preparation for absorption spectroscopy is as described in Section 5.2. An important consideration when preparing for spectroscopy is temperature. Water exhibits a significantly temperature-dependent absorption band between 700 and 800 nm (Pegau et al., 1997), so all samples and blanks should be at the same temperature and held so throughout the analysis.

Blank preparation is a significant consideration, even in situations where samples have very high levels of CDOM. It is generally necessary to carry out additional purification of deionized water, using a system such as the Milli-Q or Alpha-Q (Millipore) or Nanopure (Thermo Barnstead). High resistivity is not a sufficient indicator of water purity for the purpose of CDOM analysis, as dissolved organics may not be ionic. When choosing a water purification system it is important to include options for organics removal. Some systems include a UV light source for antimicrobial purposes that may also destroy organics in the water (see Section 2.4.3). It is important to monitor the quality of the blank water by frequent comparison to reference samples; one option is to use freshly opened commercial high-quality water preparations (e.g., Fisher Optima water). Ultrapure water for spectroscopy should not be stored. Freshly prepared ultrapure water should be allowed to degas and temperature equilibrate before use. Any change in the relationship between the reference ultrapure water and the freshly prepared blank may indicate a problem with the water purification system.

Commercial drinking water preparations can also be used in the field if high-quality laboratory water is unavailable. Charcoal-filtered drinking water (Aquafina, Dasani, and Emirates) has been found to produce good results if used immediately (Coble, unpublished data). Unused water from a container should be discarded within 4 hours of opening.

5.3.3 Instrumentation

Instruments commonly used to determine the absorption spectrum of CDOM fall into two main categories: dual-beam (or conventional) spectrophotometers and single-beam, long-path spectrophotometers. A third type of absorption meter, the point-source integrating cavity absorption meter (PSICAM), shows great promise but is in the experimental stage and is not yet commercially available (Röttgers and Doerffer, 2007). The operating principles of the PSICAM differ from those of spectrophotometers (Kirk, 1997), so discussion of their use is beyond the scope of the present chapter.

Conventional (dual-beam) spectrophotometers are laboratory instruments designed to simultaneously or near-simultaneously illuminate a sample and a reference held in identical cells (Figure 5.1a). This configuration minimizes problems due to short-term variation in the intensity of the light sources, which are usually tungsten-halogen (for visible light) or deuterium (for UV light) incandescent lamps. A common light source passes a collimated beam through both cells. In some spectrophotometers the illuminating light is passed through a monochromator with an adjustable slit before the sample. Solid-state or photomultiplier tube detectors quantify the intensity of the light passing through the sample or the reference, and absorbance is calculated electronically. In some spectrophotometers a broadband source illuminates the sample, and the light passing through the sample is diffracted by a grating to a diode array solid-state detector.

Absorbance calibration of dual-beam spectrophotometers is typically verified by comparison of instrument signals to a NIST-traceable reference material based on the NIST 930 filter set, which consists of three glass filters absorbing in visible wavelengths (Travis et al., 2000). This is usually accomplished by the manufacturer, but is rarely done by end users. NIST-traceable filter sets are available from various manufacturers. Wavelength calibration is usually accomplished automatically on start-up, by comparison to an internal standard (e.g., emission lines from the lamps).

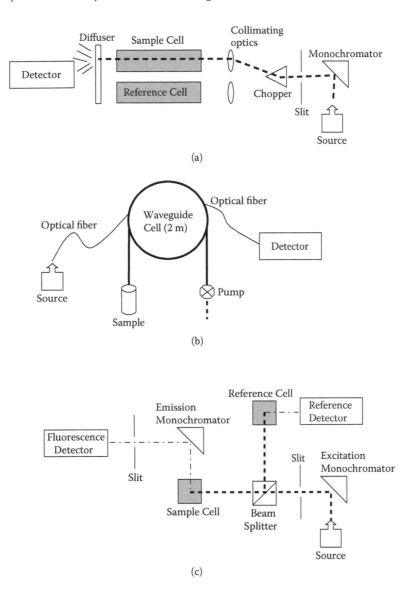

FIGURE 5.1 (a) Schematic of a dual-beam (conventional) spectrophotometer. The dashed line shows the light path through the sample cell. In this design of spectrophotometer a chopper alternates the beam through the sample and reference cells, and a diffuser behind the sample and reference cells allows the use of a single detector. (b) Schematic of a liquid waveguide single-beam spectrophotometer. (c) Schematic of a spectrofluorometer with a reference cell (to be used with quantum correction fluorescent dye).

If internal wavelength calibration is not used, a holmium oxide (HoO_4)-impregnated solid glass standard may be used to verify wavelength calibration.

Sample cells for dual-beam spectrophotometers are usually found in 1 cm, 5 cm, and 10 cm path lengths. For CDOM analysis 5 or 10 cm cells with quartz end windows are most commonly used. Absorbance is proportional to path length, so longer cells will provide a higher signal-to-noise ratio on samples from the open ocean (where overall CDOM concentration is lower) or at wavelengths longer than 400 or 500 nm (where CDOM absorption is low in all samples). It is important to consider the performance specifications of the instrument when choosing a cell. Most spectrophotometers have a linear response between 0.05 and 2 optical density units. If short-wavelength absorbance exceeds 2, then a shorter path length cell should be chosen.

In dual-beam spectrophotometers the blank spectrum is found by scanning two cells filled with blank solution. This baseline spectrum is subsequently subtracted from all subsequent scans within a session, so it is important that this initial measurement is carried out accurately. This is usually accomplished in software, and the apparent blank spectrum is not available to the user. To verify the stability of the system (including the instrument, cell, and cell holder), blanks (as samples) must be run periodically during a session.

Conventional spectrophotometers generally have a maximum cell path length of 10 cm. This constrains the use of conventional spectrophotometers for open-ocean applications, where CDOM absorption is very low. If the photometric accuracy of a spectrophotometer is 0.005 (optical density units), the corresponding absorption coefficient is over 0.1 m^{-1}. This is approximately the deep-ocean absorption coefficient of CDOM at 325 nm in the North Atlantic (Nelson et al., 2007). Surface values are lower, and of course longer wavelengths are lower as well. One way to increase the detection limit of absorption spectroscopy is to increase the path length. This has typically been accomplished by using single-beam spectrophotometers. Single-beam spectrophotometers generally work along the same principles as diode array spectrophotometers, but instead of near-simultaneous scanning of a sample and a blank, samples and blanks are run sequentially in a single optical path.

One example of the single-beam approach is the "shiny tube" absorption meter (e.g., AC-9 or AC-s, WETLabs, Philomath, Oregon), which can be deployed on a frame or CTD rosette. These instruments have cells with an optical path length of 10 cm or 25 cm, consisting of a 1 cm diameter glass tube with an air gap to provide for total internal reflection of photons (Kirk, 1992). The sample flows continuously through the cell by means of a submersible pump (as in a CTD). For measurement of CDOM, a 0.2 μm filter cartridge is installed to remove particles before pumping through the system (Twardowski et al., 2004). Blanks (air) are run before a cast to provide a zero.

A laboratory-based approach to increasing the path length for measurement of absorption in low-CDOM conditions is the liquid waveguide cell, which has been used with optical path lengths up to 2 m (e.g., UltraPath cell, World Products Intl., Sarasota, Florida). An optical waveguide consists of a long narrow tube (ca. 1 mm diameter) with fiber optics inserted at each end. One fiber optic is connected to a light source, and the other to a diode array spectrometer (Figure 5.1b). When filled with liquid the waveguide acts like a fiber optic, with near-total internal reflection of the photon flux. Samples are approximately 12 mL in volume, and are injected into the cell by way of a peristaltic pump. Typically a single cell is used, and samples are alternated with blanks.

An important feature of liquid waveguides of this nature is that the overall throughput of the cell is highly sensitive to the refractive index of the sample. Higher-index liquids, like seawater, have a higher throughput than pure water. This means that low-CDOM water typically has a negative apparent absorbance in a liquid waveguide relative to the ultrapure water customarily used for blanks. Two approaches have been used to account for this problem. Miller et al. (2002) used pure sodium chloride solutions and an optical refractometer to run samples against a blank with a similar refractive index. Nelson et al. (2007) developed an empirical correction based on salinity using artificial seawater, which was applied to absorbance spectra after acquisition. This issue limits the potential accuracy of the method to an undefined degree; nevertheless, Nelson et al. (2007) were able to make measurements of CDOM absorption coefficient down to 0.05 m^{-1}.

5.3.4 Data Processing and Quality Control

Initial data processing for CDOM analysis is straightforward. Since most spectrophotometers will directly compute absorbance, only Equation 5.1 needs to be applied to yield the absorption coefficient. This assumes that the optical path length is known and that the blank spectrum has been properly zeroed. In the case of liquid waveguide absorption spectra collected according to the methods of Nelson et al. (2007), the salinity-dependent offsets should be applied to the spectra before

Equation 5.1. Because of the exponential decay of absorption with wavelength, long wavelengths (e.g., 650 or 800 nm) should be near zero in most CDOM samples. Significant offsets from zero at these wavelengths indicate a problem with the sample or reference (contamination, dirty cuvette, unfiltered particles causing scattering of the incident beam), and these data should be discarded. Some studies using dual-beam conventional spectrophotometry (Green and Blough, 1994) have identified a small yet consistent positive offset in seawater samples relative to pure water at long wavelengths, which may be due to the refractive index differences, so this criterion should be used carefully when excluding data.

One product that is often derived from CDOM absorption spectra is referred to as the spectral slope (S, nm^{-1}), which is the parameter in a negative exponential equation describing the spectrum over a discrete wavelength range (Jerlov, 1968; Bricaud et al., 1981), according to the following equation:

$$a(\lambda) = a(\lambda_o)e^{-s(\lambda-\lambda_o)}$$

(5.2)

where $a(\lambda_o)$ is the absorption coefficient at reference wavelength λ_o.

The parameter S can be found by linear regression of log-transformed absorption spectra against the wavelength (Bricaud et al., 1981), or by a nonlinear curve fitting technique (Stedmon et al., 2000). In general the nonlinear fitting technique provides higher quality fits for S determined over longer (>50 nm) wavelength intervals, but choice of which technique to use should depend upon the application. It is important for the purpose of comparison that the parameters of the fit (wavelength interval and fitting technique) be specified. Slopes determined with differing parameters are not comparable.

Slopes for the UV and visible ranges have been observed in the range 0.014 to 0.035 nm^{-1} for oceanic samples. Typically terrestrial and coastal CDOM absorption spectra have values of S at the lower end of this range.

Careful attention must be paid to the detection limit and precision of the system in use. A typical photometric accuracy for spectrometers is 0.005 absorption units: this means in a system using 10 cm cuvettes an absorption coefficient as high as 0.06 m^{-1} may not be distinguishable from zero. On the other hand, practical experience has shown this to be overly pessimistic. Researchers have reported good results with 10 cm cuvettes down to 0.03 m^{-1} (Nelson et al., 1998) and 0.05 m^{-1} (Stedmon and Markager, 2001), based on log-linearity and stability.

Replicate samples can help establish the precision of the instrument; Nelson et al. (2007), using a liquid waveguide spectrophotometer system, found a root mean square difference in absorption coefficient at 325 nm between replicate samples (different Niskin bottles collected at the same depth) of approximately 0.013 m^{-1}. Subsequent investigations have estimated this precision as fine as 0.004 m^{-1} (unpublished data).

5.4 ANALYSIS BY SPECTROFLUOROMETRIC METHODS

5.4.1 PRINCIPLES OF OPERATION: DEFINITIONS

Fluorescence is the emission of light from a molecule as it transitions from an excited to a ground state following absorption of light at a shorter wavelength. Fluorescent compounds, called fluorophores, generally show similarities between their absorption and fluorescence excitation spectra, since absorption of photons is the first step in fluorescence. Figure 5.2 shows the difference between the absorbance, excitation, and emission spectra of CDOM. Fluorescence is not stimulated at all the wavelengths where CDOM absorbs light.

5.4.2 INSTRUMENTATION FOR FLUORESCENCE ANALYSIS

Basic components of a fluorometer include a light source, excitation optics, sample chamber, emission optics, detector, and signal processor. In the most basic systems, fluorescence intensity at a

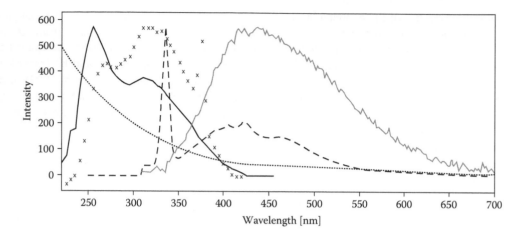

FIGURE 5.2 Absorbance and fluorescence spectra for a sample of CDOM from the Caloosahatchee River, Florida. Dotted line = absorbance; solid line = corrected excitation spectrum; plus signs = uncorrected excitation spectrum; dashed line = uncorrected emission spectrum; solid gray line = corrected emission spectrum. All intensities are relative.

single excitation/emission pair is measured. Since fluorescence is proportional to concentration, single point measurements are appropriate for solutions of known or constant composition. Single channel in situ monitoring is a valuable tool for real-time and spatial monitoring of CDOM (FDOM) and DOC in some systems. Fluorescence has been used successfully as a proxy for DOC in many freshwater systems.

More sophisticated systems allow adjustment of both excitation and emission wavelengths. Fluorescence spectrometers allow for measuring over a wide range of UV and visible excitation and emission wavelengths. In the following section, commonly used elements of fluorometers will be discussed briefly, with the primary intent of informing the user of performance and limitations of each design. A more detailed treatment of the principles of fluorescence and its measurement can be found in Lakowicz (2006).

5.4.2.1 Light Sources

Commonly used light sources for fluorescence excitation, ranging from low to high intensity, include light-emitting diodes (LEDs), xenon lamps, and lasers. The light source can be either continuous or flashing. Continuous mode is most useful for benchtop scanning spectrofluorometers, while flash lamps are useful in filter fluorometers, where excitation is controlled by a rotating filter wheel, or for in situ applications, where power, battery life, and lamp life are important design issues.

Each type of light source has distinctive wavelength emission properties and variable susceptibility to fluctuations in intensity. Most modern fluorometers have a reference channel that ratios the signal intensity to lamp intensity to correct for lamp fluctuation.

High-pressure xenon arc lamps are most commonly used in present-day commercial benchtop spectrofluorometers. These consist of an electrode and cathode in a fused quartz envelope filled with xenon under pressure. The xenon lamp is an arc lamp, and as the arc location moves during operation, this can influence optimal lamp output. The large lamps (450–1,000 W) require a high-voltage power supply to generate the arc, which generates nearly continuous emission between 250 and 700 nm. Lamp operation generates heat and causes the internal pressure to reach several atmospheres. Unless the quartz envelope is treated with a UV-absorbing material, ozone is also generated. These conditions necessitate a protective lamp housing for both cooling and sending ozone vapors into an ozone scrubber. The lamp housing is also equipped with a mechanism for altering lamp position to align the beam for maximum output to the excitation optics. Lamp alignment should be performed periodically, especially when a drop in sensitivity is noticed, or a new lamp installed. Typical lamp lifetimes are 1,500–2,000 hours, after which lamps become susceptible to explosion due to weakening of the envelope. Safety goggles and protective clothing must be worn during any tasks involving handling of the lamp, and the lamp should never be viewed directly. The outer envelope of the lamp should also not be touched, as fingerprints cause black carbon deposits to form on the outside of the lamp, decreasing light output. When handling and using a xenon lamp, all manufacturers' safety warnings should be strictly heeded.

The emission intensity of xenon lamps changes most rapidly during the initial burn-in period, but also as the lamps age, with loss of intensity especially noticeable in the UV.

Light-emitting diodes (LEDs) are semiconductor diodes that emit light in a narrow band of wavelengths. Multiple LEDs can be combined to broaden potential excitation range, or filtered to further narrow the excitation. LEDs have the advantage of low power requirement and do not generate excessive heat. The trade-off is that emission intensities are much lower than for xenon lamps. LEDs are available from the near-UV to infrared wavelengths.

Lasers and laser diodes emit monochromatic light, but with higher energy than LEDs. A configuration with multiple lasers or tunable lasers is suitable for multichannel instruments.

5.4.2.2 Optical Components

The optical components of a fluorometer can include mirrors, beam splitters, filters, and monochromators. Each of these components has reflective and absorptive properties that can affect the fluorescence results. For best sensitivity at UV wavelengths, optical components need to be made of quartz or be treated with special coatings to minimize absorbance of emission energy. Each component can affect absorption, dispersion, scatter, bandpass, and spectral biasing.

Fluorometers use a combination of filters and monochromators to select excitation and emission wavelength. Filters are available in a wide range of wavelengths, and are generally characterized by both bandpass width and wavelength of maximum transmission. Interference filters provide selection of narrower-band transmission than do color filters. High-pass filters, transmitting only light above a specified wavelength, are sometimes used to reduce stray light. Low-pass filters can be used to prevent Rayleigh and Raman scatter from reaching the dectector.

Monochromators disperse white light into a range of wavelengths. As the monochromator is pivoted from one position to another, the color of light reflected into the exit slit varies. Diffraction gratings are now used in most spectrofluorometers. The resolution, stray light, and efficiency are all properties that depend on monochromator design. Each monochromator or pair of monochromators used in an instrument will have unique correction factors associated with it. As with all optical surfaces, these should never be touched with fingers, and care should be taken not to touch the surface with any material. Dust on the surface is less detrimental than a fingerprint, which may permanently damage an optical surface.

Grating efficiency should be optimized near the excitation maximum (350 nm) and emission maximum (450 nm) of CDOM. In most cases, the desired grating can be requested with instrument purchase. UV-transmitting coatings can also be obtained to minimize loss of UV excitation and emission light below 300 nm.

Calibration of both emission and excitation monochromators should be done at the factory prior to delivery, and provided to the user for correction of fluorescence data. Calibration of the emission monochromator is performed using a standard lamp mounted in a fixed position above the instrument. Periodic check of emission wavelength accuracy can be performed by running a water Raman scan. For excitation at 275 nm, maximum Raman scatter should be at 303 nm in pure water. Periodic check of excitation wavelength accuracy is described in Section 5.4.4.1.

5.4.2.3 Sample Compartment

Conventional benchtop fluorometers use a sample compartment consisting of entrance and exit slits, entrance and exit shutters, a cuvette holder, and a cuvette (Figure 5.1c). Slit width controls bandpass of light entering and exiting the sample. Cuvette composition needs to be quartz or Suprasil (fused silica) for UV fluorescence measurement. Since temperature fluctuation causes variation in fluorescence intensity, sample temperature control is recommended. This is not as much of an issue in flow-through or in situ instruments, provided that sampling stream temperature is constant and flow-through rates are high. Samples run on benchtop instruments should be run at room temperature (18–20°C) to prevent condensation on cuvette walls.

The sample compartment may be configured for dual-emission detection. Many instruments also have a quantum counterreference detector. Excitation light passes through a beam splitter as it enters the sample compartment. A percentage of the energy is sent into the quantum counter, consisting of a cuvette filled with a saturated solution of Rhodamine B and a photomultiplier tube (PMT). The Rhodamine solution absorbs all available photons; therefore, the PMT detects only fluorescence from the Rhodamine. The quantum yield and emission maximum of Rhodamine is independent of excitation wavelength; thus, the signal on the reference PMT is solely proportional to the flux of photons from the excitation light, and the spectral bias of the reference PMT is eliminated. More modern instruments may use a solid-state reference detection system.

5.4.2.4 Detection System

Photomultiplier tubes are the most common device used to amplify and collect fluorescence emission signals. Incident photons are converted to photoelectrons, which strike a series of dynodes, with each electron generating five to twenty additional electrons at each collision. The size of the electron pulse that finally reaches the dynode is dependent on the number of photons emitted and on the voltage applied to the PMT. A nonlinear response and damage to the PMT can result from application of excessive current. The dark current of the PMT, which is the signal produced in the absence of light, can be further reduced by using a PMT cooling unit. The PMT has a unique spectral response that depends on the material used for the window as well as the material used for the photocathode.

Other types of detectors in use for spectrofluorometry include photodiode array and charge-coupled devices (CCDs). Both types are used to obtain a complete emission spectrum instantaneously, without requiring the monochromators to scan. In these instruments, the emission light is reflected off a stationary grating that spreads the wavelengths of fluorescence across the surface of the detector, and intensities of each wavelength are measured by the photodiode or CCD channel on which it is incident. Emission range is controlled by moving the monochromators across the detector, and wavelength resolution and signal intensity can be controlled by binning channels or changing the grating dispersion. It is important to use monochromators that do not distort the spatial dispersion of the signal. The CCD array is more sensitive than the photodiode, as it uses some electron amplification. Since the CCD is a two-dimensional array, it can be used to image an entire surface or to produce an entire EEM from a single pulse of white light.

5.4.3 Sample Preparation

In most cases, samples require no additional treatment prior to analysis. Samples should be allowed to reach room temperature prior to analysis. Frozen samples should be allowed to thaw in the refrigerator overnight. Gentle swirling of sample bottles prior to filling a sample cuvette is preferable to vigorous shaking, which can cause particles to form. If particles are observed in the sample bottle, filtration prior to analysis is recommended because scattering may interfere with the fluorescence signal. Causes of particle formation during storage should be investigated and remedied in future sampling, if at all possible. One common cause is freezing very concentrated samples, such as found in freshwater.

Sample cuvette (quartz or Suprasil) should be clean and rinsed three times with sample before filling the cuvette. An automatic pipetting device is recommended for sample transfer to avoid contact with the outside of the sample bottle. In no case should latex pipette bulbs be used, as particles of latex may contaminate the sample.

5.4.4 SIGNAL CORRECTION

5.4.4.1 Wavelength Calibration

There are several ways to check accuracy of excitation and emission wavelength. For excitation, the most common method is to run a lamp spectrum and check that the maximum signal is observed at 467 nm, the maximum output of the xenon lamp. For the emission, checking the positions of one or more water Raman peaks is useful. In practice, a water Raman blank should be run daily for monitoring instrument performance as well as to check the cleanliness of water used to clean cuvettes.

5.4.4.2 Rayleigh and Raman Scatter Corrections

Excitation light is scattered by the solvent molecules. Most of the photons are scattered elastically, and thus the scattered light has the same energy, or wavelength, as the incident radiation. This is called Rayleigh scatter. A smaller amount of light is inelastically scattered at higher or lower energy after causing a change in the vibrational state of the solvent molecules. This is known as Raman scatter. The wavelength at which Raman scatter is observed is a function of the solvent type and the excitation energy. For water, Raman scatter is observed at a wavenumber $3,600$ cm^{-1} lower (or higher) than the excitation wavenumber.

Rayleigh scattering can be avoided by using prefilters or adjusting the emission wavelength collection range to start 10 nm longer than the excitation wavelength. Raman scatter peaks overlap CDOM fluorescence and are not as easy to remove. In instruments with a single emission monochromator, second-order peaks (at twice the wavelength of the first-order peak) may also be observed. One procedure to remove Raman scatter is subtraction of water blanks. This method is not entirely satisfactory, especially at low CDOM levels where negative values sometimes result. Because the Raman peak results from a molecular property, its shape and intensity can also be modeled and removed mathematically.

Removal of solvent scatter peaks is not required in all applications. Blank subtraction is used to remove contamination peaks due to sample handling, if any, and to process EEMs for interpretation and presentation. If proper care is taken in sampling protocols, contamination blanks should be minimal and not of concern. Some exceptions are in cases where a fluorescent tracer has been used in part of the experimental study, as in some water mass tracing applications, and CDOM concentrations are needed independent of the dye. Process blanks should be run on ultrapure water that has been processed just as samples are processed.

Removal of Raman scatter is not as great a concern in samples with high CDOM concentrations, where the scatter peaks are small in comparison to CDOM fluorescence. However, when CDOM concentrations are low, such as in the open ocean, the scatter peak makes it difficult to view the characteristics of the CDOM signal in the EEM. Raman also interferes with determination of protein and amino acid fluorescence peaks. Removal of scatter peaks from EEMs may also be important for some, but not all, postprocessing techniques. The emission wavelength of Raman scatter is dependent on the excitation wavelength, whereas this is not true for fluorophores; therefore, fluorescence and scatter can be differentiated by parallel factors (PARAFAC) analysis.

The following procedure is used to prepare the water blank for Raman scatter removal. The water blank is run like any other sample, with the cleanest available ultrapure water in the sample cuvette. The effectiveness of blank subtraction in removing water scatter peaks is very sensitive to small variations in intensity and wavelength, which can occur from day to day. For this reason, all samples and blanks are first shifted along the x-axis to align the observed water Raman peak to a fixed position at Ex/Em 275/303 nm. In practice, the position of the water Raman usually varies by less than 1 nm. Failure to shift the data to actual Raman peak position results in large artifact peaks on either side of the Raman scatter peak.

The EEMs are then normalized to the maximum observed value for this peak observed over an extended period of measurement. The normalization step simply sets that maximum Raman scatter for water to the same value for all samples and blanks. This compensates for drift in instrument performance over time with lamp aging and loss of PMT sensitivity. In practice, the normalization value is arbitrary and can be set to some value close to what is normally observed. Each day the water blank Raman intensity is recorded and the ratio of the daily value to the normalization value is determined. This ratio is then used to normalize fluorescence intensity of every sample collected that day. The value of the daily normalization ratios is seen to drift generally downwards over the course of weeks to months. Every sample, blank, and calibration standard is normalized prior to subsequent data correction and processing.

An alternate method for removal of remnant scatter peaks after blank subtraction involves a smoothing technique (Zepp et al., 2004). This technique allows for clearer depiction of EEM spectra, but its impact upon downstream processing is unclear.

5.4.4.3　Excitation and Emission Correction

While many of the problems with stray light, scatter, lamp intensity fluctuation, and calibration of monochromators have been resolved by most modern manufacturers, excitation and emission correction factors specific to each instrument must still be applied to raw fluorescence EEMs in order for results to be compared with those from other instruments. This is not an issue with single point measurements, where only a calibration is needed. Lakowicz (2006) compares various methods for excitation and emission corrections, none of which are found to be entirely satisfactory. For the case of CDOM, several instruments produce emission scans having double peaks, whereas a properly corrected scan shows only a single peak. It has become the standard in CDOM measurement to perform rigorous wavelength correction prior to publication of results. This becomes even more critical when multicomponent data analysis techniques are used to process data for concentrations of individual fluorophores within a CDOM sample. In part, this need for correction arises from the nature of CDOM, which is a mixture of fluorophores with different quantum efficiencies, excitation maxima, and emission maxima. The properties of a pure compound, such as Rhodamine, in which emission maximum is independent of excitation wavelength, do not hold true for CDOM.

The most important first step in excitation and emission correction is to apply the manufacturer's correction spectra (Figure 5.3). Corrected spectra should then be compared to those of one or more standard substances published in the literature (Lakowicz, 2006). Quinine sulfate has been most commonly used for CDOM calibration; however, it does not cover the full range of excitation and emission wavelengths observed for CDOM. One alternative is to use prepared standards, such as those available commercially in a range of wavelengths between 220 and 450 nm excitation and 230 and 700 nm emission. If no correction factors have been provided by the manufacturer, other procedures have been described (Lakowicz, 2006).

Factory-generated emission correction factors can be used for the lifetime of the instrument, unless an optical component on the emission side is replaced. However, additional spectral correction factors may be required for excitation data. Excitation energies at wavelengths shorter than 300 nm must be monitored closely, as they decrease daily due to aging of the xenon lamp. Change in spectral output of the lamp can be corrected by routine generation of user-determined excitation correction factors. These lamp correction factors are generated by placing freshly prepared saturated Rhodamine B solution in the sample cuvette and generating an excitation scan. Results represent differences in optics between sample and reference sides of the instrument. The reciprocal

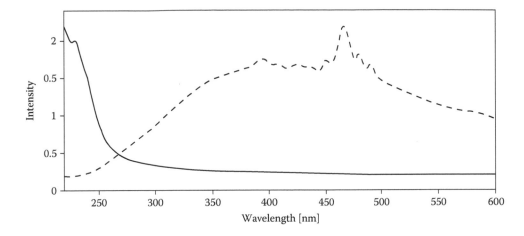

FIGURE 5.3 Excitation correction spectra. Solid line = correction factor values; dashed line = xenon lamp emission.

of this scan is then normalized to a value of 1 at the lowest fluorescence value, and these normalized values are then multiplied by corresponding sample excitation values.

5.4.4.3.1 Inner Filter Effects

Inner filter effects are caused by absorbance of light by the sample, thereby changing fluorescence intensity and spectral output. Primary inner filter effects are caused by absorbance of the excitation light, and secondary inner filter effects are caused by absorbance of the light emitted by the sample in response to excitation. Inner filter effects can be mitigated by making a dilution series until a linear relationship is observed between absorbance and fluorescence. In general, all samples that have an absorbance above 0.2 at 300 nm for a 10 cm cell will likely need to be diluted to avoid inner filter effects. Sample dilution is time-consuming and can also change the pH, salinity, and other chemical parameters of the sample matrix (Spencer et al., 2007b). An alternative procedure is correction according to an equation that accounts for absorbance of the sample at the relevant wavelengths (Ohno, 2002). Both dilution and absorbance corrections give comparable results (Spencer et al., 2007b); however, in cases where fluorescence observations are collected in situ, correction formulae are the only possible option.

5.4.5 Standards and Calibration

The two most commonly used methods of calibration are versus a quinine sulfate dilution series and normalizing to the intensity of the water Raman peak. The Raman peak is stable and easily measured; however, fluorescence to Raman ratios is dependent on the instrument design elements, including horizontal polarization of the light source and the geometry of the collection lens system (Sivaprakasam, 2002). This may not be a problem in commercial systems, which use similar designs, but the issue does need further investigation by the community before becoming accepted practice as a sole calibration standard. In the interim, we recommend collecting and reporting conversion factors between quinine sulfate and Raman units.

Quinine sulfate dilution series should be measured at least every month. Since the necessary frequency of standardization depends on instrument stability, it should be determined for each instrument. Although quinine sulfate has a limited fluorescence range compared to CDOM, it is common accepted practice to use these units for reporting CDOM fluorescence at all wavelengths. Some investigators have found the Starna standard reference blocks of Ovalene, *p*-terphenyl, tetraphenylbutadiene, Compound 610, and Rhodamine B to be useful for verification of corrections at additional wavelengths (Stedmon, personal communication); however, at this time, reporting in units of other fluorophores is not standard practice. Protein-like and amino acid-like fluorescence of peaks B and T have been reported relative to quinine sulfate, in Raman units, and also converted to protein or amino acids units (Determann et al., 1998; Mayer et al., 1999; Elliott et al., 2006).

ACKNOWLEDGMENTS

The authors thank Colin Stedmon and an anonymous reviewer for constructive comments and discussion, and Robyn Conmy, Jennifer Boehme, Carlos Del Castillo, and Chantal Swan for their contributions to technique development in our laboratories.

REFERENCES

Arrigo, K. R., and C. W. Brown, 1996. Impact of chromophoric dissolved organic matter on UV inhibition of primary productivity in the sea. *Marine Ecology Progress Series* 140:207–16.

Baker, A. 2005. Thermal fluorescence quenching properties of dissolved organic matter. *Water Research* 39:4405–12.

Blough, N. V., and R. Del Vecchio. 2002. Chromophoric DOM in the coastal environment. In *Biogeochemistry of marine dissolved organic matter*, ed. D. A. Hansell and C. A. Carlson. San Diego: Academic Press, 509–546.

Blough, N. V., O. C. Zafiriou, and J. Bonilla. 1993. Optical absorption spectra of water from the Orinoco River outflow: Terrestrial input of colored organic matter to the Caribbean. *Journal of Geophysical Research* 98:2271–78.

Bricaud, A., A. Morel, and L. Prieur. 1981. Absorption by dissolved organic matter in the sea (yellow substance) in the UV and visible domains. *Limnology and Oceanography* 26:43–53.

Cabaniss, S. E., and M. S. Shuman. 1987. Synchronous fluorescence spectra of natural waters: Tracing sources of dissolved organic matter. *Marine Chemistry* 21:37–50.

Carder, K. L., R. G. Steward, G. R. Harvey, and P. B. Ortner. 1989. Marine humic and fulvic acids: Their effects on remote sensing of ocean chlorophyll. *Limnology and Oceanography* 34:68–81.

Coble, P. G. 1996. Characterization of marine and terrestrial DOM in seawater using excitation-emission matrix spectroscopy. *Marine Chemistry* 51:325–46.

Coble, P. G. 2007. Marine optical biogeochemistry—The chemistry of ocean color. *Chemical Reviews* 107:402–18.

Coble, P. G., C. E. Del Castillo, and B. Avril. 1998. Distribution and optical properties of CDOM in the Arabian Sea during the 1995 Southwest Monsoon. *Deep Sea Research Part II: Topical Studies in Oceanography* 45:2195–2223.

Coble, P. G., S. A. Green, N. V. Blough, and R. B. Gagosian. 1990. Characterization of dissolved organic matter in the Black Sea by fluorescence spectroscopy. *Nature* 348:432–35.

Coble, P. G., C. A. Schultz, and K. Mopper. 1993. Fluorescence contouring analysis of DOC intercalibration experiment samples: A comparison of techniques. *Marine Chemistry* 41:173–78.

Conmy, R. N., P. G. Coble, R. F. Chen, and G. B. Gardner. 2004. Optical properties of colored dissolved organic matter in the Northern Gulf of Mexico. *Marine Chemistry* 89:127–44.

Cory, R. M., and D. M. McKnight. 2005. Fluorescence spectroscopy reveals ubiquitous presence of oxidized and reduced quinines in dissolved organic matter. *Environmental Science and Technology* 39:8142–49.

Del Vecchio, R., and N. V. Blough. 2004. On the origin of the optical properties of humic substances. *Environmental Science and Technology* 38:3885–91.

Determann, S., J. M. Lobbes, R. Reuter, and J. Rullkötter. 1998. Ultraviolet fluorescence excitation and emission spectroscopy of marine algae and bacteria. *Marine Chemistry* 62:137–56.

Elliott, S., J. R. Lead, and A. Baker. 2006. Characterisation of the fluorescence from freshwater, planktonic bacteria. *Water Research* 40:2075–83.

Ferrari, G. M. 2000. The relationship between chromophoric dissolved organic matter and dissolved organic carbon in the European Atlantic coastal area and in the west Mediterranean Sea (Gulf of Lions). *Marine Chemistry* 70:339–57.

Ferrari, G. M., M. D. Dowell, S. Grossi, and C. Targa. 1996. Relationship between the optical properties of chromophoric dissolved organic matter and total concentration of dissolved organic carbon in the southern Baltic Sea region. *Marine Chemistry* 55:299–316.

Green, S. A., and N. V. Blough. 1994. Optical absorption and fluorescence properties of chromophoric dissolved organic matter in natural waters. *Limnology and Oceanography* 39:1903–16.

Hu, C., F. E. Muller-Karger, and R. G. Zepp. 2002. Absorbance, absorption coefficient and apparent quantum yield: A comment on common ambiguity in use of these optical concepts. *Limnology and Oceanography* 47:1261–67.

Jerlov, N. G. 1968. *Optical oceanography.* Elsevier Oceanography Series 5. Amsterdam, The Netherlands: Elsevier.

Kalle, K. 1938. Zum Problem der Meerwasserfarbe. *Annalen der Hydrologischen und Marinen Mitteilungen* 66:1–13.

Kirk, J. T. O. 1992. Monte-Carlo modeling the performance of a reflective-tube absorption meter. *Applied Optics* 31:6463–68.

Kirk, J. T. O. 1994. *Light and photosynthesis in aquatic ecosystems.* 2nd ed. Cambridge, UK: Cambridge University Press.

Kirk, J. T. O. 1997. Point-source integrating cavity absorption meter: Theoretical principles and numerical modeling. *Applied Optics* 36:6123–28.

Lakowicz, J. R. 2006. *Principles of fluorescence spectroscopy.* Baltimore: Springer.

Mayer, L. M., L. L. Schick, and T. C. Loder. 1999. Dissolved protein fluorescence in two Maine estuaries. *Marine Chemistry* 64:171–79.

Miller, R. L., M. Belz, C. Del Castillo, and R. Trzaska. 2002. Determining CDOM absorption spectra in diverse coastal environments using a multiple pathlength, liquid core waveguide system. *Continental Shelf Research* 22:1301–10.

Mopper, K., and D. J. Kieber. 2002. Photochemistry and the cycling of carbon, sulfur, nitrogen, and phosphorus. In *Biogeochemistry of marine dissolved organic matter*, ed. D. A. Hansell and C. A. Carlson. San Diego: Academic Press, 456–508.

Mueller, J. L., G. S. Fargion, and C. R. McClain, eds. 2003. *Ocean optics protocols for satellite ocean color sensor validation.* Revision 4, Vol. V. *Biogeochemical and bio-optical measurements and data analysis methods.* NASA/TM—2003-211621/Rev4-Vol.V, NASA Goddard Spaceflight Center, Greenbelt.

Nelson, N. B., C. A. Carlson, and D. K. Steinberg. 2004. Production of chromophoric dissolved organic matter by Sargasso Sea microbes. *Marine Chemistry* 89:273–87.

Nelson, N. B., and D. A. Siegel. 2002. Chromophoric DOM in the open ocean. In *Biogeochemistry of marine dissolved organic matter*, ed. D. A. Hansell and C. A. Carlson. San Diego: Academic Press, 547–578.

Nelson, N. B., D. A. Siegel, C. A. Carlson, C. Swan, and W. M. Smethie Jr. 2007. Hydrography of chromophoric dissolved organic matter in the North Atlantic. *Deep Sea Research I* 54:710–31.

Nelson, N. B., D. A. Siegel, and A. F. Michaels. 1998. Seasonal dynamics of colored dissolved material in the Sargasso Sea. *Deep Sea Research I* 45:931–57.

Ohno, T. 2002. Fluorescence inner-filtering correction for determining the humification index of dissolved organic matter. *Environmental Science and Technology* 36:742–46.

Otero, M., A. Mendonça, M. Válega, E. B. H. Santos, E. Pereira, V. I. Esteves, and A. Duarte. 2007. Fluorescence and DOC contents of estuarine pore waters from colonized and non-colonized sediments: Effects of sampling preservation. *Chemosphere* 67:211–20.

Pegau, W. S., D. Gray, and J. R. V. Zaneveld. 1997. Absorption and attenuation of visible and near-infrared light in water: Dependence on temperature and salinity. *Applied Optics* 36:6035–46.

Repeta, D. J., N. T. Hartman, S. John, A. D. Jones, and R. Goericke. 2004. Structure elucidation and characterization of polychlorinated biphenylcarboxylic acids as major constituents of chromophoric dissolved organic matter in seawater. *Environmental Science and Technology* 38:5373–78.

Rochelle-Newall, E. J., and T. R. Fisher. 2002. Chromophoric dissolved organic matter and dissolved organic carbon in Chesapeake Bay. *Marine Chemistry* 77:23–41.

Röttgers, R., and D. Doerffer. 2007. Measurements of optical absorption by chromophoric dissolved organic matter using a point source integrating cavity absorption meter. *Limnology and Oceanography: Methods* 5:126–35.

Siegel, D. A., S. Maritorena, N. B. Nelson, and M. J. Behrenfeld. 2005. Independence and interdependencies of global ocean color properties. Reassessing the bio-optical assumption. *Journal of Geophysical Research* 110:C07011, doi:10.1029/2004JC002527.

Siegel, D. A., S. Maritorena, N. B. Nelson, D. A. Hansell, and M. Lorenzi-Kayser. 2002. Global distribution and dynamics of colored dissolved and detrital organic materials. *Journal of Geophysical Research* 107:3228, doi:10.1029/2001JC000965.

Sivaprakasam, V. 2002. UV laser induced fluorescence spectroscopic studies and trace detection of dissolved plastics (bisphenol-A) and organic compounds in water. Department of Physics, University of South Florida, Tampa.

Spencer, R. G. M., J. M. E. Ahad, A. Baker, G. L. Cowie, G. Ganeshram, R. C. Upstill-Goddard, and G. Uher. 2007a. The estuarine mixing behaviour of peatland derived dissolved organic carbon and its relationship to chromophoric dissolved organic matter in two North Sea estuaries (UK). *Estuarine Coastal and Shelf Science* 74:131–44.

Spencer, R. G. M., J. M. E. Ahad, A. Baker, G. L. Cowie, G. Ganeshram, R. C. Upstill-Goddard, and G. Uher. 2007b. Discriminatory classification of natural and anthropogenic waters in two UK estuaries. *Science of the Total Environment* 373:305–23.

Spencer R. G. M., L. Bolton, and A. Baker. 2007c. Freeze/thaw and pH effects on freshwater dissolved organic matter fluorescence and absorbance properties from a number of UK locations. *Water Research* 41:2941–50.

Stedmon, C. A., and S. Markager. 2001. The optics of chromophoric dissolved organic matter (CDOM) in the Greenland Sea: An algorithm for differentiation between marine and terrestrially derived organic matter. *Limnology and Oceanography* 46:2087–93.

Stedmon, C. A., S. Markager, and H. Kaas. 2000. Optical properties and signatures of chromophoric dissolved organic matter (CDOM) in Danish coastal waters. *Estuarine Coastal and Shelf Science* 51:267–78.

Steinberg, D. K., N. B. Nelson, C. A. Carlson, and A. Prusak. 2004. Production of chromophoric dissolved organic matter (CDOM) in the open ocean by zooplankton and the colonial cyanobacterium *Trichodesmium* spp. *Marine Ecology Progress Series* 267:45–56.

Travis, J. C., M. V. Smith, S. D. Rasberry, and G. W. Kramer. 2000. *Technical specifications for certification of spectrophotometric NTRMs*. NIST Special Publication 260-140, National Institute of Standards and Technology, Gaithersburg, MD.

Twardowski, M. S., E. Boss, J. M. Sullivan, and P. M. Donaghay. 2004. Modeling the spectral shape of absorption by chromophoric dissolved organic matter. *Marine Chemistry* 89:69–88.

Zepp, R. G., W. M. Sheldon, and M. A. Moran. 2004. Dissolved organic fluorophores in southeastern US coastal waters: Correction method for eliminating Rayleigh and Raman scattering peaks in excitation-emission matrices. *Marine Chemistry* 89:15–36.

6 Isotope Composition of Organic Matter in Seawater

Laodong Guo and Ming-Yi Sun

CONTENTS

6.1 INTRODUCTION

Marine organic matter in seawater is one of the most active carbon reservoirs on the earth's surface and plays an important role in earth's climate system. Marine organic matter is also a key component in the exchange among the biosphere, hydrosphere, and geosphere (Hedges, 1992). Therefore, knowledge of the cycling of organic matter in the marine environments is indispensable for understanding of the biogeochemistry of a variety of elements, function of ecosystems, and impact of human activities on global climate changes. Isotopic signatures, such as those of stable isotope ratios of C, N, and S, as well as radiocarbon, have been widely applied as powerful tools to study biogeochemical cycling of organic matter in marine environments (Eadie et al., 1978; Sigleo and Macko, 1985; Peterson and Fry, 1987; Altabet, 1988; Altabet et al., 1991; Cifuentes et al., 1988; Sackett, 1989; Rau et al., 1990, 1991a; Druffel and Williams, 1992; Druffel et al., 1992; Benner et al., 1997; Raymond and Bauer, 2001a; Tanaka et al., 2004; Knapp et al., 2005; Chen et al., 2006; Tagliabue and Bopp, 2008). Interactions of organic carbon with inorganic carbon in marine environments and their interchange with atmospheric CO_2 are always accompanied by variations in carbon isotope signals (Tagliabue and Bopp, 2008). Therefore, isotopic measurements provide a baseline for assessment of sources, sinks, transport, and transformation of marine organic matter, as well as the geochemical processes controlling the distribution of organic matter in the oceans (Druffel and Williams, 1992; Tumbore and Druffel, 1995; Raymond and Bauer, 2001a).

Marine organic matter is heterogeneous in terms of chemical composition (Mayer, 1994; Benner et al., 1997; Bergamaschi et al., 1997; Guo and Santschi, 1997a), isotopic signature (Druffel et al., 1992; Santschi et al., 1995; Guo et al., 1996; Wang et al., 1998; Eglinton et al., 1996; Pearson and Eglinton, 2000; Wang and Druffel, 2001), and bioavailability (Amon and Benner, 1994; Hunt et al., 2000). Organic matter in seawater includes three major forms based on its size spectrum (Figure 6.1) (Sharp, 1973; Guo and Santschi, 1997b): dissolved organic matter (DOM), colloidal organic matter (COM), and particulate organic matter (POM). These different organic matter phases are operationally defined based on specific filtration/separation techniques.

Micro- and macrofiltration, in situ pumps, and sediment traps have been used successfully to collect particulate organic matter, including suspended POM, phytoplankton, and sinking POM from marine environments for chemical and isotopic measurements (Bishop et al., 1978; Knauer et al., 1979; Altabet, 1990; Fry et al., 1991; Hedges et al., 2001; Peterson et al., 2005). Ultrafiltration, depending on pore size cutoffs of filters or membranes, has been proven to be effective for isolating colloidal organic matter (Benner et al., 1992; Guo and Santschi, 1997b; Guo and Santschi, 2007). However, sampling of DOM from seawater for chemical and isotopic characterization remains a challenging task, although recent development in methodologies has allowed direct measurements of $DO^{13}C$ and $DO^{15}N$ in seawater (Druffel et al., 1992; Clercq et al., 1998; Gandhi et al., 2004;

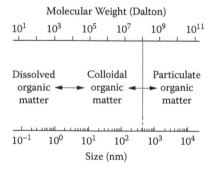

FIGURE 6.1 Size spectrum of marine organic matter, including dissolved organic matter (DOM), colloidal organic matter (COM), and particulate organic matter (POM) in seawater.

Knapp et al., 2005; Bouillon et al., 2006; Beaupre et al., 2007; Osburn and St.-Jean, 2007; Lang et al., 2007).

Biogeochemical cycling of organic matter in the ocean is complex and has not been fully understood (Hedges, 2002; Lee and Wakeham, 1992; Mopper et al., 2007), partially due to the relatively poor characterization in chemical and isotopic compositions of marine organic matter (Benner et al., 1992; Guo and Santschi, 1997b; Hansell and Carlson, 2002). This is largely caused by the difficulty in separating trace amounts of dissolved and colloidal organic matter from seawater. In general, the concentration of DOC in seawater is less than 0.5–1 mg-C/L, variable with water depth and ocean basins (Guo et al., 1995; Hansell and Carlson, 1998), while sea salt concentrations are usually in the order of 35,000 mg/L (Chester, 2003) (~35,000 times higher than those of bulk DOM in seawater).

Recent advances in separation and isolation techniques for DOM have greatly improved our ability to understand the cycling of DOM and COM and associated trace elements in marine environments (Hansell and Carlson, 2002; Guo and Santschi, 2007; Mopper et al., 2007). Indeed, a growing body of literature has reported the measurements of stable isotope and radiocarbon, especially for COM or high molecular weight DOM samples from the marine environments (Guo and Santschi, 2007, and references therein). While many DOM separation techniques/methods have been used, depending on specific research purposes, the most common methods are cross-flow ultrafiltration (Benner et al., 1992; Buesseler et al., 1996; Guo and Santschi, 2007), solid phase extraction (e.g., Louchouarn et al., 2000; Kim et al., 2003; Dittmar et al., 2008), and reverse osmosis coupled with electrodialysis (Koprivnjak et al., 2006; Vetter et al., 2007).

Major elements of marine organic matter include carbon, nitrogen, phosphorus, sulfur, oxygen, and hydrogen. So far, most stable isotope data of marine organic matter are for carbon, and measurements of N, S, and O isotopes are still scarce, especially for the DOM pool (Peterson et al., 1985; Hoefs, 2004; Fry, 2006). In this chapter, we will first describe the basic concept of isotopic fractionation, the stable isotopes of C, H, N, O, and S, as well as radiocarbon and their isotopic composition in natural marine organic matter. The focus will be shifted to sampling procedures of DOM, COM, and POM, and treatment/preparation of organic matter samples for bulk isotope, compound classes, and compound-specific isotope analyses.

6.2 STABLE ISOTOPES AND RADIOCARBON

Table 6.1 lists the common isotopes of carbon, nitrogen, sulfur, oxygen, and hydrogen, along with their isotopic mass, natural abundance (%), and standard/reference materials for isotopic measurements. Similar to carbon, nitrogen, and sulfur, phosphorus is one of the important elements that build biochemical molecules. However, phosphorus has only one stable isotope (^{31}P). Although phosphorus has many radioactive isotopes, most of them with a half-life within minutes or seconds, only ^{32}P and ^{33}P (both of them are beta-emitters) could be useful oceanic tracers (Benitez-Nelson, 2000, and references therein) due to their relatively longer decay half-lives. Therefore, phosphorus is considered to be a monoisotopic element and is not listed in Table 6.1.

In addition to traditional stable isotopes (C, N, S, H, and O), numerous stable isotope tracers with applicability to marine biogeochemical cycles have emerged in recent years. These nontraditional stable isotopes, such as Li, B, Si, Ca, Fe, Cu, Zn, Se, Mo, and Cd, can be measured via new analytical techniques, such as multiple-collector inductively coupled plasma-mass spectrometry (MC-ICP/MS), but their isotopic variations are generally very small. These nontraditional stable isotopes are not discussed in this chapter.

6.2.1 δ-NOTATION

The δ-notation is a means to express isotopic composition of a sample relative to a standard (Hoefs, 2004). The isotopic ratios in natural samples usually show a very small variance, sometimes in the range of the third to fifth decimal place. Therefore, measurements of stable isotopes at or near

TABLE 6.1
A List of Commonly Used Isotopes of Carbon, Nitrogen, Sulfur, Hydrogen, and Oxygen

Element	Isotope	Isotopic Mass	Abundance (%)	Isotope Standard
Carbon	^{12}C	12.0000	99.985	Pee Dee Belemnite (PDB)
	^{13}C	13.00335	0.015	
	$^{14}C^a$	14.00324	10^{-10}	Oxalic acid/1890 wood
Nitrogen	^{14}N	14.00307	99.63	Atmospheric N_2
	^{15}N	15.0001	0.37	
Sulfur	^{32}S	31.97207	95.00	Canyon Diablo meteorite
	^{33}S	32.97145	0.76	(CDT)
	^{34}S	33.96786	4.22	
	^{36}S	35.96708	0.014	
Hydrogen	^1H	1.00794	99.985	Standard Mean Ocean
	^2H	2.0141	0.015	Water (SMOW)
Oxygen	^{16}O	15.9949	99.758	PDB or SMOW
	^{17}O	16.9991	0.037	
	^{18}O	17.9991	0.204	

[a] ^{14}C is a radioactive isotope with a half-life of $5,730 \pm 40$ years.

natural abundance levels are usually reported in the delta (δ) notation, whereby δ is a value given in parts per thousand or per mil (‰). The delta value is dimensionless. The δ-notation for an isotope of an element X can be expressed as

$$\delta X \ (‰) = [(R_{sample} - R_{standard})/R_{standard}] \times 1,000 \tag{6.1}$$

or

$$\delta X \ (‰) = [(R_{sample}/R_{standard} - 1)] \times 1,000 \tag{6.2}$$

where R_{sample} is the relative abundance of the heavy to the light isotope of element X in the sample, and $R_{standard}$ is the isotopic ratio of standard materials (e.g., Pee Dee Belemnite (PDB) for stable carbon isotopes and atmosphere-N_2 for nitrogen isotopes). Using C isotopes as an example, the $\delta^{13}C$ is calculated as

$$\delta^{13}C = [(^{13}C/^{12}C)_{sample}/(^{13}C/^{12}C)_{standard} - 1] \times 1,000 \tag{6.3}$$

In addition to the δ-notation, other measurement notations include atom percentage and atom percent excess, among others (Hoefs, 2004).

For radiocarbon measurements, $\Delta^{14}C$ is usually used in addition to $\delta^{14}C$ (Broecker and Peng, 1982):

$$\Delta^{14}C = \delta^{14}C - 2 \ (\delta^{13}C + 25.0) \ (1 + \delta^{14}C/1,000) \tag{6.4}$$

The value of $\Delta^{14}C$ here is a normalized value of $\delta^{14}C$, taking the fractionation between ^{14}C and ^{13}C into consideration.

6.2.2 Isotope Fractionation

Isotope fractionation includes equilibrium fractionation, kinetic fractionation, and fractionation through physical processes. In equilibrium fractionation, the fractionation is temperature dependent and related to differences in thermodynamic properties (e.g., bonding properties) of molecules with different isotopes. The fractionation is normally larger at lower temperatures and declines with increase in temperature. The higher the temperature the less fractionation occurs. Heavy isotopes are typically enriched in the more condensed phases in equilibrium fractionation (Hoefs, 2004, and references therein).

Kinetic fractionation occurs as a unidirectional process that separates stable isotopes from each other by their mass. This applies to most biogeochemical reactions. Molecules with lighter isotopes react more readily. For example, lighter C and N isotopes (that is, ^{12}C and ^{14}N) are preferentially incorporated into phytoplankton cells during photosynthesis in surface ocean compared to ^{13}C and ^{15}N (Farquhar et al., 1989; Needoba et al., 2003). Meanwhile, fractionation is also related to both temperature and availability of dissolved CO_2 or inorganic N in surface ocean. For ^{14}C isotope, the assumption is that the fractionation of ^{14}C relative to ^{12}C is twice that of ^{13}C, reflecting the difference in atomic mass (Stuiver and Polach, 1977). Other fractionations include those occurring during trophic transfers or food web dynamics, and other mass-independent fractionation, such as the case for sulfur isotopes (Hoefs, 2004).

6.2.3 Isotope Mixing between End-Member Organic Components

Delta (δ) values of stable isotopes of a natural sample can usually be added linearly from different end-member components, making isotope mass balance equations straightforward (Fry and Sherr, 1984; Kwak and Zedler, 1997; Raymond and Bauer, 2001b). For example, measured isotope values (δ_S) of a specific sample, which may be composed of several different end-member sources, can be expressed as

$$\delta_{sample} = \Sigma f_i \delta_i \tag{6.5}$$

or

$$\delta_{Sample} = f_1 \delta_1 + f_2 \delta_2 + f_3 \delta_3 + \cdots \tag{6.6}$$

where f_i is the mass or molar fraction of component i, and δ_i is the isotope ratio of component i. This equation can also be used for blank corrections. According to the concept of mass balance,

$$f_1 + f_2 + \cdots = 1 \tag{6.7}$$

For a two-component system, one can calculate the contribution of each component based on the measured isotopic values of sample and components 1 and 2. For example,

$$f_1 = (\delta_{sample} - \delta_2)/(\delta_1 - \delta_2) \tag{6.8}$$

Using multiple isotopes (e.g., $\delta^{13}C$ and $\Delta^{14}C$; $\delta^{34}S$ and $\delta^{36}S$), one can use multiple equations to solve multiple unknown parameters in a multiple-component system (Bauer et al., 2001).

6.2.4 Carbon Isotopes

Carbon, along with other bioactive elements, is a building block of all biochemical molecules. Among the isotopes listed in Table 6.1, carbon isotopes are the most commonly used and measured for tracing the biogeochemical cycling of marine organic matter (see examples in Table 6.2). There are three natural carbon isotopes: ^{12}C, ^{13}C, and ^{14}C. Their average natural abundances are approximately 98.89%, 1.11%, and $10^{-10}\%$, respectively (Table 6.1). The combined measurements of ^{13}C and ^{14}C provide a more complete picture of dynamic cycling of marine organic carbon (Hedges, 1992; Trumbore and Druffel, 1995). Both ^{13}C and ^{14}C are employed extensively to study the sources, reactivity, and fates of dissolved organic carbon (DOC) and particulate organic carbon (POC) in riverine (Hedges et al., 1986; Peterson and Fry, 1987; Guo and Macdonald, 2006; Raymond et al., 2007), estuarine (Cifuentes et al., 1988; Hedges and Keil, 1999; Raymond and Bauer, 2001a, 2001b; Fry, 2006), coastal, and marine environments (Bauer et al., 1992, 2001; Druffel and Williams, 1992; Druffel et al., 1992, 1996; Santschi et al., 1995; Guo et al., 1996, 2003; Wu et al., 1999; Loh et al., 2004). In addition, the carbon isotopic signatures tend to be less affected by photochemical degradation and structural modification processes than other organic biomarkers. Thus, ^{13}C and ^{14}C are preferentially applied to study carbon biogeochemical cycles over the other organic biomarkers (Table 6.2). However, the signatures of $\delta^{13}C$ and $\Delta^{14}C$ in organic matter can also be altered by bacterial activities during mixing of water bodies (Raymond and Bauer, 2001a).

The use of carbon isotopes $\delta^{13}C$ as fingerprints to trace the contribution from different sources relies on the fact that each source has a distinct $\delta^{13}C$ signal in a DOC or POC pool. Organic matter from marine, terrestrial, and marsh environments has a $\delta^{13}C$ range from $-35‰$ to $-5‰$ (Coffin et al., 1994). Typically, organic matter from a terrestrial source (C_3 plants) is relatively depleted in ^{13}C ($-35‰$ to $-25‰$), C_4 marsh macrophytes are enriched in ^{13}C ($-18‰$ to $-8‰$), and the $\delta^{13}C$ values of C_3 marine phytoplankton are intermediate (Fry and Sherr, 1984; Hedges et al., 1986; Tan, 1987). Two major factors are responsible for the large variation in $\delta^{13}C$ of various organic matter: (1) the pathway of carbon metabolism in plants (Fry, 2006) and (2) the isotopic composition of their carbon source. In the process of carbon fixation during the Calvin cycle (C_3 pathway), the key carboxylating enzyme (ribulose bisphosphate carboxylase) discriminates further against the heavier carbon (^{13}C) than the enzyme (phosphoenolpyruvate carboxylase) produced in the Hatch–Slack cycle (C_4 pathway), resulting in a more negative $\delta^{13}C$ in the C_3 plants than in C_4 plants. On the other hand, land plants use atmospheric CO_2 ($\delta^{13}C = -7‰$) as their carbon source, while marine plants assimilate HCO_3^- ($\delta^{13}C = \sim0‰$) from seawater (Libes, 1992). Other factors also contribute to the fractionation of ^{13}C and ^{14}C in plants, including temperature (Rau et al., 1982), partial pressure of CO_2 (Rau et al., 1991b), growth rate (Laws et al., 1995), and community structure of the primary producer (Falkowski, 1991).

Because the ^{14}C is radioactive, one can use it to measure apparent ages of organic matter in marine environments (Williams and Druffel, 1987; Bauer et al., 1992; Santschi et al., 1995). The presence of ^{14}C in organic materials is the basis of the radiocarbon dating method. The half-life of ^{14}C is $5,730 \pm 40$ years. Therefore, radiocarbon can be used to trace marine biogeochemical processes with a timescale of $\sim50,000$ years. The relationship between $\Delta^{14}C$ and the age of carbon-containing materials can be expressed as

$$\text{Radiocarbon age (years BP)} = -8033 \times \ln[1 + (\Delta^{14}C/1,000)] \qquad (6.9)$$

where $-8,033$ represents the mean lifetime of ^{14}C and ln represents the natural logarithm. Table 6.3 shows examples of the radiocarbon composition of marine organic matter. One fundamental assumption for radiocarbon dating is that there is a negligible exchange of organic carbon between the target reservoir and its environment. The application of ^{14}C as a proxy of the age of organic matter is, however, complicated and ambiguous (Bauer, 2002; Mortazavi and Chanton, 2004), mainly due to mixing processes. Single-compound dating could better reflect the true age of the source material.

6.2.5 Nitrogen Isotopes

Nitrogen has two stable isotopes, ^{14}N and ^{15}N, and their natural abundances are 99.63% and 0.37%, respectively. The average abundance of ^{15}N in the atmosphere ($\delta^{15}N = 0‰$) is relatively constant at 0.366%. Atmospheric N_2 is thus used as the standard for reporting $\delta^{15}N$ values. The natural variation of the N isotopic signal of different source materials is useful for understanding food web dynamics and organic matter cycling in ecosystems, especially for tracing the source of nitrogen to aquatic systems (Schell et al., 1998; Sigman, 2000; Knapp et al., 2005; Fry, 2006). While C isotopes

TABLE 6.2
Examples of Stable Isotope Composition of Marine Organic Material

Location	Organic Fraction	$\delta^{13}C$ (‰)	$\delta^{15}N$ (‰)	Reference
Marine	DOC	−21.8	—	Eadie et al. (1978)
Pacific	DON	—	4–11	Gedeon et al. (2001)
Atlantic	DON	—	3.35–4.68	Knapp et al. (2005)
Patuxent estuary	COM	−24.8	8.5–10.8	Sigleo and Macko (1985), Sigleo (1996)
Pacific Ocean	COM	−21.63 ± 0.17	7.93 ± 0.69	Benner et al. (1997)
Atlantic Ocean	COM	−22.0 ± 0.26	8.0 ± 1.2	Benner et al. (1997)
Chesapeake Bay	COM	−24.08 ± 0.42	8.83 ± 0.19	Sigleo and Macko (2002)
San Francisco Bay	COM	−26.70 ± 0.72	7.92 ± 0.84	Sigleo and Macko (2002)
Gulf of Mexico	COM	−21.67 ± 0.76	3.87 ± 0.65	Guo et al. (2003)
Middle Atlantic Bight	COM	−21.2 ± 0.15	5.49 ± 0.77	Guo et al. (2003)
Boston Harbor	COM	−30.1 to −23.7	2.8–5.7	Zou et al. (2004)
Delaware/Chesapeake Bay	COM	−25.1 to −23.1	4.4–8.9	Zou et al. (2004)
San Francisco Bay	COM	−26.1 to −23.1	5.1–6.4	Zou et al. (2004)
Mississippi River plume	COM	−22.6 ± 1.0	4.3 ± 0.5	Guo et al. (submitted)
Atlantic	POM	—	4–11	Altabet (1988)
North Pacific	POM	−23.20 ± 0.62	3.06 ± 1.16	Wu et al. (1999)
Atlantic	POM	—	−0.8–5.4	Mino et al. (2002)
Seawater	Plankton	−21.3 ± 1.1	8.6 ± 1.0	Peterson et al. (1985)
Arctic	Zooplankton	−20.2 to −25.6	5.8–14.2	Schell et al. (1998)

Note: DOM, dissolved organic matter in <0.45 or 0.7 μm bulk fraction; COM, colloidal organic matter in the size fraction between 1 kDa and 0.2 or 0.7 μm; POM, particulate organic matter in the >0.7 μm or sediment trap samples.

vary little during trophic transfers, N isotopes undergo significant changes between trophic levels in marine environments (e.g., Schell et al., 1998). Thus, the use of multiple stable isotopes can provide insights into the biogeochemical cycling of marine organic matter.

Examples of N-isotope measurements for marine organic matter are given in Table 6.2. Most environmental $\delta^{15}N$ values vary within the range from −10‰ to +20‰ (Peterson and Fry, 1987). Although N isotopic measurements have been applied to study the effects of denitrification and assimilation in the ocean (Liu and Kaplan, 1984; Sigman, 2000), measurements of N isotopes in marine organic matter so far have been carried out mainly on phytoplankton, suspended and settling particles, and sediment samples (Wada and Hattori, 1976; Altabet, 1988; Saino and Hattori, 1987; Cifuentes et al., 1988; Benner et al., 1997; Schell et al., 1998; Chen et al., 2006). Measurements of $\delta^{15}N$ on marine dissolved organic matter are still scarce, although there are increasing reports on COM and $DO^{15}N$ measurements (Sigleo and Macko, 1985; Feuerstein et al., 1997; Guo et al., 2003; Knapp et al., 2005; Miyajima et al., 2005).

6.2.6 SULFUR ISOTOPES

Sulfur has four stable isotopes, ^{32}S, ^{33}S, ^{34}S, and ^{36}S. Their abundances are 95.02%, 0.76%, 4.22%, and 0.02%, respectively (Table 6.1). Canyon Diablo meteorite is used as a standard for sulfur isotope measurements, which are reported relative to troilite in the Canyon Diablo iron meteorite (CDT).

Sulfur is normally a minor component of natural organic matter. On average, sulfur to organic carbon ratios range from 0.014 for humic substance (Aiken et al., 1985; Thurman, 1985) to 0.037 for marine organic matter (Guo et al., 1999). In marine sediments, the organic S/C ratio varies

TABLE 6.3
Examples of Radiocarbon Composition of Marine Dissolved, Colloidal, and Particulate Organic Matter

Location	Organic Matter Type	$\Delta^{14}C$ (‰)	Apparent ^{14}C Age (Year BP)	Reference
Pacific	DOC	−150 to −540	1,300 to 5,710	Williams and Druffel (1987)
Gulf of Mexico	COM	−126 to −432	1,076 to 4,538	Santschi et al. (1995)
Atlantic	COM	−89 to −403	751–4,143	Guo et al. (1996)
Arctic	COM	−87 to −379	680–3,770	Benner et al. (2004)
Pacific	POC	43–139	> Modern	Druffel and Williams (1990)
Pacific	Sinking POC	99–136	> Modern	Druffel and Williams (1990)

from 0.01 in surface sediments to 0.11 in deeper or older sediments (Lückge et al., 2002). Natural variation in sulfur isotopes can be used for tracing sediment diagenesis, organic matter cycling, and natural and anthropogenic solute sources. However, few measurements of sulfur isotopes in marine organic matter have been made. Available data on sulfur isotopic composition are for organic matter in suspended particulate matter, sinking particles, and marine organisms (Peterson and Fry, 1987). Since the concentration of inorganic sulfate in seawater far outweighs the concentration of organic matter, separation and purification of organic sulfur from sea salts and sulfate is the first step for the measurement of its isotopes in marine organic matter, especially dissolved organic matter. Adding to this difficulty, sulfur content in organic matter is much lower than the contents of carbon and nitrogen (Guo et al., 1999); therefore, the measurements of sulfur isotopes require a much larger sample size than C and N isotope analysis.

6.2.7 HYDROGEN ISOTOPES

Hydrogen has three isotopes, two stable isotopes, ^1H and ^2H (D for deuterium), and one radioactive isotope (^3H or T for tritium). The radioactive isotope tritium (half-life $t_{1/2} = 12.3$ years) can be used for tracing water mixing with a timescale within 100 years. Although most measurements of δD or D/H ratio are used to trace water sources and hydrological processes, the D/H ratio has been widely used to study the diagenesis of sedimentary organic matter, organic geochemistry, and climate and environmental changes, since D/H ratios of organic hydrogen can preserve quantitative information about paleoenvironmental and paleoclimate conditions. In recent years, measurements of D/H ratios of carbon-bound hydrogen in individual organic compounds have considerably advanced our knowledge on climate and environmental changes (Sauer et al., 2001; Huang et al., 2004). However, measurements of hydrogen isotopes of marine organic matter are still scarce despite their great potential to trace terrestrial input to the marine environment, link food web and hydrology, and extend the carbon cycling studies (Malej et al., 1993; Fry, 2006).

6.2.8 OXYGEN ISOTOPES

Oxygen has three stable isotopes, ^{16}O, ^{17}O, and ^{18}O. Natural variation of the oxygen isotopic composition can be used for determining water mixing, precipitation sources, and evaporation effects, as well as for deciphering paleoceanographic, paleoclimate, and paleohydrologic records. Similar to hydrogen isotopes, measurements of oxygen isotopes in marine samples have been made mostly for water samples and sediment core samples. Few measurements of oxygen isotopes have been made for marine dissolved organic matter, and little is known about how oxygen isotopes are incorporated into marine organic matter.

6.3 SAMPLING OF MARINE DISSOLVED ORGANIC MATTER

The IR-MS system coupled with an elemental analyzer and combustion chamber has become a routine procedure for stable isotope measurements of bulk organic carbon and nitrogen. Many commercial stable isotope facilities or organizations/agencies have provided services for routine analyses of $\delta^{13}C$ and $\delta^{15}N$ in water samples at affordable prices (e.g., http://stableisotopefacility. ucdavis.edu/). Similarly, advances in accelerator mass spectrometry (AMS) have reduced the sample size for radiocarbon analysis to micrograms of carbon (e.g., Tuniz et al., 1998). However, direct measurements of stable isotope composition of marine dissolved organic matter remain difficult except for carbon isotopes (Beaupre et al., 2007; Osburn and St.-Jean, 2007; Lang et al., 2007). Sampling of marine organic matter is a limiting step in the measurements of isotope composition, especially for N and S isotope analyses. The development in sampling technique of particulate organic matter (suspended and sinking particles, plankton and sediment samples) has provided sufficient amounts of particulate material for the measurement of isotope composition. In this section, focus will be on the sampling of DOM and COM from seawater for stable isotope analysis.

6.3.1 BULK DOM FOR $\delta^{13}C$ AND $\Delta^{14}C$ MEASUREMENTS

There are several factors causing measurements of $\delta^{13}C$ and $\Delta^{14}C$ in seawater DOC to be difficult. First, the DOC concentrations in seawater are very low, ranging from 35–45 µM in deep waters to 60–80 µM in surface waters (Guo et al., 1995; Hansell and Carlson, 1998; see also Chapter 2). In addition, there are different ways to oxidize the DOC, including high-energy UV irradiation (Williams and Gordon, 1970; Williams and Druffel, 1987; Bauer et al., 1992; Beaupre et al., 2007), wet chemical oxidation (St.-Jean, 2003; Osburn and St.-Jean, 2007), high-temperature sealed tube oxidation of the lyophilized DOC/salt mixture (Peterson et al., 1994; Fry et al., 1996), combined UV-persulfate oxidation (Bauer et al., 1992; Clercq et al., 1998; Bouillon et al., 2006), and high-temperature catalytic oxidation of seawater DOC (Lang et al., 2007). However, some uncertainties exist in the quantitative oxidation of DOC and subsequent conversion to CO_2 (see Chapter 2).

Measurements of isotopic composition ($\delta^{13}C$ and $\Delta^{14}C$) of dissolved organic carbon (DOC) in the ocean have been reported since the 1960s (Williams et al., 1969; Williams and Gordon, 1970). In recent years, UV oxidation and high-temperature combustion are the most used techniques to convert DOC into CO_2 for isotope analysis (Bauer et al., 1992; Druffel et al., 1992; Beaupre et al., 2007; Lang et al., 2007). For low-DOC oceanic waters, a traditional vacuum line is needed to carry out large-volume seawater DOC oxidation and CO_2 purification using either UV oxidation (Figure 6.2) or high-temperature catalytic oxidation (Figure 6.3).

Seawater samples are filtered through precombusted quartz filters or glass fiber filters (GF/F, 0.7 µm) or prerinsed polycarbonate Nuclepore filters (0.45 µm) to remove particulate organic matter (see Section 2.3.1). Filters with this pore size cannot remove all bacteria from the seawater, and the DOM in samples can be decomposed microbially and photochemically. Thus, samples should be preserved or frozen immediately after sample filtration to prevent bacterial utilization of DOC (see Section 2.3.2). However, many laboratories do not recommend preservation with $HgCl_2$.

Precombusted quartz or glass bottles are used for DOC sampling and storage. Based on recommendations by the JGOFS program and GEOTRACES studies, high-density polyethylene (HDPE) bottles can also be used for DOC sampling (Dickson et al., 2007) and have been adopted by commercial stable isotope laboratories. One advantage of using HDPE bottles is that samples are safer during sample freezing and transportation than those in quartz or glass bottles.

Due to low DOC concentration in seawater, at least 50 mL of filtered seawater is needed to yield a total of 20–50 µg of DOC for C isotope analysis. With an improved low-blank UV oxidation and extraction system, this sample size is now sufficient for both $d^{13}C$ and $\Delta^{14}C$ measurements (Beaupre et al., 2007), which could not be measured previously.

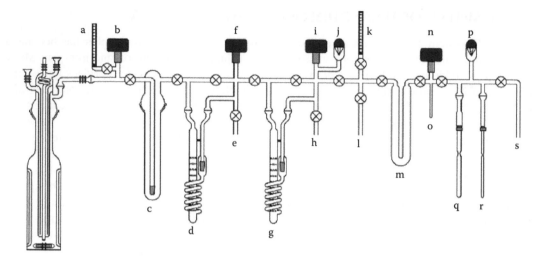

FIGURE 6.2 A schematic showing a traditional vacuum line for purifying CO_2 from the oxidation of DOC in seawater. Individual components are connected by spherical joints with Viton o-rings. (a) Flow meter 1, (k) flow meter 2, (b, f, i, n) capacitance manometers, (j, p) thermocouple pressure gauge sensors, (c) KI solution trap, (d) modified Horibe trap 1 (dry ice/isopropanol slush bath), (g) modified Horibe trap 2 (liquid nitrogen bath), (m) U-tube trap, (o) calibrated volume with a 7 mm OD, 7 cm long cold-finger, (q, r) break-seal tubes for 14C and 13C splits, respectively, secured by internally threaded o-ring adapters with Teflon bushings. (e, h, l, s) Conduits of a manifold (not shown for clarity) leading to the vacuum pump. (From Beaupre et al., 2007, with kind permission of American Society of Limnology and Oceanography.)

6.3.2 BULK DOM FOR STABLE N ISOTOPE ANALYSIS

Measurements of N isotopes have been widely conducted for nitrate or dissolved inorganic nitrogen (DIN) pools in seawater (Liu and Kaplan, 1984, 1989; Sigman et al., 2001, 2005) and marine particulate organic matter samples (Altabet, 1988; Wu et al., 1997; Schell et al., 1998; Waser et al., 2000). However, procedures for direct measurements of N isotopes in the DON pool are still lacking. Currently used methods involve multiple measurements and a number of complex preparation

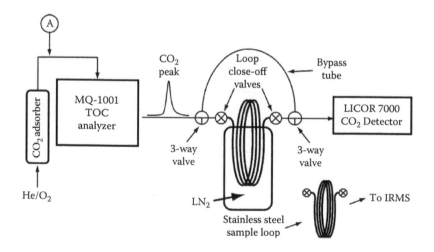

FIGURE 6.3 Schematic of oxidation and trapping system using high-temperature combustion and liquid N_2 for CO_2 trapping. (From Lang et al., 2007, with kind permission from Elsevier Limited.)

processes prior to mass spectrometric determination of the $^{15}N/^{14}N$ ratio. Recently, Miyajima et al. (2005) used gas chromatography/negative-ion chemical ionization mass spectrometry (GC-NICI-MS) to determine ^{15}N enrichment of DON in natural waters after derivatization of NO_3 with pentafluorobenzyl bromide. Knapp et al. (2005) have reported $\delta^{15}N$ isotope values for bulk DON in seawater, based on the measurements of N isotopes in the nitrate (NO^{3-}) pool coupled with the commonly used persulfate oxidation method to convert total dissolved nitrogen (TDN) into nitrate (see Section 9.4.1.2), followed by the denitrifier method for isotopic analysis of NO_3. Similar to the measurements of DON, this method also involves multiple procedures and measurements, but it provides reasonable values of $\delta^{15}N$ of bulk DON pool in seawater for the first time.

Based on Knapp et al. (2005), 12 mL of filtered seawater is used for the conversion of TDN to NO_3^-, with the addition of 2 mL of a persulfate oxidizing reagent (Solorzano and Sharp, 1980), followed by autoclaving for 55 minutes on a slow vent setting. The reagent blank is based on the measurements of NO_3^- and $\delta^{15}N$ in 12 mL of the persulfate oxidizing reagent. The concentration and isotopic composition of DON are calculated by mass balance from the analyses of $\delta^{15}N$ of TDN and NO_3^- in filtered seawater samples.

6.3.3 SEAWATER DISSOLVED ORGANIC SULFUR FOR ISOTOPE ANALYSIS

All four stable isotopes of sulfur (^{32}S, ^{33}S, ^{34}S, and ^{36}S) can be measured at the nanomole level (Ono et al., 2006). However, there are no reports for direct measurements of S isotope composition in marine dissolved organic matter samples. This is largely due to the low organic sulfur concentration and high sea salt content in seawater, although measurements of stable S isotopes in the sulfate pool have been reported previously (Rees et al., 1978).

6.4 SAMPLING OF HIGH MOLECULAR WEIGHT DISSOLVED ORGANIC MATTER FROM SEAWATER

As discussed in the previous section, measurements of stable isotopic composition in marine dissolved organic matter have been hampered by the difficulty in the quantitative extraction of DOM from seawater. Once DOM is isolated from sea salt and lyophilized into powdered DOM samples, measurements of stable isotopes will become routine.

As shown in Figure 6.1, dissolved organic matter also contains a significant portion of high molecular weight (HMW) or colloidal organic matter that can be further isolated using a number of separation and isolation techniques, such as cross-flow ultrafiltration, solid phase extraction, and reverse osmosis coupled with electrodialysis. Marine DOM is heterogeneous and contains different organic components with different molecular weights and isotope signatures, which makes the sampling of size-fractionated DOM and the measurement of their isotopes even more meaningful (Santschi et al., 1995, 1998; Guo et al., 1996; Loh et al., 2004; Wang et al., 2006). The solid phase extraction method has been proved to be simple and efficient for extracting DOM from seawater, although it requires adjustments of the pH of samples before extraction (Dittmar et al., 2008). In this section, the ultrafiltration method, solid phase extraction, and the newly developed reverse osmosis/electrodialysis method will be discussed.

6.4.1 ULTRAFILTRATION METHOD FOR ISOLATING HMW-DOM

Cross-flow ultrafiltration has been widely applied to extract DOC from seawater due to its capability in processing large volumes of seawater, and has proven to be an effective and efficient tool for deciphering chemical and isotopic composition of marine dissolved organic matter (Benner et al., 1992; Moran and Buesseler, 1992; Santschi et al., 1995; Guo et al., 1996; Guo and Santschi, 2007, and references therein). A typical ultrafiltration system is composed of a prefiltration component and ultrafiltration component (Figure 6.4). The prefiltration system is usually composed of a peristaltic pump, tubing, and a prefilter with a pore size ranging from 0.2 to 1 μm, depending on research purposes, while the ultrafiltration system consists of ultrafiltration cartridges with a molecular weight

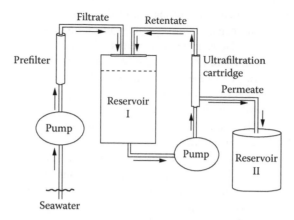

FIGURE 6.4 A diagram showing components of a typical cross-flow ultrafiltration system, including prefiltration and ultrafiltration. The prefiltration system is usually composed of a peristaltic pump, a prefilter with a pore size ranging from 0.2 to 1 μm, and Teflon tubing, while the ultrafiltration system consists of ultrafiltration cartridge(s), reservoirs for prefiltered seawater and permeate, pump, tubing, and other accessories.

cutoff ranging from 0.5 to 200 kDa, reservoirs for prefiltered seawater and permeate, a pump, Teflon tubing, and other accessories (Figure 6.4).

To avoid contamination, surface seawater is pumped peristaltically through a prerinsed cartridge prefilter (in-line setup) into an ultrafiltration reservoir. The first several tens of liters of filtrate should be discarded and used for rinsing reservoirs. Deep water collected by Niskin bottles is transferred to an enclosed reservoir by closed in-line tubing and then pumped peristaltically through a prerinsed filter cartridge into an ultrafiltration reservoir. Filtered seawater should be ultrafiltered immediately after sample collection (Guo et al., 1996).

New ultrafiltration cartridges should be thoroughly cleaned with laboratory detergent (e.g., Micro solution), NaOH, HCl solution, and large volumes of ultrapure water before sampling. Concentrations of each chemical are determined based on manufacturer recommendations for specific ultrafiltration membranes. Between cleaning solutions, large volumes of ultrapure water should be used to rinse the cartridges. To enhance the efficiency during cartridge cleaning, a recirculation mode is used for cleaning with chemical solutions (e.g., Micro, NaOH, and HCl), while a flushing mode or once-through mode (both retentate and permeate lines going into a waste) is used during ultrapure water rinsing/cleaning. Each chemical is recycled for 20–30 minutes under normal operating conditions and allowed to soak for another 20–30 minutes. Between each solution, large volumes of ultrapure water are then flushed through the ultrafiltration system, and after chemical cleaning, more ultrapure water is flushed again under normal ultrafiltration operating conditions. Once the cartridges are cleaned, a small volume of prefiltered seawater is used to condition the cartridges before ultrafiltration (Guo et al., 2000). Used ultrafiltration membranes are usually preserved in NaN_3 solution, so they should be cleaned before sampling based on the same cleaning procedure for new cartridges.

Cartridges are cleaned between samples (or sampling stations) with NaOH and HCl solutions. All ultrafiltration cartridges should be calibrated and checked for their integrity before use with macromolecules of known molecular weights based on specific cartridge cutoffs. The standard macromolecules used for calibration should have a molecular weight slightly higher than the membrane's molecular cutoff (e.g., >1 kDa standard molecule vs.1 kDa membrane, 14 kDa standard molecule vs. 10 kDa membrane, etc.). Further details about cartridge calibration procedures are described in Guo et al. (2000).

The HMW-DOC fraction isolated by cross-flow ultrafiltration is largely dependent on membrane pore size (usually ranging from 0.5 to 200 kDa), membrane materials (hydrophobic vs. hydrophilic), and operational conditions (e.g., concentration factor used). A significant fraction of low molecular weight (LMW) DOC can be retained during ultrafiltration (Guo and Santschi, 1996; Guo et al., 2000). Therefore, a high concentration factor is recommended, although the retained LMW-DOC may be flushed out during the diafiltration process (Guo et al., 2000, 2001). After ultrafiltration, the retentate of the water sample, which is highly concentrated in DOC, is desalted using diafiltration with large volumes of ultrapure water. Then, the sample is freeze-dried at –45°C to become powdered DOM for further analysis (Guo and Santschi, 1996). Due to the low concentrations of DOC in seawater and high concentrations of sea salts, diafiltration (desalting) is a necessary step in isolating COM for isotopic and chemical characterization. About 20 L of ultrapure water is sufficient for the diafiltration of a retentate solution (Guo and Santschi, 1996). After desalting, the isolated HMW-DOC may represent 5%–45% of the bulk DOC in seawater, depending on initial DOC concentration and ultrafilter cutoffs used in ultrafiltration.

6.4.2 Solid Phase Extraction Method

Solid phase extraction has been used in isolating DOM samples from seawater for chemical and isotopic characterization (Louchouarn et al., 2000; Dittmar et al., 2008, and references therein). Louchouarn et al. (2000) used C18–solid phase extraction (SPE) cartridges to determine the DOM extraction efficiency from seawater and the reusability of the C18-SPE cartridges. They found that the DOM recovery rate by the C18-SPE cartridges was comparable to that by the ultrafiltration method and higher at lower pH (1.5–4.0) conditions, while the extraction efficiencies were independent of flow rate.

Recently, Dittmar et al. (2008) presented a protocol for efficient extraction of DOM from seawater using commercially prepacked cartridges. They found a much higher DOM recovery using styrene divinyl benzene polymer type sorbents (PPL) compared to the C18 cartridges. Based on results from NMR, C/N, and $\delta^{13}C$ measurements, they found that styrene divinyl benzene polymer type sorbents (Varian PPL) could extract up to 62% of DOC as salt-free extracts, while the C18 extracts about 40% of bulk DOC. Figure 6.5 shows the seawater DOM extraction procedures using styrene

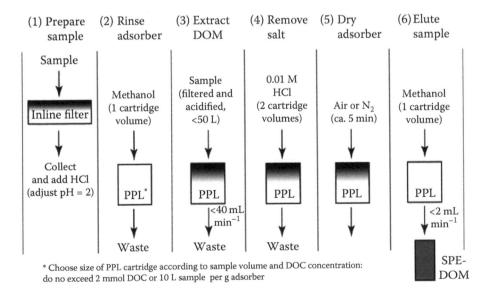

FIGURE 6.5 Procedures of DOM solid phase extraction using commercially prepacked cartridges, styrene divinyl benzene polymer type sorbents (PPL). (From Dittmar et al., 2008, with kind permission of the American Society of Limnology and Oceanography.)

divinyl benzene polymer type sorbents based on Dittmar et al. (2008). After seawater samples are filtered through precombusted GF/Fs, filtrates are acidified to pH 2 with HCl for the extraction of organic acids and phenols. To avoid overloading, less than 10 L of sample or a total of 2 mmol of DOC for 1 g of sorbent is used for extraction. The cartridges are rinsed with one column volume (CV) of methanol before extraction. To extract DOM, the seawater samples are passed through the cartridges with a peristaltic pump or gravity at a flow rate of 40 mL min^{-1} (Figure 6.5, step 3). Before DOM elution, the cartridges are rinsed with at least two CV of 0.01 M HCl to remove residual salts. DOM is eluted with one CV of methanol at a flow rate of <2 mL min^{-1} into muffled glass ampoules (Figure 6.5, step 6). The elutes are stored at –20°C until further processing and analysis.

6.4.3 Reverse Osmosis/Electrodialysis Method

Vetter et al. (2007) used the coupled reverse osmosis (RO)/electrodialysis (ED) system to separate DOC from seawater samples (Figure 6.6). This new separation method can recover up to 64%–93% of marine DOC and has greatly improved the recovery efficiency of marine DOM. This system can be used onboard research vessels. Immediately after sample collection, seawater is filtered through a 0.45 μm filter to remove particulate matter and transferred to the tank of the RO/ED system. The RO/ED system includes a commercial RO module (Dow FilmTec TW30-4021) from the Dow Chemical Company (Midland, Michigan), a Standex Procon CMP-7500 SS pump (Procon, Murfreesboro, Tennessee), and stainless steel tubing and fittings (Figure 6.6). The permeate fluxes are about 0.5–2 L min^{-1} under 200–210 PSIG. The electrodialysis component is composed of membranes (Neosepta AMX and CMX, from Ameridia) and electrodialysis stacks (Type 100, DeukumGmbH, Frickenhausen, Germany; 50 and 100 cell pairs, respectively). The anode is platinized titanium mesh, while the cathode is a stainless steel plate.

The RO system is operated at 200 PSIG pressure with a feed flow rate of 400 L per hour to the module. The flow rate through the seawater circuit of the ED system is about 640 L per hour. The flow rate in the ED concentrate circuit was roughly equal to that in the seawater circuit, while the flow rate in the ED electrode rinse was adjusted to avoid pressure gradients to the concentrate and seawater circuits. Before sampling, the RO/ED system, tanks, and tubing are first rinsed with filtered seawater. After every run, the RO module is rinsed with NaOH solution (0.01 M) manually by slowly filling and draining the RO pressure vessel several times. Electrodialysis is performed in the constant electrical current/variable voltage mode.

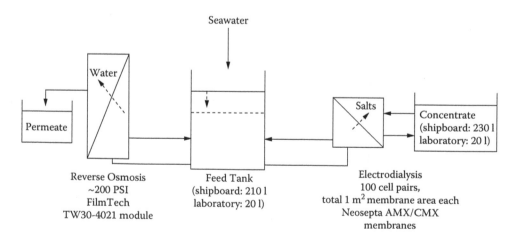

FIGURE 6.6 System for DOM separation from seawater combining reverse osmosis and electrodialysis techniques. (From Vetter et al., 2007, with kind permission from Elsevier Limited.)

6.5 SAMPLING OF PARTICULATE ORGANIC MATTER FROM SEAWATER

Conventional gravity and vacuum filtration methods have been widely used for collecting POM samples for stable isotope analysis (Sharp, 1973; Druffel and Williams, 1990). However, there are potential problems related to small-volume filtration, such as blank and DOC sorption on filters (Moran et al., 1999; Liu et al., 2005). Other practical methods for collecting marine POM samples include (see also Chapter 13):

- Multiple-unit large-volume in situ filtration system (MULVFS): This in situ filtration system can collect up to 100–200 mg of particulate samples (e.g., 1–53 μm or >1 μm) for organic and isotopic analyses, depending on membrane pore size, type of filters, and pumping time (Bishop and Edmond, 1976).
- McLane filtration system: This filtration system can collect particulate sample onto a 142 mm membrane filter (either GF/F or other membrane types) by drawing ambient seawater through the filter holder with a rotary pulse pump. This is a single-event filtration system, capable of pumping up to 25,000 L per deployment, and requires multiple systems for multiple-layer sampling (see also http://www.mclanelabs.com).
- Sediment traps: This device can collect sinking particles in situ by differential settling in a cylindrical settling trap. It could also separate different particle size fractions with different settling velocities (Peterson et al., 2005; Buesseler et al., 2007). While some preservatives used in sediment trap deployment may affect measurements of stable isotopes and other organic components, short-term sediment trap deployments may provide sufficient amounts of POM for isotope analysis.

In addition to these methods, continuous flow centrifugation, Go-Flo rosette filtration, including direct filtration (see Figure 8.3) and off-line filtration of scawater subsample, aggregate sampling devices (e.g., Lunau et al., 2004), and other in situ filtration devices are being used for collecting POM samples from seawater.

6.6 SAMPLE PREPARATION

6.6.1 SAMPLE PROCESSING FOR BULK ORGANIC SAMPLES

Sample processing for direct measurements of stable C and N isotopes in bulk DOM samples is discussed in Sections 6.3.1 and 6.3.2. For isolated COM and POM samples, carbonates are first removed before organic carbon and its isotope analyses (Guo and Macdonald, 2006). This is done through either acid fuming by using concentrated HCl in a closed desiccator or acid treatment using diluted HCl solution. Recent studies recommend the use of acid fuming for the removal of inorganic carbon from organic samples for stable isotope and radiocarbon analyses (Komada et al., 2008; see Section 2.3.3 for detailed procedure). While there is a difference in OC content and its isotope composition between acid fumigation and direct acidification, N and its isotope composition show little difference between these two removal processes.

For isolated COM samples, most of the DIC, DIN, and sulfate can be removed during diafiltration processes, and only DOM components are retained after diafiltration. Thus, there will be little inorganic substances in the desalted COM samples. However, measurements on COM samples without proper desalting or diafiltration could bias the isotope signatures of COM due to the influence of residual DIC, DIN, and sulfate in isolated COM samples.

6.6.2 ORGANIC COMPOUND CLASS SEPARATION FOR STABLE ISOTOPE ANALYSIS

Using stable C isotope as an example, methods for separating total lipid, total hydrolyzable amino acids (THAAs), total carbohydrates (TCHOs), and other organic residue fractions have been developed and used successfully in many previous studies on COM, POM, and SOM (Wang et al., 1996, 1998, 2006; Wang and Druffel, 2001; Loh et al., 2004; Zou et al., 2004). Sample size for compound class isotopic analysis can be as little as a few milligrams of organic carbon, which can be separated into several organic fractions, with 97%–103% recovery for organic C using standard compounds (Wang et al., 1996, 1998). The blank effects on $\Delta^{14}C$ and $\delta^{13}C$ measurements during sample processing and combustion are negligible (Wang et al., 1998). Separation procedures for total lipids, THAAs, TCHOs, and organic residue are described briefly below.

6.6.2.1   Lipid Extraction

Isolated marine organic matter samples are first extracted for total lipids with a 2:1 v/v mixture of methylene chloride:methanol (both high-purity grade). The extraction is repeated four times, and the combined extracts are dried by rotary evaporation. The dried sample is transferred with methylene chloride into a precombusted quartz tube and then dried with a high-purity N_2 gas stream. The extracted residue left in the centrifuge tube after lipid extraction is dried at room temperature for THAA and TCHO extractions (Wang et al., 1998).

6.6.2.2   Total Hydrolyzable Amino Acids Isolation

Isolation of the THAA fraction from marine DOM and POM requires acid hydrolysis to break down the polymer forms, such as proteins (or peptides), to the free amino acids (McCarthy et al., 2004). After lipid extraction, half of each dried DOM/POM sample is weighed directly into a 50 mL glass centrifuge tube and hydrolyzed with 6 M HCl (Ultrex pure) under N_2 at 100°C in an oven for 19 hours. After hydrolysis, samples are centrifuged and the supernatants are transferred into 100 mL pear-shaped glass flasks. The remaining solid material is rinsed twice with ultrapure water, centrifuged, and the supernatants combined with the acid hydrolysate. The hydrolysate is dried by either rotary evaporation or freeze drying.

The dried THAA fraction is dissolved in 2 mL of ultrapure water and desalted using cation exchange column chromatography, with AG 50 W-X8 resin (100–200 mesh, analytical grade, BioRad) that has been soaked in 6 M, double-distilled HCl for at least 1 week and rinsed with ultrapure water several times. Free amino acids are collected in a 1.5 N NH_4OH elute and dried by rotary evaporation or freeze drying. The dried sample is transferred with ultrapure water to a precombusted quartz tube and dried again in a desiccator *in vacuo* for final combustion (Wang et al., 1998).

6.6.2.3   Total Carbohydrate Isolation

The extraction and isolation of total carbohydrates (TCHOs) is based on the method used by Cowie and Hedges (1984). Another half of the residue in the centrifuge tube after lipid extraction is hydrolyzed with H_2SO_4 at 100°C for 3 hours. After hydrolysis, finely ground $Ba(OH)_2 \cdot 8H_2O$ is added to neutralize the acid, and the resulting $BaSO_4$ is removed by centrifuging. The solution is adjusted to a pH of 6–7 with 1.5 N NH_4OH. The hydrolysate containing free sugars is desalted on a 20 mL mixed cation/anion column packed with mixed (1:1 v/v) cation AG 50W-X8 (100–200 mesh) and anion AG 1-X8 (100–200 mesh) resins. The volume of TCHO fraction collected by elution with ultrapure water is reduced to about 2 mL by rotary evaporator at 50°C, then transferred into a precombusted quartz tube for final combustion (Wang et al., 1998). Santschi et al. (1998) used an alcohol extraction/precipitation method to extract and purify polysaccharides from marine HMW-DOM samples for radiocarbon measurements.

6.6.2.4   Acid-Insoluble Fraction

The material left in the centrifuge tube after removal of the supernatant and rinses is defined as the acid-insoluble fraction. This solid material is transferred into a precombusted quartz tube (with 1 mL ultrapure water, three or four times), and the sample is dried in a desiccator *in vacuo* for later combustion.

6.6.3 Procedures for Specific Organic Compound Separation

6.6.3.1 Fatty Acid Separation

Aliquots of freeze-dried isolated DOM, COM, or POM samples (about 10–90 mg of organic matter) are first extracted with 10 mL of methanol, followed by 3×10 mL of methylene chloride:methanol (2:1 v/v). The extracted lipids are further saponified at 100°C with 0.5 M KOH in methanol/H_2O (95:5 v/v) for 2 hours. Neutral lipids are first extracted from the solution at pH ~13, while fatty acids are extracted at pH < 2 after addition of HCl to the solution. Fatty acids are methylated with BF_3^-methanol at 100°C for 2 hours to form fatty acid methyl esters (FAMEs). The FAMEs are quantified by capillary gas chromatography (GC) with an on-column injector and a flame ionization detector. Prior to GC analysis, a known amount of internal standard (nonadecanoic acid methyl ester) is added to each extract to aid quantification. FAMEs are analyzed on a 30 m $\times$ 0.25 mm i.d. HP-5 column coated with 5% diphenyl–95% dimethylsiloxane copolymer (0.25 μm film thickness). The following GC temperature program is implemented: 60 to 150°C at 10°C min[-1], then 150 to 310°C at 4°C min[-1] and isothermally at 310°C for 5 minutes. FAMEs are identified from mass spectra obtained from a GC-MS system. The operating conditions for GC-MS are helium as carrier gas; mass range 50–610 kDa with a 0.4 second scan interval; 70 eV ionizing energy; GC temperature program 50 to 150°C at 20°C min[-1] followed by 150 to 310°C at 4°C min[-1] and a 5 minute hold at 310°C.

Fatty acid molecular stable carbon isotopic compositions are measured using a GC combustion system interfaced with an isotope ratio mass spectrometer (IR-MS). Compounds are separated with a 30 m $\times$ 0.25 mm i.d. column, and the GC temperature is programmed for 50 to 170°C at 20°C min[-1], followed by 170 to 300°C at 4°C min[-1] and a 15-minute hold at 300°C. Peaks eluting from the GC column are combusted to CO_2 over CuO/Pt wires at 850°C and online transported to the IR-MS. The isotopic composition of CO_2 peaks is measured by the IR-MS operated at 10 kV acceleration potential and by magnetic-sector mass separation. To obtain actual fatty acid isotope ratios, the $\delta^{13}C$ of FAMEs is corrected for the carbon added during methylation (Zou et al., 2006).

6.6.3.2 Amino Acids

Individual amino acid isotopic analyses on isolated DOM, COM, or POM samples are made after acid hydrolysis with 6 M HCl at 100°C for 20 hours using isopropyl-TFA derivatives. Samples are hydrolyzed in duplicate, derivatized, and analyzed on a GC-IR-MS. McCarthy et al. (2004, 2007) reported the results of $\delta^{13}C$ measurements on specific amino acid species, including alanine (Ala), aspartic acid + asparagine (Asp), glutamic acid + glutamine (Glu), glycine (Gly), valine (Val), leucine (Leu), isoleucine (Ile), lysine (Lys), and phenylalanine (Phe). Isotope variability of individual amino acid species is dependent on the sample size of marine organic matter. For example, total variability of individual AA isotope composition is usually lower in biological and particulate samples but higher in isolated DOM and COM samples (McCarthy et al., 2004). Similarly, individual AA species, which can be well separated chromatographically from other AA compounds, have a lower analytical variability.

6.6.3.3 Other Organic Compounds

Eglinton et al. (1996) first used automated preparative capillary gas chromatography (PCGC) coupled with a fraction collector to separate and recover sufficient quantities of individual target compounds for natural abundance ^{14}C analysis by accelerator mass spectrometry (AMS). Based on the PCGC approach, the natural radiocarbon abundance can be determined accurately on specific compounds in marine organic samples, which are complex and heterogeneous.

Repeta and Aluwihare (2006) measured compound-specific natural abundance radiocarbon of seven neutral sugar species in HMW-DOM from the North Pacific Ocean. The isolated HMW-DOM is hydrolyzed to release neutral sugars, which are then purified by high-pressure liquid chromatography and analyzed for radiocarbon content. The hydrolysis is conducted with 4 M HCl at 108°C for 4 hours. Following the hydrolysis, acid is removed by freeze drying and the residue is dissolved in ultrapure water and desalted using Biorex 5 anion exchange resin. The resin column is washed with 30 mL of pure water to elute the carbohydrate fraction. The carbohydrate fraction is freeze-dried, dissolved in ultrapure water, and further purified by HPLC with refractive index detection using two cation exchange columns. Monosaccharides are collected in three fractions (glucose/rhamnose, galactose/mannose/xylose, and fucose/arabinose). After freeze drying, each fraction is dissolved in ultrapure water, diluted with acetonitrile, and further purified by reverse phase HPLC (Repeta and Aluwihare, 2006). Purified sugars are collected, evaporated to dryness, dissolved in ultrapure water, and transferred to a precombusted quartz tube containing 300 to 500 mg of copper oxide. After freeze drying, the sealed tubes are combusted at 850°C to convert sugars into carbon dioxide for natural abundance radiocarbon analyses using AMS.

6.7 INSTRUMENTS FOR THE MEASUREMENT OF STABLE ISOTOPES AND RADIOCARBON

6.7.1 ISOTOPE RATIO MASS SPECTROMETER

The isotope ratio mass spectrometer (IR-MS) is the instrument used for the precise measurement of stable isotope ratios. Traditionally, isotope analysis has been carried out using off-line preparation systems followed by automated analysis. Recent advances using fully automated online preparation systems allow routine high-throughput measurements of small amounts of marine organic matter samples (a few micrograms). The mass spectrometer can be configured in both continuous-flow and dual-inlet modes, providing high precision for the determination of H/D, $^{13}C/^{12}C$, $^{15}N/^{14}N$, $^{18}O/^{16}O$, $^{34}S/^{32}S$ (from SO_2 and SF_6), and other isotope ratios. The IR-MS instrument has the following basic components (Figure 6.7): ion source, magnet, and collectors (usually an array of Faraday cups). It can measure reference and sample at the same time and provide simultaneous detection of multiple isotopes.

An elemental analyzer is used to combust organic matter in filter, powdered, or liquid samples. Evolved gases (e.g., CO_2, N_2, H_2O vapor) are carried through an oxidizing furnace tube, a reducing furnace tube, and a water trap. The mixed gases are separated based on their different affinities for the chromatography column, with nitrogen eluting from the column before carbon dioxide. The gases are sequentially transported with the helium stream to the mass spectrometer, where they are analyzed by continuous flow mode. The coupling of IR-MS and an elemental analyzer allows

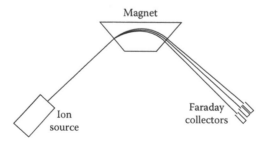

FIGURE 6.7 A diagram showing components in a typical isotope ratio mass spectrometer (IR-MS). (From Karl von Reden, Woods Hole Oceanographic Institution, with kind permission of Media Relations, Woods Hole Oceanographic Institution.)

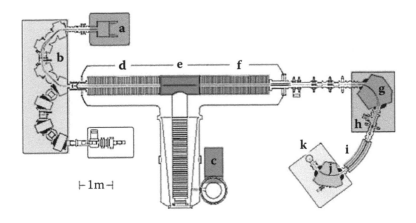

FIGURE 6.8 A diagram showing components of an accelerator mass spectrometer (AMS).

routine $\delta^{15}N$ and $\delta^{13}C$ analyses for organic samples. Although it is not as precise as the dual inlet, an elemental analyzer makes sample preparation much easier, and both $\delta^{15}N$ and $\delta^{13}C$ are measured on the same sample.

Furthermore, the IR-MS can also be interfaced with a gas chromatograph (GC) and a combustor, called GC-C-IR-MS. The GC-C-IR-MS allows measurement of isotope composition of individual compounds, such as fatty acid and amino acids (see also Section 6.6.3).

6.7.2 ACCELERATOR MASS SPECTROMETER

Radiocarbon can be measured by either beta counting or accelerator mass spectrometry (AMS) techniques (see Chapter 13). AMS has an advantage in the measurement of natural abundance radiocarbon in organic matter samples, when sample size is too small (e.g., marine DOM samples) to be accurately measured by beta counting (e.g., Tuniz et al., 1998). Components of the AMS system include ion source, injector magnet, tandem accelerator, analyzing and switching magnets, electrostatic analyzer, and gas ionization detector (Figure 6.8). The strength of AMS is its power to separate a rare isotope (e.g., ^{14}C, with an abundance in parts per trillion) from other abundant isotopes (e.g., ^{12}C, with an abundance of 98.89%). This makes it possible to detect many naturally occurring, long-lived radionuclides that can be used as tracers for organic carbon cycles, such as ^{14}C and ^{129}I (see Chapter 13). Therefore, AMS techniques, where all the ^{14}C atoms are counted directly, can outperform the decay counting techniques, where only those decaying during the counting interval are counted. Especially for isotopes with a long half-life, decay counting techniques become much disadvantaged compared to AMS techniques. The isotopic fractionation on the $^{14}C/^{12}C$ ratio is about two times that on the $^{13}C/^{12}C$ ratio due to the difference in mass. To account for differences in isotopic fractionation between ^{14}C and ^{12}C, values of $\delta^{13}C$ are needed. Depending on the detection limit and decay half-life of ^{14}C, the timescales that can be resolved using radiocarbon are higher than 50 years (the standard deviation may be larger than the age obtained for recent or young samples), but less than 50,000–60,000 years. This maximum ^{14}C age limit is encountered when the ^{14}C activity in a sample is too low to be determined compared to the background ^{14}C activity. Thus, no age is reported greater than 60,000 years.

6.7.3 GC-IR-MS

The coupling of a conventional GC with a specialized isotope ratio mass spectrometer allows the interface to convert all organic matter continuously and quantitatively into a single molecular form for isotopic measurements (e.g., Eglinton et al., 1996; Sessions, 2006). In general, if a compound of

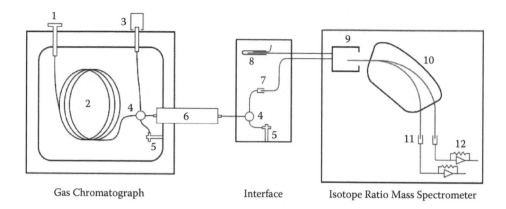

FIGURE 6.9 Schematic of a typical GC-IR-MS system configured for D/H analysis. Major components are: (1) injector, (2) analytical column, (3) flame ionization detector, (4) unions, (5) back flush valves, (6) pyrolysis reactor, (7) open split, (8) reference gas injector, (9) electron impact ionization source, (10) magnetic-sector mass analyzer, (11) Faraday detectors, and (12) analog electrometers. (From Sessions, 2006, with kind permission from Elsevier Limited.)

interest can be analyzed by GC or GC-MS, the ^{13}C isotope ratio can be measured using a specific GC column and oven program.

Figure 6.9 shows a typical GC-IR-MS system configured for D/H analysis (Sessions, 2006). Isotopes of C and N are analyzed, respectively, as CO_2 and N_2, derived from combustion of sample residues, while H and O are analyzed as H_2 and CO produced by pyrolysis.

6.8 QUALITY ASSURANCE

Common quality assurance and quality control (QA/QC) should be performed on a daily basis. Before sample analysis, standard reference materials are analyzed to verify instrument performance. If the values of the standard materials do not fall within the expected range, samples are not analyzed until the expected performance has been established. During sample analysis, blank samples, including procedural, reagent, and instrument blanks, and reference standard samples are analyzed with the samples frequently for the purpose of quality assurance and quality control. Furthermore, replicate measurements should be done for each sample to determine the reproducibility and to verify the performance of the IR-MS system. By evaluating the data from reference materials, blanks, reproducibility, and instrument drift, the data quality can be evaluated. If a problem is detected with either drift or poor performance, the samples should be reanalyzed. The standard deviation of IR-MS analyses (about ±0.4‰) can be obtained based on internal standard measurements over a few days' operation. Precision is normally 0.1‰ for $\delta^{13}C$ and 0.2‰ for $\delta^{15}N$.

6.9 FUTURE RESEARCH

Stable isotopes and radiocarbon composition of organic matter can provide useful information on biogeochemical cycles of DOM, COM, and POM in estuarine, coastal, and oceanic environments. A combination of multiple isotope measurements enhances the power to understand the complicated systems and the interlinked processes. Further research is needed to focus on:

- Improving the efficiency of DOM isolation and separation from seawater for multiple isotope analyses
- Analysis of sulfur isotopic composition in marine organic samples with low sulfur abundance, and improving direct measurements of DON-δ^{15}N in seawater samples
- Improving online coupling between compound separation techniques such as GC and HPLC, and specialized instruments for isotope and radiocarbon analyses, such as IR-MS and AMS
- Finding a specific organic compound in the marine organic matter pool that has a distinctive stable isotope signature in terrestrial, marine, atmospheric, and anthropogenic organic sources, so that the relative contribution of autochthonous and allochthonous sources can be quantified in the marine environment
- Coupling of organic size fractionation with compound class and compound specific separation techniques for the measurements of isotopic composition of marine organic matter

ACKNOWLEDGMENTS

We thank Dr. Peter Santschi and Dr. Oliver Wurl for constructive comments. The writing of this work was supported in part by the U.S. National Science Foundation (EAR 0554781, OCE 0526111, and OCE 0627820) and the University of Southern Mississippi.

REFERENCES

Aiken, G. R., D. M. McKnight, R. L. Wershaw, and P. MacCarthy. 1985. *Humic substances in soil, sediment, and water: Geochemistry, isolation, and characterization*. New York: John Wiley & Sons.

Altabet, M. A. 1988. Variations in nitrogen isotopic composition between sinking and suspended particles: Implications for nitrogen cycling and particle transformation in the open ocean. *Deep Sea Research* 35:535–54.

Altabet, M. A. 1990. Organic C, N, and stable isotopic composition of particulate matter collected on glass-fiber and aluminum oxide filters. *Limnology and Oceanography* 35:902–9.

Altabet, M. A., W. G. Deuser, S. Honjo, and C. Stiene. 1991. Seasonal and depth-related changes in the source of sinking particles in the North Atlantic. *Nature* 354:136–39.

Amon, R. M. W., and R. Benner. 1994. Rapid cycling of high-molecular-weight dissolved organic matter in the ocean. *Nature* 369:549–52.

Bauer, J. E., E. R. M. Druffel, D. M. Wolgast, and S. Griffin. 2001. Sources and cycling of dissolved and particulate organic radiocarbon in the northwest Atlantic continental margin. *Global Biogeochemical Cycles* 15:615–36.

Bauer, J. E., E. R. M. Druffel, D. M. Wolgast, and S. Griffin. 2002. Temporal and regional variability in sources and cycling of DOC and POC in the northwest Atlantic continental shelf and slope. *Deep Sea Research II* 49:4387–4419.

Bauer, J. E., P. M. Williams, and E. R. M. Druffel. 1992. ^{14}C activity of dissolved organic carbon fractions in the northcentral Pacific and Sargasso Sea. *Nature* 357:667–70.

Beaupre, S. R., E. R. M. Druffel, and S. Griffin. 2007. A low-blank photochemical extraction system for concentration and isotopic analyses of marine dissolved organic matter. *Limnology and Oceanography: Methods* 5:174–84.

Benitez-Nelson, C. R. 2000. The biogeochemical cycling of phosphorus in marine systems. *Earth Science Reviews* 51:109–35.

Benner, R., B. Benitez-Nelson, K. Kaiser, and R. M. W. Amon. 2004. Export of young terrigenous dissolved organic carbon from rivers to the Arctic Ocean. *Geophysical Research Letters* 31, L05305, doi: 10.1029/2003GL019251.

Benner, R., B. Biddanda, B. Black, and M. McCarthy. 1997. Abundance, size distribution, and stable carbon and nitrogen isotopic compositions of marine organic matter isolated by tangential-flow ultrafiltration. *Marine Chemistry* 57:243–66.

Benner, R., J. D. Pakulski, M. McCarthy, J. I. Hedges, and P. G. Hatcher. 1992. Bulk chemical characteristics of dissolved organic matter in the ocean. *Science* 255:1561–64.

Bergamaschi, B. A., E. Tsamakis, R. G. Keil, T. I. Eglinton, D. B. Montlucon, and J. I. Hedges. 1997. The effect of grain size and surface area on organic matter, lignin and carbohydrate concentration, and molecular compositions in Peru Margin sediments. *Geochimica et Cosmochimica Acta* 61:1247–60.

Bishop, J. K. B., and J. M. Edmond. 1976. A new large volume filtration system for the sampling of oceanic particulate matter. *Journal of Marine Research* 34:181–98.

Bishop, J. K. B., D. R. Ketten, and J. M. Edmond. 1978. The chemistry, biology and vertical flux of particulate matter from the upper 400 m of the Cape Basin in the southeast Atlantic Ocean. *Deep Sea Research* 25:1121–61.

Bouillon, S., M. Korntheuer, W. Baeyens, and F. Dehairs. 2006. A new automated setup for stable isotope analysis of dissolved organic carbon. *Limnology and Oceanography: Methods* 51:216–26.

Broecker, W. S., and T.-H. Peng. 1982. *Tracers in the sea*. Palisades, NY: Eldigio Press.

Buesseler, K. O., A. N. Antia, M. Chen, S. W. Fowler, W. D. Gardner, O. Gustafsson, K. Harada, A. F. Michaels, M. R. van der Loeff, M. Sarin, D. K. Steinberg, and T. Trull. 2007. An assessment of the use of sediment traps for estimating upper ocean particle fluxes. *Journal of Marine Research* 65:345–416.

Buesseler, K. O., J. Bauer, R. Chen, T. Eglinton, O. Gustafsson, W. Landing, K. Mopper, S. B. Moran, P. H. Santschi, R. Vernon-Clark, and M. Wells. 1996. An intercomparison of cross-flow filtration techniques used for sampling marine colloids: Overview and organic carbon results. *Marine Chemistry* 55:1–32.

Chen, M., L. Guo, Q. Ma, Y. S. Qiu, R. Zhang, E. Lv, and Y. P. Huang. 2006. Zonal patterns of $\delta^{13}C$, $\delta^{15}N$ and ^{210}Po in the North Pacific Ocean. *Geophysical Research Letters* 33:L04609, doi: 10.1029/2005GL025186.

Chester, R. 2003. *Marine Geochemistry*. 2nd ed. New York: Blackwell Publishing.

Cifuentes, L. A., J. H. Sharp, and M. L. Fogel. 1988. Stable carbon and nitrogen isotope biogeochemistry in the Delaware estuary. *Limnology and Oceanography* 33:1102–15.

Clercq, M., J. Van der Plicht, and H. A. J. Meijer. 1998. A supercritical oxidation system for the determination of carbon isotope ratios in marine dissolved organic carbon. *Analytica Chimica Acta* 370:19–27.

Coffin, R. B., L. A. Cifuentes, and P. M. Elderidge. 1994. The use of stable carbon isotopes to study microbial processes in estuaries. In *Stable isotopes in ecology and environment science*, ed. K. Lajitha and R. H. Michener. Oxford: Blackwell Scientific.

Cowie, G. L., and J. I. Hedges. 1984. Determination of neutral sugars in plankton, sediments, and wood by capillary gas chromatography of equilibrated isomeric mixtures. *Analytical Chemistry* 56:504–10.

Dickson, A. G., C. L. Sabine, and J. R. Christian. 2007. *Guide to best practices for ocean CO_2 measurements*. PICES Science Report 34.

Dittmar, T., B. Koch, N. Hertkorn, and G. Kattner. 2008. A simple and efficient method for the solid-phase extraction of dissolved organic matter (SPE-DOM) from seawater. *Limnology and Oceanography: Methods* 6:230–35.

Druffel, E. R. M., S. Griffin, J. E. Bauer, D. Wolgast, and X. Wang. 1996. Distribution of particulate organic carbon and radiocarbon in the water column from the upper slope to the abyssal NE Pacific Ocean. *Deep Sea Research II* 45:667–87.

Druffel, E. R. M., and P. M. Williams. 1990. Identification of a deep marine source of particulate organic carbon using bomb ^{14}C. *Nature* 347:172–74.

Druffel, E. R. M., and P. M. Williams. 1992. Importance of isotope measurements in marine organic geochemistry. *Marine Chemistry* 39:209–15.

Druffel, E. R. M., P. M. Williams, J. E. Bauer, and A. J. Ertel. 1992. Cycling of dissolved and particulate organic matter in the open ocean. *Journal of Geophysical Research* 97:15639–59.

Eadie, B. J., L. M. Feffrey, and W. M. Sackett. 1978. Some observations on the stable carbon isotope composition of dissolved and particulate organic carbon in the marine environment. *Geochimica et Cosmochimica Acta* 42:1265–69.

Eglinton, T. I., L. I. Aluwihare, J. E. Bauer, E. R. M. Druffel, and A. P. McNichol. 1996. Gas chromatographic isolation of individual compounds from complex matrixes for radiocarbon dating. *Analytical Chemistry* 68:904–12.

Falkowski, P. G. 1991. Species variability in the fractionation of ^{13}C and ^{12}C marine phytoplankton. *Journal of Plankton Research* 13:21–28.

Farquhar, G. D., J. R. Ehleringer, and K. T. Hubick. 1989. Carbon isotope discrimination and photosynthesis. *Annual Review of Plant Physiology and Plant Molecular Biology* 40:503–37.

Feuerstein, T. P., P. H. Ostrom, and N. E. Ostrom. 1997. Isotopic biogeochemistry of dissolved organic nitrogen: A new technique and application. *Organic Geochemistry* 21:363–70.

Fry, B. 2006. *Stable isotope ecology*. New York: Springer.

Fry, B., H. W. Jannasch, S. J. Molyneaux, C. O. Wirsen, and J. A. Muramoto. 1991. Stable isotope studies of the carbon, nitrogen and sulfur cycles in the Black Sea and the Cariaco Trench. *Deep Sea Research* 38 (Suppl. 2A):S1003–19.

Fry, B., E. T. Peltzer, C. S. Hopkinson, A. Nolin, and L. Redmond. 1996. Analysis of marine DOC using a dry combustion method. *Marine Chemistry* 54:191–201.

Fry, B., and E. B. Sherr. 1984. δ^{13}C measurements as indicators of carbon flow in marine and freshwater ecosystems. *Contributions to Marine Science* 27:13–47.

Gandhi, H., T. N. Wiegner, and P. H. Ostrom. 2004. Isotopic (C-13) analysis of dissolved organic carbon in stream water using an elemental analyzer coupled to a stable isotope ratio mass spectrometer. *Rapid Communication in Mass Spectrometry* 18:903–6.

Gedeon, M. L., P. H. Ostrom, N. E. Ostrom, and H. Gandhi. 2001. Simultaneous determination of dissolved organic nitrogen concentrations and isotope values in the subtropical North Pacific. In *ASLO 2001 Aquatic Science Meeting*, February 12–16, 2001, Albuquerque, NM, p. 57.

Guo, L., B. J. Hunt, and P. H. Santschi. 2001. Ultrafiltration behavior of major ions (Na, Ca, Mg, F, Cl, SO$_4$) in natural waters. *Water Research* 35:1500–8.

Guo, L., and R. W. Macdonald. 2006. Source and transport of terrigenous organic matter in the upper Yukon River: Evidence from isotope (^{13}C, ^{14}C and ^{15}N) composition of dissolved, colloidal and particulate phases. *Global Biogeochemical Cycles* 20:GB2011, doi: 10.1029/2005GB002593.

Guo, L., and P. H. Santschi. 1996. A critical evaluation of the cross-flow ultra-filtration technique for sampling colloidal organic carbon in seawater. *Marine Chemistry* 55:113–27.

Guo, L., and P. H. Santschi. 1997a. Isotopic and elemental characterization of colloidal organic matter from the Chesapeake Bay and Galveston Bay. *Marine Chemistry* 59:1–15.

Guo, L., and P. H. Santschi. 1997b. Composition and cycling of colloids in marine environments. *Review of Geophysics* 35:17–40.

Guo, L., and P. H. Santschi. 2007. Ultra-filtration and its applications to sampling and characterization of aquatic colloids. In *Environmental colloids and particles*, ed. K. Wilkinson and J. Lead. New York: John Wiley & Sons.

Guo, L., P. H. Santschi, and T. S. Bianchi. 1999. Dissolved organic matter in estuaries of the Gulf of Mexico. In *Biogeochemistry of Gulf of Mexico estuaries*, ed. T. S. Bianchi, J. R. Pennock, and R. R. Twilley. New York: John Wiley & Sons.

Guo, L., P. H. Santschi, L. A. Cifuentes, S. Trumbore, and J. Southon. 1996. Cycling of high molecular weight dissolved organic matter in the Middle Atlantic Bight as revealed by carbon isotopic (^{13}C and ^{14}C) signatures. *Limnology and Oceanography* 41:1242–52.

Guo, L., P. H. Santschi, and K. W. Warnken. 1995. Dynamics of dissolved organic carbon in oceanic environments. *Limnology and Oceanography* 40:1392–403.

Guo, L., N. Tanaka, D. M. Schell, and P. H. Santschi. 2003. Nitrogen and carbon isotopic composition of high-molecular-weight dissolved organic matter in marine environments. *Marine Ecology Progress Series* 252:51–60.

Guo, L., L. Wen, D. Tang, and P. H. Santschi. 2000. Re-examination of cross flow ultra-filtration for sampling aquatic colloids: Evidence from molecular probes. *Marine Chemistry* 69:75–90.

Guo, L., D. M. White, C. Xu, and P. H. Santschi. 2008. Chemical and isotopic composition of colloidal organic matter from the Mississippi River plume. *Marine Chemistry*, submitted.

Hansell, D. A., and C. A. Carlson. 1998. Deep-ocean gradients in the concentration of dissolved organic matter. *Nature* 395:263–66.

Hansell, D. A., and C. A. Carlson. 2002. *Biogeochemistry of marine dissolved organic matter*. New York: Academic Press.

Hedges, J. I. 1992. Global biogeochemical cycles: Progress and problems. *Marine Chemistry* 39:67–93.

Hedges, J. I. 2002. Why dissolved organic matter? In *Biogeochemistry of marine dissolved organic matter*, ed. D. A. Hansell and C. Carlson. New York: Academic Press.

Hedges, J. I., J. A. Baldock, Y. Gélinas, C. Lee, M. Peterson, and S. G. Wakeham 2001. Evidence for non-selective preservation of organic matter in sinking marine particles. *Nature* 409:801–4.

Hedges, J. I., J. R. Ertel, P. D. Quay, P. M. Grootes, J. Richey, A. H. Devol, and G. W. Farwell. 1986. Organic carbon-14 in the Amazon River system. *Science* 231:1129–31.

Hedges, J. I., and R. G. Keil. 1999. Organic geochemical perspectives on estuarine processes: Sorption reactions and consequences. *Marine Chemistry* 65:55–65.

Hoefs, J. 2004. *Stable isotope geochemistry*. Berlin: Springer-Verlag.

Huang Y., B. Shuman, Y. Wang, and T. Webb III. 2004. Hydrogen isotope ratios of individual lipids in lake sediments as novel tracers of climatic and environmental change: A surface sediment test. *Journal of Paleolimnology* 31:363–75.

Hunt, A. P., J. D. Parry, and J. Hamilton-Taylor. 2000. Further evidence of elemental composition as an indicator of the bioavailability of humic substances to bacteria. *Limnology and Oceanography* 45:237–41.

Kim, S., A. J. Simpson, E. B. Kujawinski, M. A. Freitas, and P. G. Hatcher. 2003. High resolution electrospray ionization mass spectrometry and 2D solution NMR for the analysis of DOM extracted by C18 solid phase disk. *Organic Geochemistry* 34:1325–35.

Knapp, A. N., D. M. Sigman, and F. Lipschultz. 2005. N isotopic composition of dissolved organic nitrogen and nitrate at the Bermuda Atlantic time-series study site. *Global Biogeochemical Cycles* 19:GB1018, doi: 10.1029/2004GB002320.

Knauer, G. A., J. H. Martin, and K. W. Bruland. 1979. Fluxes of particulate carbon, nitrogen, and phosphorus in the upper water column of the northeast Pacific. *Deep Sea Research* 26A:97–108.

Komada, T., M. R. Anderson, and C. L. Dorfmeier. 2008. Carbonate removal from coastal sediments for the determination of organic carbon and its isotopic signatures, d^{13}C and Δ^{14}C: Comparison of fumigation and direct acidification by hydrochloric acid. *Limnology and Oceanography: Methods* 6:254–62.

Koprivnjak, J. F., E. M. Perdue, and P. H. Pfromm. 2006. Coupling reverse osmosis with electrodialysis to isolate natural organic matter from fresh waters. *Water Research* 40:3385–92.

Kwak, T. J., and J. B. Zedler. 1997. Food web analysis of southern California coastal wetlands using multiple stable isotopes. *Oecologia* 110:262–77.

Lang, S. Q., M. D. Lilley, and J. I. Hedges. 2007. A method to measure the isotopic (13C) composition of dissolved organic carbon using a high temperature combustion instrument. *Marine Chemistry* 103:318–26.

Laws, E. A., B. N. Popp, R. R. Bidigare, M. C. Kennicutt, and S. A. Maco. 1995. Dependence of phytoplankton carbon isotopic composition on growth rate and [CO_2]$_{aq}$: Theoretical considerations and experimental results. *Geochimica et Cosmochimica Acta* 59:1131–38.

Lee, C., and S. G. Wakeham. 1992. Organic matter in the water column: Future research challenges. *Marine Chemistry* 39:95–118.

Libes, S. M. 1992. *An introduction to marine biogeochemistry*. New York: John Wiley & Sons.

Liu, K. K., and I. R. Kaplan. 1984. Denitrification rates and availability of organic matter in marine environment. *Earth and Planetary Science Letters* 68:88–100.

Liu, K. K., and I. R. Kaplan. 1989. The eastern tropical Pacific as a source of 15N-enriched nitrate in seawater off southern California. *Limnology and Oceanography* 34:820–30.

Liu, Z. F., G. Stewart, J. K. Cochran, C. Lee, R. A. Armstrong, D. I. Hirschberg, B. Gasser, and J. C. Miquel. 2005. Why do POC concentrations measured using Niskin bottle collections sometimes differ from those using in-situ pumps? *Deep Sea Research I* 52:1324–44.

Loh, A. N., J. E. Bauer, and E. R. M. Druffel. 2004. Variable ageing and storage of dissolved organic components in the open ocean. *Nature* 430:877–80.

Louchouarn, P., S. Opsahl, and R. Benner. 2000. Isolation and quantification of dissolved lignin from natural waters using solid-phase extraction and GC/MS. *Analytical Chemistry* 72:2780–87.

Lückge, A., B. Horsfield, R. Littkeand, and G. Scheeder. 2002. Organic matter preservation and sulfur uptake in sediments from the continental margin off Pakistan. *Organic Geochemistry* 33:477–88.

Lunau, M., A. Sommer, A. Lemke, H. P. Grossart, and M. Simon. 2004. A new sampling device for microaggregates in turbid aquatic systems. *Limnology and Oceanography: Methods* 2:387–97.

Malej, A., J. Faganeli, and J. Pezdi. 1993. Stable isotope and biochemical fractionation in the marine pelagic food chain: The jellyfish *Pelagia noctiluca* and net zooplankton. *Marine Biology* 116:565–70.

Mayer, L. M. 1994. Relationships between mineral surfaces and organic-carbon concentrations in soils and sediments. *Chemical Geology* 114:347–63.

McCarthy, M. D., R. Benner, C. Lee, and M. L. Fogel. 2007. Amino acid nitrogen isotopic fractionation patterns as indicators of heterotrophy in plankton, particulate, and dissolved organic matter. *Geochimica et Cosmochimica Acta* 71:4727–44.

McCarthy, M. D., R. Benner, C. Lee, J. I. Hedges, and M. L. Fogel. 2004. Amino acid carbon isotopic fractionation patterns in oceanic dissolved organic matter: An unaltered photoautotrophic source for dissolved organic nitrogen in the ocean? *Marine Chemistry* 92:123–34.

Mino, Y., T. Saino, K. Suzuki, and E. Maranon. 2002. Isotopic composition of suspended particulate nitrogen in surface waters of the Atlantic Ocean from 50 degrees N to 50 degrees S. *Global Biogeochemical Cycles* 16:GB 1059, doi: 10.1029/2001GB001635.

Miyajima, T., Y. Tanaka, and I. Koike. 2005. Determining ^{15}N enrichment of dissolved organic nitrogen in environmental waters by gas chromatography/negative-ion chemical ionization mass spectrometry. *Limnology and Oceanography: Methods* 3:164–73.

Mopper, K., A. Stubbins, J. D. Ritchie, H. M. Bialk, and P. G. Hatcher. 2007. Advanced instrumental approaches for characterization of marine dissolved organic matter: Extraction techniques, mass spectrometry, and nuclear magnetic resonance spectroscopy. *Chemical Reviews* 107:419–42.

Moran, S. B., and K. O. Buesseler. 1992. Short residence time of colloids in the upper ocean estimated from ^{238}U-^{234}Th disequilibria. *Nature* 359:221–23.

Moran, S. B., M. A. Charette, S. M. Pike, and C. A. Wicklund. 1999. Differences in seawater particulate organic carbon concentration in samples collected using small- and large-volume methods: The importance of DOC adsorption to the filter blank. *Marine Chemistry* 67:33–42.

Mortazavi, B., and J. P. Chanton. 2004. Use of Keeling plots to determine sources of dissolved organic carbon in nearshore and open ocean systems. *Limnology and Oceanography* 49:102–8.

Needoba, J. A., N. A. Waser, P. J. Harrison, and S. E. Calvert. 2003. Nitrogen isotope fractionation in 12 species of marine phytoplankton during growth on nitrate. *Marine Ecology Progress Series* 255:81–91.

Ono, S., B. Wing, D. Rumble, and J. Farquhar. 2006. High precision analysis of all four stable isotopes of sulfur (^{32}S, ^{33}S, ^{34}S and ^{36}S) at nanomole levels using a laser fluorination isotope-ratio-monitoring gas chromatography–mass spectrometry. *Chemical Geology* 225:30–39.

Osburn, C. L., and G. St.-Jean. 2007. The use of wet chemical oxidation with high-amplification isotope ratio mass spectrometry to measure stable isotope values of dissolved organic carbon in seawater. *Limnology and Oceanography: Methods* 5:296–308.

Pearson, A., and T. I. Eglinton. 2000. The origin of n-alkanes in Santa Monica Basin surface sediment: A model based on compound-specific $\Delta^{14}C$ and $\delta^{13}C$ data. *Organic Geochemistry* 31:1103–16.

Peterson, B. J., and B. Fry. 1987. Stable isotopes in ecosystem studies. *Annual Review of Ecology and Systematics* 18:293–320.

Peterson, B., B. Fry, M. Hullar, and S. Saupe. 1994. The distribution and stable carbon isotopic composition of dissolved organic carbon in estuaries. *Estuaries* 17:111–21.

Peterson, B. J., R. W. Howarth, and R. H. Grarritt. 1985. Multiple stable isotopes used to trace the flow of organic matter in estuarine food webs. *Science* 227:1361–63.

Peterson, M. L., S. G. Wakeham, C. Lee, M. A. Askea, and J. C. Miquel. 2005. Novel techniques for collection of sinking particles in the ocean and determining their settling rates. *Limnology and Oceanography: Methods* 3:520–32.

Rau, G. H., C. W. Sullivan, and L. I. Gordon. 1991a. $\delta^{13}C$ and $\delta^{15}N$ variations in the Weddell Sea particulate organic matter. *Marine Chemistry* 35:355–69.

Rau, G. H., R. E. Sweeney, and I. R. Kaplan. 1982. Plankton $^{13}C/^{12}C$ ratio changes with latitude: Differences between northern and southern oceans. *Deep Sea Research* 29:1035–39.

Rau, G. H., T. Takahashi, and D. J. DesMarais. 1991b. Latitudinal variations in plankton $\delta^{13}C$: Implications for CO_2 and productivity in past oceans. *Nature* 341:516–18.

Rau, G. H., J. I. Teyssie, F. Rassoulzadegan, and S. W. Fowler. 1990. $^{13}C/^{12}C$ and $^{15}N/^{14}N$ variations among size-fractionated marine particles: Implications for their origin and trophic relationships. *Marine Ecology Progress Series* 59:33–38.

Raymond, P. A., and J. E. Bauer. 2001a. Use of ^{14}C and ^{13}C natural abundances for evaluating riverine, estuarine, and coastal DOC and POC sources and cycling: A review and synthesis. *Organic Geochemistry* 32:469–85.

Raymond, P. A., and J. E. Bauer. 2001b. DOC cycling in a temperate estuary: A mass balance approach using natural ^{14}C and ^{13}C isotopes. *Limnology and Oceanography* 46:655–67.

Raymond, P. A., J. W. McClelland, R. M. Holmes, A. V. Zhulidov, K. Mull, B. J. Peterson, R. G. Striegl, G. R. Aiken, and T. Y. Gurtovaya. 2007. Flux and age of dissolved organic carbon exported to the Arctic Ocean: A carbon isotopic study of the five largest arctic rivers. *Global Biogeochemical Cycles* 21:GB4011, doi: 10.1029/2007GB002934.

Rees, C. E. 1978. Sulphur isotope measurements using SO_2 and SF_6. *Geochemica et Cosmochemica Acta* 42:377–81.

Repeta, D. J., and L. I. Aluwihare. 2006. Radiocarbon analysis of neutral sugars in high-molecular-weight dissolved organic carbon: Implications for organic carbon cycling. *Limnology and Oceanography* 51:1045–53.

Sackett, W. M. 1989. Stable carbon isotope studies on organic matter in the marine environment. In *Handbook of environmental isotope geochemistry*, ed. P. Fritz and J. Ch. Fontes. Amsterdam, The Netherlands: Elsevier Science.

Saino, T., and A. Hattori. 1987. Geographical variation in the water column distribution of suspended particulate organic nitrogen and its ^{15}N natural abundance in the Pacific and its marginal seas. *Deep Sea Research I* 34:807–27.

Santschi, P. H., E. Balnois, K. Wilkinson, J. Buffle, and L. Guo. 1998. Fibrillar polysaccharides in marine macromolecular organic matter as imaged by atomic force microscopy and transmission electron microscopy. *Limnology and Oceanography* 43:896–908.

Santschi, P. H., L. Guo, M. Baskaran, S. Trumbore, J. Southon, T. S. Bianchi, B. D. Honeyman, and L. Cifuentes. 1995. Isotopic evidence for contemporary origin of high-molecular weight organic matter in oceanic environments. *Geochimica et Cosmochimica Acta* 59:625–31.

Sauer, P. E., T. I. Eglinton, J. M. Hayes, A. Schimmelmann, and A. L. Sessions. 2001. Compound-specific D/H ratios of lipid biomarkers from sediments as a proxy for environmental and climate conditions. *Geochimica et Cosmochemica Acta* 65:213–22.

Schell, D. M., B. A. Barnett, and K. A. Vinette. 1998. Carbon and nitrogen isotope ratios in zooplankton of the Bering, Chukchi and Beaufort Sea. *Marine Ecology Progress Series* 162:11–23.

Sessions, A. L. 2006. Isotope-ratio detection for gas chromatography. *Journal of Separation Science* 29:1946–61.

Sharp, J. H. 1973. Size classes of organic carbon in seawater. *Limnology and Oceanography* 18:441–47.

Sigleo, A. C. 1996. Biochemical components in suspended particles and colloids: Carbohydrates in the Potomac and Patuxent estuaries. *Organic Geochemistry* 24:83–93.

Sigleo, A. C., and S. A. Macko. 1985. Stable isotope and amino acid composition of estuarine dissolved colloidal material. In *Marine and estuarine geochemistry*, ed. A. C. Sigleo and A. Hattori. Chelsea, MI: Lewis Publishers.

Sigleo, A. C., and S. A. Macko. 2002. Carbon and nitrogen isotopes in suspended particles and colloids, Chesapeake and San Francisco estuaries, USA. *Estuarine, Coastal and Shelf Sciences* 54:701–11.

Sigman, D. M. 2000. The $\delta^{15}N$ of nitrate in the Southern Ocean: Nitrogen cycling and circulation in the ocean interior. *Journal of Geophysical Research—Oceans* 105:19599–614.

Sigman, D. M., K. L. Casciotti, M. Andreani, C. Barford, M. Galanter, and J. K. Bohlke. 2001. A bacterial method for the nitrogen isotopic analysis of nitrate in seawater and freshwater. *Analytical Chemistry* 73:4145–53.

Sigman, D. M., J. Granger, P. J. DiFiore, R. Ho, C. Cane, and A. van Geen. 2005. Coupled nitrogen and oxygen isotope measurements of nitrate along the eastern North Pacific margin. *Global Biogeochemical Cycles* 19:GB4022.

Solorzano, L., and J. H. Sharp. 1980. Determination of total dissolved nitrogen in natural waters. *Limnology and Oceanography* 25:754–58.

St.-Jean, G. 2003. Automated quantitative and isotopic (^{13}C) analysis of dissolved inorganic carbon and dissolved organic carbon in continuous flow using a total organic carbon analyzer. *Rapid Communications in Mass Spectrometry* 17:419–28.

Stuiver, M., and H. A. Polach. 1977. Discussion: Reporting of ^{14}C data. *Radiocarbon* 19:355–65.

Tagliabue, A., and L. Bopp. 2008. Towards understanding global variability in ocean carbon-13. *Global Biogeochemical Cycles* 22:GB1025, doi: 10.1029/2007GB003037.

Tan, F. C. 1987. Discharge and carbon isotope composition of particulate organic carbon from the St. Lawrence River, Canada. In *Transport of carbon and minerals in major world rivers*, ed. E. T. Degens, S. Kempe, and W. Gan, 64, 301–10. Mitt. Geol-Paläont. Inst. Univ. Hamburg, SCOPE/UNEP Sonderbd.

Tanaka, T., L. Guo, C. Deal, N. Tanaka, T. Whitledge, and A. Murata. 2004. Nitrogen deficiency in a well oxygenated cold bottom water over the Bering Sea Shelf: Influence of sedimentary denitrification. *Continental Shelf Research* 24:1271–83.

Thurman, E. M. 1985. *Organic geochemistry of natural waters*. Dordrecht, The Netherlands: Kluwer Academic.

Tumbore, S. E., and E. R. M. Druffel. 1995 Carbon isotopes for characterizing sources and turnover of non-living organic matter. In *Role of nonliving organic matter in the earth's carbon cycle*, ed. R. G. Zepp and Ch. Sonntag. New York: John Wiley & Sons.

Tuniz, C., J. R. Bird, D. Fink, and G. F. Herzog. 1998. *Accelerator mass spectrometry. Ultrasensitive analysis for global science*. Boca Raton, FL: CRC Press.

Vetter, T. A., E. M. Perdue, E. Ingall, J.-F. Koprivnjak, and P. H. Pfromm. 2007. Combining reverse osmosis and electrodialysis for more complete recovery of dissolved organic matter from seawater. *Separation and Purification Technology* 56:383–87.

Wada, E., and A. Hattori. 1976. Natural abundance of $\delta^{15}N$ in particulate organic matter in the North Pacific Ocean. *Geochimica et Cosmochimica Acta* 40:249–51.

Wang, X. C., J. Callahan, and R. F. Chen. 2006. Variability in radiocarbon ages of biochemical compound classes of high molecular weight dissolved organic matter in estuaries. *Estuarine, Coastal and Shelf Science* 68:188–94.

Wang, X.-C., and E. R. M. Druffel. 2001. Radiocarbon and stable carbon isotope compositions of organic compound classes in sediments from the NE Pacific and Southern Oceans. *Marine Chemistry* 73:65–81.

Wang, X. C., E. R. M. Druffel, S. Griffin, C. Lee, and M. Kashgarian. 1998. Radiocarbon studies of organic compound classes in plankton and sediment of the Northeast Pacific Ocean. *Geochimica et Cosmochimica Acta* 59:1787–97.

Wang, X. C., E. R. M. Druffel, and C. Lee. 1996. Radiocarbon in organic compound classes in particulate organic matter and sediment in the deep Northeast Pacific Ocean. *Geophysical Research Letters* 23:3583–86.

Waser, N. A., W. G. Harrison, E. J. H. Head, B. Nielson, V. A. Lutz, and S. E. Calvert. 2000. Geographic variations in the nitrogen isotope composition of surface particulate nitrogen and new production across the North Atlantic Ocean. *Deep Sea Research I* 47:1207–26.

Williams, P. M., and E. R. M. Druffel. 1987. Radiocarbon in dissolved organic matter in the central North Pacific Ocean. *Nature* 330:246–48.

Williams, P. M., and L. I. Gordon. 1970. Carbon-13:carbon-12 ratios of dissolved and particulate organic matter in the sea. *Deep Sea Research* 17:19–27.

Williams, P. M., H. Oeschger, and P. Kinney. 1969. Natural radiocarbon activity of dissolved organic carbon in north-east Pacific Ocean. *Nature* 224:256–58.

Wu, J., S. E. Calvert, and C. S. Wong. 1997. Nitrogen isotope variations in the subarctic northeast Pacific: Relationships to nitrate utilization. *Deep Sea Research I* 44:287–314.

Wu, J., S. E. Calvert, C. S. Wong, and F. A. Whitney. 1999. Carbon and nitrogen isotopic composition of sedimenting particulate material at Station Papa in the subarctic northeast Pacific. *Deep Sea Research II* 46:2793–832.

Zou, L., M. Y. Sun, and L. Guo. 2006. Temporal variations of organic carbon inputs into the upper Yukon River: Evidence from fatty acids and their stable carbon isotopic compositions in dissolved, colloidal and particulate phases. *Organic Geochemistry* 37:944–56.

Zou, L., X. C. Wang, J. Callahan, R. A. Culp, R. F. Chen, M. A. Altabet, and M. Y. Sun. 2004. Bacterial roles in the formation of high-molecular-weight dissolved organic matter in estuarine and coastal waters: Evidence from lipids and the compound-specific isotopic ratios. *Limnology and Oceanography* 49:297–302.

7 Determination of Marine Gel Particles

Anja Engel

CONTENTS

7.1 INTRODUCTION

Gel particles have gained increasing attention in the marine scientific community. They have been examined with respect to abiotic particle formation (Chin et al., 1998; Passow, 2000; Kerner et al., 2003; Verdugo et al., 2004; Engel et al., 2004), organic matter coagulation and sedimentation/export processes (Logan et al., 1995; Engel, 2000; Passow et al., 2001; Beauvais et al., 2006; Kahl et al., 2008), carbon cycling (Mari, 1999; Engel and Passow, 2001; Mari et al., 2001), absorbance of trace elements and nutrients (Quigley et al., 2002; Guo et al., 2002; Azetsu-Scott and Niven, 2005), climate change feedback (Engel, 2002; Arrigo, 2007), habitat structuring, and food web dynamics in benthic and pelagic ecosystems (Krembs and Engel, 2000; Mari et al., 2004; Wild et al., 2004; Shackelford and Cowen, 2006; Grossart and Simon, 2007; Heinonen et al., 2007).

Marine gel particles are hydro gels, which means that water, specifically seawater, is the fluid penetrating a solid three-dimensional network of polymers, resulting in a semisolid or jelly-like texture of the gel. In general, the reactivity and rigidity of a gel depends on its molecular composition and on the type of binding and interactions between the individual polymer chains, such as covalent binding, cation bridging, physical entangling, and hydrophobicity. The size of the polymer network representing the gel can range from a single macromolecule-containing colloid (~1 nm) to macro gels spanning several millimeters. Micro and macro gels can aggregate with each other and thereby further increase in size, up to the centimeter scale. The size continuum of gel particles (Figure 7.1)

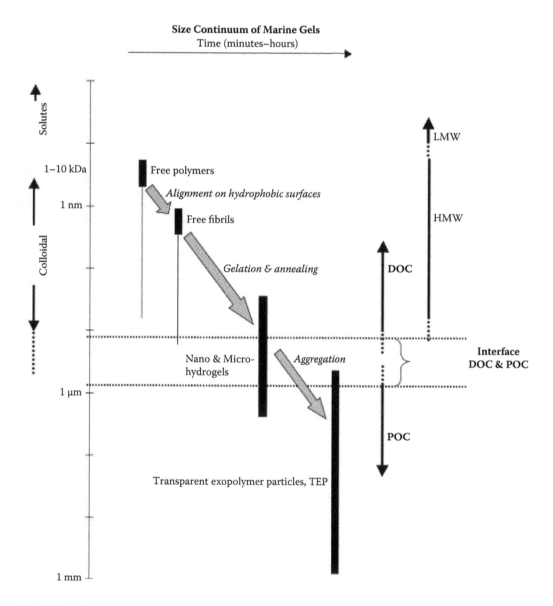

FIGURE 7.1 Sketch of the size continuum of marine gel particles illustrating the size distribution of gels, and the processes governing gel formation. Organic molecules originating from biological activity can adsorb and align to each other to form colloids. Aggregation of colloids leads to gel particles, such as transparent exopolymer particles (TEP), which favor micro- and macroaggregate formation due to their high stickiness. (From Verdugo et al., 2004, with kind permission of Elsevier Limited.)

and the role of gels for organic matter partitioning in the ocean have been reviewed by Verdugo et al. (2004).

Overall, the study of gel particles in marine systems is still in its infancy. Among the types of gel particles that have been examined more closely are mucopolysaccharide gels. Mucopolysaccharides are heteropolysaccharides that contain acidic sugars, such as glucuronic and galacturonic acid, and aminosugars, such as glucosamin. They are released by various organisms, for example, bacteria, phytoplankton, macro algae, corals, or bivalves (Decho, 1990). Mucus substances also serve as building material for biological tissues, such as cell coatings, stalks, tubes, apical pads, and adhering films (Hoagland et al., 1993). Depending on their nutrient status, marine phytoplankton cells release between 3% and 40% of the carbohydrates that they assimilate during photosynthesis via exudation (Baines and Pace, 1991). In general, polysaccharides represent the largest fraction of phytoplankton exudates (Ittekkot, 1981; Myklestad et al., 1989; Biddanda and Benner, 1997), and are a major constituent of the high molecular weight fraction of dissolved organic matter (DOM) in the surface ocean (Aluwihare et al., 1997). The monomeric composition of marine polysaccharides is diverse (see Decho, 1990; Hoagland et al., 1993; Leppard, 1995) and varies with both species and physiological stage (Myklestad, 1977; Grossart et al., 2007). Myklestad et al. (1972) observed that exopolysaccharides of the diatom *Chaetoceros affinis* consisted mainly of fucose, rhamnose, and arabinose. Similar results were obtained by Mopper et al. (1995) and Zhou et al. (1998), who found that surface-active polysaccharides and mucopolysaccharide gels that formed by bubble coagulation were enriched in the desoxysugars fucose and rhamnose. Little is known on the acidic sugar content of extracellular polysaccharides. About 20% of the total sugar content of dissolved polysaccharides that are released by the coccolithophore *Emiliania huxleyi* is represented by the acidic sugar D-galacturonic acid (De Jong et al., 1979). Compared to phytoplankton exudates, bacterial exopolysaccharides are enriched in acidic sugars (20%–50%; Kennedy and Sutherland, 1987). Bacteria may also be involved in the chemical modification of dissolved polysaccharides (Giroldo et al., 2003).

Mucopolysaccharides represent the major fraction of extracellular polymeric substances (EPS). A special type of particulate EPS, that is, EPS that are retained on a filter with a pore size of 0.4 μm, is classified as transparent exopolymer particles (TEP; Figure 7.2a) (Alldredge et al., 1993). These are discrete polysaccharide particles that contain acidic sugars. TEP do not include mucopolysaccharide gels that are or were part of an organism, as, for instance, cell coatings. Another type of particulate EPS is of proteinaceous origin and described as Coomassie stained particles

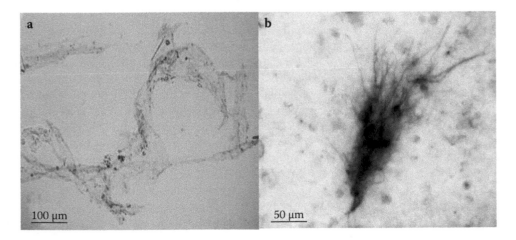

FIGURE 7.2 *(A color version of this figure follows page 132.)* Microscopic images of large web-like TEP stained with (a) the dye Alcian Blue and (b) the dye Coomassie Brilliant Blue G.

(CSP; Figure 7.2b) (Long and Azam, 1996). It has been observed that CSP are closely associated to TEP, and so far it is not known to what extent CSP and TEP represent different chemical subunits, for example, proteins and polysaccharides, of the same gel particle. Discrete gel particles, such as TEP and presumably also CSP, are generated abiotically from interactions of dissolved and colloidal precursors during bubble adsorption (Mopper et al., 1995; Zhou et al., 1998; Mari, 1999), or laminar and turbulent shear (Passow, 2000; Engel and Passow, 2001). Thereby, the mucopolysaccharide gel structure is formed due to divalent cation (Ca^{2+}, Mg^{2+}) and half-ester sulfate (OSO_3^-) bridging between acidic monomers of the individual molecules. While nano gels may assemble spontaneously from free exopolymers (Chin et al., 1998; Kerner et al., 2003), larger particles (>0.4 μm), such as TEP, are formed by aggregation processes (Engel et al., 2004). As gel particles originate from the pool of dissolved organic matter, their size distribution shows an exponential decrease, $d(n)/d(d) = ad^{-\delta}$, with δ generally ranging between 2 and 3 (Long and Azam, 1996; Mari and Burd, 1998; Harlay et al., 2008).

Gel particles such as TEP and CSP are generally transparent unless stained by a specific dye. CSP are determined with the amino acid-specific dye Coomassie Brilliant Blue G (Long and Azam, 1996). Mucopolysaccharides contained in EPS and TEP are stained with Alcian Blue (Decho, 1990; Alldredge et al., 1993), a cationic copper phthalocyanine dye that complexes carboxyl (COO^-) and half-ester sulfate (OSO_3^-) reactive groups of acidic polysaccharides. The blue-stained gel particles can be viewed microscopically. The amount of Alcian Blue adsorption to mucopolysaccharides depends on the anion density of the exopolymer and varies among different polysaccharides, such as agar, carrageneen, or natural algal exopolymers. Ramus (1977) showed that the amount of adsorbed Alcian Blue can be directly related to the weight of a specific polysaccharide. Passow and Alldredge (1995) modified the method of Ramus (1977) for quantifying TEP colorimetrically. Their colorimetric TEP method is a semiquantitative procedure that uses the relationship between Alcian Blue staining capability and weight of a polysaccharide. Here, a weight equivalent for the Alcian Blue adsorption is given by standardization with the polysaccharide Gum Xanthan.

Polysaccharides in gel particles can also be visualized by epifluorescent or confocal laser scanning microscopy using molecular probes conjugated with fluorescent dyes. The lectin concanavalin A (Con-A) conjugated with fluorescein isothiocyanate (FITC) has often been applied to detect the carbohydrates species D(+)-glucose, N-acetyl-D-glucosamine, D(+) mannose, and methyl-α-D-mannopyranoside in marine snow and mucilages (Baldi et al., 1997; MacKenzie et al., 2002).

The temporal and spatial distributions of TEP in the ocean have been reviewed by Passow (2002) and Bhaskar and Bhosle (2005). Numerical abundances of TEP usually are in the order of 10^6 L^{-1}, but values up to 10^8 L^{-1} have also been observed. The abundance of CSP is similar to that of TEP and reaches 10^6 to 10^8 L^{-1} in coastal waters (Long and Azam, 1996). In terms of Gum Xanthan weight equivalence (Xeq.), TEP concentrations in the order of 10^3 μg Xeq. L^{-1} can occur during times of phytoplankton blooms, but are generally lower in oligotrophic waters, below the surface mixed layer, and out of vernal seasons. TEP and CSP can accumulate in the sea surface microlayer and are transported to the atmosphere by aerosols (Kuznetsova et al., 2005). Average carbon concentrations due to TEP in the upper water column of different marine systems range from 27 μg C L^{-1} during fall in the open Atlantic to 769 μg C L^{-1} in spring in the Adriatic (Engel and Passow, 2001), implying that the contribution of TEP to particulate organic carbon (POC) can be at times significant. Thus, TEP, and presumably also CSP, establish a bridge between the DOM and the particulate organic matter (POM) pool, and influence the biogeochemical composition of particles qualitatively by selective enrichment of carbon and nitrogen, respectively. Recent studies indicated that the production of TEP is sensitive to changes in seawater pH (Engel, 2002; Engel et al., 2004; Mari, 2008). Hence, ocean acidification due to increasing anthropogenic CO_2 concentration may change the role of marine gel particles for particle dynamics, trophic interactions, and marine biogeochemistry in the future ocean.

Based on the current knowledge, marine gel particles are numerically important, ubiquitous, and of biogeochemical and ecological significance. There have been an increasing number of publications on marine gels during the last 20 years. Due to their chemical heterogeneity, continuous size distribution, and unique physicochemical properties, the analysis of gel particles is difficult. This chapter gives a detailed description of how to implement the most commonly applied methods for gel particle analysis in routine lab work. These methods are primarily based on the work of Decho (1990) for EPS; Alldredge et al. (1993), Passow and Alldredge (1995), Mari (1999), and Engel and Passow (2001) for TEP; and Long and Azam (1996) for CSP. More elaborate methods that investigate marine gel particles using atomic force microscopy (AFM) (Higgins et al., 2003) or dynamic laser scattering (DLS) (Chin et al., 1998) will not be discussed.

7.2 DEFINITIONS

7.2.1 EXTRACELLULAR POLYMERIC SUBSTANCES (EPS)

EPS are defined as any dissolved or particulate macropolymer organic substances that are excreted by the cell external to the biomembrane. Marine EPS have a high content of gelling agents, such as mucopolysaccharides. EPS that are retained on a membrane or glass fiber filter are therefore classified as gel particles. The term EPS includes TEP and CSP.

7.2.2 TRANSPARENT EXOPOLYMER PARTICLES (TEP)

TEP are operationally defined as discrete exopolymer particles (>0.4 μm) that stain with the cationic copper phthalocyanine dye Alcian Blue at a pH of 2.5 (Alldredge et al., 1993). TEP are a special class of mucopolysaccharide gels and a subcategory of EPS.

7.2.3 COOMASSIE BLUE STAINABLE PARTICLES (CSP)

CSP are discrete extracellular proteinaceous gel particles (>0.4 μm) that are transparent and stain blue with the amino acid-specific dye Coomassie Brilliant Blue G at a pH of 7.4 (Long and Azam, 1996). CSP represent a subcategory of EPS.

7.3 SAMPLING AND STORAGE

7.3.1 SAMPLING

Sampling of seawater can be undertaken by a variety of different procedures and depends on the purpose, field site, sampling depth, and volumes to be sampled. Approved systems include stainless steel samplers, Niskin or Go-Flo bottle samplers, and vacuum and rotary pumps. Because gel particles are prone to aggregation, peristaltic pumps should be avoided when information on size frequency distributions is to be obtained. The sample volume depends on the concentration of gel particles and the method applied, and may range between one and several hundred milliliters. Samples should be processed immediately, within 2 h of sampling, or preserved for storage.

7.3.2 STORAGE

For storage purposes, gel samples are chemically preserved through addition of formalin to a final concentration of 2%. Preserved samples should be kept cold (4°C) and retreated with formalin if stored for more than 2 months. Fixation should be avoided if sensitive organisms, such as colonies of *Phaeocystis*, are present that may disintegrate in the presence of formalin (Passow and Alldredge, 1995). Formalin fixation and storage may change the size frequency distribution of gel particles, due to continued interaction and aggregation of the particles. Storage of stained gel particles on filters is described below.

7.4 METHODS

7.4.1 PREPARATION OF STAINING SOLUTIONS

7.4.1.1 Alcian Blue

An aqueous solution of Alcian Blue (AB) is used for staining of the mucopolysaccharides of particulate EPS and TEP.

An AB stock solution is prepared from:

- 97 mL ultrapure water
- 3 mL acetic acid
- 1 g Alcian Blue (8GX, Sigma)

The stock solution is diluted with ultrapure water in a ratio of 1:50 (vol:vol), yielding a working solution of 0.02% Alcian Blue at a pH of 2.5. Both the stock and working solution are stored in the dark and cold (4°C) and can be used for several months. Prior to usage, the working solution is filtered through 0.2 μm disposable membrane filters, for example, syringe filters, to remove dye particles that may form spontaneously in the solution.

7.4.1.2 Coomassie Brilliant Blue G

An aqueous solution of Coomassie Brilliant Blue G (CBBG) is used for staining of proteinaceous particles of particulate EPS and CSP.

A CBBG stock solution is prepared from:

- 100 mL ultrapure water
- 1 g Coomassie Brilliant Blue (G-250, Serva)

The CBBG stock solution (1%) is diluted with 0.2 μm filtered seawater in a ratio of 1:25 (vol:vol), giving a working solution of 0.04% CBBG at a pH of approximately 7.4. The working solution is prepared daily before use. Prior to usage, the working solution is filtered through 0.2 μm disposable syringe filters to remove dye particles that may form spontaneously in the solution.

7.4.2 FILTRATION AND STAINING

For the filtration of gel particles, it is recommended to use a filtration unit that allows for vacuum release, for example, by use of a three-way valve. The filtration unit can be made of stainless steel, glass, or polysulfone. The filter support screen should be made of stainless steel or polysulfone, but not of glass fiber, as glass fiber frits adsorb the stain and are difficult to clean after the filtration. The methods described below can also be used to analyze the mucopolysaccharide or proteinaceous components of particulate EPS, respectively.

7.4.2.1 TEP

Depending on the concentration of TEP, variable amounts of seawater (1–200 mL) are filtered onto 25 mm polycarbonate filters (Nuclepore, Poretics) at low and constant vacuum (<150 mmHg). Filters with 0.4 μm pore size are commonly used, but depending on the purpose of analysis, filtrations over 0.2 to 1 μm pore sizes have also been applied. Immediately after the filter falls dry, the vacuum is released. Then 1 mL of AB working solution is added to the filter. Others have used 0.5 mL staining solution, but depending on the nature and amount of particles stained, 0.5 mL may be insufficient to fully cover the material on the filter. The stain is allowed to react with the material on the filter for approximately 4 seconds. Then the vacuum pressure is reestablished to remove the excess stain, and the filter is rinsed with a few milliliters of ultrapure water. Between filtration of samples, the filter holder and funnel are rinsed thoroughly with ultrapure water. The stained filter containing the particles can be either analyzed directly or stored frozen at −20°C for several weeks.

7.4.2.2 CSP

Depending on CSP concentration, variable amounts of seawater (1–200 mL) are filtered onto 25 mm polycarbonate filters (Nuclepore, Poretics) at low and constant vacuum (<150 mmHg). Depending on the purpose of analysis, filters with pore sizes from 0.2 to 1 μm pore have been used. The Nuclepore filter is backed by two 0.45 μm HA filters (Millipore). Immediately after the filter falls dry, the vacuum is released. Then 1 mL of CBBG working solution is added to the filter. Some authors have used 0.5 mL of CBBG, or less. However, depending on the nature and amount of particles stained, a volume of 0.5 mL may be insufficient to fully cover the material on the filter. The stain is allowed to react with the material on the filter for approximately 30 seconds. Then the vacuum pressure is reestablished to remove the excessive stain, and the filter is rinsed with a few milliliters of ultrapure water. Between samples, the filter holder and funnel are rinsed thoroughly with ultrapure water. The stained filter containing the particles can be either analyzed directly or stored frozen at −20°C for several weeks.

7.4.3 THE COLORIMETRIC METHOD FOR ANALYZING TEP

For the colorimetric analysis of TEP, the stained filter is transferred into 10 mL acid-resistant tubes, for example, polypropylene tubes. Then, 6 mL of sulfuric acid (H_2SO_4, 80%) is added and the tube is tightly sealed. The tubes are gently agitated to ensure that the whole filter is covered by the acid. The filters are incubated in sulfuric acid for a period of >2 hours and <12 hours. During the incubation, AB is released from its binding sites, and the POM is decomposed. About two or three times during incubation, the tubes are agitated gently. Finally, the absorbance of the amount of AB in the solution is determined at 787 nm in a 1 cm cell against distilled water. It is recommended to prepare at least three replicate filters for each sample and three blank filters.

Blanks are prepared from either 1–2 mL of ultrapure water or 1–2 mL of 0.2 μm filtered seawater, processed like the samples. The advantage of using 0.2 μm filtered seawater as a blank is that the final TEP concentration is corrected for dissolved acidic polysaccharides that may eventually adsorb onto the filter. However, as the amount of acidic polysaccharides may vary between samples, especially for samples from vertical profiles, the seawater blank has to be prepared for each sample individually, which is more laborious and often not feasible.

TEP concentrations are given in micrograms of Xanthan equivalents per liter (μg Xeq. L^{-1}) and calculated using

$$TEP\,(\mu g\ Xeq.L^{-1}) = \frac{(E_{787} - C_{787})}{V} \times F(x) \tag{7.1}$$

where E_{787} is the absorption of the sample at 787 nm, C_{787} is the absorption of the blank at 787 nm, V is the volume filtered (in L), and F (in μg) is the calibration factor determined for the standard polysaccharide Gum Xanthan (see below).

7.4.3.1 Calibration of the Alcian Blue Solution

The procedure for determining the calibration factor $F(x)$ is the most critical stage of the colorimetric method and has to be performed carefully. To calibrate the AB working solution the exact concentration of stain is determined using the standard polysaccharide Gum Xanthan. Gum Xanthan is produced by a pure culture of the bacterium *Xanthomonas campestris*. It is a microbial heteropolysaccharide that contains D-glucuronic acid. The amount of Alcian Blue absorbed is directly related

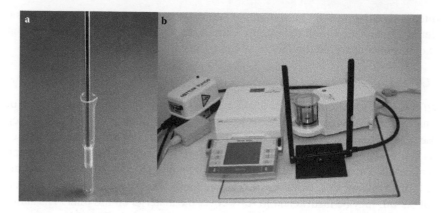

FIGURE 7.3 Photographs of the instruments used to calibrate the amount of Alcian Blue in the staining solution for TEP according to the colorimetric method of Passow and Alldredge (1995): (a) tissue grinder and (b) microbalance equipped with an ionizer.

to the weight of the standard; that is, the slope of the linear relationship between the weight of the standard polysaccharide and the amount of stain absorbed yields the calibration factor. Using the calibration factor, the equivalent mass of Gum Xanthan is calculated from the sample absorption. The results are expressed in micrograms of Gum Xanthan equivalents per liter (μg Xeq. L^{-1}).

The standard solution is prepared by adding 15 mg of Gum Xanthan (Sigma G-1253) to 200 mL of ultrapure water. The polysaccharide in solution is allowed to swell for about 15 minutes. To break up larger particles of Gum Xanthan, and to obtain a size distribution similar to what is expected for TEP in the sample, the standard solution is treated with a tissue grinder (Figure 7.3a). About 20–30 mL is filled into the grinder and the pistil is lowered and raised a few times. The procedure is repeated until the whole solution has been processed. The solution is set to rest again for 15 minutes and the grinding procedure repeated once more.

To test whether the solution is suitable for calibration, two test filters are prepared: 0.5 and 4 mL of the standard solution each are filtered onto 0.4 μm polycarbonate filters and stained as described above. The two volumes should be filtered in less than 10 minutes. The stained areas should appear homogeneous and clearly differ in staining intensity from each other. Sometimes blue-stained particles are visible on the filter by the naked eye. This indicates that the size of the Gum Xanthan particles is still too large. In this case, the grinding procedure needs to be repeated. If the filtration is very quick and the staining intensity does not differ clearly between the two volumes, this could be an indication that the grinding created too small particles that passed through the filter. In this case, a new solution needs to be prepared with less intensive grinding. If the larger volume takes too long to be filtered, the maximum volume can be lowered to 3 mL.

A 5-point calibration is performed using volumes of 0.5, 1, 1.5, 2, and 4 mL of standard solution. Triplicate filters of each volume are stained and analyzed according to the colorimetric method as described above. At least three blank filters are prepared for determining Alcian Blue adsorption to the filter.

To determine the mass of Gum Xanthan retained on the filter, identical volumes of standard solution are filtered onto preweighed 0.4 μm Nuclepore filters. Five replicate filters are prepared for each volume of standard solution. Filters are dried at 60°C for 12 hours, stored, and cooled to room temperature in a desiccator. It is recommended that the filters equilibrate for about 60 minutes with temperature and moisture in the room where the weighing of the filters is performed. Weight determination of Nuclepore filters requires a balance with an accuracy of 1 μg, such as the Mettler Toledo UMX-2 microbalance (METTLER Corp., Germany) (Figure 7.3b). Because dry Nuclepore filters have electrostatic charges, they need to be neutralized prior to weighing on

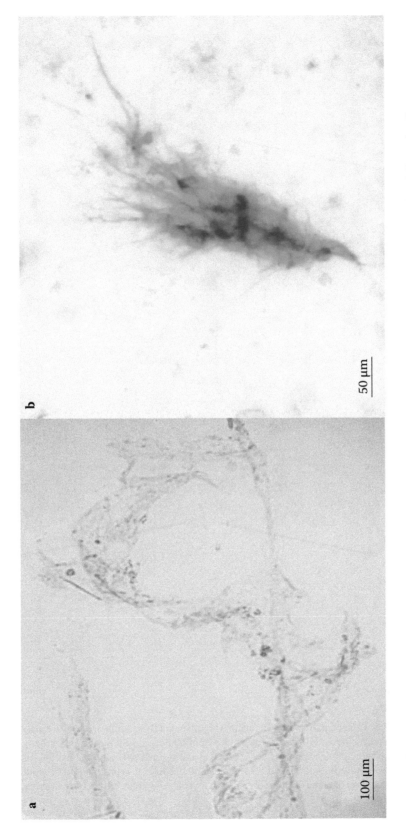

COLOR FIGURE 7.2 Microscopic images of large web-like TEP stained with (a) the dye Alcian Blue and (b) the dye Coomassie Brilliant Blue G.

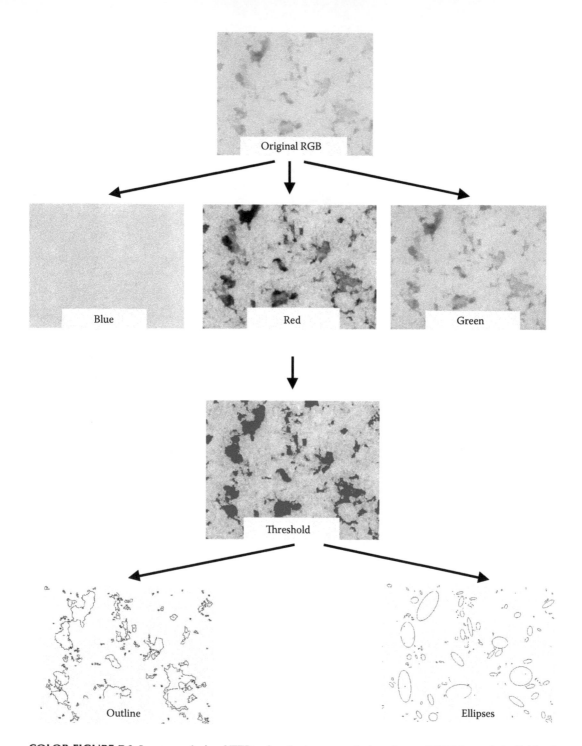

COLOR FIGURE 7.6 Image analysis of TEP, using the image analysis software WCIF ImageJ. A digitized microscopic RGB picture, here a TEP sample at 200× magnification, is split into its red, green, and blue channel components. The red channel enhances the contrast of the stained particles against the filter background and is used for further analysis. A threshold value is set, and particles that are marked in red are further analyzed. The outline encloses the area of the individual gel particle, whereas ellipses are fitted to determine the major and minor sizes.

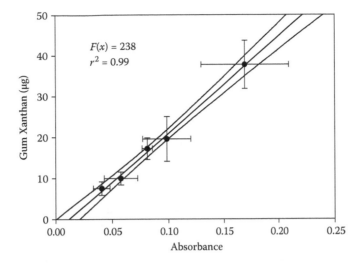

FIGURE 7.4 Linear regression of weight versus absorbance of the standard polysaccharide Gum Xanthan. The slope of the regression gives the conversion factor $F(x)$, used to calculate the Gum Xanthan equivalent weight of TEP.

an electronic balance, for example, with an ionization system (HAUG Corp., Germany). Each of the empty and standard solution-containing filters is weighed three times, and the average weight and standard deviation per filter are determined. At least five filters are prepared for blank weight determination. To determine the weight (W_i) of Gum Xanthan retained on the filter for each volume filtered, the average weight of the empty filters (W_{fe}) is subtracted from the average weight of the filters containing the standard (W_{fi}):

$$W_i = (W_{fi} - W_{fe}) - (W_{bl1} - W_{bl2}) \tag{7.2}$$

The difference between the two weight measurements is corrected for potential other changes in blank filter weight ($W_{bl1} - W_{bl2}$), for example, due to differences in the moisture content of filters. The standard deviation of W_i is calculated by addition of the standard deviations of W_{fi} and W_{fe}, obtained from the repeated weighing. During storage and drying, each filter is kept in combusted glass petri dishes to minimize contamination.

The calibration factor $F(x)$ of the standard is calculated from the slope of a regression of weight (µg) against the corresponding absorbance (ABS) (Figure 7.4):

$$f(x) = \frac{\Delta W}{\Delta ABS} \tag{7.3}$$

Values for $F(x)$ typically range between 50 and 300 µg. The calibrated staining solution can be used for up to 3 months when kept in the dark at 4°C. Over a longer time, the staining intensity and hence the calibration factor may change, for example, to about 25% within the first year. Therefore, the working solution should be recalibrated after 6 months.

7.4.4 MICROSCOPIC DETERMINATION OF GEL PARTICLES

7.4.4.1 Slide Preparation

For microscopy of EPS, TEP, and CSP, slides are prepared from samples filtered onto 0.4 µm Nuclepore filters and stained with 1 mL of AB or CBBG working solution, respectively, as described above. The sample volume should be low enough to ensure that the particles on the filter do not abut or even overlay. For field samples, volumes ranging from 10 to 20 mL are often sufficient. Lower volumes are generally used for phytoplankton cultures (i.e., 0.5–5 mL). There are two different methods for slide preparation, the filter-transfer-freeze (FTF) technique (Hewes and Holm-Hansen, 1983) and the CytoClear slide technique (Logan et al., 1994).

For the FTF technique, the stained and still damp filter is placed upside down on a glass slide. If the filter is too dry, the slide should be moisturized by either placing a drop of water below the filter or breathing upon the cold glass slide. The latter will give a more even distribution of moisture on the slide. The slide, filter side up, is placed on a block of dry ice and frozen for a couple of minutes. Then the filter is peeled off, leaving behind the particles on the slide. The particles are covered by a thin film of gelatine mixture that holds the particles in place and makes a semipermanent mount. The gelatine mixture (about 100 g) is prepared by adding 2.5 g of gelatine to 80 mL boiled distilled water. Once the gelatine is dissolved, 20 mL of liquid glycerine is added. Before use, the solution has to cool down, until it becomes viscous enough to form a film on a tin or copper wire bended to a loop up. Then, the gelatine mixture is taken up with the wire loop and transferred onto the slide so that all particles are covered. The sample is allowed to thaw and solidify for about 30 minutes at room temperature. All slides are prepared in duplicate and can be stored in the cold (4°C) for up to 3 weeks until analysis.

The advantage of using CytoClear slides (PORETICS Corp.) compared to the FTF technique is that the particles can be viewed directly on the filter. CytoClear slides are glass slides that are glazed on one side. The purpose of the glaze is to remove interference with the filter pores under bright field and epifluorescence microscopes. The filter containing the stained gel particles is placed, sample side up, onto the slide with a drop of 0.2 μm filtered immersion oil underneath. Another drop is placed on top of the filter and the filter is then covered by a glass slip. All slides are prepared in duplicate and can be stored in the freezer for 2 to 3 months until analysis. Two blank filters per filtration series are prepared from Nuclepore filters moistened with ultrapure water and stained with 1 mL of AB or CBBG working solution, respectively. Stained blank filters are processed like sample filters according to the FTF or CytoClear technique described above.

7.4.4.2 Image Analysis of Gel Particles

Slides containing the stained gel particles are transferred to a compound light microscope, and analyzed either directly or with the aid of image analysis software. For image analysis of gel particles, the filter is screened by a digital color camera at 100–400× magnification. Photos are taken from thirty frames per filter area, chosen in a cross section, that is, fifteen photos along the horizontal and fifteen photos along the vertical axis. Because gels have irregular shapes, it is recommended to analyze them with an image analysis program that can automatically count and size individual particles (Figure 7.5), such as WCIF ImageJ, a public domain program developed at the U.S. National Institutes of Health (http://www.uhnresearch.ca/facilities/wcif/fdownload.html, courtesy of Wayne Rasband, National Institute of Mental Health, Bethesda, Maryland). To automatically analyze particles, the ImageJ software requires a binary image. Therefore, a threshold range is chosen, including all pixels that are comprised by the particle. Since AB and CBBG stain gel particles, the color information can be used to enhance the contrast between the particles and the background. Therefore, the composite RGB picture is split into the individual red, green, and blue channels (Figure 7.6). The picture of the red channel accentuates the stained gel particles and is used for setting the threshold range. The original photo and the threshold image are compared visually to ensure that the threshold includes the stained gels but no other particles. Gel particles of a selected minimum area size, that is, >0.2 μm^2, are enumerated and sized for individual area and size. Size information can be gained from the major or minor axis of the best-fitted ellipse enclosing the gel particle, or calculated from the area of the particle (A_p) as the equivalent spherical diameter of the individual particle

---

Commands for analyzing gel particles with WCIF ImageJ

→*File/ Open* 'Picture-file'

→*Analyze/ Set scale*

The pixel size has to be set prior to analysis. This can be done by drawing a line of known distance on the picture, e.g. 100 µm: The distance is then filled in under *known distance*.

→ *Image/Color/RGB Split*

Continue with the 'Picture' (red)

→ *Analyze/Set measurements*

Individual choice; the minimum information required is area and *fit ellipse*. The latter will give the *minor* and *major* size of the particle. Mark 'limit to threshold' and redirect to 'Picture' (red).

→ *Image/Adjust/threshold*

All pixels with an intensity above the threshold value are marked and will be treated as TEP. This should be carefully validated by the user.

→ *Analyze/Analyze Particles*
Settings in 'Analyze Particles' window:

Minimum size: (individually chosen)
Circularity: 0-1
Mark '*display results*' and '*exclude on edges*'

Results will be displayed in a separate window. Here, the file can be saved in xls. or wks format.

→ *File/Save Results*

FIGURE 7.5 Flowchart of the commands used to analyze gel particles such as TEP and CSP with the image analysis program WCIF ImageJ. The size scale needs to be set prior to analysis, for example, using a micrometer scale that is photographed under the microscope at the same magnification as the sample slides. For more information, see text.

(ESD_P), assuming the geometry of a sphere, with

$$ESD_P = \sqrt{\frac{A_P}{\pi}} \times 2 \qquad (7.4)$$

The number of gel particles per volume of sample (N in L^{-1}) is determined from

$$N(L^{-1}) = -\frac{A_F \times n}{b \times M \times V} \qquad (7.5)$$

where A_F is the total filter area (mm^2) stained with AB or CBBG solution, n the total number of gel particles counted, b the number of fields of view examined, M the area size of one field of view

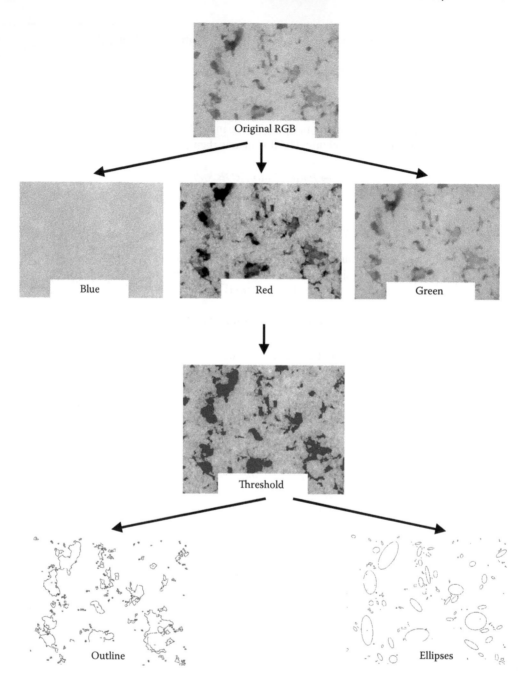

FIGURE 7.6 *(A color version of this figure follows page 132.)* Image analysis of TEP, using the image analysis software WCIF ImageJ. A digitized microscopic RGB picture, here a TEP sample at 200× magnification, is split into its red, green, and blue channel components. The red channel enhances the contrast of the stained particles against the filter background and is used for further analysis. A threshold value is set, and particles that are marked in red are further analyzed. The outline encloses the area of the individual gel particle, whereas ellipses are fitted to determine the major and minor sizes.

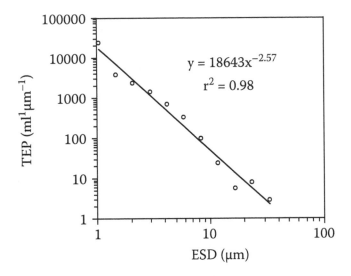

FIGURE 7.7 Example of the size spectrum of TEP in a field sample, harvested in surface waters of the Bay of Biscay during spring 2006. Linear regression of data on a log-log scale yields the characteristic exponent δ.

(mm²), and *V* the volume (L) of sample. The total area of gel particles, that is, EPS, TEP, or CSP, per liter volume (A_P, L^{-1}) is calculated accordingly.

To gain information on the size frequency distribution, gel particles are classified according to their size, that is, area, major axis, or ESDs, into logarithmic size classes. The size distributions of gel particles can then be described by a power law relationship:

$$\frac{dn}{d(d_p)} = kd_P^\delta \tag{7.6}$$

where *dn* is the number of particles per unit volume in the size range d_p to ($d_p + d(d_p)$). The constant *k* depends on the concentration of particles, and the spectral slope δ (usually δ < 0) describes the size distribution. Both constants can be obtained from linear regression of log($dn/d(d_p)$) versus log(d_p) (Figure 7.7), where *k* is the y-intercept of the regression line and δ is the slope. The exponent δ is related to the slope of the cumulative size distribution, $N = ad_p^\beta$ by δ = β + 1.

7.4.5 CARBON CONTENT OF GEL PARTICLES

So far, only the carbon content of TEP (TEP-C) has been determined empirically. Two different approaches have been used to estimate TEP-C concentrations; one is based on a direct relationship between carbon and colorimetrically determined TEP concentrations (TEP-C$_{color}$) (Engel and Passow, 2001), and the other derives the carbon content of individual TEP from the relationship between carbon and TEP size distribution as determined by the microscopic method (TEP-C$_{micro}$) (Mari, 1999).

Both methods generate TEP from dissolved polysaccharide precursors. This has the advantage of separating TEP from other particulate matter, primarily phytoplankton cells and bacteria. About 2 to 4 L of culture media or field sample is required to produce enough material for each TEP-C measurement. The media or sample is filtered first through GF/F filters (0.7 μm, Whatman) in order to remove diatom cells, and subsequently through 0.2 μm polycarbonate filters (Nucleopore) to remove all remaining particles, including almost all bacteria. TEP can be generated under laminar and turbulent shear using either Couette chambers or tangential flow filtration

(TFF) devices (Engel and Passow, 2001), or by bubbling with air (Mari, 1999). When using the Couette chamber procedure, the filtrate is rotated for 24 to 64 hours at a shear rate of $G = 8$ to $12\ s^{-1}$. When using the TFF procedure, the filtrate is cycled through a TFF chamber (e.g., Filtron Ultrasette, with a 0.16 μm membrane) for 24 hours. During TFF, substances larger than the pore size of the membrane are trapped within the turbulent motion of the fluid. TEP precursors coagulate during the TFF procedure and accumulate in the retentate fraction. Engel and Passow (2001) cycled the filtrate together with the retentate for ~20 hours and concentrated TEP for 4 hours to a final volume of 2 L. After generating TEP, the sample is analyzed for both carbon and TEP concentration. The procedure has to be repeated to yield a minimum of five different observations on TEP and carbon.

For the elemental (CHN) analysis of TEP, silver filters with 0.45 μm pore size (Poretics) and GF/F filters can be used. The advantage of silver filters is that their pore size is close to the pore size of the polycarbonate filters (0.4 μm) that are used in the standard protocol for TEP determination. Carbon concentration is determined by CHN analyses with all filters prepared at least in duplicate. It is important to avoid carbon contamination during the whole assay. Thus, all materials in contact with the sample shall be either autoclaved or acid (10% HCl) rinsed. Blanks for silver or GF/F filters need to be prepared for each filtration series (see Chapter 2).

TEP are quantified using either the colorimetric method or the microscopic method. For TEP-C_{color}, the carbon concentrations of generated TEP are directly related to the colorimetrically determined TEP concentrations. The slope of the regression gives a first-order approximation of the increase of carbon with an increase in TEP, and is used as a conversion factor to calculate TEP-derived carbon. For TEP prepared from dissolved precursors produced by diatoms and coccolithophores, Engel (2004) estimated a conversion of

$$\text{Carbon}\ [\mu g\ L^{-1}] = (0.63 \pm 0.03)\ [\text{TEP};\ \mu g\ \text{Xeq.}\ L^{-1}] \tag{7.7}$$

$$(n = 47,\ r^2 = 0.80,\ p < 0.001)$$

For diatoms solely, a higher conversion factor was obtained by Engel and Passow (2001):

$$\text{Carbon}\ [\mu g\ L^{-1}] = (0.75 \pm 0.05)\ [\text{TEP};\ \mu g\ \text{Xeq.}\ L^{-1}] \tag{7.8}$$

$$(n = 28,\ r^2 = 0.92,\ p < 0.001)$$

For the coccolithophore *Emiliania huxleyi*, a lower conversion factor was obtained by Engel et al. (2005):

$$\text{Carbon}\ [\mu g\ l^{-1}] = 0.40 \pm 0.08\ [\text{TEP},\ \mu g\ \text{Xeq.}\ l^{-1}] \tag{7.9}$$

$$(n = 11,\ r^2 = 0.73,\ p < 0.005)$$

Assuming that the volume of individual TEP is proportional to r^D, where r is the equivalent spherical radius in μm and D is the fractal scaling dimension, TEP-C_{micro} can also be determined from TEP size (Mari, 1999):

$$\text{TEP-C}_{micro}\ [\mu g\ L^{-1}] = 0.25 \times 10^{-6}\ r^D \tag{7.10}$$

with $D = -2.55$. For estimating the scaling dimension, a semiempirical relationship has been proposed by Burd and Jackson (unpublished), as referenced in Mari and Burd (1998):

$$D = (64 - \delta)/26.2 \tag{7.11}$$

It has to be emphasized that the carbon content of TEP very likely depends on the source of poly-saccharides involved in TEP formation, and on the environmental processing of TEP, in particular microbial diagenesis and decomposition. Carbon conversion factors should therefore be applied with caution. For TEP of unknown origin, a direct determination of the carbon content is preferred.

7.4.6 GEL PARTICLES IN SEDIMENT TRAP MATERIAL

Due to their high water content, gel particles are almost neutrally buoyant. However, gel particles aggregate with solid particles, which undergo gravitational settling. Within aggregates, gel particles contribute to the vertical flux of organic matter in the ocean and have been observed in material collected from sediment traps or recovered from the sediment. In order to measure TEP in trap material, Passow et al. (2001) have modified the colorimetric method for TEP. Therefore, a defined subfraction of trap material is diluted 1:10 and 1:20 with 0.2 μm filtered seawater of the same salin-ity as the trap water. Four replicates each of different aliquots (i.e., 1 to 2 mL) of the two dilutions are filtered onto 0.4 μm Nuclepore filters and stained with AB solution as described above. Two rep-licates of each aliquot are filtered onto 0.4 μm Nuclepore filters and remain unstained. The stained as well as the unstained filters are processed according to the protocol of the colorimetric method for TEP (see above). The unstained filters are used as an estimate for absorption due to turbidity and are subtracted from the sample absorption. Four replicate blank filters are prepared from 0.2 μm filtered seawater used to dilute the trap material and are subtracted from the sample absorption value as well.

The corrected absorption of the samples is graphed against the filtration volume, and should yield a linear relationship. From the slope of the regression the absorbance of TEP per liter of sample is calculated and multiplied with the calibration factor to give TEP concentration (μg Xeq. L^{-1}).

7.5 TROUBLESHOOTING

The exact chemical composition of gel particles, such as TEP and CSP, is unknown, but presum-ably they are heteropolymers. Consequently, the weight concentration of monomers that undergo complex binding with the stain, that is, the colorability of gels, may vary depending on the origin of precursor material, and on microbial and chemical diagenesis. This is a particular problem when the amount of stain adsorbed is used for quantification of gels, as in the case of the colorimetric TEP method. Here, a weight equivalent for TEP is given, relative to a standard polysaccharide, that is, Gum Xanthan. If samples are compared with respect to TEP concentration determined in this way, differences may not necessarily be due to different quantities, that is, different weights per vol-ume, but to different chemical composition. Hence, this method is sensitive to the amount of acidic sugars as well as to the amount of particles. This should also be kept in mind when results from the colorimetric method are compared with those of the microscopic method. The microscopic method quantifies the number and size of gel particles, but cannot account for the chemical composition. A potential source of error is the volume estimation of gel particles from area measurements obtained by microscopy. Due to their high flexibility and semisolid consistency, gels flatten during the filtra-tion process, which likely leads to an overestimation of their size and of the derived volume. The choice of methods therefore depends on the scientific purpose. For example, if properties of gel par-ticles are examined that are related to the amount of acidic sugars, that is, trace metal binding and stickiness, the colorimetric method is a good choice, whereas if interactions between gel particles and organisms are considered, the microscopic method may be more appropriate.

Another difficulty in the quantification of gels lies in the separation of individual types of par-ticles. TEP often cannot be separated from mucopolysaccharides that are contained in other EPS, like adhering films or extracellular gels as part of organisms. This is in particular not possible when TEP are quantified by the colorimetric approach. In order to estimate TEP concentration solely, the amount of AB adsorption due to the staining of cells and organisms themselves has to be determined, and subtracted from the results. Caution also needs to be taken when gel particles are quantified by different methods, that is, staining with AB or CBBG. Since a gel can contain poly-saccharides as well as amino acids, TEP and CSP may sometimes represent the same gel particle. Hence, the total amount of gel particles is not necessarily given by the sum of TEP and CSP.

7.6 SUMMARY AND PERSPECTIVES

Gel particles play an enormous role within biological, physical, and chemical interactions in marine systems. They are a chemically diverse group of particles and cannot easily be determined and quantified. This chapter gives a detailed description of the currently most often applied methods for mucopolysaccharide and proteinaceous gel particles, that is, TEP and CSP, analysis. One purpose of this chapter is to give a standardization for these methods in order to enhance the comparability of scientific findings. Moreover, limitations of the methods, as well as potential sources of error, are discussed in order to support a correct interpretation of results. Application of common methods for gel particle analysis in field and laboratory investigations will help to better understand gel particle dynamics, and to improve their chemical analysis. Potential fields for improving gel particle analysis include a simpler and less error-prone calibration of staining solutions, the development of a colo-rimetric method for CSP determination, a better three-dimensional representation of gel particles during microscopy, and most importantly, the development of a method to directly determine the chemical composition of gel particles. The latter is a particular challenge because it requires the separation of the chemical analysis of gel particles from that of coexistent bulk particles.

REFERENCES

Alldredge, A. L., U. Passow, and B. E. Logan. 1993. The abundance and significance of a class of large, trans-parent organic particles in the ocean. *Deep-Sea Research* 40:1131–40.

Aluwihare, L. I., D. J. Repeta, and R. F. Chen. 1997. A major biopolymeric component of dissolved organic carbon in surface sea water. *Nature* 387:166–69.

Arrigo, K. R. 2007. Marine manipulations. *Nature* 450:491–92.

Azetsu-Scott, K., and S. E. H. Niven. 2005. The role of transparent exopolymer particles (TEP) in the transport of ^{234}Th in coastal water during a spring bloom. *Continental Shelf Research* 25:1133–41.

Baldi, F., A. Minacci, A. Saliot, L. Mejanelle, P. Mozetic, V. Turk, and A. Malej. 1997. Cell lysis and release of particulate polysaccharides in extensive marine mucilage assessed by lipid biomarkers and molecular probes. *Marine Ecology Progress Series* 153:45–47.

Baines, S. B., and M. L. Pace. 1991. The production of dissolved organic matter by phytoplankton and its importance to bacteria: Patterns across marine and freshwater systems. *Limnology and Oceanography* 36:1078–90.

Beauvais, S., M. L. Pedrotti, J. Egge, K. Iversen, and C. Marrase. 2006. Effects of turbulence on TEP dynam-ics under contrasting nutrient conditions: Implications for aggregation and sedimentation processes. *Marine Ecology Progress Series* 323:47–57.

Bhaskar, P. V., and N. B. Bhosle. 2005. Microbial extracellular polymeric substances in marine biogeochemical processes. *Current Science* 88:45–53.

Biddanda, B., and R. Benner. 1997. Carbon, nitrogen, and carbohydrate fluxes during the production of particu-late dissolved organic matter by marine phytoplankton. *Limnology and Oceanography* 42:506–18.

Chin, W., M. V. Orellana, and P. Verdugo. 1998. Spontaneous assembly of marine dissolved organic matter into polymer gels. *Nature* 391:568–72.

Decho, A. W. 1990. Microbial exopolymer secretions in the ocean environments: Their role(s) in food webs and marine processes. *Oceanography and Marine Biology Annual Reviews* 28:73–153.

De Jong, E., L. van Rens, P. Westbroek, and L. Bosch. 1979. Biocalcification by the marine alga *Emiliania huxleyi* (Lohmann) Kamptner. *European Journal of Biochemistry* 99:559–67.

Engel, A. 2000. The role of transparent exopolymer particles (TEP) in the increase in apparent particles sticki-ness (α) during the decline of a diatom bloom. *Journal of Plankton Research* 22:485–97.

Engel, A. 2002. Direct relationship between CO_2-uptake and transparent exopolymer particles (TEP) produc-tion in natural phytoplankton. *Journal of Plankton Research* 24:49–53.

Engel, A. 2004. Distribution of transparent exopolymer particles (TEP) in the northeast Atlantic Ocean and their potential significance for aggregation processes. *Deep-Sea Research I* 51:83–92.

Engel, A., A. Benthien, L. Chou, B. Delille, J.-P. Gattuso, J. Harley, C. Heemann, L. Hoffmann, S. Jacquet, J. Nejstgaard, M.-D. Pizay, E. Rochelle-Newall, U. Schneider, A. Terbrueggen, I. Zondervan, and U. Riebesell. 2005. Testing the direct effect of CO_2 concentration on a bloom of the coccolithophorid *Emiliania huxleyi* in mesocosm experiments. *Limnology and Oceanography* 50:493–507.

Engel, A., and U. Passow. 2001. Carbon and nitrogen content of transparent exopolymer particles (TEP) in relation to their Alcian Blue adsorption. *Marine Ecology Progress Series* 219:1–10.

Engel, A., S. Thoms, U. Riebesell, E. Rochelle-Newall, and I. Zondervan. 2004. Polysaccharide aggregation as a potential sink of marine dissolved organic carbon. *Nature* 428:929–32.

Giroldo, D., A. A. H. Vieira, and B. Paulsen. 2003. Relative increase of dexoysugars during microbial degradation of an extracellular polysaccharide released by a tropical freshwater *Thalassiosira* sp. (Bacillariophyceae). *Journal of Phycology* 39:1109–15.

Grossart, H.-P., A. Engel, C. Arnosti, C. De La Rocha, A. Murray, and U. Passow. 2007. Microbial dynamics in autotrophic and heterotrophic seawater mesocosms. III. Organic matter fluxes. *Aquatic Microbial Ecology* 49:143–56.

Grossart, H.-P., and M. Simon. 2007. Interactions of planktonic algae and bacteria: Effects on algal growth and organic matter dynamics. *Aquatic Microbial Ecology* 47:163–76.

Guo, L., C. C. Hung, P. H. Santschi, and I. D. Walsh. 2002. [234]Th scavenging and its relationship to acid polysaccharide abundance in the Gulf of Mexico. *Marine Chemistry* 78:103–19.

Harlay, J., C. de Bodt, A. Engel, S. Jansen, Q. d'Hoop, J. Piontek, N. van Oostende, S. Groom, K. Sabbe, and L. Chou. 2008. In situ Abundance and size distribution of transparent exopolymer particles (TEP) in a coccolithophore bloom in the northern Bay of Biscay. *Deep-Sea Research I*, accepted.

Heinonen, K. B., J. E. Ward, and B. A. Holohan. 2007. Production of transparent exopolymer particles (TEP) by benthic suspension feeders in coastal systems. *Journal of Experimental Marine Biology and Ecology* 341:184–95.

Hewes, C. D., and O. Holm-Hansen. 1983. A method for recovering nanoplankton from filters for identification with the microscope: The filter-transfer-freeze (FTF) technique. *Limnology and Oceanography* 28:389–94.

Higgins, M. J., P. Molino, P. Mulvaney, and R. Wetherbee. 2003. The structure and nanomechanical properties of the adhesive mucilage that mediates diatom-substratum adhesion and motility. *Journal of Phycology* 39:1181–93.

Hoagland, K. D., J. R. Rosowski, M. R. Gretz, and S. C. Roemer. 1993. Diatom extracellular polymeric substances: Function, fine structure, chemistry and physiology. *Journal of Phycology* 29:537–66.

Ittekkot V., U. Brockmann, W. Michaelis, and E. T. Degens. 1981. Dissolved free and combined carbohydrates during a phytoplankton bloom in the northern North Sea. *Marine Ecology Progress Series* 4:299–305.

Kahl, L. A., A. Vardi, and O. Schofield. 2008. Effects of phytoplankton physiology on export flux. *Marine Ecology Progress Series* 354:3–19.

Kennedy, A. F. D., and I. W. Sutherland. 1987. Analysis of bacterial exopolysaccharides. *Biotechnology and Applied Biochemistry* 9:12–19.

Kerner, M., H. Hohenberg, S. Ertl, M. Reckermann, and M. Spitzy. 2003. Self-organization of dissolved organic matter to micelle-like microparticles in river water. *Nature* 422:150–54.

Krembs, C., and A. Engel. 2001. Abundance and variability of microorganisms and transparent exopolymer particles across the ice-water interface of melting first-year sea ice in the Laptev Sea (Arctic). *Marine Biology* 138:173–85.

Kuznetsova M., C. Lee, and J. Aller. 2005. Characterization of the proteinaceous matter in marine aerosols. *Marine Chemistry* 96:359–77.

Leppard, G. G. 1995. The characterization of algal and microbial mucilages and their aggregates in aquatic ecosystems. *Science of the Total Environment* 165:103–31.

Logan, B. E., H.-P. Grossart, and M. Simon. 1994. Direct observation of phytoplankton, TEP and aggregates on polycarbonate filters using brightfield microscopy. *Journal of Plankton Research* 16:1811–15.

Logan, B. E., U. Passow, A. L. Alldredge, H.-P. Grossart, and M. Simon. 1995. Rapid formation and sedimentation of large aggregates is predictable from coagulation rates (half-lives) of transparent exopolymer particles (TEP). *Deep-Sea Research II* 42:203–14.

Long, R. A., and F. Azam. 1996. Abundant protein-containing particles in the sea. *Aquatic Microbial Ecology* 10:213–21.

MacKenzie, L., I. Sims, V. Beuzenberg, and P. Gillespie. 2002. Mass accumulation of mucilage caused by dino-flagellate polysaccharide exudates in Tasman Bay, New Zealand. *Harmful Algae* 1:69–83.

Mari, X. 1999. Carbon content and C:N ratio of transparent exopolymer particles (TEP) produced by bubbling of exudates of diatoms. *Marine Ecology Progress Series* 33:59–71.

Mari, X. 2008. Does ocean acidification induce an upward flux of marine aggregates? *Biogeosciences Discussion* 5:1631–54.

Mari, X., S. Beauvais, R. Lemee, and M. L. Pedrotti. 2001. Non-Redfield C:N ratio of transparent exopoly-meric particles in the northwestern Mediterranean Sea. *Limnology and Oceanography* 46:1831–36.

Mari, X., and A. Burd. 1998. Seasonal size spectra of transparent exopolymeric particles (TEP) in a coastal sea and comparison with those predicted using coagulation theory. *Marine Ecology Progress Series* 163:63–76.

Mari, X., F. Rassoulzadegan, and C. P. D. Brussaard. 2004. Role of TEP in the microbial food web structure. II. Influence on the ciliate community structure. *Marine Ecology Progress Series* 279:23–32.

Mopper, K., J. Zhou, K. S. Ramana, U. Passow, H. G. Dam, and D. T. Drapeau. 1995. The role of surface-active carbohydrates in the flocculation of a diatom bloom in a mesocosm. *Deep-Sea Research II* 42:47–73.

Myklestad, S. 1977. Production of carbohydrates by marine planktonic diatoms. II. Influence of the N/P ratio in the growth medium on the assimilation ratio, growth rate and production of cellular and extracel-lular carbohydrates by *Chaetoceros affinis* var Willei (Gran) Hustedt and *Skeletonema costatum* (Grev) Cleve. *Journal of Experimental Marine Biology and Ecology* 29:161–79.

Myklestad, S., A. Haug, and B. Larsen. 1972. Production of carbohydrates by the marine diatom *Chaetoderos affinis* var Willei (Gran) Hustedt II: Preliminary investigation of the extracellular polysaccharide. *Journal of Experimental Marine Biology and Ecology* 9:137–44.

Myklestad, S., O. Holm-Hansen, K. M. Varum, and B. E. Volcani. 1989. Rate of release of extracellular amino acids and carbohydrates from the marine diatom *Chaetoceros affinis*. *Journal of Plankton Research* 11:763–73.

Passow, U. 2000. Formation of transparent exopolymer particles, TEP, from dissolved precursor material. *Marine Ecology Progress Series* 192:1–11.

Passow, U. 2002. Transparent exopolymer particles (TEP) in aquatic environments. *Progress in Oceanography* 55:287–333.

Passow, U., and A. L. Alldredge. 1995. A dye-binding assay for the spectrophotometric measurement of trans-parent exopolymer particles (TEP) in the ocean. *Limnology and Oceanography* 40:1326–35.

Passow, U., R. F. Shipe, A. Murray, D. K. Pak, M. A. Brzezinski, and A. L. Alldredge. 2001. Origin of transpar-ent exopolymer particles (TEP) and their role in the sedimentation of particulate matter. *Continental Shelf Research* 21:327–46.

Quigley, M. S., P. H. Santschi, C.-C. Hung, L. Guo, and B. D. Honeyman. 2002. Importance of acid polysac-charides for [234]Th complexation to marine organic matter. *Limnology and Oceanography* 47:367–77.

Ramus, J. 1977. Alcian Blue: A quantitative aqueous assay for algal acid and sulfated polysaccharides. *Journal of Phycology* 13:445–48.

Shackelford, R., and J. P. Cowen. 2006. Transparent exopolymer particles (TEP) as a component of hydrother-mal plume particle dynamics. *Deep-Sea Research* 10:1677–94.

Verdugo, P., A. Alldredge, F. Azam, D. Kirchman, U. Passow, and P. H. Santschi. 2004. The oceanic gel phase: A bridge in the DOM-POM continuum. *Marine Chemistry* 92:67–85.

Wild, C., M. Huettel, A. Klueter, S. G. Kremb, M. Y. M. Rasheed, and B. B. Joergensen. 2004. Coral mucus functions as an energy carrier and particle trap in the reef ecosystem. *Nature* 428:66–70.

Zhou, J., K. Mopper, and U. Passow. 1998. The role of surface active carbohydrates in the formation of trans-parent exopolymer particles by bubble adsorption of seawater. *Limnology and Oceanography* 43:1860–71.

8 Nutrients in Seawater Using Segmented Flow Analysis

Alain Aminot, Roger Kérouel, and Stephen C. Coverly

CONTENTS

8.1 INTRODUCTION

Seawater, with its high concentration of ions and the presence of Ca and Mg, whose salts are insoluble at high pH, represents a sample matrix that is many times more complex than that of typical freshwater samples. The nutrients that form the basis of life in the ocean are present at concentrations up to 100 million times lower than the typical salt content, and this presents a great challenge to the analyst. Measuring low levels of dissolved nutrients requires careful attention to detail in cleanliness, sample handling, laboratory technique, and instrument design.

Continuous flow analysis (CFA) systems are well suited to this task. In such systems a peristaltic pump delivers samples and reagents into a long tube (≤ 2 mm internal diameter [ID]), which acts as the reaction vessel. The specific product formed during the reaction with the species to be determined is subsequently detected optically.

Two forms of CFA are in common use. In flow injection analysis (FIA) the flow is not air segmented and dispersion in the tubing mixes the components. In contrast, in segmented flow analysis (SFA) air bubbles are regularly introduced into the flow to minimize dispersion and maintain a steady state. Although these two principles have been applied to nutrient analysis in seawater, FIA, in which detection occurs in a transition phase, is not as sensitive or precise as SFA (Zhang, 2000). Optical perturbations due to refractive index changes are major problems in FIA, whereas they can be ignored in modern segmented flow analyzers. In situ analyzers will normally use FIA, which is not altered by pressure, but in the laboratory SFA provides more precise data, and this technology is the choice of most international laboratories. Many CFA methods for nutrient determination are available in the literature, including hybrid versions such as those described by Hansen and Grasshoff (1983) and Hansen and Koroleff (1999), which are partly segmented and partly not, and omit intersample wash.

This chapter describes aspects of modern instrumental design and discusses precautions needed to protect samples against changes or contamination before they are analyzed. This is followed by a description of some of the technical aspects of nutrient analysis by SFA. Finally, we describe methods that have been used for many years to analyze oceanic, coastal, and estuarine waters, and successfully proven in international collaborative studies (Aminot and Kérouel, 1995, 1997; Aminot and Kirkwood, 1995; Aminot et al., 1997). These methods are not specifically dedicated to nanomolar ranges, but still can detect concentrations close to 1 nmol L^{-1} for some nutrients.

8.2 INSTRUMENT DESIGN

8.2.1 SAMPLER

The sampler introduces samples and wash water (i.e., the baseline water) successively into the manifold at a frequency and sample/wash ratio suitable for the analyses. Analyses that run with different frequencies and sample/wash ratios require separate samplers. However, the performance of modern analyzers generally allows multichannel work using a single sampler. When there is only one sampler for several channels, the total flow is split by separators to feed each individual channel. If a channel needs a special wash water, this should be compatible with all the channels fed by the same sampler (e.g., ammonium). During the movement of the probe from sample to wash water and vice versa, the small air bubble (the intersample air bubble) that separates the two liquids is essential to minimize their mixing. The probe movement should be rapid enough to avoid producing an intersample bubble that is too large. In most cases this bubble alters the hydraulic conditions in the manifold and is therefore normally removed after the pump. However, when there is an intersample air compression (ISAC) effect (see Section 8.2.2), the bubble is removed before the pump. Exceptionally, with a low sample flow, when no perturbation occurs, there is no need to remove this bubble.

To analyze nutrients in seawater using an AutoAnalyzer and to limit contamination risks, it is strongly recommended to pump aliquots directly from the sample bottle placed on the sampler tray (Kérouel and Aminot, 1987). Compact autosamplers, able to receive various kinds of sample bottle, are now available. The sample probe should be made from material that is not affected by acid and does not interact with phosphate or ammonium ions, and a recent development that meets these requirements and provides rigidity is stainless steel coated inside and out with a thin film of polytetrafluoroethylene (PTFE). An alternative is a rigid plastic such as polyether ether ketone (PEEK). Probes up to 1.5 mm internal diameter are available for modern samplers and higher flow rates. Probes with a smaller diameter are used for low sample flow, for example, single-channel analysis.

8.2.2 PUMP

Regular hydraulic flow is essential for precise measurements, and the peristaltic pump can therefore be considered the heart of any SFA system. Consistency in the composition of each liquid segment, that is, the exact proportion of reagent and sample, is a critical factor in system reproducibility, and hence baseline noise and detection limit. The volume of reagents and sample in each liquid segment must be equal. As the output of a peristaltic pump is not constant, the only way to ensure constant composition is to synchronize the bubble injection with the pump rollers. Early pumps used a cam and lever to activate the air injection. This is effective in new pumps, but the precision deteriorates as the mechanical parts wear out. More advanced designs use an optical switch coupled to solenoid valves, and a recent development is programmable air injection frequency, which allows the amount of air in the system to be reduced for methods with a long reaction time.

The intersample air bubble is compressed when it passes through the pump tube. This ISAC effect momentarily modifies the sample flow, and hence the reaction conditions in one or more segments. The ISAC effect may pass unnoticed since its position and magnitude depend on various characteristics of the manifold, but it can also alter the peak (Figure 8.1) in such a way that the computerized automatic height measurement sometimes fails. Removing the intersample bubble just before the pump eliminates the effect.

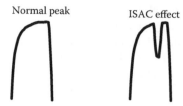

FIGURE 8.1 Schematic example of peak alteration by the ISAC effect (see text).

8.2.3 MANIFOLD

The manifold, which is the reaction vessel, is an assembly of various components such as debubblers, injectors, mixing and reaction coils, and heating baths. Glass is preferred for its inert properties, smooth surface, which reduces intersegment carryover, and long life. Interference on silicate from borosilicate glass is negligible under the acidic conditions of the reaction. Platinum or sapphire inserts are recommended as connectors for the addition of reagents as they have the advantage of inertness and long life. Heating baths are included in the manifolds to increase the reaction rate in order to enhance sensitivity, to ensure a consistent temperature for the reaction, and to ensure completeness of reaction in case of low laboratory temperature (e.g., onboard ships) or low nutrient concentration or reagent quality (e.g., phosphate; see Section 8.8.1).

Air segmentation allows fast reagent/sample mixing in glass reaction coils and minimizes carryover in order that peaks reach their plateau as quickly as possible, hence increasing the analytical rate and precision. Satisfactory conditions are achieved when the bubble length is 1.5 to 2 times the tube diameter (e.g., 3–4 mm in macroflow) with two to four segments per coil turn. The internal diameter (ID) of any tubing carrying the air-segmented stream should be kept close to that of the glass manifold: if the ID is too large, the bubbles will not be large enough to prevent carryover from one liquid segment to the next; if the ID is too small, bubbles may be elongated so much that they break. Air bubbles should not contact the inner surface of the tubing but be separated from it by a thin film of liquid; otherwise, the force needed to push them through the tubing becomes higher and nonconstant. An easy way to check this is to observe the shape of the bubbles in motion through the tubing. If the tubing surface is wetted, as it always should be with glass, the bubbles are rounded at the front and back. If it is not wetted, the trailing edge of the bubble appears to be straight.

If there is a carryover effect (Section 8.4.4), it will result in peak tailing, such that a peak's height is affected by the residual signal from the tail of the preceding peak. The longer the time between peaks, the less the interference will be. When seawater alternates with freshwater, the carryover characteristics of sample and wash may differ, so that the sample/wash ratio may influence carryover. This should not be ignored if standard sampling rates are increased. Manifold design should minimize carryover as follows: sample line as short as possible; sample probe, T-pieces, and nipples having the correct size according to sample flow rate; good connections between glass fittings, with no dead volume; and manifold cleaned properly after analysis (Section 8.3.4). Debubblers should be selected to have the minimum dead volume that is compatible with their function. For that reason, debubblers of microflow systems (see below) are sometimes suggested in the macroflow manifolds presented here.

The methods presented in this chapter are designed for macroflow systems with a glass manifold of an internal diameter of 2 mm. Unless otherwise mentioned, given references correspond to Bran+Luebbe/SEAL AutoAnalyzer3 (AA3) material, with first-generation colorimeters. Microflow systems, with an internal diameter of 1 mm, offer potential advantages of higher sampling rates, lower reagent consumption, and smaller instrument size. However, the smaller volume of each liquid segment makes a microflow system more sensitive to disturbances due to dissolved air in reagents, which may come out of solution and enter the manifold with the reagent stream. The influences of dead volume and the ISAC effect (Section 8.2.2) are also proportionately higher. For these reasons, microflow systems need more careful attention to detail when preparing reagents and constructing or rebuilding the manifold, if good results are to be obtained.

8.2.4 DETECTORS: COLORIMETER AND FLUOROMETER

Filter-based detectors are generally preferred for their sensitivity and robustness since measured compounds normally exhibit wide spectral bands. LEDs, which are unaffected by vibration, should replace krypton or tungsten-halogen lamps, particularly for onboard ship analyses.

Because of nonideal optical characteristics in flowcell design, variations in sample refractive index can cause false absorbance differences in colorimetry, even though the true absorbance may be constant. Such differences occur in CFA when seawater samples are separated by pure water as the wash solution. A transient signal, the Schlieren effect, is caused by refraction anomalies in the transmitted light beam when the unstable sample/wash mixing zone passes through the cell. The Schlieren effect is intense when the segmenting air bubbles are removed with a debubbler before the flowcell. It is prevented by passing the air bubbles through the flowcell, a technique that also allows a significant increase of the analysis rate by reducing carryover (Figure 8.2). After the transient phase the liquid is homogeneous in the flowcell, and the refractive anomaly of the light beam remains in the steady state. The corresponding constant absorbance difference, in general positive, is called the refractive index blank (RIB). Loder and Glibert (1977) and Froelich and Pilson (1978) describe detailed studies of the RIB in sea and estuarine waters (note that some wetting agents can increase the RIB). Figure 8.2 shows samples with a salinity of 35, separated by a pure water wash, passing through flowcells with and without debubbling. The RIB is a function of detector and flowcell design, and it is minimized in SFA analyzers, such as the one described here, by using an optical system that generates a near-parallel light beam through the flowcell and onto the detector. The RIB with the AA3 new-generation high-resolution (HR) colorimeters appears to be reduced to the point where it is practically negligible.

Fluorometry offers several advantages over colorimetry. It has high sensitivity, no RIB, and is almost unaffected by sample turbidity. Since few SFA fluorometers are commercially available,

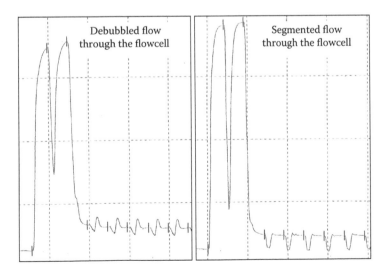

FIGURE 8.2 Peak shape with debubbled flow (left) and segmented flow (right) through the flowcell. Samples are seawater; sampler wash solution is ultrapure water. When the bubbles pass through, the cell carryover (tailing peaks) is markedly reduced and the Schlieren effect (visible on low peaks) is almost removed.

HPLC instruments can be used with minor modifications. Cell capillary metal tubing should be replaced with PTFE tubing (~0.8 mm ID) protected from external light with an opaque material over 5–10 cm in order to avoid signal alteration (optic fiber effect). Temperature-stabilized fluorometers are preferred since fluorescence is more temperature sensitive than colorimetric measurement.

8.2.5 SOFTWARE

To measure very low concentrations accurately the software must be designed to take an average reading of the steady-state plateau of peaks, where the sample is at its highest concentration in the flowcell, starting and stopping the peak data collection to measure only the plateau, so that the wash and intersample disturbances are ignored. All modern computer controlled analytical systems are able to correct for baseline drift. For the highest accuracy, the baseline can be measured before and after every peak, rather than interpolated over several samples.

The contributions of the RIB and residual nutrients in the natural or synthetic low-nutrient seawater used for calibration standards are rarely negligible. This means that blank corrections for standards and samples are usually not equal. It follows that the software should not compute sample concentrations directly from the calibration curve equation ($ax + b$), but only use the slope (a) to obtain the conversion factor, then correct for the sample blank as necessary (see Section 8.4.3). Note that no software can precisely compute concentrations in samples of varying salinity if the RIB is not negligible and if the calibration slope is altered by a salt effect; in such cases, additional manual corrections for RIB and slope at any salinities may be required.

8.3 FROM SAMPLING TO ANALYSIS: PRECAUTIONS AGAINST CONTAMINATION

8.3.1 SAMPLING, SUBSAMPLING, AND FILTRATION

Determination of dissolved nutrients should be performed on a sample conventionally filtered through a 0.45 μm pore size filter. However, filtering the samples requires great care to avoid contamination. For this reason, it is generally omitted for oceanic water where the amount of suspended matter is very low and cannot interfere with the analyses. In areas where particles (including plankton) may slightly interfere, direct online rough prefiltration is usually appropriate, as described below (Figure 8.3). In turbid areas conventional filtration is generally required. Full explanations about these procedures and related materials can be also found in Aminot (1995).

A primary raw sample is taken from the water column using a water sampling bottle: Niskin bottles or similar devices with stoppers are recommended. These bottles are operated by means of either a hydrowire, to which they are individually attached (e.g., onboard small ships for coastal studies), or using rosettes fitted with many bottles together. To avoid contamination, never grasp the water sampling bottles by the opening or touch the stoppers with bare hands. Between sampling stations the water sampling bottles should be kept closed in a clean place.

The sample to be analyzed is conveniently collected in screw-capped sample bottles made of plastic (e.g., HDPE). Glass bottles can also be used, except for silicate determination. Caps should be thoroughly leakproof, and unlined caps are recommended. Liners, if present, should not be separable from the cap itself. Bottles should not be too small in order to minimize potential contamination: 50–100 mL is recommended. If the analyses of all nutrients cannot be performed simultaneously, as many samples as the number of analytical runs necessary for the determination of all nutrients should be collected. Before the first use, bottles must be filled with 1 M HCl and stored tightly closed for a few hours for cleaning. Then, between cruises, bottles may be machine washed using a phosphate-free detergent and stored capped in a clean, dry place, in darkness.

When filtration is not necessary, the sample bottle is filled directly from the water sampling bottle without any treatment. The sample bottle is rinsed two or three times (with the cap screwed

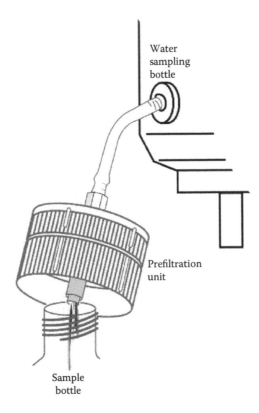

FIGURE 8.3 Prefiltration system fitted to the water sampling bottle (Niskin or Go-Flo bottle).

on each time), then filled with the sample. If conventional filtration is required, a subvolume of the raw water sample is collected in a bottle of the same material as the sample bottle, and which has been cleaned similarly. Then filtration should be done as quickly as possible and the filtered sample, which will be analyzed, is poured into the usual sample bottle with rinsing as described previously. When a prefiltration is appropriate to remove part of the suspended matter, it is easily performed by connecting a low dead-volume online filter holder (e.g., Millipore Swinnex; Figure 8.3) equipped with a medium pore size filter (e.g., 10 μm hydrophilic polypropylene Gelman membrane). As this unit is always rinsed together with the sample bottle, its cleanliness is permanently ensured. If the samples are to be frozen, the sample bottle should be filled only to three-quarters of its volume and closed tightly to prevent leakage.

Since skin is a major source of contamination in nutrient measurements (Kérouel and Aminot, 1987), it is recommended to wear disposable gloves (not powdered, latex, or vinyl) from sampling to analysis.

8.3.2 STORAGE AND PRETREATMENT OF SAMPLES

Whenever possible, nutrients in seawater samples should be determined as quickly as possible after collection. This is essential for oligotrophic oceanic waters in the nanomole per liter range to obtain the highest accuracy. However, onboard small vessels, for coastal and estuarine studies, immediate analyses are often not possible and samples have to be stored until analysis in a shore-based laboratory. Since freezing has been shown to be reliable for sample storage over several months (Clementson and Wayte, 1992; Gordon et al, 1993; Avanzino and Kennedy, 1993; Aminot, 1995; Dore et al., 1996), it is the method recommended here. Chemical preservatives either have poor

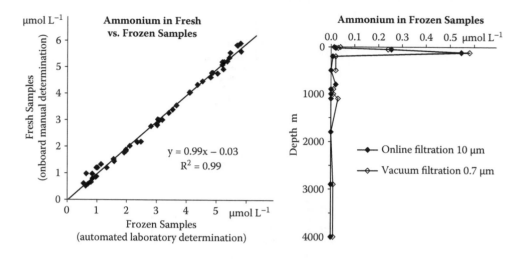

FIGURE 8.4 Tests on the stability of ammonium concentrations in frozen samples. Left: Coastal waters, fresh samples (manual method) vs. frozen samples (automated method), all of them prefiltered through 10 μm. Right: Frozen offshore waters, 10 μm prefiltered samples stored in HDPE bottles compared with 0.7 μm filtered samples stored in glass bottles.

efficiency (chloroform; Kirkwood, 1992), or are hazardous substances (mercury), or modify analytical conditions (mercury, acids).

Storage by freezing can be validated using ammonium as a reference nutrient with regard to its high biological reactivity and its sensitivity to poor handling conditions. Experimental data show a high correlation between fresh and frozen coastal waters, with no statistical difference between samples (Figure 8.4, left), and very consistent results in duplicates of frozen offshore waters (Figure 8.4, right). The small excess of ammonium (0.015 ± 0.01 μmol/L) in samples treated by conventional filtration shows that it should be avoided whenever possible.

When samples are analyzed on the day of collection they can be stored in a refrigerator. All other samples should be immediately placed in a freezer (in an upright position and at $-25°C$ or lower for complete solidification). Note that some silicate samples should not be frozen (see Section 8.8). If a freezer is not available, samples must be cooled in a refrigerator or in an icebox containing sufficient cooling capacity, then analyzed or frozen within hours of sampling. Freezers, refrigerators, or iceboxes should be strictly devoted to the storage of water samples (never use them to store chemicals, reagents, food, or biological samples).

Sample thawing should be done with care (Aminot, 1995; Aminot and Kérouel, 2004): maintain bottles in an upright position, tightly capped during the whole process; avoid warming the samples; mix the bottle contents from time to time (wearing gloves); when no ice remains, thoroughly mix the samples before analysis.

Until analysis, keep the samples cool and in the dark. If samples contain suspended matter, let the particles settle down while stored in a fridge or, preferably, centrifuge the bottles (without transfer) for a few minutes in a refrigerated centrifuge.

8.3.3 Risk of Sample Contamination During Analysis

When the sample is poured into a cup of a few milliliters volume, contamination of the transferred aliquot may occur from spilling on the skin and dripping into the cup. Even external wetness of the cup allows contaminant substances to migrate into the aliquot. Wearing disposable gloves (Section 8.3.1) is therefore essential. Overall, nutrient contamination from skin decreases

TABLE 8.1
Typical Contaminations from Fingerprints

Nutrient	Contamination µmol L⁻¹		Typical Marine Concentration µmol L⁻¹	Relative Contamination	
	100 mL Bottle	5 mL Cup		100 mL Bottle	5 mL Cup
Ammonium	0.45 ± 0.05	No cups	0.5	90%	—
Nitrite	<0.002	0.005 ± 0.002	0.5	<0.4%	1%
Nitrate	0.5 ± 0.1	6.2 ± 0.8	5	10%	120%
Primary amines (in N)	1 ± 0.3	No cups	0.05	2,000%	—
N-urea	1.2 ± 0.3	83 ± 16	0.5	250%	17,000%
Phosphate	0.09 ± 0.05	1.0 ± 0.3	0.5	20%	200%

Note: Average (± standard deviation) of two to four independent assays from one or two people (Kérouel and Aminot, 1987; Aminot and Kérouel, 2007).

in the following order, primary amines > urea >> ammonium > nitrate, phosphate >> nitrite (see Table 8.1). Contamination from volatile substances present in the laboratory atmosphere (ammonia, nitric acid) can be a major problem, even in so-called clean laboratories. Figure 8.5 shows ammonia contamination during analysis of samples in various cups placed on an autosampler tray. To decrease this contamination source, increase the volume of the aliquot submitted for analysis, but avoid filling the sample cups or tubes completely. The use of a sampler able to receive the sample bottles directly is recommended (no transfer, whole volume handled). As a final precaution, for nitrate and ammonium determination at trace levels, only a few sample bottles at a time should be placed together on the sampler.

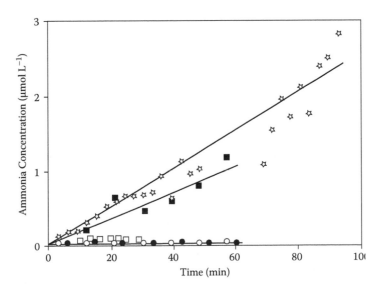

FIGURE 8.5 Contamination by ammonia from air: series of low-ammonium seawater samples simultaneously placed on the sampler and successively sampled. Symbols: ☆ = full 5 mL cups; ■ = full 10 mL tubes; □ = half-full 10 mL tubes; ○ = 60 mL PE bottles (3/4 full); ● = 125 mL PE bottles (3/4 full).

8.3.4 GENERAL PROCEDURE FOR THE ANALYSIS OF A SAMPLE SERIES

The day before the analysis:

- Make sure that sufficient amounts of reagents are available and that their age is within the allowable stability period. Before the analysis begins the reagents should be at room temperature.
- If the AutoAnalyzer has not been used for some time, make sure that the pump tubes are in good condition (normally with less than 200 hours of use) and that the AutoAnalyzer is working normally. This can be checked by pumping the RIB reagents (see Sections 8.5 to 8.9), or alternatively water + surfactant at the same concentration as in the reagents, through the reagent lines with pure water through the sample. The baseline under these conditions should be a perfect straight line, with no noise or drift.
- Replace the RIB reagents by the normal reagents and record the reagent absorbance, making sure that this is within allowable limits (if not, renew the reagents).
- Check the baseline stability, then the peak shape with a few typical samples or standards for all the analyses to be run the next day.
- Samples that have been frozen may be placed in a refrigerator to begin thawing.

On the day of analysis:

- Start the AutoAnalyzer and check the baseline stability and the peak shape, as previously, before completely thawing the samples.
- Prepare calibration standards and analyze them (subsequently, keep them in a refrigerator if they are intended to be used again).
- Finish thawing the samples, but keep them cool (Section 8.3.2).
- As soon as the calibration is valid, analyze the series of samples, including all necessary reference materials and any useful control waters as blank.
- If required, once all sample peaks have been recorded, replace normal reagents by those designed for RIB determination, wait for baseline stability, and measure RIBs.
- After RIB determination, record the baseline for some time, then pump water for sufficient time to clean the manifold before removing the pump platen.

8.4 TECHNICAL ASPECTS OF NUTRIENT DETERMINATION IN SEAWATER

8.4.1 SPECIFICITY OF MARINE WATER ANALYSIS

Seawater analysis differs from freshwater analysis insofar as some chemical reactions can be affected by the salinity of marine waters (the so-called salt effect). In continuous flow colorimetry, variation of the refractive index from fresh to seawater generates optical perturbations. Adsorption on the manifold walls, predominantly in fresh and low-salinity waters, enhances peak tailing and the carryover effect for some parameters. In coastal and estuarine waters these alterations may require corrections as a function of sample salinity. In the case of oceanic surface water, the concentrations of some nutrients are so low that great precautions are necessary to prevent contamination of the sample, and analytical techniques may need special methodology and equipment in order to achieve the highest sensitivity. It follows that high precision and low detection limits in seawater analysis rely on manifolds and methods that have been optimized to maximize reproducibility and to minimize corrections (especially for carryover and drift).

In CFA, the reagent blank is ignored in the calculation since it is part of the baseline signal. However, it must be lowered as far as possible since high reagent absorbance will make the baseline level more sensitive to salt effects (Alvarez-Salgado et al., 1992) and usually increase its variability, and hence the detection limit. Low reagent absorbance is achieved by using analytical-grade chemicals or better and high-quality, freshly prepared (not stored) ultrapure water. To avoid contamination of the reagents through the transfer between containers, we recommend preparing them directly in the storage bottle, which can be simply marked at the desired volume (only standards require highly precise volumes).

In addition to the constraints of CFA, such as carryover or drift, nutrient determination in seawater should take several other points into account:

1. The sample blank, that is, the RIB (Section 8.2.4), varies from one analytical system to another.
2. The baseline sometimes differs from the zero concentration level.
3. The calibration water matrix can contain a measurable amount of the nutrient to be determined.
4. The salt effect has to be checked under the specific analytical conditions of each laboratory.

Taking the above points into consideration, the steps involved in calculating sample concentrations are:

- Identify the baseline level.
- Measure the peak heights of standards.
- Determine the calibration factor.
- Compute the raw concentrations.
- Correct for the carryover effect.
- Correct for the sensitivity drift.
- Correct for the baseline blank.
- Correct for the RIB.
- Correct for the salt effect.

Computations are usually done after peak heights have been converted into concentrations, but it is equally valid to use peak heights and convert into concentrations at the final stage. Modern CFA analyzers have software that can perform most of the above steps automatically.

8.4.2 Baseline, Peak Measurement, Zero Concentration Level

The baseline is recorded when the intersample wash water is sampled for a time long enough to obtain a stable signal. The baseline is the reference level from which peak heights are measured. The height is measured near the end of the plateau, or at its maximum for sharp peaks without Schlieren effects. The time interval between the determinations of two successive peak heights should correspond, on average, to the sampling period (= sample + wash). When the baseline tends to drift, it should be recorded at shorter intervals to ensure precise interpolation of its level for all peak evaluations. This is realized by programming longer wash periods from time to time or adding wash cups in the series. When the baseline drift is not constant, the most accurate results are obtained by measuring it before and after each peak. Baseline resolution can be achieved between sample peaks when the manifold is optimized and the sampling rate and sample/wash ratio are not too high. If the software does not allow this mode of calculation, data can be obtained by printing the chart for manual peak measurement.

The sampler wash water may contain low levels of nutrients. In such cases, the baseline is not at the zero concentration level and a correction is required for the samples. Except when very low concentrations produce negative peaks, contamination of the baseline often passes unnoticed. Checking the zero level is therefore important, and the only valid reference is high-purity, freshly produced 18 Mohm ultrapure water, for example, from a Millipore or similar system (distilled water is not a reliable reference). Hence, bottles with freshly prepared ultrapure water should be occasionally analyzed in a series to check the baseline blank (the difference between the baseline and ultrapure water analyzed as sample) to be added to peak heights. In most instances, when the wash water is prepared with care (using ultrapure water, protected from direct contact with ambient air), the baseline is the actual reference level, and no correction is required.

8.4.3 CALIBRATION

Primary (concentrated) standards are prepared using analytical-grade salts and ultrapure water. However, working standards should not be prepared using ultrapure water. The matrix can be surface oceanic water, nutrient-depleted natural seawater (NDSW), or artificial seawater. NDSW can be prepared by collecting a bulk volume of clean coastal seawater when nutrient concentrations have been lowered by phytoplankton growth. Once filtered through a nylon net with a pore size of 50 µm to remove zooplankton (but not phytoplankton), the water is stored in a nonopaque carboy at the ambient temperature and light of the laboratory. Usually, the nutrient concentrations become very low after about 2 months, then remain so as long as the water is not filtered. For use in making the calibration standards, the necessary volume is conventionally filtered just before use, taking care to avoid contamination.

In a linear calibration, the working standard heights (H, in user's units), measured from the baseline as reference, are related to their concentrations (C) by the equation

$$H = a \times C + h_0 \tag{8.1}$$

The constant h_0 (the calibration blank) is the signal of the nonspiked matrix (RIB + background concentration, if any). This signal is usually neither equal to zero nor equal to the sample RIB. Consequently, the full equation should not be used to calculate sample concentrations (Section 8.2.5). The calibration factor is $F = 1/a$, from which all raw sample concentrations are obtained:

$$C_{sample} = F \times H_{sample} \tag{8.2}$$

8.4.4 CARRYOVER AND SENSITIVITY DRIFT CORRECTIONS

8.4.4.1 Carryover

The carryover effect (Section 8.2.3) increases each peak by a proportion P of the peak that precedes it. The effect should be negligible beyond one peak. P can be calculated by two methods (Figure 8.6). It is recommended to determine P at several points in a run and average the values.

   Method 1: Analyze one high standard several times in succession. $H1$ is the height of the first peak and $H2$ that of the following peaks; it follows that $P = (H2 - H1)/H1$.
   Method 2: Analyze one high standard followed by two identical analyses of a low standard. H, $L1$, and $L2$ are the heights of the three successive standards; it follows that $P = (L1 - L2)/(H - L1)$.

The carryover correction is applied to all raw concentrations as follows: with C_n the raw concentration of sample n, C_{n-1} the raw concentration of sample $n - 1$, and P the relative carryover effect defined above, the corrected concentration $C_n(cor)$ for sample n is

$$C_n(cor) = C_n - (C_{n-1} \times P) \tag{8.3}$$

8.4.4.2 Sensitivity Drift

Sensitivity drift (in other words, a calibration factor drift) occurs when the height of a given, stable standard or a reference material varies throughout the analysis of a sample series. The proportional

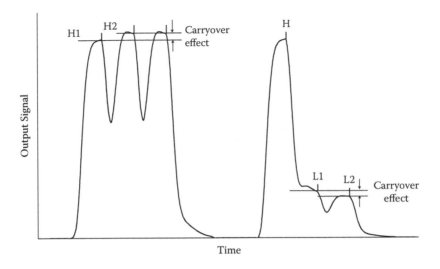

FIGURE 8.6 Effect of carryover on peak heights.

variation of the calibration factor relative to the initial calibration is determined by interpolation for each sample, and a correction applied accordingly.

8.4.5 REFRACTIVE INDEX BLANK (RIB) CORRECTION

In CFA, the RIB (Section 8.2.4) is the actual sample blank. When not optically corrected, it should be separately measured and subtracted from the sample concentrations. The RIB is the signal that a zero concentration sample would produce, which is similar in composition to the analyzed sample. In practice, samples are reanalyzed, but preventing any color development by using special RIB reagents. RIB reagents are prepared like normal reagents, but omitting a chemical in order that the chemical reaction cannot proceed. If the samples are not turbid, RIBs are linearly related to salinity and only a few selected samples have to be reanalyzed for a precise determination. Figure 8.7 shows an example of RIB determination. RIB heights (h_{RIB}) are converted into concentrations using the calibration factor F: $c_{RIB} = F \times h_{RIB}$. RIBs must be used to correct the sample concentrations. Note that when RIBs vary from sample to sample (if salinity varies, for example), software is usually unable to make an automatic correction.

For a given salinity, the RIB varies according to the analyzed nutrient and the optical design of the colorimeters. In terms of concentration, it rarely exceeds 0.1 µmol L^{-1} in seawater, for all nutrients. In estuaries, when concentrations are significantly higher than in oceanic waters, RIBs may be ignored.

8.4.6 SALT EFFECT CORRECTION

The salt effect of an automated method may differ from that of a corresponding manual method. It should therefore be determined with each analytical technique, in each laboratory. It is assumed to remain the same as long as the method is unchanged, but it is recommended to check it occasionally.

The salt effect is determined by measuring the same concentration of nutrients added in nutrient-depleted seawaters of different salinities, prepared by a series of dilutions, by mass, of seawater with ultrapure water. Artificial seawater can be used provided it mimics the buffer capacity of natural seawater (hence containing at least Na^+, Ca^{2+}, Mg^{2+}, Cl^-, SO_4^{2-}, and bicarbonate). A simple NaCl solution is not artificial seawater. The salt effect, sometimes nonlinear, should be determined over

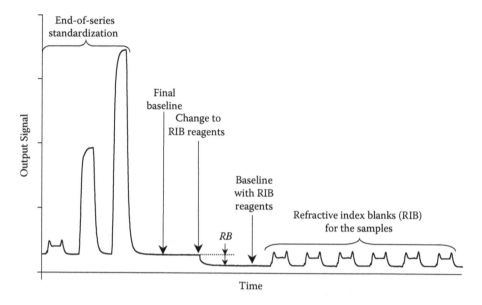

FIGURE 8.7 Example of the end of a series (nitrite, AutoAnalyzer II colorimeter), including replacement of normal reagents by RIB reagents, then RIB determination. (Note: RB is the reagent blank if the baseline is the zero concentration level.)

the whole range of the sample salinities. Because of possible adsorption below a salinity of ~2, the salt effect at zero salinity will be obtained by extrapolation.

For the same concentration of the added nutrients, the height is $H_{S(C)}$ at the salinity corresponding to the calibration matrix $S(C)$ of the analytical series, and H_S at any other salinity S. Given $K_S = H_{S(C)}/H_S$, all concentration values obtained after the previous correction steps are multiplied by K_S to obtain the final concentrations.

8.4.7 Specific Manifold Fittings

Figure 8.8 shows various manifold modifications that may be useful for optimizing hydraulic conditions and increasing the performance in any medium, from sea- to fresh water.

If an ISAC effect is present, it can be completely eliminated by removing the intersample bubble just before the pump (Figure 8.8A and B). To minimize the effect on carryover, the length of the debubbled flow must be kept as short as possible.

The concentration ranges can be expanded up to estuarine concentrations by increasing sample dilution in the reaction medium, in various ways. When the initial sample flow is low compared with the total flow, it can often be reduced while keeping satisfactory reaction conditions. If this is not sufficient, a dilution stage can be added before mixing the sample with reagents (Figure 8.8C). When the initial sample flow is relatively high it may be suitable to dilute the sample online (no repumping), keeping the total flow of sample + dilution water as near constant as possible (Figure 8.8D).

8.5 DETERMINATION OF NITRITE

8.5.1 Overall Description of the Method

The method is based on the Griess reaction, adapted to seawater by Bendschneider and Robinson (1952). Nitrite ions react first with sulfanilamide to form a diazo compound, which then combines with *N*-naphthyl-ethylenediamine (NED) in acid conditions (pH < 2, strictly) to produce a final pink-colored complex.

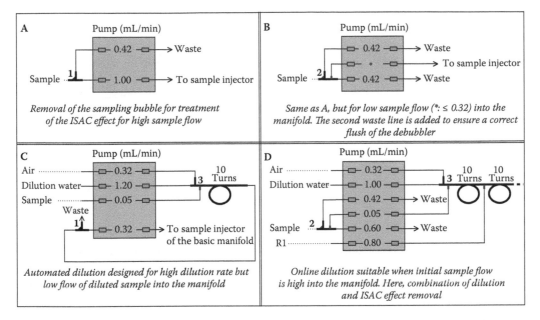

FIGURE 8.8 Examples of manifold options (suitable for the nutrient methods described) designed to remove the ISAC effect or to automatically dilute the sample. 1 = microflow debubbler (e.g., Alpkem A303-0103-01 or possibly SEAL 188-G017-02, with a smaller chamber); 2 = microflow debubbler-injector (e.g., Alpkem A303-0104-03); 3 = injector SEAL 116-0489-01.

In the automated method the reaction is accelerated, at ambient temperature, by increasing the reagent concentrations in the reaction medium, while still being within the optimum range, as described by Benschneider and Robinson (1952). (For a comparison of manual and automated versions, see Aminot and Kérouel, 2007). Reagent solutions for automated analysis are typically more diluted to avoid very low reagent flow rates. Below 15°C the reaction may become incomplete, hence adding delay coils, or a 37°C heating bath may be useful to preserve maximum sensitivity. The manifold is designed so that nitrite and nitrate can be analyzed using the same reagents.

Depending on colorimeter design, such as light path and signal amplification, a linear range from 0.001 to 10 µmol L^{-1} is obtained for undiluted samples. The detection limit near 1 nmol L^{-1} is comparable to more complex methods (Garside, 1982; Mikuska et al., 1995; Pai et al., 1996; Yao et al., 1998; Zhang, 2000). Normally, there is no interference between the chemistry and the matrix of seawater (Aminot and Kérouel, 2004). The formation of the colored complex is not altered by salt, but at salinities lower than 10 the pink compound is slightly adsorbed onto the manifold glass walls and the peaks become sharper and lower.

Sampling, subsampling, and storage procedures are described in Section 8.3. Pasteurization may be a reliable storage method for filtered samples (Aminot and Kérouel, 1998).

8.5.2 MANIFOLD AND ANALYTICAL CONDITIONS

To prevent hydraulic perturbation resulting from the high sample flow, the intersample bubble is removed by a debubbler placed just before the air segmentation injector (Figure 8.9). The debubbler type and the debubbling flow may need to be adjusted depending on the sampler dimensions and speed, which determine the intersample bubble size. The sample probe diameter should be of a larger diameter. The sampler wash receptacle is fed with freshly prepared ultrapure water (Section 8.4.2).

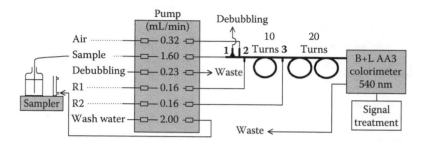

FIGURE 8.9 Standard manifold for nitrite determination. 1 = CO debubbler; 2 = injector 116-0489-01; 3 = A10 injector. The debubbler (1) and the injector (2) must be assembled without a gap between.

The characteristics and settings of the AA3 colorimeter used for the presented data are: standard 540 nm filter (bandpass 20 nm) and standard krypton lamp or LED light source; standard detector; 1 cm light path; gain = 200, smoothing factor = 4.

A sampling rate of 30 h^{-1} and a sample:wash ratio of 1:1 provide the highest precision. Higher frequencies are suitable for routine analysis. It is recommended to place only a few samples on the autosampler tray in order to keep them cool for as long as possible, thus reducing biological activity. Noncontaminating disposable gloves should be worn.

8.5.3 Reagents and Standard Solutions

Brown glass bottles are preferred for reagent storage.

> **Reagent R1:** Dilute 50 mL of hydrochloric acid (HCl 37%) into ~500 mL of ultrapure water. Add and dissolve 5 g of sulfanilamide ($C_6H_8N_2O_2S$; M = 172 g mol^{-1}), then make up to 1,000 mL with ultrapure water and mix. Add 400 µL of Brij 35 (in 30% solution). Store at 5°C in the dark and renew each month, imperatively.
> *Optional*: If the peaks become sharper at salinities below 10, add 100 g L^{-1} of analytical-grade NaCl to reagent R1 to restore normal peak shape.
>
> **Reagent R2.** Dissolve 0.25 g of *N*-(1-naphthyl)-ethylenediamine hydrochloride (NED; $C_{12}H_{14}N_2$, 2HCl; M = 259 g mol^{-1}) in 1,000 mL of ultrapure water. Store cool and in the dark. Renew the reagent each month or if it becomes brownish.
>
> **RIB reagents:** Use reagent R1 unchanged and replace reagent R2 by ultrapure water.
>
> **Nitrite calibration standards:** Use analytical-grade sodium nitrite ($NaNO_2$; M = 69.00 g mol^{-1}). If the purity differs from 100% but is certified, increase the mass to be weighed proportionally. Do not use an old product (Hansen and Koroleff, 1999).
> *Primary nitrite standard* (5,000 µmol L^{-1}): Dry sodium nitrite (105°C, 1 hour), then let it cool in a desiccator. Weigh 0.345 g for 1,000 mL of solution prepared in ultrapure water, in a volumetric flask. When the salt is completely dissolved, mix the solution and transfer it to a clean glass or plastic bottle. Store at ambient temperature, in the dark, and renew each month, imperatively. Never add acid or mercury as a preservative, because they accelerate nitrite loss (Aminot and Kérouel, 1996).
> *Secondary nitrite standard* (500 µmol L^{-1}): Dilute the primary nitrite standard exactly ten times with freshly prepared ultrapure water to obtain the secondary standard, 1 mL of which contains 0.5 µmol of nitrite. Store at ambient temperature, in the dark, and renew each week.

8.5.4 PROTOCOL

Start the AutoAnalyzer and follow the general instructions listed in Section 8.3.4.

Prepare a series of calibration standards, corresponding to expected concentrations, by diluting the secondary nitrite standard (500 µmol L^{-1}) in NDSW (Section 8.4.3). Ultrapure water should not be used because adsorption onto the manifold walls may alter the results (Section 8.5.1). As an example, the typical range 0.1–0.2–0.5–1.0 µmol L^{-1}, prepared in 500 mL volumetric flasks, requires 100–200–500–1,000 µL of the secondary standard solution.

If the calibration is satisfactory, the samples can be analyzed as previously described (Section 8.3.4). Then, replace normal reagents by RIB reagents and determine the sample RIBs. Using AA3 lamp colorimeters, the RIB is usually around 0.01–0.02 µmol L^{-1} at a salinity of 35. It may be lower with an LED light source.

The calculation sequence is described in Section 8.4.1. Under optimized working conditions, the RIB correction is the only one that has to be applied to raw concentrations.

8.5.5 METHOD PERFORMANCE

Figure 8.10 shows a typical record of calibration standards prepared in NDSW (salinity ~ 35), with ultrapure water as the baseline.

Under normal conditions the signal stability is about 0.05% at the level of 1 µmol L^{-1}, corresponding to a resolution of at least 0.001 µmol L^{-1} of nitrite in the range 0–1 µmol L^{-1}.

Method precision was assessed using various reference materials prepared for international intercomparison programs (Aminot and Kérouel, 1995, 1997). Within the range of most natural seawater concentrations (<0.5 µmol L^{-1}), the repeatability is about 1 nmol L^{-1}. Reproducibility (which includes calibration variability over 2 years) is better than 4 nmol L^{-1} up to a concentration of about 1 µmol L^{-1}.

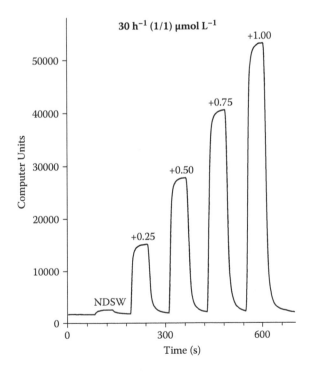

FIGURE 8.10 Typical record of a series of calibration standards for nitrite in nutrient-depleted seawater (NDSW). AA3 colorimeter settings as described in the text.

From repeatability data, the detection limit, calculated according to Taylor (1990; three times the standard deviation of the blank or a very low concentration), is 1–2 nmol L^{-1}.

8.6 DETERMINATION OF NITRATE

8.6.1 OVERALL DESCRIPTION OF THE METHOD

Nitrate is determined after its reduction to nitrite by passing the sample over copperized cadmium packed in a column (Wood et al., 1967). A copperized cadmium column prepared with care converts 100% nitrate into nitrite (Garside, 1993). Nevertheless, reduction remains the critical step, even though automation makes it more reliable. The resulting nitrite is determined in-line following Benschneider and Robinson's (1952) colorimetric method (Section 8.5). Thus, the result is the sum of nitrate + nitrite in the sample, so that if the nitrate concentration alone is required, nitrite should be separately measured and subtracted from the total result.

Several types of reducing column described in the literature (cadmium granules, wire, or tube) have been tested. Cadmium tubes are convenient for microflow systems, but their limited surface area makes them less suitable for macroflow. Stainton (1974) described the use of a cadmium wire. Cadmium in granules is here preferred for its high and stable reduction efficiency and its longevity. The grain size, as well as the column diameter, is smaller than in manual methods. This enhances the reduction speed and allows a flow rate of up to about 5 cadmium volumes per minute through the column, four times higher than Nydahl's (1976) recommendation. The ammonium chloride buffer stabilizes the chloride concentration in the reaction medium (6 ± 2 g L^{-1}), hence eliminating salt effects for a wide range of sample flow rates. The NH$_4$Cl solution should be buffered at a pH close to 8.5 to ensure a high, stable reductor efficiency while preventing reduction of nitrite itself (Nydahl, 1976). To overcome the risk of sample contamination by ammonia vapor from the buffer, the latter is prepared inside the manifold by precisely combining NH$_4$Cl and NaOH solutions.

Since air bubbles should not pass through the column, the flow is debubbled just before the column and segmented again immediately after. It is essential to minimize dead volumes and shorten the nonsegmented section as far as possible. Of various hydraulic configurations described in the literature (Tréguer and Le Corre, 1975; Whitledge et al., 1981), the greatest stability was obtained by repumping the debubbled buffered sample into the column since the flow remains constant, even when granule lumps and particle accumulation partially clog the column. When excess clogging finally occurs, the flow drop causes sensitivity drift, then overpressure causes the column connection to burst and data are lost. It is therefore essential to check the column appearance before each run and to verify that the flow freely passes through it: this can be estimated by observing the bubble shapes and moves in the segmented flow just after the column.

Garside (1993) studied the error resulting from poor nitrate reduction efficiency in the presence of nitrite in the sample. If the reductor efficiency is assumed to be 100% but it is not, the absolute error on nitrate concentration (using Garside's symbols) will be

$$\text{Nitrate concentration error} = (1-c) \times \frac{S_2 N_3 - S_3 N_2}{(N_2 + cN_3)} \tag{8.4}$$

N_3 and N_2 are the concentrations of nitrate and nitrite in the standard, S_3 and S_2 the concentrations of nitrate and nitrite in the sample, and c the reductor efficiency of nitrate into nitrite.

It follows that the error cancels only in two particular cases: either if $c = 100\%$ or for $c < 100\%$, if the nitrate/nitrite ratio is the same in standards and samples. For example, using a combined standard with [NO$_3$]/[NO$_2$] = 10 and expecting the nitrate error within the ±1% range, nitrite must remain <30% of nitrate if $c = 95\%$ and <20% of nitrate if $c = 90\%$.

It is therefore essential (1) to obtain a reductor efficiency as close to 100% as possible, (2) to check this efficiency regularly, and (3) to calibrate with standards containing nitrate and nitrite in proportions close to those of the samples.

Nitrate concentrations can be measured on a linear range from less than 0.010 up to 50 µmol L^{-1}, depending on colorimeter design, such as light path and signal amplification. The range can be expanded up to ~1,000 µmol L^{-1} (for estuaries or rivers) with automated dilution of the samples requiring minor modifications of the manifold. In order to lower the detection limit in oligotrophic areas, the manifold can be adapted from the version of Raimbault et al. (1990), which includes a lower dilution of the sample flow. Note: With some colorimeters of previous generations or using a long flowcell, a nonlinear response may occur beyond 25 and 30 µmol L^{-1}.

Few interferences have been documented. Olson (1980) observed that, with aged columns, even low concentrations of phosphate alter the reductor efficiency. For the working conditions described here, no effect was detected up to 10 µmol L^{-1} of phosphate with a column used for hundreds of samples. The method is free from salt effects.

Sampling, subsampling, and storage procedures are described in Section 8.3. Pasteurization may be a reliable storage method for filtered samples (Aminot and Kérouel, 1998).

8.6.2 MANIFOLD AND ANALYTICAL CONDITIONS

The manifold described in Figure 8.11 is suitable for oceanic and coastal waters. For very low concentrations, sensitivity may be enhanced by using a sample pump tube of 0.42 mL min^{-1} instead of 0.23 mL min^{-1}, while reducing the sample flow to 0.05 mL min^{-1} expands the range up to 200 µmol L^{-1}. These changes in the sample flow, which occur in a preliminary stage, have no influence on column reduction, reaction conditions, or relative performance. The sample probe should have a diameter not greater than that of the sample pump tube unless a multichannel manifold is used requiring higher sample flow rates. The sampler wash cuvette is fed with freshly prepared ultra-pure water (Section 8.4.2). To expand the range further (up to 1,000 µmol L^{-1}), add a dilution stage (Figure 8.8C, Section 8.4.7).

Since the wetting agent has been shown to decrease the reductor efficiency (Tréguer and Le Corre, 1975), it is not added to reagents introduced before the cadmium column. Correct segmented flow under these conditions relies on the glass manifold components being clean, and it follows that the hydraulic behavior may be improved by pumping 1 mol L^{-1} NaOH or a diluted phosphate-free laboratory detergent through any new manifold (but not through the column) for about 10 minutes.

It is important to minimize carryover by shortening the nonsegmented section of the manifold as far as possible, so that (1) the low dead-volume debubbler before the column is directly connected

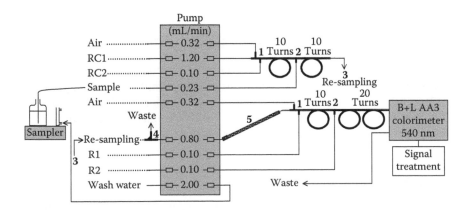

FIGURE 8.11 Standard manifold for nitrate determination. 1 = injector 116-0489-01; 2 = A10 injector; 3 = Tygon tubing (not PTFE), ID 1.3 mm; 4 = microflow debubbler (same as 1 in Figure 8.8); 5 = reduction column. Note: Mixing the two waste effluents prevents NH$_3$ release (use the shortest possible tubes). See manifold options in Section 8.4.7.

to the resampling pump tube; (2) similarly, the column is connected as close as possible to the other end of the pump tube, and (3) the flow is resegmented immediately after the column. The column assembly described here achieves optimum conditions. Alternatively, operation is made easier if a four-port two-way valve is fitted to the manifold to switch the column into and out of the reagent flow, provided a low dead-volume valve is used to prevent excess carryover.

The characteristics and settings of the AA3 colorimeter used for the presented data are: standard 540 nm filter (bandpass 20 nm) and standard krypton lamp or LED light source; standard detector; 1 cm light path; gain = 130, smoothing factor = 4. Amplifying the signal (gain = 1,000 and smoothing factor = 8) enables the analysis of oceanic water in the range 0–100 nmol L^{-1} with a repeatability of about 3 nmol L^{-1}.

A sampling rate of 30 h^{-1} and a sample:wash ratio of 1:1 provide the highest precision. Higher frequencies are suitable for routine analysis. It is recommended to place only a few samples on the autosampler tray in order to keep them cool for as long as possible, thus reducing biological activity. Disposable gloves should be worn.

8.6.3 REAGENTS AND STANDARD SOLUTIONS

Reagent for column reduction RC1: Dissolve 11.0 g (0.206 mol) of anhydrous NH_4Cl in ultrapure water and make up to 1,000 mL. Store at ambient temperature.

Reagent for column reduction RC2: Dissolve 9.0 g (0.225 mol) of sodium hydroxide NaOH in ultrapure water and make up to 1,000 mL. Store in a plastic bottle at ambient temperature.

Reagent R1: Same as nitrite reagent R1 (Section 8.5.3).

Reagent R2: Same as nitrite reagent R2 (Section 8.5.3).

RIB reagents: Same as nitrite RIB reagents (Section 8.5.3).

Nitrite standard at 5,000 μmol L^{-1}: Same as the primary nitrite standard (Section 8.5.3).

Nitrate standard at 5,000 μmol L^{-1}: Dry (105°C, 1 hour) analytical-grade potassium nitrate (KNO_3; M = 101.10 g mol^{-1}), then let it cool in a desiccator. Weigh 505.5 mg for 1,000 mL of solution prepared with ultrapure water, in a volumetric flask (note: if the KNO_3 purity differs from 100% but is certified, increase the weighted mass proportionally). When the salt is completely dissolved, mix the solution and transfer it to a clean glass or plastic bottle: 1 mL of this solution contains 5 μmol of nitrate. Stored at ambient temperature and in the dark, the solution is stable for at least 1 year provided no evaporation occurs (Aminot and Kérouel, 1996).

8.6.4 PREPARATION AND MAINTENANCE OF THE REDUCTION COLUMN

A well-prepared reduction column, used daily and regularly reconditioned, exhibits stable properties for weeks and allows hundreds, or even thousands, of samples to be analyzed. The column (Figure 8.12) is filled with fine cadmium granules, such as Aldrich 41.489-1 (30–80 mesh, i.e., 0.18–0.6 mm) or Merck 102088 (20–50 mesh, i.e., 0.3–0.8 mm), and treated as in the following procedure. Since nitric acid is used, do not perform this in the laboratory where nitrate is determined.

8.6.4.1 Cadmium Washing

In an Erlenmeyer flask shake ~2 g of cadmium with a small amount of acetone, then rinse thoroughly with ultrapure water. Pour out the water, cover the Cd with HCl 2 mol L^{-1} and shake for about 30 seconds, then rinse copiously with ultrapure water. Pour out the water, cover the Cd with HNO_3 0.3 mol L^{-1} and shake for a few seconds, then rinse again copiously with ultrapure water. Repeat washing with HCl, and then rinse again copiously with ultrapure water.

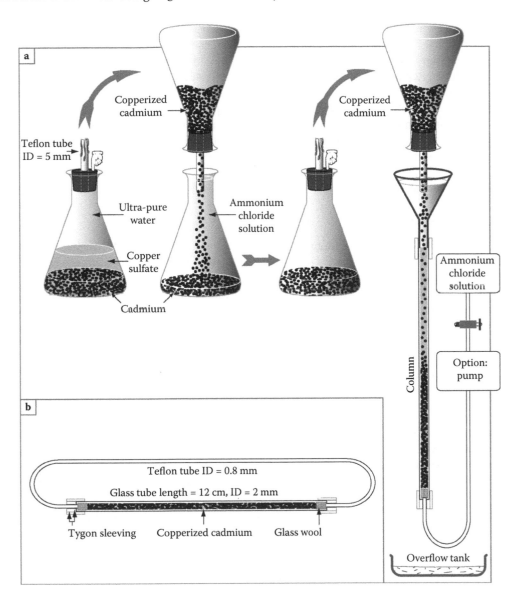

FIGURE 8.12 (a) A technique to transfer copperized cadmium into ultrapure water, then to fill the column, without any risk of contact with air. (b) Reduction column filled with copperized cadmium and stored protected from air.

8.6.4.2 Copperization

Without delay, pour out the water, add 30 mL of copper sulfate solution (20 g L^{-1} of CuSO$_4$, 5H$_2$O), and wait 5 minutes, but not longer. Transfer the copperized cadmium to another Erlenmeyer containing reagent RC1 (Section 8.6.3) as shown in Figure 8.12 (this removes most of the Cu solution without bringing the reductor into contact with air).

8.6.4.3 Column Filling

As shown in Figure 8.12, fit a small funnel to the top of the column and introduce a little glass wool (3–4 mm) at its end. Feed the column at low flow, and let it overflow, with reagent RC1. Then fill the column as shown, up to about 5 mm from the top. Increase the NH_4Cl flow for a short time to eliminate residual Cu solution and Cd dust. Stop the flow, and tap the column up and down to pack down the cadmium. Then add copperized Cd up to 3–4 mm from the top and insert glass wool to fill the dead volume. Replace the funnel by a PTFE tube (0.8 mm ID) and restart the NH_4Cl flow to fill it. Carefully disconnect the column from the NH_4Cl inlet and immediately connect the free end of the PTFE tube at the other end of the column without trapping any air (Figure 8.12). The column is ready for use and can be stored in this way at ambient temperature for a few months. Excess copperized cadmium can be stored for a few weeks in a bottle completely filled with NH_4Cl solution and stoppered.

Caution: Drying out at its ends indicates a leak at the sleeving and makes the column inoperative.

8.6.4.4 Maintenance

During analysis, dissolution and alteration of the reductor slowly occurs. When loss or clogging becomes obvious, the column can be regenerated as long as bright granules are not present over more than a quarter of the column length. Connect the column to NH_4Cl solution as for the initial filling. Using a small hooked needle, remove the glass wool. Then, under high solution flow, gently break up the granule lumps at the top of the column and drain impurities by overflowing. Stop the flow, pack up the cadmium, and fill the column with copperized Cd. Insert a little glass wool and close the column with the PTFE tube as before.

8.6.5 Activating and Checking a Reduction Column

Every new reduction column must be activated, then checked for efficiency as follows. Start the analyzer, with all reagents, but without the reductor column. When the baseline is stable, connect the column (Section 8.6.6) and continuously sample a 50 $\mu mol\ L^{-1}$ nitrate solution for 5–10 minutes to activate the reductor. Then wait for resolution of the baseline and analyze several 10 $\mu mol\ L^{-1}$ nitrite standards followed by several 10 $\mu mol\ L^{-1}$ nitrate standards. Compare the peak heights to obtain the efficiency. This should be stable and exceed 95%; otherwise, the column may not be well activated, or is defective and must be replaced.

Efficiency should be determined by comparing a nitrate standard reduced by the column to a nitrite standard of the same concentration analyzed without the column. However, because the hydraulic conditions are modified by the column, the nitrite standard is typically analyzed with the column installed, with the assumption that nitrite itself is not reduced. In general, using columns prepared as described, nitrite and nitrate peaks exhibit exactly the same heights, which can occur only if the efficiency in nitrite out of the column is 100% for the two nutrients (no nitrite reduction). If the nitrate and nitrite peak heights differ, check nitrite transfer efficiency with and without the column, setting the sampling time long enough to prevent a possible carryover effect. If nitrite is reduced, either the reductor is defective or the buffer conditions have to be reexamined. Recent versions of software allow the user to include a nitrate and nitrite standard sequence in a run to automatically calculate the reduction efficiency.

8.6.6 Operating Protocol

Start the AutoAnalyzer and follow the general instructions listed in Section 8.3.4. Prepare a 10 $\mu mol\ L^{-1}$ nitrite standard by diluting 1,000 μL of the 5,000 $\mu mol\ L^{-1}$ nitrite standard in a 500 mL volumetric flask. Use the same matrix as for the nitrate calibration (below).

Prepare a calibration series, corresponding to the expected concentrations, by diluting the nitrate standard (5,000 $\mu mol\ L^{-1}$) preferably in NDSW (Section 8.4.3). Ultrapure water is suitable only if it is compatible with other nutrients when mixed standards are used (e.g., problem of salt effect). As an example, the typical range 0.5–1 … 10–20 $\mu mol\ L^{-1}$, prepared in 500 mL volumetric flasks, requires

50–100 … 1,000–2,000 µL of the standard solution. If the reductor efficiency may depart from 100%, standards should contain nitrate and nitrite in proportions close to those of the samples (Section 8.6.1).

As soon as the flow becomes stable, switch the reduction column into the flow (if connected to a valve) or install it manually as follows. Protect the pump with absorbing paper and wear gloves. Disconnect the end of the resampling pump tube from the manifold. Disconnect the PTFE tube from the column entrance, maintained downward to prevent air from entering, while connecting the column to the resampling pump tube. Finally, remove the PTFE tube from the column and connect the column to the manifold.

Start standardizing when the baseline is stable again. When the calibration is satisfactory, the samples can be analyzed as previously described (Sections 8.3.2 to 8.3.4). Then, if necessary, replace normal reagents by RIB reagents and determine the sample RIBs. Using AA3 lamp colorimeters, the RIB usually lies around 0.01–0.03 µmol L^{-1} at a salinity of 35. It may be lower with an LED light source.

Before rinsing the manifold, remove RC2, wait 5 to 10 min, then switch the column out of the flow with the valve or remove it in the opposite way to how it was installed, taking care not to trap air by closing the column with the PTFE tube.

Calculation of concentrations is described in Section 8.4.1. The RIB correction may often be omitted in most coastal and estuarine waters. No salt effect correction may be required. Even if the reductor efficiency is 100% all the time, occasional minor drift (attributed to flow variation in the column) may require a correction (Section 8.4.4). Under the normal working conditions described in Section 8.6.1 for seawater, no reductor efficiency correction is required (even if the reductor efficiency is lower than 100%) and the results are the sum of nitrate and nitrite in the samples.

8.6.7 Method Performance

Figure 8.13 shows a typical record of calibration standards prepared in nutrient-depleted seawater (salinity ~35), with ultrapure water as the baseline.

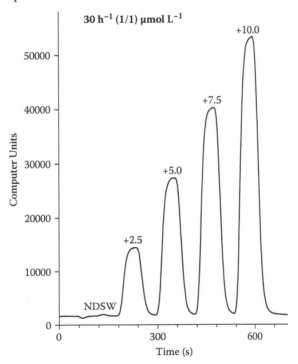

FIGURE 8.13 Typical record of a series of calibration standards for nitrate in nutrient-depleted seawater (NDSW). AA3 colorimeter settings as described in the text.

Under normal conditions the signal stability is about 0.05% at 10 μmol L^{-1}, corresponding to a resolution of a few nanomoles per liter of nitrate (+ nitrite) in the range 0–10 μmol L^{-1}.

Method precision was assessed using various reference materials prepared for international intercomparison programs (Aminot and Kérouel, 1995, 1997). Within the range of most natural surface seawater concentrations (<15 μmol L^{-1}), the repeatability, assessed from different bottles of the same sample, is generally better than 0.02 μmol L^{-1} overall and 0.005–0.01 μmol L^{-1} below concentrations of 2 μmol L^{-1}. When assessed from the same bottle a repeatability of 3–4 nmol L^{-1} was obtained in the range 0–1,000 nmol L^{-1}. Reproducibility (which includes calibration variability over 2 years) was similar to the intersample repeatability.

From repeatability data, the detection limit, calculated according to Taylor (1990; three times the standard deviation of the blank or a very low concentration), is ~10 nmol L^{-1}.

8.7 DETERMINATION OF AMMONIUM

8.7.1 OVERALL DESCRIPTION OF THE METHOD

The fluorometric method offers many advantages over indophenol blue colorimetry: simplicity, very high sensitivity, stable reagents with low toxicity, no refractive index blank, mostly insignificant fluorescence blank from natural substances, almost unaltered by sample turbidity, and a low salt effect. Both methods produce equivalent results (Kérouel and Aminot, 1997). While a detection limit of 5 nmol L^{-1} is claimed by Li et al. (2005), using long-capillary-cell colorimetry, fluorometry coupled with SFA might allow ammonium determination at subnanomole levels if contamination problems could be solved at every stage of sample handling and analysis.

Ammonium reacts with ortho-phthaldialdehyde (OPA) in the presence of a sulfurous reductor and in a slightly alkaline medium. Developed for primary amines by Roth (1971), the reaction was made specific for ammonium in freshwater by Genfa and Dasgupta (1989), who replaced 2-mercaptoethanol with sulfite. Using a borate (instead of phosphate) buffer the method was reexamined for its application to seawater (Kérouel and Aminot, 1997). A maximum sensitivity is obtained at 75°C ± 5°C for 3 minutes 40 seconds. While the reaction temperature could be lower, these conditions preserve a high stability at a wide range of ambient temperature conditions.

Depending on fluorometer settings, a linear range is obtained from 0.001 to 12 μmol L^{-1}. The detection limit of ~2 nmol L^{-1} is actually restricted by difficulties in controlling sources of contamination. The range may be expanded up to 250 μmol L^{-1} with automated sample dilution (Section 8.7.2).

Interferences are negligible in natural unpolluted waters (Kérouel and Aminot, 1997). Since the response drops by 5% to 80% at mercury concentrations between 10 and 40 mg L^{-1}, this preservative should not be used. Sulfur decreases the signal at a linear rate of 1% per 4 μmol L^{-1} of S. In seawater, in contrast with Li et al.'s (2005) assertion, fluorescence from natural dissolved substances produces only a very low blank (natural fluorescence blank [NFB]) of about 1–3 nmol L^{-1} expressed in ammonium concentration. The salt effect is less than 3% over the whole salinity range.

Sampling, subsampling, and storage procedures are described in Section 8.3. Remember that the determination of ammonium is very sensitive to poor sampling and handling conditions, especially contamination from surrounding air, handling without gloves, and pouring in general.

8.7.2 MANIFOLD AND ANALYTICAL CONDITIONS

As shown in Figure 8.14, the intersample bubble is removed just before the sample pump tube. This prevents hydraulic perturbations resulting from the high sample flow, as well as the ISAC effect (Section 8.2.2). The debubbler should be selected and the debubbling flow adjusted as a function of the bubble size. The sample probe diameter should be at least 1 mm. In order to minimize adsorption on tubing walls and improve the wash characteristics, the sampler wash cuvette is fed with slightly saline water (Section 8.7.3).

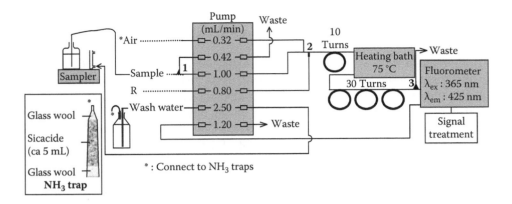

FIGURE 8.14 Standard manifold for ammonium determination. 1 = microflow debubbler (same as 1 in Figure 8.8); 2 = injector 116-0489-01; 3 = A2 debubbler. Volume of heating bath coil is 7.7 ml. See manifold options in Section 8.4.7.

Important: At the nanomolar level, the baseline stability can be improved by feeding the sampler cuvette by gravity, instead of pumping the wash water as usual.

Protection against ammonia contamination from surrounding air is important, at least for the segmentation air and wash water (less critical for reagents). The common method of bubbling air through H_2SO_4, which is cumbersome and may alter hydraulics, is replaced by tubes containing Sicacide (Merck, 719) (Figure 8.14), to be renewed when the Sicacide becomes moist.

The manifold includes a thirty-turn coil to cool the flow after the heating bath and minimize temperature variations inside the fluorometer cuvette.

With minor modifications of the manifold, an automated dilution of the sample can be undertaken to expand the measuring range up to 250 µmol L^{-1} (see Section 8.4.7).

The JASCO FP-2020 detector is equipped with a standard 16 µL cuvette and set as follows: excitation wavelength λ_{ex} = 365 nm (bandpass = 18 nm); emission wavelength λ_{em} = 425 nm (bandpass = 40 nm); gain = 1; attenuation between 16 and 128, depending on concentration range; response time = standard.

A sampling rate of 20 h^{-1} and a sample:wash ratio of 1:1 provide the highest precision. Higher frequencies are suitable for routine analysis: at a frequency of 40 h^{-1}, the carryover remains lower than 1%. It is important to avoid sample transfer, especially into small-volume cups. Therefore, the sampler should directly accommodate the sample bottles. Only a small number of samples should be placed on the autosampler tray in order to keep them cool for as long as possible, thus reducing biological activity. Disposable gloves should be worn (Section 8.3.1).

8.7.3 Reagents and Standard Solutions

Borate stock solution: In a transparent plastic bottle dissolve 30 g of disodium tetraborate decahydrate ($Na_2B_4O_7$, $10H_2O$; M = 381.4 g mol^{-1}) in 1,000 mL of ultrapure water. The bottle is tightly closed and stored at ambient temperature. If particles are visible in a new or aged solution, filter it before use (glass fiber, 1–2 µm pore size).

OPA stock solution in ethanol: Use standard-grade *o*-phthaldialdehyde (OPA; $C_8H_6O_2$; M = 134.1 g mol^{-1}), not more than 2 years old. Dissolve 5 g of OPA in 125 mL high-purity ethanol, by stirring for several minutes in the dark. Store in a refrigerator in a tightly capped glass bottle (e.g., cap with a PTFE-faced silicone septum) and always protect from light. This solution, initially light yellow, then rapidly colorless, can be used for at least 1 year.

Sulfite stock solution: Dissolve 200 mg of sodium sulfite (Na_2SO_3; $M = 126$ g mol^{-1}) in 25 mL of ultrapure water. Since the stability of this solution seems unexpectedly variable, it is recommended to prepare it daily.

Working reagent R: Use a 1 L bottle in opaque material (e.g., black PTFE). Rinse the bottle with ultrapure water and drain. Pour 1,000 mL of borate stock solution into the bottle, add 20 mL of the OPA stock solution, mix, add 2 mL of the sulfite stock solution, mix, then add 500 µL of Brij 35 (in 30% solution) and mix. In a tightly capped bottle, this reagent can be stored for 1 month at ambient temperature (do not cool, to prevent borate precipitation).

NFB reagent (nanomolar range only): Prepare like R but without OPA.

Wash water: Spike freshly prepared ultrapure water with either 5 mL L^{-1} of clean, ammonium-free seawater (NDSW) or 0.2 g L^{-1} of analytical-grade NaCl.

Ammonium calibration standards: Use analytical-grade ammonium sulfate (($NH_4)_2SO_4$; $M = 132.14$ g mol^{-1}), prefered to NH_4Cl, which is slightly hygroscopic; note that ammonium sulfate contains two ammonium groups per molecule. If its purity differs from 100% but is certified, increase the mass to be weighted proportionally.

Primary ammonium standard (10,000 µmol L^{-1} of NH_4): Dry ammonium sulfate (105°C, 1 hour), then let it cool in a desiccator. Weigh 660.7 mg for 1,000 mL of solution prepared with ultrapure water, in a volumetric flask. When the salt is completely dissolved, mix the solution and transfer it to a clean glass or plastic bottle (PE or PP). This standard is stable at least 1 year at ambient temperature (Aminot and Kérouel, 1996).

Secondary ammonium standard (1,000 µmol L^{-1} of NH_4): Dilute the primary ammonium standard exactly ten times with freshly prepared ultrapure water to obtain the secondary standard, 1 mL of which contains 1 µmol of ammonium. Storage and stability are similar to that of the primary standard.

8.7.4 PROTOCOL

Start the AutoAnalyzer and follow the general instructions listed in Section 8.3.4.

Prepare a series of calibration standards, corresponding to expected concentrations, by diluting the secondary ammonium standard (1,000 µmol L^{-1}) in NDSW (Section 8.4.3). Never use ultrapure water or water of salinity lower than 2 because NH_4 adsorption onto glass walls would alter the calibration. As an example, the typical range 0.5–1–2–4–8 µmol L^{-1}, prepared in 500 mL volumetric flasks, requires 250–500–1,000–2,000–4,000 µL of the secondary standard solution. Since tranfer of solutions is a source of contamination, preparation of standards by directly weighing into the bottles that will be placed on the sampler is a good alternative approach (Aminot and Kérouel, 2007). Note that for estuary work, calibration in water of salinity 15–20 will restrict the salt effect to within a range of ±1.5%.

When the calibration is satisfactory, the samples can be analyzed as previously described (Sections 8.3.2 to 8.3.4). It is worth remembering how strict handling precautions should be, especially to analyze samples at the nanomolar level. For example, a good approach is to analyze directly from the sample bottle and one at a time. After ammonium is measured in the samples, NFB can be measured if required (nanomolar range or organic rich waters). Thus, replace reagent R by the NFB reagent and determine sample NFBs.

If required in estuaries, the salt effect should be determined on each instrument, then checked from time to time (Section 8.4.6).

The calculation sequence is described in Section 8.4.1. Under usual working conditions in seawater, no correction is required. Only the NFB correction might be required at the nanomolar level. In estuaries, the salt effect correction should be considered. Based on calibration in seawater (S = 35), error remains lower than ~3%, while a calibration in water of 15–20 salinity restricts errors within ±1.5%.

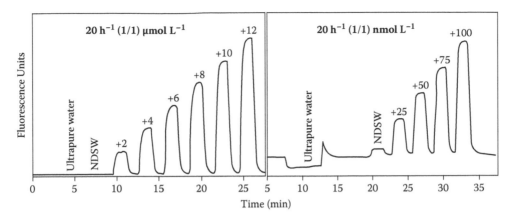

FIGURE 8.15 Typical record of a series of calibration standards for ammonium in nutrient-depleted seawater (NDSW). Left: Usual range for most marine waters. Right: Nanomolar range. JASCO FP-2020 fluorometer settings as described in the text.

8.7.5 Method Performance

Figure 8.15 shows typical records of calibration standards prepared in nutrient-depleted seawater (salinity ~35), using ultrapure water spiked with 5 mL L^{-1} of ammonium-free seawater as the baseline.

In fluorometry, sensitivity is not easy to express given the various combinations of settings. For example, with gain = 10, 100 nmol L^{-1} of ammonium produces 650 mV at the computer output. The signal stability allows a resolution of 0.3–0.5 nmol L^{-1}.

Reliable precision data are difficult to obtain since instability of ammonium in the samples does not allow long-term assessment. The repeatability from the same seawater sample is 0.005 μmol L^{-1} at a level of 3 μmol L^{-1}. Tested on thirty-eight samples (S ~34), duplicated at 1-hour intervals, a repeatability of 0.01 μmol L^{-1} was achieved for a concentration range of 0.5–5 μmol L^{-1}. In the nanomolar range, duplicate analyses of twelve samples at 3- and 30-minute intervals led to repeatabilities of 0.7 and 1.5 nmol L^{-1}, respectively.

From the above data, the detection limit, calculated according to Taylor (1990; three times the standard deviation of the blank or a very low concentration), is ~2 nmol L^{-1}.

8.8 DETERMINATION OF PHOSPHATE

8.8.1 Overall Description of the Method

The phosphomolybdic complex formed by phosphate with molybdate in acid conditions (pH ~0.8) in the presence of antimony is reduced by ascorbic acid to molybdenum blue (Murphy and Riley, 1962). Excess acid slows the reaction, while the acid:molybdenum ratio plays a major role in silicate interference. The following method follows as closely as possible the reaction conditions recommended by Murphy and Riley (1962) and confirmed by Pai et al. (1990) and Drummond and Maher (1995). It is therefore recommended to closely control the reaction conditions to ensure they remain valid when phosphate is determined in acid-hydrolyzed samples (organic P or polyphosphates; see Chapter 9).

Antimony not only plays a significant catalytic role in the reaction, but it also takes part in the phosphomolybdic complexation and makes it colloidal. As a consequence, the complex may adsorb more or less onto the manifold walls, including the colorimeter cell, so that the peaks become sharper and tailed and the baseline drifts. Some analysts therefore recommend removing antimony or reducing its concentration, but the resulting reaction rate is so low that heating up to 70°C is required, with the risk that organic phosphorus may be partially hydrolyzed. The option with antimony is considered to be the preferred approach, provided the manifold is cleaned as soon as alteration of the signal occurs (rarely more than a few times a year). The low-temperature heating bath (37°C) prevents hydrolysis, but ensures a fast reaction in any circumstances: acidity variations, very low phosphate concentrations (Sjösten and Blomqvist, 1997), and unexpectedly poor molybdate salt quality may all affect the reaction rate. The high sample flow enhances rinsing of the debubbler and improves peak shapes.

An unexplained, slightly nonlinear calibration might occur when measuring concentrations at the near-infrared absorption maximum (880 nm), but this does not happen at 820 nm, which is the closest available commercial light filter.

Depending on colorimeter design such as light path and signal amplification, the range is linear from about 0.003 to 50 µmol L^{-1}. The detection limit is 2–3 nmol L^{-1}. Normally, there is little interference in unpolluted seawater (Aminot and Kérouel, 2004). Silicate does not interfere below 2000 µmol L^{-1}. Arsenate reacts like phosphate, but interference can be eliminated (Hansen and Koroleff, 1999) or usually ignored since it rarely exceeds 0.03 µmol L^{-1} in seawater (Karl and Tien, 1992).

The determination of phosphate by the manual method is assumed free of salt effect (<1%, according to Murphy and Riley, 1962). However, in the automated method, a significant salt effect may occur (plus or minus several percent), mainly attributed to the wetting agent and aged antimony tartrate. Since additional factors (reaction temperature, phosphate concentration, detector wavelength) apparently influence the salt effect, it is recommended that laboratories working with estuarine samples check their method under their actual working conditions.

Sampling, subsampling, and storage procedures are described in Section 8.3.

8.8.2 MANIFOLD AND ANALYTICAL CONDITIONS

To prevent hydraulic perturbation resulting from the high sample flow, the intersample bubble is removed by a debubbler placed just before the air segmentation injector (Figure 8.16).

The debubbler type and the debubbling flow rate may need to be adjusted as a function of the sampler probe dimensions and speed, which determine the intersample bubble size. Use a

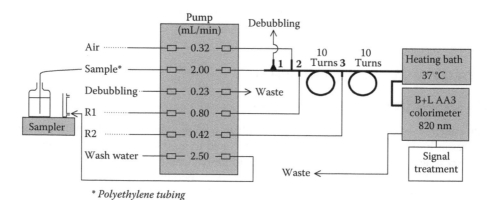

FIGURE 8.16 Standard manifold for phosphate determination. 1 = C0 debubbler; 2 = injector 116-0489-01; 3 = A10 injector. The debubbler (1) and the injector (2) must be assembled without a gap. The whole manifold should be in glass from injector (2) up to the colorimeter cell, with a minimum (and edge to edge) of connections.

large-diameter sampling probe. The sampler wash cuvette is fed with freshly prepared ultrapure water (see Section 8.4.2).

The characteristics and settings of the AA3 colorimeter used for the presented data are: standard 820 nm filter (bandpass 20 nm) and standard krypton lamp or LED light source; standard detector; 1 cm light path; gain = 300, smoothing factor = 8.

A sampling rate of 30 h^{-1} and a sample:wash ratio of 1:1 provide the highest precision. Higher frequencies are suitable for routine analysis. Since phosphate cannot undergo contamination from air, many samples can be placed on the sampler, provided they remain cool to reduce biological activity. In contrast, contamination from skin may be high; thus, it is strongly recommended to wear noncontaminating disposable gloves.

As mentioned in Section 8.8.1, when the peak shape begins to change due to adsorption of the complex onto the manifold, the latter should be cleaned by continuously sampling a hypochlorite solution (5% available Cl) for ~30 minutes, with water pumping through the reagent lines.

8.8.3 Reagents and Standard Solutions

Sulfuric acid stock solution (1.5 mol L^{-1})**:** In a 1 L borosilicate glass bottle, add 800 mL of ultrapure water, then, while continuously stirring, slowly add 85 mL of concentrated sulfuric acid (δ = 1.84 kg L^{-1}), then dilute further to 1,000 mL after cooling. If silicate is also determined, transfer the solution to a plastic bottle. It can be stored indefinitely.

Ammonium molybdate stock solution: In a 1 L plastic (e.g., HDPE) bottle, dissolve 40 g of ammonium heptamolybdate ($(NH_4)_6Mo_7O_{24}$, $4H_2O$; M = 1236 g mol^{-1}) in about 800 mL of ultrapure water and dilute further to 1,000 mL. This solution can be stored for several weeks at ambient temperature in the dark, but should be replaced if precipitation occurs.

Antimony tartrate stock solution: A product of more than 2–3 years old must not be used. Dissolve 1.5 g of potassium antimony(III) oxide tartrate hemihydrate ($K(SbO)C_4H_4O_6$, $\frac{1}{2}H_2O$; M = 334 g mol^{-1}) in 500 mL of ultrapure water. This solution can be stored for 2 months in a refrigerator, in either a glass or plastic bottle.

Reagent R1: Dissolve 2 g of ascorbic acid ($C_6H_8O_6$; M = 176 g mol^{-1}) in 500 mL of ultrapure water. Add 0.5 mL of the wetting agent Aerosol 22 (Sigma A-9753) or, alternatively, 200 mg of pure-grade sodium dodecyl sulfate. Prepare this reagent daily, in either a plastic or a glass bottle.

Reagent R2: In a brown glass bottle, add 800 mL of the sulfuric acid stock solution. Add 150 mL of molybdate stock solution and mix. Add 50 mL of antimony tartrate stock solution and mix. This reagent can be stored for about 1 month in a refrigerator.

RIB reagents: Use reagent R1 unchanged. Replace reagent R2 by sulfuric acid 1.2 mol L^{-1} (800 mL of the H_2SO_4 stock solution + 200 mL of ultrapure water).

Phosphate calibration standards: Use analytical-grade potassium dihydrogenphosphate (KH_2PO_4; M = 136.09 g mol^{-1}). If its purity differs from 100% but is certified, increase the mass to be weighed proportionally.

Primary phosphate standard (5,000 µmol L^{-1}): Dry potassium dihydrogenphosphate (105°C, 1 hour), then let it cool in a desiccator. Weigh 680.5 mg for 1,000 mL of solution prepared with ultrapure water, in a volumetric flask. When the salt is completely dissolved, mix the solution and transfer it to a clean glass or plastic bottle. This standard is stable for at least 1 year at ambient temperature (Aminot and Kérouel, 1996).

Secondary phosphate standard (500 µmol L^{-1}): Dilute the primary phosphate standard exactly ten times with freshly prepared ultrapure water to obtain the secondary standard, 1 mL of which contains 0.5 µmol of phosphate. Storage and stability are similar to that of the primary standard.

8.8.4   PROTOCOL

Start the AutoAnalyzer and follow the general instructions listed in Section 8.3.4.

Prepare a series of calibration standards, corresponding to expected concentrations, by diluting the secondary phosphate standard (500 µmol L^{-1}) in NDSW (Section 8.4.3). Do not use ultrapure water or water of salinity lower than 2 because phosphate adsorption onto glass walls would alter the calibration. As an example, the typical range 0.1–0.2–0.5–1.0 µmol L^{-1}, prepared in 500 mL volumetric flasks, requires 100–200–500–1,000 µL of the secondary standard solution.

When the calibration is satisfactory, the samples can be analyzed as previously described (Sections 8.3.2 to 8.3.4). Even though turbidity of the samples is low, it is important to centrifuge them to remove particulate phosphate that may react like dissolved phosphate (Kérouel and Aminot, 1987). Once phosphate is measured in the samples, replace normal reagents by RIB reagents and determine the sample RIBs. Using AA3 lamp colorimeters, the RIB is usually around 0.10–0.13 µmol L^{-1} at a salinity of 35. It may be lower, especially with an LED light source.

Each laboratory should test for the presence of a salt effect if required (Sections 8.8.1 and 8.4.6).

The calculation sequence is described in Section 8.4.1. Under optimal working conditions, the RIB correction is the only one that has to be applied to raw concentrations. In estuaries, correct for the salt effect if required.

8.8.5   METHOD PERFORMANCE

Figure 8.17 shows a typical record of calibration standards prepared in nutrient-depleted seawater (salinity ~35), with ultrapure water as the baseline.

Under normal conditions, the signal stability is better than 0.1% at levels of 1 µmol L^{-1}, resulting in a resolution of about 0.001 µmol L^{-1} of phosphate in the range 0–1 µmol L^{-1}.

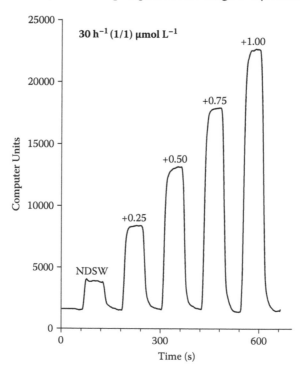

FIGURE 8.17 Typical record of a series of calibration standards for phosphate in nutrient-depleted seawater (NDSW). Concentration of NDSW here is below detection limit of phosphate, and thus equal to the RIB. AA3 colorimeter settings as described in the text.

Method precision was assessed using various reference materials prepared for international inter-comparison programs (Aminot and Kérouel, 1995, 1997). Within the range of most natural surface seawater concentrations (<1 µmol L^{-1}), the repeatability is about 1 nmol L^{-1}. The long-term reproducibility is actually better than the 0.02 µmol L^{-1} obtained in the programs mentioned since it was altered by slow phosphate release from the bottle walls.

From repeatability data, the detection limit, calculated according to Taylor (1990; three times the standard deviation of the blank or a very low concentration), is 2–3 nmol L^{-1}.

8.9 DETERMINATION OF SILICATE

8.9.1 Overall Description of the Method

The β-silicomolybdic complex formed by silicate with molybdate in acid conditions (pH ~1.5) is then reduced by ascorbic acid to a blue compound. Transformation of the beta complex into the alpha form is prevented if the pH < 1.8. Reaction conditions have been widely described in the literature (Aminot and Kérouel, 2004). The present method is mainly based on Truesdale and Smith's work (1976), with a slightly lower pH (~1.6) to accelerate the reaction. Phosphate forms the phosphomolybdic complex, but it is decomposed by oxalic acid under these reaction conditions. Sulfuric acid is added to the oxalic acid, in contrast with Hansen and Koroleff (1999), but in agreement with other authors (Strickland and Parsons, 1972; Tréguer and Le Corre, 1975; Truesdale and Smith, 1976), since it stabilizes the baseline after a seawater sample. The high ascorbic acid concentration allows 98% reduction in 1 minute; thus, 3-minute delay coils have been used instead of a heating bath. The tenfold dilution of the sample enables the whole range of seawater concentrations (0–200 µmol L^{-1}) to be measured, and additionally reduces the salt effect, the RIB, and the effect of high alkalinity.

Silicate is measured at a wavelength of 820 nm. Depending on colorimeter design, such as light path and signal amplification, the range is linear from 0.01 to 200 µmol L^{-1} of Si (detection limit ~ 0.03 µmol L^{-1}). There is no interference in unpolluted seawater (Aminot and Kérouel, 2004). The response is about 4% lower in sea- than in fresh water (linear effect).

8.9.2 Specific Contamination and Storage Problems

General sampling, subsampling, and storage procedures are described in Section 8.3, but silicate determination requires some specific precautions, as follows.

Contamination of seawater samples by silicate occurs mainly from contact with glassware (containers, volumetric flasks, filtration units, and glass fiber [GF] filters) and from particulate biogenic silica dissolution. Dissolution from glassware increases as a function of salinity; hence, standards in seawater prepared in classical glass volumetric flasks run the risk of being slightly overestimated. Removal of biogenic silica by filtration should be performed through polymer membranes only (not GF filters). Contamination of the baseline water may result from storage of the water in a glass container or release by aged demineralization resins.

Concerning storage, freezing is reliable only for oceanic water, provided the concentration does not exceed 120 µmol L^{-1} (Dore et al., 1996). Indeed, freezing causes polymerization of silicate, in proportion that increases with decreasing salinity (Burton et al., 1970; Macdonald and McLaughlin, 1982; Macdonald et al., 1986). On thawing, depolymerization slowly occurs and samples should be maintained at ambient temperature for at least 1 hour if salinity S > 33 and up to 24 hours in the S range ~20–33. For S < 20, depolymerization is often incomplete, in which case nonconservative mixing of silicate in estuaries might be erroneously assumed. Therefore, coastal and estuarine waters should be filtered, just refrigerated, and analyzed preferably within 2 weeks.

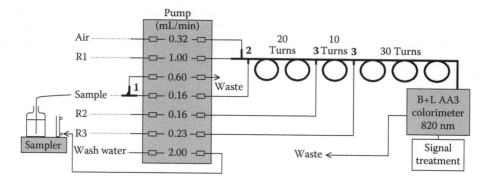

FIGURE 8.18 Standard manifold for silicate determination: 1 = microflow debubbler (same as 1 in Figure 8.8); 2 = injector 116-0489-01; 3 = A10 injector.

8.9.3 MANIFOLD AND ANALYTICAL CONDITIONS

As shown in Figure 8.18, the intersample bubble is removed just before the sample pump tube to prevent an ISAC effect (Section 8.2.2). The relatively high debubbling flow helps to flush the sample line and debubbler. The sample probe diameter should be at least 1 mm. The sampler wash station is fed with ultrapure water stored in a plastic bottle and periodically checked against Si-free water (Section 8.9.4). If another baseline matrix must be used, check its silicate content.

The characteristics and settings of the AA3 colorimeter used for the presented data are: standard 820 nm filter (bandpass 20 nm) and standard krypton lamp or LED light source; standard detector; 1 cm light path; gain = 250, smoothing factor = 4.

A sampling rate of 30 h^{-1} and a sample:wash ratio of 1:1 provide the highest precision. Higher frequencies are suitable for routine analysis. No particular precautions are required.

8.9.4 REAGENTS AND STANDARD SOLUTIONS

The reagents are stored in plastic bottles unless stated otherwise.

Si-free water: Because of its low ionization rate, silicate is released in ultrapure water by aged deionization resins, even though other nutrients are 100% retained. Water distilled using a metal unit and collected in plastic is the only recourse to produce Si-free water to periodically check the ultrapure water currently used. If the ultrapure water is Si contaminated, replace the cartridges of the water purification unit as soon as possible.

Sulfuric acid stock solution: Same as for phosphate determination (Section 8.8.3).

Ammonium molybdate stock solution: Same as for phosphate determination (Section 8.8.3).

Reagent R1: In a 500 mL plastic (e.g., HDPE) bottle, add 12 mL of sulfuric acid stock solution. Add 120 mL of molybdate stock solution, dilute to 500 mL with ultrapure water, and mix. Prepare daily. If a precipitate appears on the bottle wall, it can easily be removed with an alkaline solution (e.g., NaOH 0.1 mol L^{-1}).

Reagent R2: In a PTFE bottle, add 800 mL of ultrapure water. Under continuous mixing, slowly add 100 mL of concentrated sulfuric acid (δ = 1.84 kg L^{-1}). Add 30 g of oxalic acid dihydrate ((COOH)$_2$, 2H$_2$O; M = 126.1 g mol^{-1}), stir to dissolve, let cool, dilute further to 1,000 mL, and mix. Store at ambient temperature. This reagent is stable for 1 month.

Reagent R3: Dissolve 7.5 g of ascorbic acid (C$_6$H$_8$O$_6$; M = 176 g mol^{-1}) in 250 mL of ultrapure water. Add 0.5 mL of the wetting agent Aerosol 22 (Sigma A-9753), or alternatively 200 mg of pure-grade sodium dodecyl sulfate. Prepare daily.

RIB reagents: Replace reagent R1 with sulfuric acid 0.036 mol L^{-1} (12 mL of the H$_2$SO$_4$ stock solution diluted to 500 mL with ultrapure water). Use reagent R2 and R3 unchanged.

Silicate calibration standard (5,000 µmol L^{-1}): Use analytical-grade sodium hexafluosilicate (Na$_2$SiF$_6$; M = 188.06 g mol^{-1}) in a fine powder of purity ≥ 99% (e.g., Carlo Erba 480005 or Fluka 71596). If its purity differs from 100% but is certified, increase the mass to be weighed proportionally.

Dry sodium hexafluorosilicate (105°C, 1 hour), then let it cool in a desiccator. Weigh 940.3 mg and transfer it using ultrapure water into a 1,000 mL plastic volumetric flask. Add about 800 mL of ultrapure water and leave under magnetic stirring to ensure complete dissolution (up to several hours at ~20°C). Remove the stirrer magnet with care while properly rinsing it, then adjust the volume, mix the solution, and transfer it to a plastic bottle. In this solution, 1 mL contains 5 µmol silicate. In a tightly closed bottle, this standard is stable for several years at ambient temperature (Aminot and Kérouel, 1996).

8.9.5 PROTOCOL

Start the AutoAnalyzer and follow the general instructions listed in Section 8.3.4. Take into account that analysis may be postponed if samples have been frozen (Section 8.9.2).

Prepare a series of calibration standards, corresponding to expected concentrations, by diluting the silicate standard (5000 µmol L^{-1}) in NDSW (Section 8.4.3) or in ultrapure water. As an example, the typical range 0.5–1–2–5–10–20 µmol L^{-1}, prepared in 500 mL plastic volumetric flasks, requires 50–100–200–500–1,000–2,000 µL of the standard solution. Note that, for estuarine samples, calibration in water with a salinity of 17–18 will restrict the salt effect to within a range of ±2%.

When the calibration is satisfactory, the samples can be analyzed as previously described (Sections 8.3.2 to 8.3.4). If samples have been frozen, they should be kept at ambient temperature for some time before analysis to allow depolymerization to occur (Section 8.9.2). Once silicate is measured in the samples, replace normal reagents by RIB reagents and determine the sample RIBs. Using AA3 lamp colorimeters, RIBs usually stand around 0.07–0.10 µmol L^{-1} at a salinity of 35. It may be lower with an LED light source. If required, the salt effect should be determined by each laboratory, then checked from time to time (Section 8.4.6).

The calculation sequence is described in Section 8.4.1. Under optimized working conditions, results have to be corrected for the RIB only (if accuracy better than 0.1 µmol L^{-1} is required). For estuarine samples, a salt effect correction should be considered.

8.9.6 METHOD PERFORMANCE

Figure 8.19 shows a typical record of calibration standards prepared in nutrient-depleted seawater (salinity ~ 35), with ultrapure water as the baseline. Note that silicate concentrations in NDSW often remain between a few tenths and several micromoles per liter.

Under normal conditions the signal stability is better than ~0.15% at the level 10 µmol L^{-1}, corresponding to a resolution of about 0.02 µmol L^{-1} of silicate in the range 0–10 µmol L^{-1}.

Method precision was assessed using reference materials prepared for a French monitoring network intercomparison program (Daniel et al., 2006; Daniel and Kérouel, 2007). The intersample repeatability, assessed from different bottles of the same sample, was 0.005–0.01 µmol L^{-1} at a level of 2 µmol L^{-1} and 0.02–0.05 µmol L^{-1} between 12 and 17 µmol L^{-1}. Reproducibility at the same levels (which includes calibration variability over 3-4 months) was ~0.02 and ~0.07 µmol L^{-1}, respectively.

From the repeatability data, the detection limit, calculated according to Taylor (1990; three times the standard deviation of the blank or a very low concentration), is below 0.03 µmol L^{-1}. No attempt was made to reach a lower detection limit since silicate concentrations in marine waters are rarely below a few tenths of a micromole per liter.

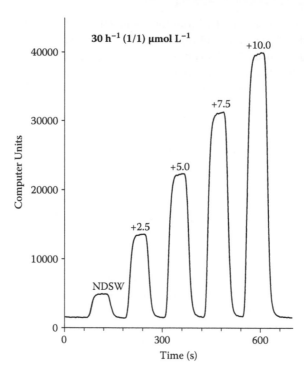

FIGURE 8.19 Typical record of a series of calibration standards for silicate in nutrient-depleted seawater (NDSW). AA3 colorimeter settings as described in the text.

REFERENCES

Alvarez-Salgado, X. A., F. Fraga, and F. F. Pérez. 1992. Determination of nutrient salts by automatic methods both in seawater and brackish water: The phosphate blank. *Marine Chemistry* 39:311–19.

Aminot, A. 1995. Quality assurance of pre-determination steps for dissolved nutrients in marine samples. In *Quality assurance in environmental monitoring, sampling and sample pretreament*, ed. P. Quevauviller. Weinheim, Germany: Wiley-VCH.

Aminot, A., and R. Kérouel. 1995. Reference material for nutrients in seawater: Stability of nitrate, nitrite, ammonia and phosphate in autoclaved samples. *Marine Chemistry* 49:221–32.

Aminot, A., and R. Kérouel. 1996. Stability and preservation of primary calibration solutions of nutrients. *Marine Chemistry* 52:173–81.

Aminot, A., and R. Kérouel. 1997. Reference material for nutrients for the QUASIMEME laboratory performance studies 1993–1996. *Marine Pollution Bulletin* 35:78–83.

Aminot, A., and R. Kérouel. 1998. Pasteurisation as an alternative method for preservation of nitrate and nitrite in seawater samples. *Marine Chemistry* 61:203–8.

Aminot, A., and R. Kérouel. 2004. *Hydrologie des écosystems marins: paramètres et analyses*. France: Editions Ifremer.

Aminot, A., and R. Kérouel. 2007. *Dosage automatique des nutriments dans les eaux marines*. France: Editions Ifremer.

Aminot, A., and D. S. Kirkwood. 1995. Report on the results of the fifth ICES intercomparison exercise for nutrients in sea water. *ICES Cooperative Research Report* 213:79.

Aminot, A., D. S. Kirkwood, and S. Carlberg. 1997. The QUASIMEME laboratory performance study (1993–1995): Overview of the nutrient section. *Marine Pollution Bulletin* 35:28–41.

Avanzino, R. J., and V. C. Kennedy. 1993. Long-term storage of stream water samples for dissolved orthophosphate, nitrate plus nitrite, and ammonia analysis. *Water Resources Research* 29:3357–62.

Bendschneider, K., and R. J. Robinson. 1952. A new spectrophotometric method for the determination of nitrite in sea water. *Journal of Marine Research* 11:87–96.

Burton, J. D., T. M. Leatherland, and P. S. Liss. 1970. The reactivity of dissolved silicon in some natural waters. *Limnology and Oceanography* 15:472–76.

Clementson L. A., and S. E. Wayte. 1992. The effect of storage of open-ocean sea water samples on the concentration of dissolved phosphate and nitrate. *Water Research* 26:1171–76.

Daniel, A., and R. Kérouel. 2007. Rapport de synthèse de l'essai interlaboratoire pour la mesure des sels nutritifs en milieu marin—Essai du 27/02/07. IFREMER report, Dyneco/Pelagos/07.04.

Daniel, A., R. Kérouel, A. Aminot, and A. Youenou. 2006. Rapport de synthèse de l'essai interlaboratoire pour la mesure des sels nutritifs en milieu marin—Essai du 14/02/06. IFREMER report, Dyneco/Pelagos/06.03.

Dore, J. E., T. Houlihan, D. V. Hebel, G. Tien, L. Tupas, and D. M. Karl. 1996. Freezing as a method of sample preservation for the analysis of dissolved inorganic nutrients in seawater. *Marine Chemistry* 53:173–85.

Drummond, L., and W. Maher. 1995. Determination of phosphorus in aqueous solution via formation of the phosphoantimonylmolybdenum blue complex: Re-examination of optimum conditions for the analysis of phosphate. *Analytica Chimica Acta* 302:69–74.

Froelich, P. N., and M. E. Q. Pilson. 1978. Systematic absorbance errors with Technicon Autoanalyzer II colorimeters. *Water Research* 12:599–603.

Garside, C. 1982. A chemiluminescence technique for the determination of nanomolar concentrations of nitrate and nitrite in seawater. *Marine Chemistry* 11:159–67.

Garside, C. 1993. Nitrate reductor efficiency as an error source in seawater analysis. *Marine Chemistry* 44:25–30.

Genfa, Z., and P. K. Dasgupta. 1989. Fluorometric measurement of aqueous ammonium ion in a flow injection system. *Analytical Chemistry* 61:408–12.

Gordon, L. I., J. C. Jennings Jr., A. A. Ross, and J. M. Krest. 1993. A suggested protocol for continuous flow automated analysis of seawater nutrients (phosphate, nitrate, nitrite and silicic acid) in the WOCE Hydrographic Program and the Joint Global Ocean Fluxes Study. Methods Manual WHPO 91-1, WOCE Hydrographic Program Office.

Hansen, H. P., and K. Grasshoff. 1983. Automated chemical analysis. In *Methods of seawater analysis*, ed. K. Grasshoff, M. Ehrhardt, and K. Kremling. Weinheim, Germany: Verlag Chemie.

Hansen, H. P., and F. Koroleff. 1999. Determination of nutrients. In *Methods of seawater analysis*, ed. K. Grasshoff, K. Kremling, and M. Ehrhardt. Weinheim, Germany: Wiley-VCH.

Karl, D. M., and G. Tien. 1992. MAGIC: A sensitive and precise method for measuring dissolved phosphorus in aquatic environments. *Limnology and Oceanography* 37:105–16.

Kérouel, R., and A. Aminot. 1987. Procédure optimisée hors-contaminations pour l'analyse des éléments nutritifs dissous dans l'eau de mer. *Marine Environmental Research* 22:19–32.

Kérouel, R., and A. Aminot. 1997. Fluorimetric determination of ammonia in sea and estuarine waters by direct segmented flow analysis. *Marine Chemistry* 57:265–75.

Kirkwood, D.S. 1992. Stability of solution of nutrient salts during storage. *Marine Chemistry* 38: 151–164.

Kirkwood, D., A. Aminot, and M. Perttilä. 1991. ICES report on the results of the fourth intercomparison exercise for nutrients in sea water. *ICES Cooperative Research Report* 174:83.

Li, Q. P., J. Z. Zhang, F. J. Millero, and D. A. Hansell. 2005. Continuous colorimetric determination of trace ammonium in seawater with a long-path liquid waveguide capillary cell. *Marine Chemistry* 96:73–85.

Loder, T. C., and P. M. Glibert. 1977. *Blank and salinity corrections for automated nutrient analysis of estuarine and sea waters*. UNH Sea Grant UNH-5G-JR-101 and WHOI 3897.

Macdonald, R. W., and F. A. McLaughlin. 1982. The effect of storage by freezing on dissolved inorganic phosphate, nitrate and reacyive silicate for samples from coastal and estuarine waters. *Water Research* 16:95–104.

Macdonald, R. W., F. A. McLaughlin, and C. S. Wong. 1986. The storage of reactive silicate samples by freezing. *Limnology and Oceanography* 31:1139–42.

Mikuska, P., Z. Vecera, and Z. Zdrahal. 1995. Flow-injection chemiluminescence determination of ultra low concentrations of nitrite in water. *Analytica Chimica Acta* 316:261–68.

Murphy, J., and J. P. Riley. 1962. A modified single solution method for the determination of phosphate in natural waters. *Analytica Chimica Acta* 27:31–36.

Nydahl, F. 1976. On the optimum conditions for the reduction of nitrate to nitrite by cadmium. *Talanta* 23:349–57.

Olson, R. J. 1980. Phosphate interference in the cadmium reduction analysis of nitrate. *Limnology and Oceanography* 24:758–60.

Pai, S. C., S. W. Chung, T. Y. Ho, and Y. J. Tsau. 1996. Determination of nano-molar levels of nitrite in natural water by spectrophotometry after pre-concentration using Sep-Pak C(18) cartridge. *International Journal of Environmental Analytical Chemistry* 62:175–89.

Pai, S. C., C. C. Yang, and J. P. Riley. 1990. Effects of acidity and molybdate concentration on the kinetics of the formation of the phosphoantimonylmolybdenum blue complex. *Analytica Chimica Acta* 229:115–20.

Raimbault, P., G. Slawyk, B. Coste, and J. Fry. 1990. Feasibility of using an automated colorimetric procedure for the determination of seawater nitrate in the 0 to 100 nM range: Examples from field and culture. *Marine Biology* 104:347–51.

Roth, M. 1971. Fluorescence reaction for amino acids. *Analytical Chemistry* 43:880–82.

Sjösten, A., and S. Blomqvist. 1997. Influence of phosphate concentration and reaction temperature when using the molybdenum blue method for determination of phosphate in water. *Water Research* 31:1818–23.

Stainton, M. P. 1974. Simple, efficient reduction column for use in the automated determination of nitrate in water. *Analytical Chemistry* 46:1616.

Strickland, J. D. H., and T. R. Parsons. 1972. *A practical handbook of seawater analysis*. Fisheries Research Board of Canada.

Taylor, J. K. 1990. *Quality assurance of chemical measurements*. Boca Raton, FL: Lewis Publishers.

Tréguer, P., and P. Le Corre. 1975. *Manuel d'analyse des sels nutritifs dans l'eau de mer*. Brest, France: Université de Bretagne Occidentale.

Truesdale, V. W., and C. J. Smith. 1976. The automatic determination of silicate dissolved in natural fresh water by means of procedures involving the use of either α- or β-molybdosilicic acid. *Analyst* 101:19–31.

Whitledge, T. E., S. C. Malloy, C. J. Patton, and C. D. Warick. 1981. *Automated nutrient analyses in seawater*. BNL Report 51398, Brookhaven National Laboratory, Upton, NJ.

Wood, E. D., F. A. J. Armstrong, and F. A. Richards. 1967. Determination of nitrate in sea water by cadmium copper reduction to nitrite. *Journal of the Marine Biological Association of U.K.* 47:23–31.

Yao, W., R. H. Byrne, and R. D. Waterbury. 1998. Determination of nanomolar concentrations of nitrite and nitrate in natural waters using long path length absorbance spectroscopy. *Environmental Science and Technology* 32:2646–49.

Zhang, J. Z. 2000. Shipboard automated determination of trace concentrations of nitrite and nitrate in oligotrophic water by gas-segmented continuous flow analysis with a liquid waveguide capillary flow cell. *Deep-Sea Research Part I* 47:1157–71.

9 Dissolved Organic and Particulate Nitrogen and Phosphorous

Gerhard Kattner

CONTENTS

9.1 INTRODUCTION

In all ecosystems nitrogen and phosphorous are continuously cycled. These two elements are the major nutrients for plant growth on earth and similarly in the marine environment for algae. The inorganic forms become part of organic molecules during primary production and through growth of organisms, forming the particulate material together with nonliving matter. During decay and active release, dissolved organic material is formed, some of which becomes highly recalcitrant, but finally a major fraction is remineralized to inorganic carbon, nitrogen, and phosphorus. Mineralization rates differ between the elements. A detailed study has shown high spatial and seasonal variability of particulate and dissolved organic matter in the North Sea from near-shore to open oceanic waters (Brockmann and Kattner, 1997). An excellent overview of the worldwide distribution of marine dissolved organic matter (DOM) has been compiled by Hansell and Carlson (2002). Dissolved organic nitrogen and phosphorus are still often missing from budget and global model calculations and have considerable shortcomings in the formulation of organic matter cycling in models (e.g., Toggweiler, 1989; Najjar et al., 2007), since methods are less well developed and less reliable than those for dissolved organic carbon (DOC; see Chapter 2).

Although the composition of organic molecules in living cells and organisms is quite well known, the molecular composition of DOM is still widely unknown. A small portion of the dissolved organic nitrogen, usually less than 10%, is composed of hydrolyzable combined amino acids. The remaining part is almost completely unidentified. Even less is known about dissolved organic phosphorous (DOP). The dominant compound classes of marine DOP in the high molecular weight fraction have been determined by nuclear magnetic resonance (NMR) (e.g., Kolowith et al., 2001), but this is still far away from a characterization at the molecular level.

Total organic nitrogen and phosphorous comprise the particulate plus the dissolved fraction. Differentiation between the particulate and dissolved form is usually achieved by filtration through glass fiber filters. The filter with the residue is used for the determination of the particulate fraction and the filtrate for the dissolved fraction.

The measurement of particulate organic nitrogen (PON) is a well-established method and is usually combined with the measurement of particulate organic carbon (POC; see Chapter 2) using a carbon hydrogen nitrogen (CHN) analyzer. The term *total particulate nitrogen* (TPN or PN) is more correct because methods do not differentiate between organic and inorganic nitrogen, but the latter is negligible and can be ignored in the marine particulate matter. The same holds for particulate organic phosphorous (POP) since methods determine the amount of *total particulate phosphorous* (TPP or PP). The most common abbreviations are, however, still PON and POP. The determination of TPP is more complicated, and various methods have been published. Inorganic phosphorous, as part of the particulate phosphorous, may play a role in freshwater and estuarine regions, where particulate matter is dominated by mineral sediments with a low contribution from organic matter. Phosphate is very particle active and adsorbs onto suspended matter and sediments (Froelich et al., 1982). In open oceanic regions, however, inorganic phosphorous contributes insignificantly to TPP (Kolowith et al., 2001; Faul et al., 2005).

Dissolved organic nitrogen (DON) and phosphorous (DOP) are much more complicated to determine because (1) the conversion of the partially recalcitrant compounds to inorganic nitrogen or phosphate must be complete, and (2) in almost all methods for the measurement of the organic fraction, inorganic nitrogen and phosphorous are included. These methods determine the amount of total dissolved nitrogen (TDN) or phosphorous (TDP), if samples are filtered. Thus, DON and DOP are always the difference of TDN minus total dissolved inorganic nitrogen (DIN) or of TDP minus total dissolved inorganic phosphorous (DIP) (Figure 9.1). The analyst therefore also needs a very good and reliable method for the determination of inorganic nutrients (see Chapter 8). If the method applied allows, it is important to perform the determination of the organic and inorganic fractions on the same analytical system, for example, an autoanalyzer, in the same laboratory, and at the same time to reduce the analytical error, which comprises the error of the TDN/TDP and the nutrient determinations. At very low concentrations of DON and DOP, which are usually found in oligotrophic waters, in the deep sea and in many surface waters, the precision of the DON and DOP measurements must be extremely high, because small errors in the measurements of the organic and inorganic components create a very large overall error. The errors become even larger if inorganic nitrogen and phosphorous are the major constituents.

The concentration of DOM is determined either by the high-temperature catalytic oxidation (HTCO) or by wet oxidation with persulfate or UV oxidation alone or in combination with persulfate oxidation. Several intercalibration studies have been performed during the last 10 to 20 years for establishment of the best method for the determination of DOC and DON with special focus on DOC (Hedges et al., 1993; Williams, 1993; see also Chapter 2 for DOC). Whereas DOC can be measured with good precision by using international standard material, DON measurements are still far from this status and need further intercalibration exercises (Sharp et al., 2002). Until now there has been no agreement on reference material for DON. The situation for DOP has been still worse until now. There have been only small group attempts at intercalibration (e.g., HELCOM, 1991; Sharp, 2002).

The HTCO method and wet oxidation with persulfate are to date the most established methods for DON. However, phosphorous can only be determined with the wet oxidation method. Several variations of both methods for the determination of the dissolved organic fractions in seawater have been described. Raimbault and Slawyk (1999), for example, applied the alkaline wet oxidation method with persulfate to measure dissolved organic carbon, nitrogen, and phosphorous

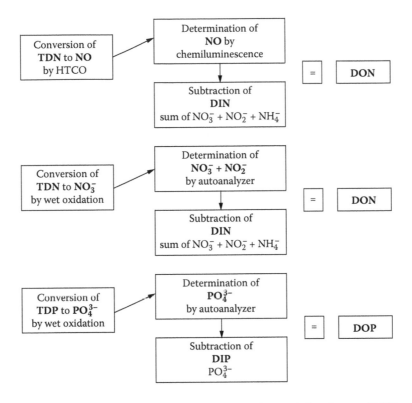

FIGURE 9.1 Steps for calculating dissolved organic nitrogen (DON) and phosphorous (DOP).

simultaneously using an autoanalyzer system. While the precision is good for coastal waters, the precision for oligotrophic oceanic water is unsatisfactory (10%–25%), especially for DOP, since values are often around the detection limit. For all phosphorous fractions, from particulate phosphorous to inorganic phosphate, the method of choice for the determination is still the colorimetric molybdenum-blue method introduced by Murphy and Riley (1962).

9.2 PARTICULATE NITROGEN

Total particulate nitrogen (TPN) is the amount of nitrogen in suspended material, which is composed of living cells, cell fragments, detritus, aggregates, and other particles and some colloids usually retained on glass fiber filters with a nominal pore size of 0.7 μm, or on 0.2 μm membrane filters. The molar ratios of carbon to nitrogen vary widely from low ratios of living material (5 to 8) to high ratios in dead material or land-derived particulate organic matter in near-shore areas (8 to >20) (e.g., Biddanda and Benner, 1997; Verity et al., 2000).

9.2.1 OVERVIEW OF METHODS

TPN is mostly determined together with carbon by high-temperature combustion using a CHN or CN analyzer. If the concentration of nitrogen is small compared to carbon, it might be necessary to repeat the determination by measuring only TPN. This, however, needs a second filter with particulate material.

The determination of TPN is also possible by the digestion via persulfate wet oxidation of suspended matter on filters (Raimbault et al., 1991). The use of unfiltered seawater samples for the measurement of TPN (digestion procedure outline in Section 9.4.1.2) cannot be recommended because of too many uncertainties, which are (1) the small volume of sample used for the determination of total nitrogen being unrepresentative for TPN, and (2) the subtraction of the amount of dissolved nitrogen, which is difficult to determine (see Section 9.4) and of course also requires filtration of the samples.

9.2.2 SAMPLING AND ANALYSIS

Seawater samples for the determination of TPN are usually taken with Niskin bottles. For sample collection from the Niskin bottles, clean (at best new) polyethylene bottles should be used. Alternatively, used polyethylene bottles have to be carefully cleaned in 1 M HCl and rinsed several times with ultrapure water. Ultrapure water with <2 µmol TOC L^{-1} is produced with a Milli-Q Gradient A10 system (Millipore) and is used for all methods. Samples are carefully poured into the polyethylene bottles, which have been washed three times with the sample just before use. Samples are filtered through glass fiber filters, usually Whatman GF/F with a nominal pore size of about 0.7 µm (e.g., UNESCO, 1994). Filters have to be precombusted for 4–5 hours at 450°C–500°C. Several filters can be stored together in aluminum foil. The amount of seawater to be filtered depends on the concentration of the suspended material in the sample and can range from 50 ml in surface water to several liters in deep water. A slight vacuum is used for filtration ($20*10^3$–$30*10^3$ Pa, i.e., residual pressure below the filter $70*10^3$–$80*10^3$ Pa). This is important to avoid destruction of the cells. Filters are not allowed to dry out during filtration. Two methods to store the filters are recommended: (1) filters are stored deep-frozen, preferably at –80°C, until analysis, or (2) filters are dried at 60°C immediately after sampling and stored frozen at –20°C until analysis. Several unused filters are stored to determine blank values.

The usually acid-treated filters (compare Chapter 2 for POC) are transferred into precleaned tin vials and compressed into small pellets. Calibration is performed with a standard of acetanilide, which has a C:N ratio of 8, similar to the average ratio of naturally occurring material. The standard is weighed directly into the tin vials (usually 100–200 µg). Pellets and standards are automatically introduced onto the combustion column of a CHN analyzer. The combustion temperature is 1,050°C. This temperature rises to 1,800°C during the flash combustion of the tin vial. For further details refer to POC (Chapter 2).

CHN analyzers are available from various companies. There are small differences in the instrumental properties, but the method is widely independent and can easily be adapted to many instruments. All instruments are equipped with a column reactor for combustion, a reduction column, a separation system comparable to gas chromatography, and a detector unit. With the oxidizing and catalyzing reagents, nitrogen compounds are converted to nitrogen and nitrogen oxides, and the latter to nitrogen by elemental copper in the reduction column. The end products, carbon dioxide and nitrogen, are separated by gas chromatography (e.g., Ehrhardt and Koeve, 1999). Peaks are automatically integrated. The data have to be corrected for the blank, which is determined with the unused filters collected during sampling. The largest possible error is the contamination of the filter itself or the tin vials. Instrumental errors are small, being <1% within the calibration range.

9.3 PARTICULATE PHOSPHOROUS

9.3.1 OVERVIEW OF METHODS

Total particulate phosphorous (TPP) is analyzed by the digestion of suspended material collected on filters. It is also possible to oxidize the sample with and without filtration. The difference between the two measurements gives the amount of TPP. These data are, however, the most unreliable, because the amount of particles in a 50 ml sample, or with even smaller sample volumes in some other protocols, is usually not representative for particulate material.

The more precise method is to filter the sample and determine TPP in the residue. The filter with the residue is either dried and combusted at high temperature (450°C–550°C; Solorzano and Sharp, 1980a) or decomposed and oxidized in concentrated sulfuric acid (Kattner and Brockmann, 1980).

Both procedures oxidize TPP to phosphate. A manual or automated method for the determination of the resulting ortho-phosphate can be applied, which is based on the colorimetric molybdenum-blue method by Murphy and Riley (1962). The recommended method, using sulfuric acids for digestion, can be modified to the individual lab method used for the phosphate determination, but it must be ensured that the acid concentration in the digested TPP sample is the same as that in the phosphate method. The major difference between the methods for TPP and inorganic phosphate (see Chapter 8) is that the *sample* contains the sulfuric acid concentration necessary for the phosphate determination instead of the *reagent*, which is used for the inorganic phosphate determination.

9.3.2 SAMPLING AND ANALYSIS

The samples are collected and stored as described for particulate carbon and nitrogen. The frozen glass fiber filter (2.4 cm diameter or smaller) with the residue is placed in a glass vial in a digester block and oxidized with sulfuric acid. For this, 2.15 ml of concentrated sulfuric acid (95%–97%) and 0.5 ml of hydrogen peroxide are added to the filter, so that the filter is fully covered by the digestion reagent. The samples are heated to 275°C for 2 hours uncovered under a fume hood. The vials are removed from the digester block, and after cooling, 38 ml of ultrapure water is added to ensure a 0.91 M H_2SO_4 solution necessary for the following ortho-phosphate measurement (about 0.15 ml H_2SO_4 is lost during the digestion procedure). The amount of acid and water can be adjusted for a different filter size, which may increase or decrease the sensitivity of the determination, depending on the dilution.

The oxidized residue can now be analyzed with a manual or automatic method based on the phospho-molybdenum complex method (Kattner and Brockmann, 1980). Particles of the glass fiber filter, which is destroyed during the digestion procedure, settle out after some minutes, and the supernatant is transferred into the autoanalyzer cups. Remaining particles can be removed with a small quartz wadding filter behind the autoanalyzer needle; both must be made of glass or acid-resistant material because the samples are acidic.

The manifold for the automatic determination is shown in Figure 9.2. The method using an autoanalyzer requires the following reagents (numbers are the same as in Figure 9.2). Note that the sample already contains the acid necessary for the phosphate determination:

1. A solution of 6 g ammonium heptamolybdate tetrahydrate $((NH_4)_6Mo_7O_{24}\ 4H_2O)$ dissolved in 1,000 ml of ultrapure water.
2. A solution of 8 g ascorbic acid dissolved in 1,000 ml ultrapure water, and finally 5 ml sodium dodecyl sulfate solution (10 g in 100 ml ultrapure water) is added (as wetting agent; also 0.5 ml of aerosol22 can be used, but it lowers sensitivity of the method.

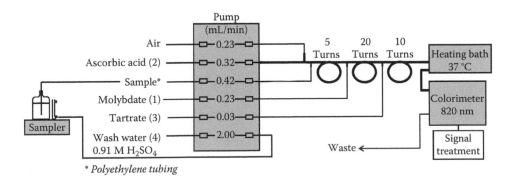

FIGURE 9.2 Manifold for the determination of total particulate phosphorous (TPP). (From Kattner, G., and U. H. Brockmann. 1980. *Fresenius Zeitschrift für Analytische Chemie* 301:15. With kind permission of Springer Science+Business Media.)

3. A solution of 1.0 g potassium antimony tartrate (K(SbO)C₄H₄O₆ H₂O) dissolved in 1,000 ml ultrapure water.
4. The washing solution is a 0.91 M H_2SO_4 solution consisting of 50 ml sulfuric acid (95%–97%) in 1,000 ml ultrapure water.

The standard phosphate solution is prepared from potassium dihydrogen phosphate (KH_2PO_4) in 0.91 M H_2SO_4 solution using a stock standard. The stock standard should be the same as for the nutrient determination. The measuring range is from 0.1–5 μmol P L^{-1}, which can be changed using a less sensitive method (Kattner and Brockmann, 1980). Due to the enrichment of the samples by filtration, the lower limit of the POP determination is about 0.01 μmol L^{-1}.

9.4 DISSOLVED ORGANIC NITROGEN

In the subsurface and deep ocean, concentrations of DON are reported from about 2 to 8 μmol N L^{-1}. In coastal and high-productivity regions DON values are usually much higher, often being the dominant component of TDN. A detailed table of worldwide concentrations of TDN and DON has been compiled by Bronk (2002). Oceanic DON values are low compared to the high nitrate concentrations at depths that exhibit, for example, values of more than 30 μmol L^{-1} in Antarctic waters. Thus, it is easy to understand that small inaccuracies in the determination of both total dissolved nitrogen and inorganic nitrogen will result in a large error. During a recent methods comparison for DON (Sharp et al., 2002) no differences were found between the UV, persulfate, and high-temperature catalytic oxidation (HTCO) methods. Five samples were distributed, and twenty-nine sets of TDN analyses were performed (Figure 9.3). The community comparability (precision) for analyzing TDN was in the 8%–28% range (coefficient of variation). However, if DIN was subtracted from TDN (the same independent set of DIN values), the comparability was only in the range of 19%–46%. The reproducibility of the deep-sea sample was the worst, because of the high nitrate and low DON values. If individually determined DIN data are subtracted from TDN, the variability of the data might become even larger.

9.4.1 OVERVIEW OF METHODS

The determination of DON has a long history, because it is an important component of the total nutrient concentration in natural waters. The major challenge for all DON methods is the

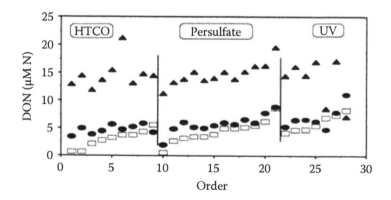

FIGURE 9.3 DON data from the intercalibration study compiled by Sharp et al. (2002). The analyses of three different samples using three different methods are compared. Order numbers belong to the participating laboratories. Samples are ordered in increasing concentrations; open squares are deep sea, closed circles are surface shelf, and closed triangles are lower estuary samples. (From Sharp et al. 2002. *Marine Chemistry* 78:171–84. With kind permission of Elsevier Limited.)

complete digestion of the often recalcitrant components dissolved in seawater. The huge number of organic compounds containing nitrogen has to be transferred into one final product, which can be easily, precisely, and accurately determined. A limited number of procedures, followed by various modifications, have been published, comprising UV oxidation, Kjeldahl digestion, wet oxidation with potassium peroxodisulfate, and the modern HTCO method. The UV and wet oxidation methods result in an oxidation of the dissolved nitrogen to nitrate (NO_3^-), which is measured after reduction as nitrite (NO_2^-) manually or with an autoanalyzer (see Chapter 8). The HTCO method converts all nitrogen to nitric oxide (NO), which is then measured by chemiluminescence detection (Badr et al., 2003).

Several intercalibrations have been performed comparing different methods and samples of various origins (Walsh, 1989; Hedges et al., 1993; Bronk et al., 2000; Sharp et al., 2002). However, it is still difficult to decide which method offers the most accurate and precise results. HTCO and wet oxidation with persulfate are recommended and presented for the determination of DON.

9.4.1.1 High-Temperature Catalytic Oxidation

Using the HTCO method, nitrogenous components are converted to NO at temperatures varying between 680°C and 720°C according to the manufacturer's instructions. NO is finally converted to the radical NO_2^* species that chemiluminesce upon decay to NO_2 and is determined by chemiluminescence detection. An advantage of this method is that TDN and DOC can be determined simultaneously from the same sample. This avoids additional work and thus contamination. A disadvantage is that the HTCO method for the determination of TDN and the method for nitrate determination are totally different. This might result in a lower precision because of additive errors of the two methods. This fact is controversially discussed and shows the importance of improving the TDN determinations. In addition, two sets of samples are needed, one in glass ampoules (stored at –20°C) and another for the nutrient determination in polyethylene bottles.

9.4.1.1.1 *Filtration and Analysis*

Filtration must be carried out carefully to avoid any contamination. Filtration devices should be made entirely of glass and carefully cleaned. The filtrate produced by filtering samples for particulate matter can be used. The filtrate is collected in a glass or plastic (if silicate is determined from the same filtrate) flask, and aliquots are taken for the various measurements of chemical constituents. For the TDN determination with the HTCO method, filtered samples (15 ml) are stored frozen at –20°C until analysis in sealed 20 ml glass ampoules (precombusted at 450°C–500°C for 5 hours). The volume stored depends on the size of the autosampler vials. Care must be taken that ampoules do not crack; they should be stored canted during freezing. Alternatively, samples can be acidified to pH 2 by adding 1 M HCl (1 ml HCl to 100 ml sample) and stored at 4°C in a fridge. Both storage methods can also be used for the determination of DOC (Badr et al., 2003; see Chapter 2). Another alternative is autoclaving of samples and storage in a fridge (Sharp et al., 2002), which is, however, not suitable for field sampling.

The nitrogen calibration stock solution (10 mmol N L^{-1}) is prepared by dissolving 25.275 g potassium nitrate (101.10 g mol^{-1}) in 250 ml acidified (0.05 M HCl) ultrapure water. From the nitrogen stock solution the calibration solution can be prepared (e.g., from 10 to 100 µmol N L^{-1}), which also needs to be acidified with 1 M HCl for conservation. In some instruments the stock standard is automatically diluted to the working standard solution. The calibration is external, and the calibration solution has to be dissolved depending on the expected total N concentrations of the samples.

Samples are filled into vials that have been well rinsed with ultrapure water and precombusted (at 450°C–500°C for 5 hours). An injection volume of 50 or 100 µL is recommended, but can vary depending on the system used. The oxidation quartz column in the TOC/TN analyzer contains a platinum mesh at the bottom, and is filled with 1 cm of quartz wool, 9 cm of platinum catalyst (e.g., Shimadzu, support: 5/64-inch platinum-impregnated (3 mass%) alumina balls, 2–2.5 mm diameter), and 1 cm of quartz wool on top to minimize destruction of the catalyst during sample injection and for collection of salt. After about 200 samples, which comprise about 600–800 injections, the catalyst has to be exchanged (the number of injections is dependent on the repetitions, which vary between three and five times, depending on the reproducibility). An indication to change the catalyst is when the shape of the peaks on the recording is asymmetric or with shoulder. The quartz column is cleaned with concentrated HCl and rinsed several times with ultrapure water. The catalyst is also put into concentrated HCl overnight and carefully washed with ultrapure water; this procedure is repeated three times. Thereafter, the catalyst is dried at 500°C overnight. Broken Al balls and catalyst dust have to be removed, for example, by sieving. The better method is to use a guided aluminum sheet (about 1 m length with a flange on each side) with a narrow tip, on which intact balls can roll down into a collecting flask, whereas broken balls and dust remain on the sheet.

As TDN and DOC are determined simultaneously, the samples are automatically acidified to pH 2 and purged with high-purity oxygen (5.0 grade) in the autosampler vials with a flow rate of 40 ml min^{-1}, or directly in the injection syringe. The samples are injected directly onto the oxidation column, which is heated to 680°C. The flow rate of the carrier gas (ultrapure oxygen, 5.0 grade) is 120 ml min^{-1}. Total nitrogen is quantified by the chemiluminescence detector. Before starting a run with samples, one run with at least five blanks of ultrapure water is required to equilibrate to the baseline of the TOC/TN analyzer. After every five samples one blank and one standard are measured for quality control. Every sample, standard, and blank is measured at least three times. The blank of ultrapure water is about 1.5 ± 0.3 μmol L^{-1}.

9.4.1.2 Wet Oxidation with Potassium Peroxodisulfate

The wet oxidation method with potassium peroxodisulfate for the simultaneous determination of nitrogen and phosphorous compounds was introduced by Koroleff on intercalibration workshops (Koroleff, 1977, 1983). The method has been modified by several analysts (e.g., Solorzano and Sharp, 1980b). Compared to the HTCO method, wet oxidation needs relatively simple instrumentation for the digestions, and the resulting nitrate and phosphate can be analyzed with flow analysis (AutoAnalyzer; see Chapter 8) or manually. The method can be used in the field, and it is possible to measure DON and PON as well as nutrients immediately with the same methods and directly after sampling, for example, during research cruises. This possibility may increase the precision of the DON determination because the same standards and the same analytical method are used. Another advantage of this method is that the digested sample can also be used for the determination of both organic nitrogen and phosphorous (Section 9.5). A disadvantage of the wet oxidation compared to the HTCO method is that the handling of the samples is more laborious, which may increase the risk of contamination.

The digestion procedure is described using a microwave or alternatively a presser cooker or an autoclave (the autoclave should be preferred over the cooker because of better temperature and pressure regulation).

9.4.1.2.1 *Sampling and Analysis*

For the wet oxidation method samples should be directly oxidized and measured after sampling and filtration. Alternatively, about 50 ml filtered samples (see Section 9.4.1.1.1) can be stored in 50 ml polyethylene bottles. The bottles must be new and unused and have to be carefully cleaned with the sample without any precleaning. One hundred fifty microliters of mercuric chloride solution (35 g HgCl$_2$ in 1,000 ml ultrapure water) is added, which corresponds to a final concentration of about 0.4 mM (105 μg HgCl$_2$ ml^{-1}) in the sample. The poisoned samples must be closed carefully, shaken, and stored at 4°C in the dark until analysis (as proposed by Kattner (1999) for storage of samples for nutrient determinations). This storage method allows the measurement of TDN and TDP as well as nutrients from the same sample. A disadvantage of this poisoning is that the cadmium reduction column used for the nitrate determination needs to be regularly checked and reactivated more frequently. Alternatively, samples may be frozen at $-20°C$ in polyethylene bottles, but our unpublished lab comparisons have shown that this method is less reliable than the poisoning with $HgCl_2$ of the samples. Poisoned samples have to be correctly disposed of after analysis.

For the digestion of the seawater sample, the digestion reagent is prepared by dissolving 15 g NaOH (sodium hydroxide, 0.375 M), 30 g H_3BO_3 (boric acid), and 50 g $K_2S_2O_8$ (potassium peroxodisulfate) in 1,000 ml ultrapure water. It is recommended that the reagent is freshly prepared each day (Eberlein and Kattner, 1987), but it may also be possible to store the reagent for some weeks in the dark, in a fridge ($<8°C$; Hansen and Koroleff, 1999).

For the digestion procedure a high-performance microwave digestion unit (MLS GmbH, Microwave Laboratory Systems, Germany) is used. The microwave Teflon vessels are cleaned at least twice with 10 ml of 0.1 M HCl for 30 minutes at $150°C$ in the microwave and rinsed after cooling with ultrapure water. The vessels must be carefully leak-proof sealed. Twenty-five milliliters of sample and 3 ml of digestion reagent are filled into the precleaned vessels and heated for 2.5 hours at $150°C$ (350 W; including heating-up time) under temperature-controlled conditions. If larger volumes are necessary, sample volume and reagent must be combined in the same ratio as above. This can be necessary for the manual determination of the resulting nitrate, or if TDN and TDP are determined from the same digested sample.

After complete oxidation, vessels are allowed to cool to room temperature until they are opened shortly before the manual or automatic measurement of the resulting nitrate. Losses of sample should be checked by weighing the vessels before and after oxidation. The pH value (about 9 at the beginning and 4–5 at the end) should be regularly checked in a subsample.

For the determination of blanks only the reagent (3 ml or the same volume as in the sample) is treated as the samples. After cooling, artificial seawater is added, prepared in ultrapure water (25 ml or the same volume as the sample).

Instead of the microwave, an autoclave or a pressure cooker (easier to handle onboard a research vessel, but temperature is usually only about $115°C$) can be used for the digestion procedure. Preferably, Teflon bottles should be used for the digestion, but glass bottles with Teflon sealing or polyethylene bottles are a more economical alternative. Polyethylene bottles, in particular, must be carefully checked for nitrogen and phosphorous contamination and also for deformation during the digestion. Bottles (precleaned with 0.1 M HCl for 30 minutes at $120°C$ in the autoclave or cooker) of 25, 50, or 100 ml volume are used. They are filled with the sample (not more than about half of the bottle volume), and the corresponding amount of the digestion reagent is added, mixed (see microwave digestion), carefully sealed, and cooked for 2 hours at $120°C$. Some analysts regard 30 minutes as sufficient, but a longer digestion period is recommended for a more complete oxidation. Blanks are prepared as for the microwave digestion.

Each sample and blank should be determined at least twice. An aliquot of the digested sample is used for the determination of nitrate (see Chapter 8), and the results represent TDN. For calibration, the same nitrate standards are used for the digested samples as for the inorganic nitrate determination (see Chapter 8). By subtracting the nitrate concentration in the undigested sample from the TDN concentration, the concentration of DON is obtained. The results must be corrected for dilution and blank. It is controversially discussed whether the digestion of the nitrate standards results in a more precise calibration than without digestion. I recommend calibration of DON with the same calibration procedure as applied for the nitrate determination, hence without digestion. Alternatively, organic nitrogen standards can be used, for example, disodium-EDTA. Triplicates of these standards have to be processed in the same manner as for the samples (Hansen and Koroleff, 1999). There is no clear improvement of the calibration by this more labor-intensive procedure. However, the oxidation efficiency of various nitrogen-containing compounds can be checked (Badr et al., 2003), which is certainly not fully representative because of the still unknown composition of DON (e.g., Benner, 2002).

9.5 DISSOLVED ORGANIC PHOSPHOROUS

Concentrations of DOP are low in the open ocean and usually higher in highly productive coastal regions (e.g., Karl and Björkman, 2002; van der Zee and Chou, 2005). Since the measurement of TDP includes the amount of inorganic phosphate, it is difficult to determine DOP accurately at low concentrations. The subtraction of phosphate from total dissolved phosphorous often results in values near zero or even below.

9.5.1 Overview of Methods

Several possible methods exist for the determination of DOP (reviewed by Karl and Björkman, 2002, but without detailed descriptions of the analytical procedures). These methods seem to be less developed and are less internationally compared than the methods for the determination of organic carbon and nitrogen. The use of the same oxidation methods as for nitrogen (UV and wet oxidation with persulfate) was introduced in the 1960s. Digestion and oxidation with high temperature with and without acid hydrolysis (HTC method; 450°C–500°C for 2 hours) for dried dissolved and particulate samples was proposed by Solorzano and Sharp (1980a), which is, however, quite laborious. For all digestion methods the final determination of the resulting phosphate is still based on the colorimetric molybdenum-blue method by Murphy and Riley (1962). Some small group comparison studies have been performed, showing that persulfate oxidation and the HTC method are best suited to fully oxidize seawater samples (Kérouel and Aminot, 1996; Monaghan and Ruttenberg, 1999). Since inorganic polyphosphates are also oxidized by this method, DOP might be marginally overestimated by the subtraction of ortho-phosphate from total dissolved phosphorous. In most natural waters and especially in open-ocean waters the concentrations of polyphosphates are extremely low, and thus this error is obviously far below the overall analytical error (for the determination of polyphosphates refer to Hansen and Koroleff, 1999).

The persulfate wet oxidation method is recommended because it requires only one digestion procedure to measure TDP (or DOP) and TDN (or DON) from the same sample. This allows the determination of both TDP and TDN immediately onboard a research vessel, if equipment for nutrient measurements is available. Another wet oxidation method is acid persulfate oxidation (Hansen and Koroleff, 1999).

9.5.1.1 Analysis

The digestion method described for the simultaneous determination of nitrogen and phosphorous is recommended. The reagents and digestion procedure are the same as described in Section 9.4.1.2. For calibration, potassium dihydrogen phosphate (KH_2PO_4), the same as for the ortho-phosphate determination, is recommended, without digestion. The determination of phosphate is given in Chapter 8. The results from the phosphate determination have to be corrected by the dilution and blank values. By subtracting the phosphate concentration in the undigested sample from the TDP concentration, the concentration of DOP is obtained.

REFERENCES

Badr, E.-S. A., E. P. Achterberg, A. D. Tappin, S. J. Hill, and C. B. Braungardt. 2003. Determination of dissolved organic nitrogen in natural waters using high-temperature catalytic oxidation. *TrAC Trends in Analytical Chemistry* 22:819–27.

Benner, R. 2002. Chemical composition and reactivity. In *Biogeochemistry of marine dissolved organic matter*, ed. D. A. Hansell and C. A. Carlson. San Diego: Academic Press, 59–90.

Biddanda, B., and R. Benner. 1997. Carbon, nitrogen, and carbohydrate fluxes during the production of particulate and dissolved organic matter by marine phytoplankton. *Limnology and Oceanography* 42:506–18.

Brockmann, U., and G. Kattner. 1997. Winter-to-summer changes of nutrients, dissolved and particulate organic material in the North Sea. *German Journal of Hydrography* 49:229–42.

Bronk, D. A. 2002. Dynamics of DON. In *Biogeochemistry of marine dissolved organic matter*, ed. D. A. Hansell and C. A. Carlson. San Diego: Academic Press, 153–247.

Bronk, D. A., M. W. Lomas, P. M. Glibert, K. J. Schukert, and M. P. Sanderson. 2000. Total dissolved nitrogen analysis: Comparisons between persulfate, UV, and high temperature oxidation methods. *Marine Chemistry* 69:163–78.

Eberlein, K., and G. Kattner. 1987. Automatic method for the determination of ortho-phosphate and total dissolved phosphorus in the marine environment. *Fresenius Zeitschrift für Analytische Chemie* 326:354–57.

Ehrhardt, M., and W. Koeve. 1999. Determination of particulate organic carbon and nitrogen. In *Methods of seawater analysis*, ed. K. Grasshoff, K. Kremling, and M. Ehrhardt, 437–44. Weinheim, Germany: Wiley-VCH Verlag.

Faul, K. L., A. Paytan, and M. L. Delaney. 2005. Phosphorus distribution in sinking oceanic particulate matter. *Marine Chemistry* 97:307–33.

Froelich, P. N., M. L. Bender, N. A. Luedtke, G. R. Heath, and T. De Vries. 1982. The marine phosphorus cycle. *American Journal of Science* 282:474–511.

Hansell, D. A., and C. A. Carlson, eds. 2002. *Biogeochemistry of marine dissolved organic matter*. San Diego: Academic Press.

Hansen, H. P., and F. Koroleff. 1999. Determination of nutrients. In *Methods of seawater analysis*, ed. K. Grasshoff, K. Kremling, and M. Ehrhardt. Weinheim, Germany: Wiley-VCH Verlag, 159–228.

Hedges, J. I., B. A. Bergamaschi, and R. Benner. 1993. Comparative analyses of DOC and DON in natural waters. *Marine Chemistry* 41:121–34.

HELCOM. 1991. Third Biological Intercalibration Workshop. Baltic Sea Environment Proceedings 38, Baltic Marine Environment Protection Commission, Helsinki.

Karl, D. M., and K. M. Björkman. 2002. Dynamics of DOP. In *Biogeochemistry of marine dissolved organic matter*, ed. D. A. Hansell and C. A. Carlson. San Diego: Academic Press, 250–366.

Kattner, G. 1999. Storage of dissolved inorganic nutrients in seawater: Poisoning with mercuric chloride. *Marine Chemistry* 67:61–66.

Kattner, G., and U. H. Brockmann. 1980. Semi-automated methods for the determination of particulate phosphorus in the marine environment. *Fresenius Zeitschrift für Analytische Chemie* 301:14–16.

Kérouel, R., and A. Aminot. 1996. Model compounds for the determination of organic and total phosphorus dissolved in natural waters. *Analytica Chimica Acta* 318:385–90.

Kolowith, L. C., E. D. Ingall, and R. Benner. 2001. Composition and cycling of marine organic phosphorus. *Limnology and Oceanography* 46:309–20.

Koroleff, F. 1977. Simultaneous persulphate oxidation of phosphorous and nitrogen compounds in water. In *Report on the Baltic Intercalibration Workshop*, ed. K. Grasshoff (Compiler), Kiel, Germany.

Koroleff, F. 1983. Total and organic nitrogen. In *Methods of seawater analysis*, ed. K. Grasshoff, M. Ehrhardt, and K. Kremling. Weinheim, Germany: Verlag Chemie, 162–168.

Monaghan, E. J., and K. C. Ruttenberg. 1999. Dissolved organic phosphorus in the coastal ocean: Reassessment of available methods and seasonal phosphorus profiles from the Eel River Shelf. *Limnology and Oceanography* 44:1702–14.

Murphy, J., and J. P. Riley. 1962. A modified single solution method for the determination of phosphate in natural waters. *Analytica Chimica Acta* 27:31–36.

Najjar, R. G., X. Jin, F. Louanchi, et al. 2007. Impact of circulation on export production, dissolved organic matter, and dissolved oxygen in the ocean: Results from Phase II of the Ocean Carbon-cycle Model Intercomparison Project (OCMIP-2). *Global Biogeochemical Cycles* 21:GB3007, doi: 10.1029/2006GB002857.

Raimbault, P., W. Pouvesle, F. Diaz, N. Garcia, and R. Sempere. 1999. Wet-oxidation and automated r.b. colorimetry for simultaneous determination of organic carbon, nitrogen and phosphorus dissolved in seawater. *Marine Chemistry* 66:161–69.

Raimbault, P., and G. Slawyk. 1991. A semiautomatic, wet-oxidation method for the determination of particulate organic nitrogen collected on filters. *Limnology and Oceanography* 36:405–8.

Sharp, J. H. 2002. Analytical methods for total DOM pools. In *Biogeochemistry of marine dissolved organic matter*, ed. D. A. Hansell and C. A. Carlson. San Diego: Academic Press, 35–58.

Sharp, J. H., K. R. Rinker, K. B. Savidge, et al. 2002. A preliminary methods comparison for measurement of dissolved organic nitrogen in seawater. *Marine Chemistry* 78:171–84.

Solorzano, L., and J. H. Sharp. 1980a. Determination of total dissolved phosphorus and particulate phosphorus in natural waters. *Limnology and Oceanography* 25:754–58.

Solorzano, L., and J. H. Sharp. 1980b. Determination of total dissolved nitrogen in natural waters. *Limnology and Oceanography* 25:751–58.

Toggweiler, J. R. 1989. Is the downward dissolved organic matter (DOM) flux important in carbon transport? In *Productivity of the ocean: Present and past*, ed. W. H. Berger, V. S. Smetacek, and G. Wefer. New York: Wiley, 65–83.

UNESCO. 1994. *IOC/SCOR*. UNESCO Manuals and Guides 29.

Walsh, T. W. 1989. Total dissolved nitrogen in seawater: A new high temperature combustion method and comparison to photo-oxidation. *Marine Chemistry* 26:295–311.

Williams, P. J. LeB. 1993. DOC subgroup report. *Marine Chemistry* 41:11–21.

Van der Zee, C., and L. Chou. 2005. Seasonal cycling of phosphorus in the Southern Bight of the North Sea. *Biogeosciences* 2:27–42.

Verity, P. G., S. C. Williams, and Y. Hong. 2000. Formation, degradation, and mass:volume ratios of detritus derived from decaying phytoplankton. *Marine Ecology Progress Series* 207:53–68.

10 Pigment Applications in Aquatic Systems

Karen Helen Wiltshire

CONTENTS

10.1 INTRODUCTION TO PHOTOSYNTHETIC PIGMENTS IN SEAWATER

Pigmented microalgae are pivotal organisms in marine aquatic systems, ranging from clear open-ocean waters to turbid sediment-laden coastal waters. Indeed, these pigmented organisms (only 1% of the total plant biomass on earth) are primarily responsible for approximately 40% of global photosynthesis (Falkowski and Raven, 1997). Their productivity is of increased interest in times of global warming, due to their ability to take up CO_2, making their accurate determination in the world's aquatic systems a priority. Light drives the photosynthetic reaction and pigments are required by the organisms to absorb the light. These pigments, particularly chlorophyll a, are used as chemical markers for the microalgae, and thus their accurate determination is imperative.

Photosynthetic pigments are those molecules in microalgae that can absorb light. Light that is not absorbed is transmitted, resulting in the vivid colors characteristic of these pigments. Pigments are fascinating in microalgae and macroalgae in aquatic systems because of their diversity. Figure 10.1 shows the different color schemes (due to the transmitted light) into which different aquatic algal groups can be placed, and the main pigments resulting in this coloration.

There are two main groups of pigment-protein complexes in microalgae in aquatic systems. The first is the chlorophyll-protein complex, which usually also contains carotenoids. The second group consists of the phycobiliproteins. The portion of a pigment molecule that absorbs light is known as a chromophore and is a result of a series of chemical double bonds. The basic chemical units of the four main types of chromophores found in algae are depicted in Figure 10.2. The pigments are mainly closed tetrapyrroles and their derivatives, for example, the porphyrin chlorophyll c and chlorins such as chlorophyll a, open tetrapyrroles (phycobilins), and the carotenoids such as β-carotene and fucoxanthin. The covalent π bonds within these molecules are largely responsible for the absorption characteristics and therefore for the absorption of light.

Although they have been studied in the literature since the early nineteeth century (see Pelletier and Caventou, 1818; Berzelius, 1838; Stokes, 1864; Sorby, 1873, 1877), the exact roles and functions of all the individual pigments in algae and plants are still not truly understood. In general,

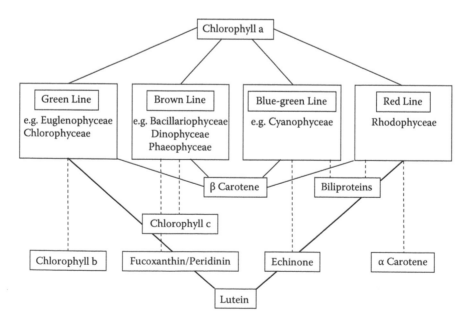

FIGURE 10.1 Simplified color/pigment scheme of the algae.

Chlorophylls & Phycobilins

Carotenoids

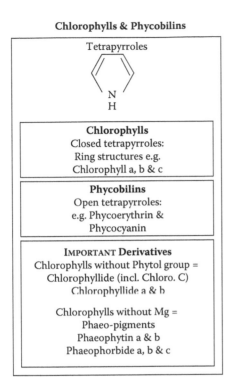

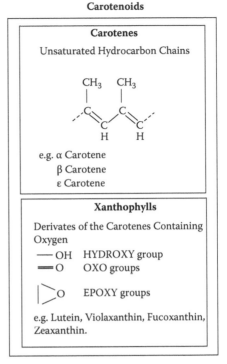

FIGURE 10.2 Structure of basic chemical units of the pigment chromophores.

the functions can be placed in four categories: light absorption to drive photosynthesis, either (1) as pigments directly incorporated into the light absorbing complexes or (2) as accessory pigments absorbing light at bands enhancing photosystem light usage; (3) photochemical protection pigments preventing photooxidation of the chlorophylls; (4) storage products. The different pigment categories are briefly described below, and major pigments found in algae are summarized in Table 10.1.

10.1.1 THE CHLOROPHYLLS

In 1818 the term *chlorophyll* was given to the green pigments involved in photosynthesis in higher plants (Pelletier and Caventou, 1818). The chlorophylls are the backbone pigments of photosynthesis. Chlorophyll a is the most important, and it usually accounts for the bulk of pigments in algae.

The chlorophylls are cyclic tetrapyrrole compounds with a Mg atom in the center of the ring. Specifically they are 13^2 methylcarboxylates of Mg-phytoporphyrins and Mg-phytochlorins (Moss, 1987). Scheer (1991) has detailed the structure of over fifty chlorophylls. Although the number found in nature is about ten, the most frequent chlorophyll is chlorophyll a, followed by b, c, d, protochlorophyll, and bacteriochlorophylls. Most of the chlorophylls have two main absorption bands in the visible spectrum. The so-called Soret bands absorb in the blue/blue-green spectrum, and the second bands absorb in the red (Q-bands). As chlorophyll c is actually a porphyrin and not a chlorin, it has only a weak Soret band.

10.1.1.1 Chlorophyll a

Chlorophyll a is the central pigment in photosynthesis and generally makes up 0.3% to 5% of the dry weight of a microalgal cell (Meeks, 1974; Kirk, 1983). The ratio of chlorophyll to carbon is roughly between 1:20 and 1:100 in natural mixed phytoplankton populations (Kirk, 1983).

TABLE 10.1
The Main Pigments in Selected Algae

Pigments	Cyanophyta	Prochlorophyta	Glaucophyta	Rhodophyta	Chrysophyceae	Synurophyceae[1]	Xanthophyceae	Eustigmatophyceae	Bacillariophyceae	Raphidophyceae	Dictyochophyceae	Phaeophyceae	Haptophyta	Dinophyceae	Cryptophyta	Chlorarachniophyta	Euglenophyta	Prasinophyceae[2,4,7,8]	Chlorophyceae	Micromonas pusilla[2-5]	Acaryochloris marina[9,11]	Notes
Chlorophylls																						
Chlorophyll a	X	X	X	X	X	X	X	X	X	X	X	X	X	X	X	X	X	X	X	X	X	
Chlorophyll b		X												s		X	X	X	X	X		s = present when endosymbionts are green algae
Chlorophyll c_1					X		x		X	X		X	X	s								s = present when endosymbionts are Chrysophytes
Chlorophyll c_2					X	X	X		X	X	X	X	X	X	X							Synurophyceae are characterized by not having chlorophyll c_1
Chlorophyll c_3					t							x^5	x^5									c_1 may be replaced by c_3
Chlorophyll ccs-170																				X^5		c_1 may be replaced by c_3
Chlorophyll c-like											X^6							$X^{7,8}$				Chlorophyll c* pigment or MgDVP[7,8]
Chlorophyll d				X^{10}																	X^{11}	In Rhodophyta still a matter of dispute[10]
Divinylchlorophyll a		X^5																				Replaces other chlorophylls in Prochlorococcus[5]
Divinylchlorophyll b		X^5																				Replaces other chlorophylls in Prochlorococcus[5]
MgDVP		x^5																$X^{7,8}$		X^5		Chlorophyll c* pigment or MgDVP[7,8]
Carotenes																						
β,ε Carotene(α)	x	x		X	x								x	s	X		t	t	t	x	x	s = present when endosymbionts are Chrysophytes
β,β Carotene(β)	x	x	X	X	X	X	X	X	X	X	X	X	X	X	t		x	x	x	x	x	t = trace or very rarely
β,ψ Carotene(g)																		t	t			t = trace or very rarely
ε,ε Carotene(e)					x				x				t		t							t = trace or very rarely
Ψ,Ψ Carotene(lycopene)																	t	t				t = trace or very rarely

Xanthophylls

Pigment	Data	Notes
Alloxanthin	X^{14}　x x	
Antheraxanthin	t　x x　x t^{12}　x　t t t	t = very rarely; Lohr and Wilhelm found accumulation under high to light stress[12]
Astaxanthin	x　t t	t = very rarely
19′-butanoyloxyfucoxanthin	X　X s	s = present when endosymbionts are Haptophytes
Canthaxanthin	x x　x x　t	t = very rarely
Crocoxanthin	x　x	
a-Cryptoxanthin	t x　t t	
b-Cryptoxanthin	t t　x　t	t = very rarely
Diadinoxanthin	f^{13} X　x x x X　f	f = freshwater species
Diatoxanthin	f^{13} X　x x x x　f	f = freshwater species
Dinoxanthin	x　x x	
Echinenone	X^{14} x　x x　X　t t t	Mainly in freshwater cyanophyta; t = very rarely[14]
Fucoxanthin	X X　m^6 X X　X X s　t t t	m = marine species; s = when endosymbionts are Haptophytes[6]
Heteroxanthin	X　x	
19′-hexanoloxyfucoxanthin	x X　X s	s = present when endosymbionts are Haptophytes
Isocryptoxanthin	x x　t	t = very rarely, e.g., in some Cladopherales and *Ulva* sp.
Loroxanthin	x　x x	
Lutein	x　x　x X	
Micromonal	X X	
Monadoxanthin	x　X	
Myxoxanthophyll	X	
Neofucoxanthin	x　x s　s	s = present when endosymbionts are Chrysophytes
Oscillaxanthin	X	
Peridinin	X X	
Peridininol	x	
Prasinoxanthin	X^4　x　X	Many Prasinophytes just have normal chlorophyte pigments and no prasinoxanthin[4]
Pyrrhoxanthin	x	
Siphonaxanthin	t t	
Siphonein	t t t	

(Continued)

TABLE 10.1 (CONTINUED)
The Main Pigments in Selected Algae

Pigments	Cyanophyta	Prochlorophyta	Glaucophyta	Rhodophyta	Chrysophyceae	Synurophyceae[1]	Xanthophyceae	Eustigmatophyceae	Bacillariophyceae	Raphidophyceae	Dictyochophyceae	Phaeophatyceae	Haptophyta	Dinophyceae	Cryptophyta	Chlorarachniophyta	Euglenophyta	Prasinophyceae[2,4,7,8]	Chlorophyceae	Micromonas pusilla[2-5]	Acaryochloris marina[9,11]	Notes
Uriolide																				X		
Vauchariaxanthin							X	X														
Violaxanthin				t	f[13]		X	X	t[12]	m[6]	x	X						X	X			f = freshwater; m = marine; Lohr and Wilhelm found accumulation under high to light stress[12,13]
Zeaxanthin	X	X	X	X	x	x			t[12]	m[6]		t			x		t[3]	X	X		X	m = marine; Lohr and Wilhelm found accumulation under high to light stress[12]

Sources: 1, Andersen and Mulkey, 1983; 2, Egeland and Liaaen-Jensen, 1993; 3, Egeland et al., 1995; 4, Ricketts, 1970; 5, Jeffrey et al., 1996; 6, van den Hoek et al., 1997; 7, Wilhelm et al., 1986; 8, Jeffrey and Wright, 1987; 9, Miyashita et al., 1997; 10, Manning and Strain, 1943; 11, Miyashita et al., 1996; 12, Lohr and Wilhelm, 1999; 13, Withers et al., 1981; 14, Nichols, 1973.

Note: Those algae contributing to the phytoplankton are Cyanophyta, Prochlorophyta, Glaucophyta (freshwater), Rhodophyta, Chrysophyceae, Synurophyceae, Xanthophyceae (mostly freshwater rare in marine), Eustigmatophyceae (mostly freshwater, one genus in marine), Bacillariophyceae, Raphidophyceae, Dictyophyceae (marine, rare), Dinophyceae, Cryptophyta, Euglenophyta, Prasinophyceae, and Chlorophyceae.

However, it must be emphasized that chlorophyll a is found in variable amounts in algae depending on environmental conditions such as nutrients (Yentsch and Vaccaro, 1958; Steele and Baird, 1962, 1965) and light (Steele and Baird, 1965; Kirk, 1983). In environments rich in nutrients such as estuaries, the chlorophyll contents are generally much higher than in those with depletion in nutrients. The highest concentrations of chlorophyll a are achieved under low-light and nitrogen-rich culture conditions (see Kirk, 1983), for example, turbid systems. Additionally, when dealing with aquatic systems it is important to note that the chlorophyll a content of cells is most dependent on light climate. At both extremes, that is, at very high and low light conditions, the chlorophyll a concentrations are reduced. The chlorophyll a per cell can increase by a factor of 5 to 10 with a reduction in light levels (Falkowski, 1980; Kirk, 1983; Post, 1986; Foy, 1993; Falkowski and Raven, 1997). Nicklisch and Woitke (1999) have shown that fluctuating light also causes increases in chlorophyll a content. Changes in chlorophyll a concentrations can occur in the order of minutes to hours. Chlorophyll a is easily destroyed at light-intensive conditions, with cells appearing bleached.

10.1.1.2   Chlorophyll b

Chlorophyll b is found in the green algae (see Table 10.1). It has two major absorption bands—one in the blue spectral region and one in the red region. The chemical difference between chlorophyll a and b is the presence of a formyl group rather than the methyl group on ring 2, resulting in a shorter-wavelength absorption in the red region (= higher energy) and a longer one in the blue region. Thus, the lowest singlet excited state of a chlorophyll b molecule is at a higher energy level than that of chlorophyll a, facilitating the direct energy transfer from chlorophyll b to a, and reflecting the function of chlorophyll b as an accessory light-absorbing pigment.

The molar ratio of chlorophyll a to chlorophyll b is around 3 for freshwater algae and between 1 and 2.3 in marine algae (Meeks, 1974; Kirk, 1983). However, this is dependent on the growing conditions and the algal species. For example, there seems to be a general tendency for the chlorophyll a:chlorophyll b ratio to decrease as light intensity decreases (Falkowski and Owens, 1978). It has been shown that as algal cultures become nutrient depleted, the ratio of chlorophyll a to chlorophyll b also decreases (Beutler, 1998).

10.1.1.3   Chlorophyll c

Chlorophyll c is used as an indicator of the so-called brown algal line (Table 10.1). Chlorophyll c is, chemically speaking, not a true chlorophyll, as its structure makes it a porphyrin. It has only a weak red (Q) absorption band and, as is the case for chlorophyll b, is at a higher energy stage than chlorophyll a. Consequently, it also transfers absorbed energy to chlorophyll a in algal cells. Usually chlorophyll c is a mixture of two or three compounds, chlorophyll c_1, chlorophyll c_2, or chlorophyll c_3, which are slightly structurally different from one another. Most algae containing chlorophyll c have the c_1 and c_2 combination. Some diatoms, Haptophytes and Chrysophytes, have a c_2 and c_3 combination (see Jeffrey and Vesk, 1997; Table 10.1). These pigments are usually present in roughly the same proportion as chlorophyll a to chlorophyll b. The molar ratios seem to vary from 1.5 to 5.5, with an average value of about 3 (Kirk, 1983).

10.1.1.4   Chlorophyll d

Chlorophyll d was found to be a major pigment in an oxygenic photosynthetic prokaryote (oxyphotobacterium), *Acaryochloris marina*. This organism is a symbiont in the ascidian *Lissoclinum patella* (Miyashita et al., 1996, 1997). This is a rare occurrence of chlorophyll d, and otherwise, this pigment often counts only for a minor fraction found as an accessory pigment in some red algae (Manning and Strain, 1943) or to be an artifact of extraction. Chemically speaking, it is a chlorophyll closely related to chlorophyll a. It is found in highly variable amounts, and it has not been found yet in planktonic microalgae. Thus, its analysis is not further described here.

10.1.1.5 Other Indicator Chlorophylls

Apart from the chlorophylls already discussed above, many other chlorophyll-related compounds exist (see also Table 10.1). Some Prasinophytes, for example, have the pigment Mg-2,4-divinyl pheoporphyrin a_5 monomethyl ester (MgDVP) (Ricketts, 1970; Wilhelm et al., 1986; Egeland et al., 1995). *Prochlorococcus*, the picoplankter in the Prochlorophytes, also contains MgDVP (see Goericke and Repeta, 1992; Jeffrey et al., 1997). In addition to MgDVP, *Prochlorococcus* contains the pigments divinyl chlorophyll a and divinyl chlorophyll b (Gieskes and Kraay, 1983b; Goericke and Repeta, 1992, 1993) as its major chlorophyll pigments.

In *Micromonas pusilla*, Jeffrey et al. (1997) have found what is suggested to be a propionate derivative of chlorophyll c_3, called chlorophyll c_{cs-170}. Currently many new chlorophyll structures are being discovered and elucidated every year. For example, Helfrich et al. (1999) found that (8-vinyl)-protochlorophyllide a is present in *Prochloron* and *Micromonas pusilla*, and that it could play an important role in photosynthesis. Recently Garrido et al. (2000a) and Zapata et al. (2001) have shown that the main nonpolar chlorophyll c from *Emiliania huxleyi* is a chlorophyll c(2)-monogalactosyldiaglyceride ester as well as a novel marker pigment for *Chrysochromulina* species (Haptophyta), respectively.

The bacteriochlorophylls associated with photosynthetic bacteria, although not considered here in detail, are also of relevance in both sedimentary and pelagic environments (Goericke, 2002; Koblížek et al., 2007; Lami et al., 2007) and should not be forgotten when dealing with marine systems. They can be analyzed using adaptations (see Van Heukelem and Thomas, 2001) of the high-performance liquid chromatography (HPLC) methods described below and elsewhere in the literature (Koblížek et al., 2007; Lami et al., 2007).

10.1.2 The Carotenoids

Over 600 carotenoids have been described (Straub, 1976), of which approximately 30 are important in algae. The distribution of the main carotenoids in algae is summarized in Table 10.1. Those which are of importance here are those found in phytoplankton and benthic algal communities. The carotenoids are composed of the carotenes and the xanthophylls. The basic structure of the carotenoids consists of an 18-carbon chain with conjugated double bonds with two unsaturated cyclic end groups. Unlike the carotenes, the xanthophylls have oxygen bound to their cyclic end groups, making them an epoxide or alcohol. The photosynthetically most important carotenoids, for example, diadinoxanthin, diatoxanthin, fucoxanthin, lutein, violaxanthin, peridinin, and zeaxanthin, absorb light in the blue/blue-green region. These bands partially overlap with the Soret bands of the chlorophylls; thus, their role is in the transfer or removal of excitation energy from chlorophylls. In general, the carotenoids that exhibit excitation at longer wavelengths can absorb energy from chlorophylls. In addition, they protect cells from photochemical reactions.

Perhaps one of the most important carotenoid roles in aquatic systems—protecting the chlorophylls from overexcitation—is carried out by the xanthophyll cycle. The chemical interconversions are epoxidation and de-epoxidation reactions, where the pigment zeaxanthin is converted into antheraxanthin and violaxanthin under low light (epoxidation), and antheraxanthin and violaxanthin are converted back to zeaxanthin under high light conditions (de-epoxidation). It is important to note that these adaptations can occur in the order of minutes in algae. With the correct sampling and extraction techniques (see below), analyses of these pigments are useful and instantaneous indicators of algal reaction to incident light.

Some of the main carotenoids are useful as marker pigments in marine systems. For example, the carotenoid peridinin is highly specific for the Dinophyceae. 19′-Butanoyloxyfucoxanthin and 19′-Hexanoyloxyfucoxanthin are markers for Haptophyta. Alloxanthin is a marker for the Cryptophyta. The carotenoid fucoxanthin can be used as a marker pigment for Bacillariophyceae, Chrysophyceae, Dinophyceae, and Prymnesiophyceae. The pigment lutein is a marker pigment for higher plants and chlorophytes and to some extent the Rhodophytes. Phycobiliproteins are important in the Cyanophyceae, Rhodophyceae, and Cryptophyceae. However, exceptions to the rule are being found continuously (see Egeland et al., 1995). The pigment contents of algae, in particular in relation to one another, can be highly variable, and therefore interpretations must be made cautiously regarding the presence or absence of pigments in algal communities.

10.1.3 THE PHYCOBILIPROTEINS

The phycobiliproteins are the second major group of light-absorbing pigment-protein complexes in algae. They occur in the Cyanophyceae, Rhodophyceae, and Cryptophyceae. The chromophore in the phycobiliproteins is a bile pigment (bilin) that consists of a series of open tetrapyrroles. These are covalently bound by a sulfide bridge to a protein. Because of the strength of this bond the phycobiliproteins do not relinquish their proteins in nonpolar solutions and therefore are water soluble. This is in contrast to the chlorophylls and carotenoids, which easily relinquish their protein connections in nonpolar solvents. As with the other pigments discussed above, the light absorption properties of the chromophore are due to double bonds in the tetrapyrroles. Three main chromophores (phycobilins) are found in phycobiliproteins. These are phycoerythrobilin, phycocyanobilin, and phycourobilin (marine algae). The phycocyanobilins absorb orange light, whereas phycoerythrobilin and phycourobilin absorb green light. Thus, these pigments fill the optical gap between the two major visible absorption bands of chlorophyll.

10.1.4 SUMMARY

Algae contain pigments, which are generally related to the photosynthetic units. The two main groups of pigment-protein complexes are the chlorophyll-protein complex, which usually contains carotenoids, and phycobiliproteins. The algal pigment composition has been shown to be specific to certain groups of algae used as an indictor in studies of algal communities. However, increasingly exceptions to the rule are being found. Moreover, the functions of pigments are for direct or accessory light absorption to drive photosynthesis, as photoprotective pigments preventing photooxidation of the chlorophylls, and as storage products. Thus, it is clear that the pigment contents of algae, in particular in relation to one another, can be highly variable.

10.2 METHODS

10.2.1 INTRODUCTION: GENERAL REMARKS

Photosynthetic pigments in seawater are, with the rare exception of water-soluble phycobiliproteins leached from algal cells, confined to plant cells, photosynthetic bacteria, chloroplasts, and other particulate matter found in the water column. Thus, the accurate determination of pigments in the large variety of substrates in aquatic systems is a considerable challenge. Table 10.2 shows the broad spectrum of substrates that one might encounter in work on photosynthetic pigments in aquatic systems. These range from pure algal cultures, through algal mixtures, particulate matter with algae, sediments with algae, sediments with old algae, to ancient sediments with pigments. Tswett (1906) first extracted and separated the pigments from plastids in 1906, and since then many methods have been described in the literature for the extraction and separation of pigments from all sorts of substrates (Riley and Wilson, 1965; Strain and Svec, 1966; Tett et al., 1975; Gieskes and Kraay, 1983a; Mantoura and Llewellyn, 1983; Bowles et al., 1985; Deventer, 1985; Daemen, 1986; Wright et al., 1997; Zapata and Garrido, 1991; Wilhelm et al., 1991, 1992, 1998a; Pinckney et al., 1994; Van Heukelem and Thomas, 2001; Garrido and Zapata, 2006). However, despite this volume of work, problems associated with the extraction, analyses, and determination of pigments from pure algal cultures to complex substrates remain and have to be considered by the marine analytical chemist each time pigments are to be analyzed.

TABLE 10.2
Broad Spectrum of Substrates and Pigments Encountered in Pigment Work in Aquatic Systems

	Chlorophylls	Phaeopigments	Porphyrins	Carotenes	Xanthophylls	Phycobilins
Algal cultures	X	X	X	X	X	X
Phytoplankton	X	X	X	X	X	X
Microphytobenthos	X	X	X	X	X	X
Decaying algae	X	X	X	X	X	X
Suspended matter	X	X	X	X	X	X
Sediment (oxic)			X		X	
Sediment (anoxic)	X		X	X	X	
Zooplankton				X	X	
Zooplankton and algae	X	X	X	X	X	X
Benthic fauna				X	X	
Benthic fauna and algae	X	X	X	X	X	X
Fossil sediments			X	X	X	
Fecal pellets		X	X	X	X	

The three main problem areas are:

- Representative sampling and preservation of substrates for pigment analysis
- Complete extraction of pigments from algae and particulate matter
- Accurate analysis of pigments

10.2.2 Sampling and Preservation of Particulate Matter for Pigment Analyses

The first criterion when working on pigments in seawater is the representative sampling and the preservation of the substrates to be analyzed. With the term *sampling*, I do not mean to include statistical methods of sampling, as these must be readjusted for each system and must be highly flexible in order to deal with the different systems one may encounter. The physical methods for sampling of suspended matter and sediment surfaces are of more importance here. Filtration of a certain volume of seawater is, for example, required to retain the substrate for the analysis unless in situ measurements are conducted using fluorescence instruments.

10.2.2.1 Water: Suspended Matter Sampling and Preservation

The sampling techniques used for obtaining sufficient substrate from the water column are described here only briefly, and the physical apparatus available for water sampling are well thought out and described in detail elsewhere (Sournia, 1978). Current techniques range from sampling devices such as Niskin bottles to Kerner's funnels. The methods available are more than adequate for sampling the water column. In the shallow intertidal where one cannot deploy sampling devices, water samples can be taken by hand using wide-necked bottles or small buckets.

To fulfill the requirement of representative sampling, replicate samples should be collected. The sampling bottle should be well, but not aggressively, mixed (so as not to cause cell rupture) before collecting the water subsample for analysis.

It should be noted here that it is almost impossible to take samples from the water column whereby the pigments of the xanthophyll cycles are preserved in exact relationships. The reason for this is that these relationships are highly sensitive to changes in light, often requiring only a period of minutes before conversions occur, and this can hardly be avoided during sampling and preservation. With laboratory cultures of algae this is easier to control, as one can sample directly under the same light conditions and also subsequently filter the algae out at the same light intensity as they were conditioned to.

In phytoplankton studies, seawater samples generally are (pre-)filtered. Size fractionations over different filters with decreasing pore sizes are used to separate different size classes of phytoplankton organisms and to allow distinct examinations. Filters can be used for further investigations in microscopic or chromatographic (for example, HPLC) methods, while the filtrates can be used for fluorometric measurements as well as for microscopic investigations. Which filter should chosen for a study often depends on the filter pore size, the costs, and in particular for HPLC measurements, its chemical compatibility. Especially nowadays, where the importance of picophytoplankton organisms (0.2–3 µm) is realized, the effectiveness of commercial filters in phytoplankton research studies has to be considered.

Knefelkamp et al. (2007) compared the chlorophyll a retention by HPLC as well as fluorescence and nutrient content of the filtrates after filteration through filters of different material and with various pore sizes. Although Whatman GF/C and GF/F filters are historically preferred in phytoplankton studies, it was reported that nylon membrane filter of 0.2 µm pore size yielded the best results in chlorophyll a retention in seawater samples. Filtering either through membrane filters or GF/F glass fiber filters increases the error by 20%–50%. Nylon filters give better results. These filters also have the advantage that they have even pore structures and retain almost no salt. This is important, as salt retained on filters can interfere with chromatographic gradients.

The analyst, when filtering a water sample, must ensure that the size fraction of interest (e.g., miniscule photosynthetic bacteria vs. gigantic diatom chains) is retained on the filter and that enough water is filtered to ensure a representative sample. Thus, one has to differentiate between oligotrophic oceanic and eutrophic coastal waters. Water samples from the former have only trace amounts of pigments, and one has to filter large amounts (10–20 L), and for the latter, with more pigmented organisms, smaller amounts (200–1,000 mL) need to be filtered.

The formation of pigment breakdown products is almost eliminated by putting the filters straight into acetone and instantly freezing them in liquid nitrogen with subsequent storage at –70°C. Another alternative method is to immediately freeze filters in acetone in the dark and then to store them air tight at –20°C. It is not advisable to keep the filters for longer than a year before analyses.

Because of inaccuracies that occur with filtering (pouring of uneven quantities and uneven pump suction) as well as problems associated with the extraction of suspended matter on filters, it is best to avoid it where possible. Thus, when dealing with samples with relatively large amounts of particulate matter (e.g., estuaries, intertidal, eutrophic lagoon systems) or algal cultures, it is best to centrifuge the sample and immediately freeze-dry the pellet straight into acetone, and then instantly freeze in liquid nitrogen and with subsequent storage at –70°C. Stored samples should also be analyzed immediately upon thawing, and preferably injected into an HPLC via a cooled autosampler. The protocol developed for dealing with suspended matter in the water column is given in Tables 10.3 and 10.4.

10.2.2.2 Sediment Sampling and Preservation

The accurate sampling of surface sediments has proven to be perhaps the most challenging aspect of method development for pigment analysis over the last few years (Wiltshire et al., 1997, 1998b, 2000; Brotas et al., 2007). Sampling of the upper sediment layers without distortion is very difficult, particularly when accuracy on scales of a millimeter or less is required. With unconsolidated muddy sediments, this becomes even more problematic. Such sediments are easily compacted; even when sampling with large corers (diameters larger than 10 cm), there is always a certain amount of compaction of the sediments (Parker, 1991). Where the properties of interest are at scales greater than

TABLE 10.3
Suspended Matter Sample Preparation Pigment Analysis:
Low Concentrations of Particulate Matter
(e.g., Open-Ocean Samples)

1. Work in dimmed light.
2. Filter as much water as possible (e.g., 5–20 L) through 45 mm diameter; nylon filters, 0.2 μm pore size.
3. Freeze filter instantly in liquid nitrogen.
4. Freeze-dry in the dark overnight.
5. Add 2–5 mL of 100% HPLC-grade acetone; if green algae are to be expected, consider using 2–5 mL of 100% HPLC-grade methanol.
6. Store at < −20°C and in the dark.

centimeters, this distortion can be seen as negligible, for example, in cores with diameters of upwards of 10 cm. This is not the case when dealing with the vertical distribution of physical properties or materials on a micrometer-millimeter scale. Thus, initially the scale of resolution needs to be defined as required. Here the resolution is divided up into millimeter-centimeter and micrometer-millimeter scales; protocols are given in Tables 10.5 and 10.6.

10.2.2.2.1   Millimeter-Centimeter Sediment Depth Resolution

The most commonly used method for sampling the upper millimeters of intertidal sediments is the syringe coring method (Joint et al., 1982). The sampler was constructed from a micrometer screw and the sawn-off barrel of a 10 mL disposable plastic syringe (see Figure 1 in Joint et al., 1982). The syringe plunger is pulled up via the micrometer screw, creating a counter vacuum, while the syringe device is slowly pushed into the sediment, theoretically minimizing vertical distortion. In practice, this only works for sandy sediments, and even for these there may be distortion on a millimeter scale. Parker (1991) discussed the problems associated with larger sediment coring devices (not involving freezing) and found that these are highly deficient, particularly when it comes to preserving the structure of muddy sediments even at large scales. However, in order to get a rough estimate of trends the syringe method is useful.

TABLE 10.4
Suspended Matter Sample Preparation Analysis: Medium
to High Concentrations of Particulate Matter
(e.g., Estuaries, Intertidal, etc.)

1. Work in dimmed light.
2. Centrifuge as much water as possible (e.g., 1,000 mL) at 2,000–3,000 rpm for 15 minutes, cool, and subsequently decant water from pellet. Alternatively, filter as much as possible (e.g., 200–1,000 mL) through 47 mm diameter nylon filter, 0.2 μm pore size.
3. Freeze pellet instantly in liquid nitrogen. Alternatively, freeze filter instantly in liquid nitrogen.
4. Freeze-dry in the dark overnight.
5. Add 2–5 mL of HPLC-grade acetone. If green algae are to be expected, consider using 2–5 mL of 100% HPLC-grade methanol.
6. Store at < −20°C and in the dark.

TABLE 10.5
Syringe Sample Preparation for Subsequent Pigment Analysis

1. Work in dimmed light.
2. Press sediment sample up from below and cut off in millimeter to centimeter slices with thin blade. Place in petri dish.
3. Mix sample in petri dish with spatula for 3 minutes on a cooled base.
4. Remove subsample.
5. Place sediment subsample in preweighed brown plastic vial.
6. Weigh (for determination of water content).
7. Freeze instantly in liquid nitrogen.
8. Freeze-dry overnight.
9. Remove from freeze drier and determine dry weight.
10. Put 1 mL of acetone on sample, mix, and extract for 72 hours in the dark and cold for subsequent pigment analyses. If green algae are to be expected, consider using 1 mL of 100% HPLC-grade methanol.

10.2.2.2.2 Micrometer-Millimeter Sediment Depth Resolution

It is often necessary to work on the profiles of substances such as chlorophyll, nutrients, and oxygen in the upper millimeters of aquatic sediments. This can involve the need for the sectioning of the sediments with depth. However, because of the cohesive nature of sediments, slicing of sediments on a scale of less than a millimeter, with depth, is very difficult. The only way to section the upper micrometer-millimeter of sediments with accuracy and precision is to freeze the sediments and then cut them using a freezing microtome (Wiltshire et al., 1997; Wiltshire, 2000). Thus, apart from the compaction problems associated with sampling, we have the problem of additional structural distortion caused by subsequent freezing of a sample. Rutledge and Fleeger (1988) have shown conclusively that it is not possible to freeze a muddy sediment core, while conserving its microstructure. Such problems remained even when fast-freezing techniques, such as plunging the core into liquid nitrogen, were used. The negative effects of freezing are always the same, regardless of the conductivity of the core tubing (i.e., plastic vs. copper). The pore waters freeze faster on the outer edge of the core than at the center, and the result is a conical protrusion of pore water upwards.

TABLE 10.6
Cryolander Sample Preparation for Subsequent Pigment Analysis

1. Cut frozen Cryolander sample into 1 cm³ blocks, with diamond lapidary blade or hand saw.
2. Place frozen block onto a freezing (–40°C) microtome stage, so that the sediment surface is absolutely horizontal. Use water at the base, to freeze block into position.
3. Lower down microtome blade until it rests on surface of sediment.
4. Cut a slice between 50 and 200 μm thick.
5. Wipe sediment off the blade with preweighed 3 mm² piece of glass fiber filter paper.
6. Place sediment and paper in preweighed brown plastic vial.
7. Weigh (for determination of water content).
8. Freeze instantly in liquid nitrogen.
9. Freeze-dry overnight.
10. Remove from freeze drier and determine dry weight.
11. Put 1 mL of acetone on sample, mix, and extract for 72 hours in the dark and cold for subsequent pigment analyses. If green algae are to be expected, consider using 1 mL of 100% HPLC-grade methanol.

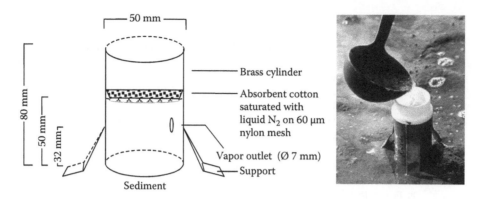

FIGURE 10.3 The Cryolander. (Redrawn after Wiltshire et al., 1997. Photo courtesy of N. Aberle. Used with permission.)

A way around the distortional problems with sampling and freezing surface sediments is the use of a simple liquid nitrogen surface sediment sampler, the Cryolander, depicted in Figure 10.3 (Wiltshire et al., 1997). In brief, the method functions as follows: The Cryolander is placed on a sediment surface and some liquid nitrogen dribbles on to the absorbent cotton in it. As this cotton is at ambient temperature, the liquid nitrogen instantly vaporizes. This vapor freezes the immediate sediment surface without distortion on a micrometer scale. After the surface is frozen, more liquid nitrogen (approximately 100 mL) is poured into the Cryolander and evenly through the Cryolander mesh. Once the fluid liquid nitrogen reaches the sediment, it freezes quickly into an approximately 2 cm thick disc of sediment (equivalent area of 20.4 cm^2). This can then be lifted up and stored in liquid nitrogen at –70°C until further use. When the sediments are required for analyses of vertical profiles of substances such as pigments, the frozen samples can be cut up in the laboratory into 1 cm^2 blocks using a diamond lapidary saw.

10.3 PIGMENT EXTRACTION

10.3.1 INTRODUCTION

The factors of primary importance with extraction of pigments from specific substrates are: (1) the interaction of solute and solvent to form solvated particles, (2) the reactions of solvent and solute to form a new substance that becomes solvated, and (3) the tendency of physical and chemical systems to attain the maximum state of disorder. Under ideal conditions, solid substances may dissolve in a solvent rapidly and totally in a short time period. An extraction procedure should ideally extract above 90% of the pigments regardless of their chemistry and ratios. However, this cannot be assumed the case with complex mixtures of substances, particularly with substances in algal cells bound to membranes and associated with protein complexes. For maximum extraction efficiency, extraction methods must always be tailored to suit the substrate and type of samples (Wiltshire et al., 2000).

General criteria for the extraction of pigments from algae and algae-particulate matter mixtures are:

- Conservation of extracted substances
- Maximum extraction efficiency of all pigments in a single step
- Use of solvents of low toxicity
- Ease of handling

There are many methods that one could theoretically apply to the extraction of pigments from algae. Indeed, the literature describes such methods and intercomparisons of extraction techniques intensively (e.g., Wright et al., 1997; Hooker et al., 2000). Extraction methods include simple solution in solvents (e.g., Scor-Unesco, 1966), grinding (Gieskes et al., 1988), and ultrasonification (Chang and Rossmann, 1982). Theoretically, it is irrelevant how the extraction occurs so long as it is complete, reproducible, user-friendly, and conserves the substances. However, the literature can be very confusing and the methods often seem to ignore basic chemical extraction principles. They are often inadequate when it comes to the extraction of certain algae (see Porra, 1991), in particular green algae.

10.3.1.1 Extraction Criteria: Conservation of Extracted Substances

The most accepted literature (see review by Porra, 1991; Svec, 1991; Wright et al., 1997) on the analyses of pigments, in particular for the chlorophylls, points out that a primary criterion in the extraction of pigments is the preservation of these. Alteration should not result from the extraction. For example, in the case of the chlorophylls the formation of phaeopigments, chlorophyl-lides, chlorophyll allomers, and carotenoid alteration products needs to be avoided. Alcohols (methanol) are generally better extractants than acetone, but these are known to promote the formation of allomers of chlorophyll (Strain and Svec, 1966; Tett et al., 1975; Bowles et al., 1985; Porra, 1991). Alteration of pigments can be prevented by using 100% acetone. Extractions should be carried out in the dark and in the cold (on ice or in the fridge) mainly because the production of chlorophyllides and phaeopigments from chlorophyll (for example) are a common problem in the extraction of algae. No matter which extraction method is applied to substrates containing pigments, this is a problem that should always be addressed and checked. Perhaps the most difficult to extract are certain green algae due to latex-like cell walls containing sporopollenin. When dealing with samples containing green algae or even culture samples, when possible these should be centrifuged down to a pellet and subsequently ground and, before extraction, have water re-added.

10.3.1.2 Extraction Criteria: Maximum Extraction Efficiency of All Pigments in One Step

Generally, the question addressing the remaining amounts of pigments in a sample after a single extraction step seems to be largely ignored in the literature. Although many authors have been intent on developing effective extraction methods (e.g., Wright et al., 1997), it is often assumed that one extraction treatment is sufficient without control procedures on completeness of extraction, particularly in routine lab work. Similarly, the fact that not all pigments are extracted at the same rate is generally assumed also to be of negligible importance. Of course, when developing an extraction method, ease of use (i.e., one extraction step) must be balanced against efficient extraction. Multiple extractions of pigments from algae are rarely reported in the phycological literature and, when discussed, are often inconclusive and confusing.

10.3.1.3 Extraction Criteria: Use of Solvents of Low Toxicity

The choice of solvent is perhaps one of the most controversial issues with regard to the extraction of pigments, in particular chlorophylls. Many different solvents have been used for the extraction of pigments in marine algae (Wright et al., 1997). The conclusion has generally been that extraction into dimethylformamide (DMF) with sonication was the most efficient combination. However, due to the toxicity of DMF, it cannot be recommended as a routine technique. Although alcohols (methanol) are generally better extractants than acetone, these are known to promote the formation of allomers of chlorophylls (Strain and Svec, 1966; Tett et al., 1975; Bowles et al., 1985; Porra, 1991; Wiltshire et al., 2000). One hundred percent acetone is generally a preferred solvent over acetone-water mixes, as the presence of water has been shown to be a problem with regard to the breakdown of chlorophylls via chlorophyllase (Wright et al., 1997). However, green algae are sometimes impossible to extract in the absence of water, and thus, analysts often resort to DMF/methanol.

10.3.1.4   Extraction Criteria: Ease of Handling

The step that is most important in the extraction of pigments from algae is undoubtedly the lysis of the cell membranes and walls. Many different extraction techniques have been compared in the literature for the extraction of pigments in algae (Porra, 1991; Svec, 1991; Wright et al., 1997; Wiltshire et al., 2000). Techniques such as sonification, hand grinding, and use of the French cell press have been applied. The presence of water has been shown to be a problem with regard to the breakdown of chlorophylls via chlorophyllase (Wright et al., 1997). Hence, the samples should generally be handled in a manner minimizing the water content during extraction. For completeness, the main procedures that have been tested are summed up to help the analyst.

Slightly better cell lysis has been observed when the pellet and filters were disrupted while still wet. However, at a later stage, during the pigment analyses, more breakdown products have been found in the wet samples. It made no difference to the cell lysis if the samples were shock-frozen in liquid nitrogen at $-197°C$ or frozen slowly to $-10°C$ (i.e., the size of the water crystals in the cells did not aid lysis). Consequently, because of the danger of pigment conversion during a slow-freezing procedure, the liquid nitrogen technique is the method of choice. The frozen samples were then freeze-dried in the dark for 24 hours. Although the hand cell grinding method, with a pestle and mortar, was superior to the mechanical method and resulted in relatively high cell lysis percentages (40%–60%), it is laborious and not suited to routine work. One problem encountered is the cooling of the sample, which could only be reasonably achieved by adding frozen CO_2 pellets. Other methods, such as the French press and cell disrupters, can also be considered, but these are also generally not recommended because of the difficulty of keeping the sample cold.

Simply adding quartz sand (i.e., Merck, particle size 10–30 µm) and solvent to freeze-dried algal material/suspended matter and subsequent extraction in an ultrasound bath for 90 minutes resulted in excellent extraction of pigments and fatty acids even with really difficult cell material (Wiltshire et al., 2000). No alteration or breakdown products were observed with this method. The method was also found to be as effective for *Scenedesmus obliquus* (Chlorophyceae), *Cryptomonas erosa* (Cryptophyceae), *Cyclotella meneghiniana* (Bacillariophyceae), *Microcystis aeruginosa* (Cyanophyceae), and *Staurastrum paradoxum* (Chlorophyta, Zygnematophyceae, Desmidiales) and is thus applicable to a wide range of algae. Tables 10.7 and 10.8 provide details of the recommended extraction steps for microalgal cell cultures and particulate matter as well as sediments, respectively (see Wiltshire et al., 2000).

TABLE 10.7
Extraction of Pigments from Cultured Algae and Water Samples

1. Algae culture sample is spun down to a pellet in a cooled centrifuge at 2,000–3,000 rpm for 15 minutes; subsequently decant water from pellet.
   Alternatively, filter culture of water sample through 47 mm diameter nylon filter, 0.2 µm pore size.
2. Pellet/filter is frozen in liquid nitrogen and subsequently freeze-dried in the dark.
3. Pellet/filter is weighed and 1 g of quartz sand is added.
4. 2–5 mL of 100% acetone is added.
5. Sample is extracted at $-4°C$ (on a bath of salt water and ice) in an ultrasound bath for 90 minutes in the dark.
6. A minimum of three samples from each batch should be subjected to a sequential extraction to check the extraction efficiency for all pigments.
7. Sample is filtered through a cellulose syringe filter to remove particles.
8. Filtrate is injected into HPLC.
9. Note: Relevant controls (with only sand and solvent) should also be analyzed.

TABLE 10.8
Extraction of Aquatic Sediments

1. Sample is frozen in liquid nitrogen and subsequently freeze-dried overnight in the dark.
2. Sample is weighed.
3. 0.2–5 mL of 100% acetone is added, depending on sediment quantities.
4. 0.5 g of quartz sand is added to the sediment-solvent mixture in the vial.
5. Sample is then extracted at –4°C (on a bath of salt water and ice) in an ultrasound bath for 90 minutes in the dark.
6. A minimum of three samples from each batch should be subjected to a sequential extraction to check the extraction efficiency for all pigments.
7. Sample is filtered through a cellulose syringe filter to remove particles.
8. Filtrate is injected into HPLC.
9. Note: Relevant controls (with only sand and solvent) should also be analyzed.

10.4 ANALYTICAL METHODS FOR PIGMENTS

10.4.1 WET CHEMICAL METHODS WITH SPECTROPHOTOMETRY AND FLUOROMETRY

In the past the main methods recommended for chlorophyll analyses were simple extraction methods using ethanol or methanol, often with an acidification step for phaeophytin analysis, coupled with spectrophotometry and fluorometry (Yentsch and Menzel, 1963; SCOR-UNESCO, 1966, 1980; AWWA, 1985, 1989). Indeed, managers, particularly in Europe, still follow some of these methods for chlorophyll a analyses, in the belief that these methods allow accurate determination. However, common problems include overestimation of chlorophyll a in waters with dying algal populations or higher concentrations of humic substances, and negative chlorophyll a concentrations when acidification steps are performed to determine phaeophytin concentrations. The problems with analytical accuracy can be summarized as follows:

- Chlorophyll degradation products or derivatives, often found in high concentrations in turbid waters, absorb light at the same wavelengths as chlorophyll and can cause overestimations. This is a common problem in coastal systems and sediments.
- The absorption and emission spectra of the true chlorophylls (e.g., a and b) overlap. Thus, in systems with a lot of chlorophyll b, for example, one can get inaccurate results for chlorophyll a.

These problems are well known, and resultant inaccuracies in algal biomass estimations as well as in incorrect assessment of the chlorophyll pigments, and in particular their breakdown products in aquatic systems, have been described over and over again by many authors (e.g., SCOR-UNESCO, 1980; Jacobsen, 1982; Coveney, 1982; Gieskes and Kraay, 1983a, 1983b; Schanz and Rai, 1988). Often trends might be similar when one compares different methods, including HPLC (SCOR-UNESCO, 1980; Wiltshire et al., 1998a); however, absolute values can diverge substantially. There is a clear relationship between methodological congruity and "cleanliness" of sample. The less additional substances and breakdown products (e.g., a fast-growing culture of diatoms vs. a decaying phytoplankton bloom in a coastal system), the better HPLC methods seem to correlate with older wet chemical methods.

Thus, for the exact quantification of chlorophyll in any natural system, HPLC is recommended (see also Section 10.4.3 on fluorescence). This technique has the advantage that not only the chlorophylls can be measured, but also their derivatives and the carotenoids.

10.4.2 HPLC Methods

10.4.2.1 General Remarks

By the early 1990s many HPLC methods had been described in the literature for the measurement of the chlorophylls and carotenoid pigments, and every year new methods and versions of older methods are published (e.g., Gieskes and Kraay, 1983a; Mantoura and Llewellyn, 1983; Daemen, 1986; Deventer, 1985; Wiltshire, 1992; Zapata and Garrido, 1991; Wilhelm et al., 1991; Pinckney et al., 1994; Wiltshire et al., 2000; Airs et al., 2001; Van Heukelem and Thomas, 2001; Garrido and Zapata, 2006).

There are many suitable HPLC devices commercially available. However, two-pump systems, with a column oven and a UV-VIS detector or fluorescence detector, are the minimum requirement. Rather than injecting by hand, these systems are easier and produce better results for use with an autosampler with injector. For the analysis of a larger number of pigments (chlorophylls and breakdown products, carotenoids and carotenes) in complex matrices such as muddy water rich in organic material, a photodiode array detector is recommended because during chromatographic separation peak overlapping can be partially eliminated through detection at multiple wavelengths. If one is just interested in chlorophylls at very low concentrations, a fluorescence detector is recommended for increased sensitivity.

Settings of HPLC parameters depend much on the application, for example, whether the pigments are to be analyzed in oceanic, limnic, or estuarine algae/suspended matter/sediments, or whether other specific pigments or compounds are to be analyzed at the same time. Regardless of the author's specific preferences, most HPLC methods published separate the different chlorophylls and their breakdown products, allowing the quantitative assessment of the chlorophylls. However, HPLC methods do usually require work on optimization and adjustments to ensure accurate, efficient, and clean separation of the substances to be analyzed in different substrates (e.g., algae vs. sediments) and with varying instrumentation. The literature has many good methods for specific applications; for example, perfectly tweaked methods for chlorophyll c(2)–monogalactosyldiaglyceride ester or 2,4-divinyl phaeoporphyrin a_5 monomethyl ester identification exist—useful marker pigments for *Chrysochromulina* and *Prochlorococcus* or for divinyl-chlorophyll a (e.g., Goericke and Repeta, 1992, 1993; Garrido et al., 2000a; Zapata et al., 2001). The range of solvents, gradients, and column types offered in the literature is wide, and each method presented has it merits; a good intercomparison is given by Claustre et al. (2004), and an overview of newer methods by Garrido and Zapata (2006). The reversed phase columns used are usually C_8 and C_{18}, and the choice of solvents varies accordingly. It is important that the analyst seriously evaluates what she or he is looking for before embarking on pigment analyses. Two methodological approaches are suggested below that are good for starting out in using C_{18} methods, the first of which avoids the use of very toxic solvents.

The first is for simple differentiation of the chlorophylls a, b, c, 19′-Butanoyloxyfucoxanthin, fucoxanthin, 19′-Hexanoyloxy-fucoxanthin, lutein, and zeaxanthin. It includes most of the carotenoids important in photosynthetic pigment analysis in seawater as well as the carotenes as a single peak with an affordable RPC_{18} column and user-friendly solvents, within 40 minutes. The second method requires less time and provides a similar differentiation of the relevant pigments, and although it is more straightforward, it involves more toxic solvents.

10.4.2.2 First Method: Long Reversed Phase C_{18} Method

HPLC methods using a reversed phase C_{18} column are recommended for separating the major pigments in particulate matter and sediments (Mantoura and Llewellyn, 1983; Wiltshire et al., 2000). The main pigments separated are usually chlorophylls a, b, phaeophorbides a, b, phaeophytin, chlorophyllides, chlorophyll c (as a single peak of c1 and c2), fucoxanthin, antheraxanthin, violaxanthin, diatoxanthin, diadinoxanthin, lutein, and the carotenes. The method uses a binary gradient with solvent A of a methanol:water:ammonium acetate solution (80:10:10, v:v), and solvent B of a

TABLE 10.9
Gradient of Solvents Devised for Simple HPLC Analysis of Pigments in Algae and Particulate Matter

Time (min)	%A	%B	%C
0	100	0	0
5	50	50	0
10	50	50	0
15	0	100	0
25	0	100	0
27	100	0	0
29	0	0	100
32	0	0	100
35	100	0	0

Note: Solvent mixture A is 80:10:10 of methanol:water: ammonium acetate solution, and B is 90:10 of methanol: acetone. Optional solvent mixture C is 10:7.7 of methanol: propanol.

methanol:acetone mixture (90:10, v:v). A third solvent, C, can be introduced as a methanol:propanol mix in a ratio of 10:7.7. It can support the separation of the carotenes more sharply on some columns. All solvents should be of HPLC grade and be degassed prior to use. The recommended flow rate is 1 mL min^{-1}. Suitable columns are reversed phase $5C_{18}$, 25 cm long, and packed with Nucleosil or equivalent. The gradient and protocol for this method are given in Tables 10.9 and 10.10, respectively. Before injection all extracts are filtered through a 0.45 μm pore size cellulose syringe filter. For a sharper resolution of the peaks at the front of the chromatogram the sample can be packed in between two water plugs, that is, by filling, for example, 20 μL water, 60 μL sample, 20 μL water into the injection loop and injecting as a block into the system. Samples should at least be injected in duplicate via a cooled autosampler directly into the HPLC system.

TABLE 10.10
HPLC Analysis of Pigments in Algae and Particulate Matter for HPLC with Autosampler and Injector, Long C_{18} Method

1. Extracts should be filtered using a 0.45 μm pore size cellulose (syringe) filter.
2. 60 μL of sample is packed in between two 20 μL water plugs (after Villerius et al., unpublished) injected in duplicate via a cooled autosampler straight into an HPLC system.
3. Commercial standards of chlorophyll a and b in 100% acetone (e.g., 0.01 mg L^{-1}, 0.1 mg L^{-1}, 1 mg L^{-1}, and 5 mg L^{-1}) should be used in a 5-point calibration every 100 samples. A 2-point calibration can be carried out every 15 samples, in order to check for possible system anomalies.
4. The optimal flow rate is 1 mL min^{-1}.
5. Column recommended is a reversed phase $5C_{18}$ column, 25 cm long. This column should preferably be kept thermostatized at 15°C in the column oven.
6. Solvent gradient as described in Table 10.9. All solvents should be of HPLC grade and degassed.

Commercial standards of chlorophyll a and b in 100% acetone (for example, 0.01, 0.1, 1, and 5 mg L^{-1}) can be used in a 5-point external calibration every one hundred samples to check the regression coefficient is in a range of 0.98 and 0.99 for the concentration range of 0.01 and 5 mg L^{-1}. A blank and 2-point calibration should be carried out every fifteen samples, in order to check for possible system anomalies. The standards used for the carotenoids and carotenes originate from a variety of sources. Often only lutein, α-carotene, and ß-carotene are available for reasonable costs. These can be calibrated as external standards, similar to the procedure with the chlorophyll standards, or if preferred, as an internal standard.

A calibration can be carried out as follows:

- The absorbance of the standard substance is measured in either ethanol or acetone on a sensitive spectrophotometer. The positions and heights of the absorbance maxima are checked against literature values (e.g., Jeffrey et al., 1997).
- Using the Lambert–Beer law and extinction coefficients from the literature (Jeffrey et al., 1997), the exact molar concentration of the stock solution is determined.
- This stock solution is then diluted volumetrically to allow 5-point calibration of the HPLC instrument. These standards are checked spectrophotometrically for their concentrations.
- Upon injection into the HPLC instrument, the absorbance of each substance at all wavelengths, the maxima, and the retention times are used as identification and, if possible, inserted in the reference library of the instrument.
- The measurements of most plant pigments of interest here can be made at 430 and 668 nm for a general differentiation or other specific wavelengths, depending on the maximum absorbance of the pigment to be identified.
- The areas of the chromatographic peaks are calculated using a valley-to-valley integration method with a peak threshold of 50, a peak width of 30, a minimum peak area of 5, and a minimum peak height of 1,000 absorbance units. Integrations and calculations of peak areas are typically included in the software of the HPLC instrument.
- If necessary, where peak bases overlap, peak purities are calculated (usually integrated into peak analyses software; see Jeffrey et al., 1997).

When the calibration of pigments is not possible, due to a lack of suitable standards, the algal pigment complements can be expressed in absolute absorbance units relative to the ubiquitous pigment chlorophyll a, which has been calibrated and used as reference for retention times of other pigments. Algal cultures with known pigment complements can be used as reference standard to verify peak identifications.

An example of the typical chromatographic separation that can be expected for a variety of algae (diatoms and green algae) using a C_{18} column is given in Figure 10.4a, and an example for a sediment (with diatoms and euglenids) in Figure 10.4b. The retention times and absorbance wavelengths are given in Table 10.11.

10.4.2.3   Second Method: Short Reversed Phase C_{18} Method

The second method is faster in the separation with an equal performance, but more suitable for cleaner samples, such as culture or oceanic samples (Garrido et al., 2000b). The C_8 methods of Zapata et al. (2000, 2001) are excellent for separating out specific pigments, but they also involve pyridine, which might be prohibitive under certain work conditions.

This second method involves a tertiary solvent gradient with solvent A as 100% methanol and solvent B as 0.025 mol L^{-1} aqueous pyridine solution (adjusted to pH 5 with acetic acid). The third solvent, C, is 100% acetone. All solvents are HPLC grade and degassed. The recommended flow rate is 4 mL min^{-1}. The column typically used is a monolithic reversed phase C_{18} column (100×4.6 mm). The gradient and protocols for this method are summarized in Tables 10.12 and 10.13, respectively.

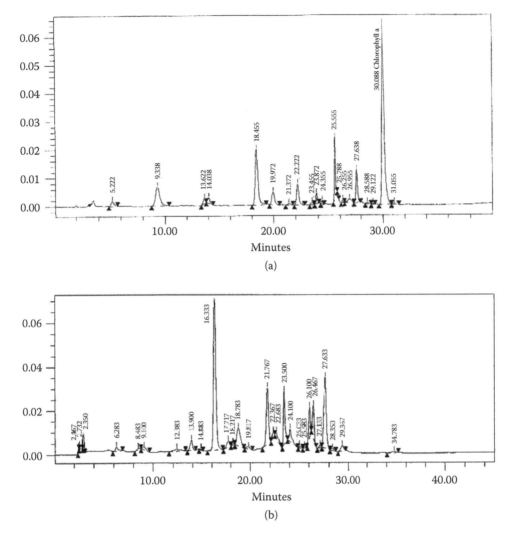

FIGURE 10.4 (a) Typical chromatogram of algal pigments, simple C_{18} method. (b) Chromatogram of pigments in sediments, simple C_{18} method.

See Garrido et al. (2000b) for specifics on retention times of pigments. First, all extracts were filtered using a 0.45 μm pore size nylon membrane filter (Whatman) (Knefelkamp et al., 2007). Then 100 μL of sample was packed in between two 15 μL water plugs (after Villerius, personal communication) injected in duplicate via a cooled autosampler (4°C) straight into an HPLC system.

The procedure for calibration is the same as described in Section 10.4.2.1.

10.4.3 IN SITU FLUORESCENCE METHODS

10.4.3.1 Introduction

The assessment of algal populations has vastly improved since more accurate methods (HPLC) for measuring pigments such as chlorophylls have become more widespread. However, the determination of algal biomass and exact species composition is still very problematic. Most current methods, although quite accurate, involve rather time-consuming enumeration to species or major taxonomic groups using counting chamber methods (Utermöhl, 1958; Sournia, 1978; Duarte et al., 1990).

TABLE 10.11

Retention Times and Absorbance Wavelength Maxima of the Different Pigments from Figure 10.4a and b, Using the C_{18} Column in Sediments and Algae

Compound Name	Sediment Retention Time (min)	Algae Retention Time (min)	Maxima (Eluent) (nm)
Chlorophyllide a	6.283	5.222	432.2, 667.2
Chlorophyllide b	6.483		469, 652
Chlorophyll c_1	8.483	13.622	442.9, 577.1, 636
Chlorophyll c_2	9.100	14.038	449.1, 582.8, 632.9
Neoxanthin	12.383		411.7, 435.8, 463.6
Fucoxanthin	16.333	18.883	452.4
Violaxanthin	18.783	19.972	415.3, 438.2, 468.5
Diadinoxanthin	21.767	22.222	420, 446.3, 476.8
Diatoxanthin	23.500	23.872	426, 453.6, 480.4
Lutein	24.100	25.555	421, 444.3, 474.3
Zeaxanthin		25.788	428, 452.4, 482.9
Phaeophorbide a	26.100		410.0, 666.0
Chlorophyll b	26.467	27.638	468.5, 652.5
Chlorophyll a	27.633	30.088	432.2, 664.7
Phaeophytin a	34.783		409.2, 665.5

Perhaps the greatest problem is that these methods are usually ex situ, and not suited to instant assays in situ. Often, aspects such as patchiness, vertical mixing, and algal migration are not detected because of the problems outlined above. Although it seems that not much has changed since 1890, when Haeckel (1890) considered phytoplankton counting to be a task that could not be accomplished without "ruin of mind and body," in situ methods do exist nowadays. Such new technologies include the differentiation of algal populations via fluorescence, algal color, and optical properties.

10.4.3.2 General Fluorescence Methods

Fluorescence emission measured at 685 nm is generally accepted as a measure of the chlorophyll contents of algae in aquatic environments. Indeed, depth profiling of chlorophyll fluorescence in aquatic systems has been carried out since the early 1970s (Kiefer, 1973). There are countless fluorescence devices in the form of bench-top instruments, in situ sensors, and fiber optic sensors available

TABLE 10.12

Tertiary Gradient Method Where Solvent A Consisted of 100% Methanol and Solvent B Is a 0.025 M Aqueous Pyridine Solution (Adjusted to pH 5 with Acetic Acid) (The Third Solvent, C, Is 100% Acetone)

Time (min)	%A	%B	%C
0	80	20	0
3	80	0	20
5	80	0	20
5.1	80	20	0

TABLE 10.13
HPLC Analysis of Pigments in Algae and Particulate Matter
for HPLC with Autosampler and Injector, Short C$_{18}$ Method

1. Extracts should be filtered using a 0.45 µm pore size cellulose (syringe) filter.
2. 100 µL of sample is packed in between two 15 µL water plugs after Villerius et al., unpublished) injected in duplicate via a cooled autosampler straight into an HPLC system.
3. Commercial standards of chlorophyll a and b in 100% acetone (e.g., 0.01 mg L^{-1}, 0.1 mg L^{-1}, 1 mg L^{-1}, and 5 mg L^{-1}) should be used in a 5-point calibration every 100 samples. A 2-point calibration can be carried out every 15 samples, in order to check for possible system anomalies.
4. The optimal flow rate is 4 mL min^{-1}.
5. Column recommended is a reversed phase 5C$_{18}$ column, 100 × 4.6 mm. This column should preferably be kept thermostatized at 15°C in the column oven.
6. Solvent gradient as described in Table 10.11. All solvents should be of HPLC grade and degassed.

(e.g., bbe Moldaenke, Chelsea Instruments, Hansatech, Perkin-Elmer, Hitachi, PSI Instruments, WET Labs, WALZ). The fluorescence from illuminated algae can be detected with such sensors that separate the red fluorescence emission from ambient light. The technology involved is basically either continuous excitation methods or modulated fluorescence methods. The former allows measurements of fast fluorescence kinetics (Kautsky effect) from dark-adapted algae. Optical filters separate excitation light used to illuminate the sample from the fluorescence signal. The sample must be measured in a light-proof environment to prevent ambient light from interfering with the measurement. In the modulated fluorescence method pulsed light is used to induce a pulsed fluorescence signal from the sample. Such measurements of the fluorescence can be made under ambient light conditions (see Turner, 1985; Schreiber, 1986; Falkowski et al., 1988; Boyd et al., 1997). These highly sensitive sensors are useful for the determination of chlorophyll and for analyzing the basic parameters of photosynthetic activity (see, for example, Falkowski et al., 1988; Dau, 1994; Kolber et al., 1990; Schreiber, 1998; Gilbert et al., 2000). However, ecological interests often go beyond this in that the differentiation of algal groups in the phytoplankton communities is of great importance.

10.4.3.3 Multialgal Color Differentiation Fluorescence Methods

Some attempts have been made in the past to distinguish different groups of phytoplankton using their fluorescence properties (Yentsch and Yentsch, 1979; Yentsch and Phinney, 1985; Kolbowski and Schreiber, 1995; Desiderio et al., 1997). However, mainly due to technical constraints, none of these methods was particularly successful, and they did not allow in situ real-time depth profiles. Through the advent of better LED, fiber optic, computer, and photomultiplier technology in recent years, these ideas have been revived.

In 1997–1998 Beutler et al. (1998) developed a new in situ measurement device, based on the fact that the color of algae is influenced by the pigments of the photosynthetic apparatus, and that different taxonomic groups differ significantly in their fluorescence excitation spectrum. The property of color is also a useful taxonomic criterion. Fluorescence is emitted mainly by the chlorophyll a of the photosystem II (PS II) antenna system, which consists of the evolutionarily conserved chlorophyll a core antenna and specific peripheral antennae. The association results in spectral differences in the fluorescence excitation spectra. Based on the fact that this pigment association is a fluorescence-determining factor, algal groups were designated according to similar fluorescence excitation spectra to distinguish between spectral signature groups (Beutler et al., 1998). Four spectral groups can

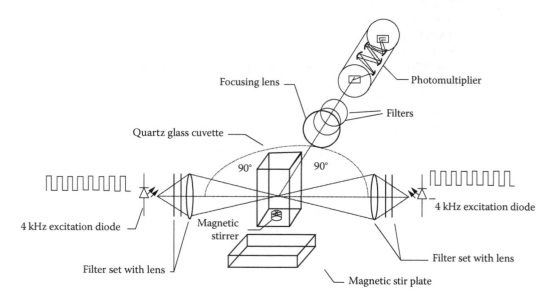

FIGURE 10.5 Schematic representation of multispectral fluorescence cuvette instrument. (Reprinted from Beutler, 1998.)

be distinguished using five distinct excitation wavelengths: (1) green, algae containing chlorophyll a/b; (2) blue, algae containing phycobilisomes rich in phycocyanin; (3) brown, algae containing chlorophyll a/c and green light-absorbing xanthophylls; and (4) mixed, algae containing chlorophyll a/c and phycoerythrin. The mean excitation probabilities per chlorophyll a concentration (in calibrated instrument dependent units) are calculated and used as norm spectra. Thus, it is possible to quantitatively determine the algal population distribution (i.e., given in terms of chlorophyll a per spectral algal group) within a phytoplankton sample. This technology can be applied in a cuvette stand-alone laboratory instrument similar to that conceptualized in Figure 10.5 or as submersible instruments for in situ measurements (Figure 10.6). The optics and electronics of the probe are mounted in a stainless steel housing (l = 45 cm, $\varnothing$ = 14 cm), allowing application down to a water depth of 100 m.

Chlorophyll a is excited with light from five LEDs (emission wavelengths: 450, 525, 570, 590, and 610 nm). The LEDs are switched alternately by a microcontroller with 4 kHz. Data can be stored in the probe or transferred to a PC. The resultant PS II fluorescence emission peak (685 nm) is detected rectangularly using an optimized optical bandpass-filter combination and a red-sensitive miniature photomultiplier. An external probe cover prevents the incidence of direct sunlight, which could cause perturbation of the measurement. High sensitivity and dynamic range enable measurement of fluorescence excitation spectra at low chlorophyll concentrations (0.02 µg L^{-1}). Spectra are recorded automatically at an integration time of 1 second. An iterative Gaussian fit weighted with the standard deviations of the norm spectra facilitates the estimation of the distribution of the spectral groups. Such fluoroprobes and in situ flow cytometers are really useful because they allow the differentiation of phycobilins, in particular between phycocyanin and phycoerythrin, providing differentiation of cyanobacteria.

10.4.3.4 The Analysis of Phycobilins Using Fluorescence and HPLC

The quantitative analysis of the phycobilin pigments has started to play a role. The reason for this is that the wet chemical methods available for the investigations of these pigments ex situ result in extremely variable results for algae and particulate matter substrates.

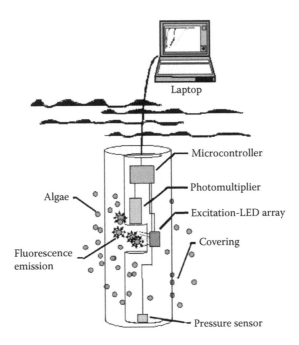

FIGURE 10.6 Schematic representation of multispectral fluorescence fluoroprobe, in situ instrument. (After Beutler et al., 1998. With permission.)

These methods generally utilize the absorbance spectra of crude aqueous extracts of the substrate (e.g., macro algae, phytoplankton, and particulate matter). They are highly susceptible to extraction errors (Sampath-Wiley and Neefus, 2007), and the formulae used to analyze the troughs and peaks in absorbance spectra often result in erroneous data. Even the most up-to-date methodology regarding phycobilin estimation shows clearly that many inaccuracies ensue in the extractions alone, and then thereafter also in the wet chemical analyses and calculations of concentrations.

There are HPLC methods available for the analyses of some of the subunits of the phycobilins and phycoerythrin, in particular phycourobilin. Isailovic et al. (2004) has used a C_4 analytical column and a binary gradient with 0.1% trifluoroacetic acid (TFA) in water and 0.1% TFA in acetonitrile as solvents. However, the sample preparation is laborious, as trypsin and pepsin digestions have to precede the injection of the pigments onto the column. These digests, however, leave not so easily determinable peptides on the chromophore.

Currently there are no practical and thus recommendable methods for the everyday analytical determination of phycobiliproteins from the complex phytoplankton matrices in aquatic systems. However, the multispectral in situ fluorescence determination (i.e., with the above-described fluoroprobe) can provide valuable monitoring data and an idea of in situ trends, even though it is impossible to calibrate beyond using laboratory cultures that have been counted as biomass.

10.4.3.5 The Reliability of Fluorescence Measurements

All fluorescence methods are as good as their calibration to algal biomass or chlorophyll, etc., and, for example, the cleanliness of the optical windows. It should be noted that old enumeration and HPLC methods of pigments cannot be entirely replaced by either the delayed fluorescence instruments or the fluoroprobe. Such analyses are vital to the calibration of in situ fluorescence measurements. Every new situation recorded by a fluorescence probe should be checked with an alternative method (e.g., HPLC). It should also be noted that the fluorescence signal of microalgae, especially when combined with measurements of photosynthetic activity, is extremely dependent on the nutrient status and health of the organisms. The success of the newer instruments is based on the differentiation of the vertical distribution of algae related to pigments and biomass in different water bodies. It is also based on the wide variety of other instruments that have recently become available, allowing algal differentiation. Such technology represents an important tool in aquatic ecology and for supervision of aquatic resources.

10.5 FUTURE PERSPECTIVES

Photosynthetic pigments are used in many areas of aquatic research (see also Millie et al., 1993). Although many of the ideas attached to these new applications may not be entirely new, the applications have largely only become possible through the advent of new technologies. For example, we are, through the advances in chromatographic systems (HPLC), now truly able to differentiate between chlorophylls and their breakdown products, thus eliminating all the problems associated with the erroneous inclusion of these products in more classical chlorophyll assessments. This was an important move forward, particularly benefiting photosynthesis research and vital work in which chlorophyll is used as a monitoring proxy of changes in algal biomass in oceanic climate research.

Research on and evaluation of microalgal biomass in aquatic systems has clearly profited from the development of new methods and technology. In the early 1990s we had to accept the separation of only the chlorophylls from one another. Now new methods, particularly HPLC and fluorescence, allow really high resolution of carotenoids, carotenes, chlorophylls, and derivatives of these in innumerable substrates. This is very useful, as it prevents errors in the classical chlorophyll and biomass estimations. It allows the separation of algal marker pigments in small amounts of substrates of complex matrices such as sediments. If required, a combination with techniques involving mass spectrometry would allow exact differentiation of unknown substances. This form of resolution has, however, not yet been required for most aquatic pigment work, but would be necessary in the analyses of pigments in ancient sediments, or the geological record (see work of Vallentyne, 1954; Gorham, 1960; Fogg and Belcher, 1961; Baker, 1966; Sanger, 1988; Callot, 1991; Repeta, 1989). This resolution may also be important in studies on degradation of algae in the water column and sediments, a scenario that would also be relevant in studies of feeding on phytoplankton (see Sanger, 1988; Repeta and Gagosian, 1984; Strom and Welschmeyer, 1991; Abele, 1988; Abele-Oeschger, 1991).

It is important to remember when analyzing aquatic pigments that all pigment complements in particulate matter (algae) are governed by environmental conditions, ranging from self-shading in large algal colonies to grazing and light conditions.

The main problem with high-resolution chemical analysis of algal material from aquatic systems is that it is retrospect. This is when in situ optical techniques become important. The ideas behind the investigation of optical differentiation of phytoplankton populations can also be applied to the development of measuring devices for algal populations on sediments. Similar to the PHYTOPAM (see http//www.walz.com) and the Scuba fluorometer of Gorbunov et al. (2000), optical methods can be conceptualized for differentiating algae in or on sediments. This would be useful for algal biomass estimations and "ground truth" measurements for remote sensing applications.

With all the new optical methods available, questions arise as to what kind of future developments will be feasible. It may be possible that in situ measurements will use chip technology using genetic markers for specific algae. It might even become possible to do broad-spectrum antibody tests for organisms in situ. However, it is unlikely that classical counting techniques, HPLC techniques, and the newer optical techniques will disappear in the future. All of these methods have their own merits, and that is unlikely to change. A combination of methods is usually the best and most useful approach, and an adjustment of the methods to the question at hand the most practical. At present, exact taxonomic differentiation can only be carried out by counting and biovolume measurement. Reliable quantitative pigment determination can only be carried out using HPLC. However, future developments of in situ techniques for determining algal components in aquatic systems are likely to be based on methods like the Fluoroprobe, PHYTOPAM, and chip technology. Classical photometric and fluorescence methods should be relegated to those studies only involving mere trends in algal populations.

REFERENCES

Abele, D. 1988. Carotinoide als biogene Marker für benthische Makroalgen im Sediment der Kieler Bucht. *Berichte aus dem Institut fuer Meereskunde Kiel* 183:1–116.

Abele-Oeschger, D. 1991. Potential of some carotenoids in two recent sediments of Kiel Bight as biogenic indicators of phytodetritus. *Marine Ecology Progress Series* 70:83–92.

Airs, R. L., J. E. Atkinson, and B. J. Keely. 2001. Development and application of a high resolution liquid chromatography method for the analysis of complex pigment distributions. *Journal of Chromatography* 917A:167–77.

American Water Works Association (AWWA-APHA). 1985. *Standard methods for the examination of water and waste water.* 16th ed.

American Water Works Association (AWWA-APHA). 1989. *Standard methods for the examination of water and waste water.* 17th ed.

Andersen, R. A., and T. J. Mulkey. 1983. The occurrence of chlorophyll C1 and C2 in the Chrysophyceae. *Journal of Phycology* 19:289–94.

Baker, E. W. 1966. Mass spectrometric characterization of petroporphyrins. *Journal of the American Chemical Society* 88:2311.

Berzelius, J. 1838. Untersuchungen des Blattgrüns (Chlorophylls). *Annalen der Pharmazie* 27:296–318.

Beutler, M. 1998. Entwicklung eines Verfahrens zur quantitativen Bestimmung von Algengruppen mit Hilfe computergestützter Auswertung spektralaufgelöster Fluoreszenzanregungsspektren. Diploma thesis, Christian-Albrechts-Universität Kiel.

Beutler, M., B. Meyer, K. H. Wiltshire, C. Moldaenke, and H. Dau. 1998. Rapid depth-profiling of the distribution of 'spectral groups' of microalgae in lakes, rivers and the sea. The method and a newly-developed submersible instrument which utilizes excitation of chlorophyll fluorescence in five distinct wavelength ranges. In *Photosynthesis: Mechanisms and effects*, ed. G. Garab. Vol 5. Dordrecht, The Netherlands: Kluwer Academic Publishers.

Bowles, N. D., H. W. Pearl, and J. Tucker. 1985. Effective solvents and extraction periods employed in phytoplankton carotenoid and chlorophyll determination. *Canadian Journal of Fisheries and Aquatic Sciences* 42:1127–31.

Boyd, P. W., J. Aiken, and Z. Kolber. 1997. Comparison of radiocarbon and fluorescence based (pump and probe) measurements of phytoplankton photosynthetic characteristics in the northeast Atlantic Ocean. *Marine Ecology Progress Series* 149:215–26.

Brotas, V., C. R. Mendes, and P. Cartaxana. 2007. Microphytobenthic biomass assessment by pigment analysis: Comparison of spectrophotometry and high performance liquid chromatography methods. *Hydrobiologia* 587:19–24.

Callot, H. J. 1991. Geochemistry of chlorophylls. In *Chlorophylls*, ed. H. Scheer. Boca Raton, FL: CRC Press.

Chang, W. Y. B., and R. Rossmann. 1982. The influence of phytoplankton composition on the relative effectiveness of grinding and sonication for chlorophyll extraction. *Hydrobiologia* 88:245–49.

Claustre, H., S. B. Hooker, L. Van Heukelem, J. F. Berthond, R. Barlow, J. Rasa, H. Sessions, C. Targad, C. S. Thomas, D. van der Linde, and J. C. Marty. 2004. An intercomparison of HPLC phytoplankton pigment methods using in situ samples: Application to remote sensing and database activities. *Marine Chemistry* 85:41–61.

Coveney, M. N. 1982. Elimination of chlorophyll b interference in the fluorometric determination of chlorophyll a and phaeopigments. *Archiv für Hydrobiologie* 16:77–90.

Daemen, E. A. 1986. Comparison of methods for the determination of chlorophyll in estuarine sediments. *Netherlands Journal of Sea Research* 20:21–28.

Dau, H. 1994. Molecular mechanisms and quantitative models of variable photosystem II fluorescence. *Photochemistry and Photobiology* 60:1–24.

Desiderio, R. A., C. Moore, C. Lantz, and T. J. Cowles. 1997. Multiple excitation fluorometer for in situ oceanographic applications. *Applied Optics* 36:1289–96.

Deventer, B. 1985. Methodische Untersuchungen zur Chlorophyllbestimmung im Gewässer. Diploma thesis, Faculty of Biology, University of Hamburg.

Duarte, C. M., C. Marrase, D. Vaque, and M. Estrada. 1990. Counting error and the quantitative analysis of phytoplankton communities. *Journal of Plankton Research* 12:295–304.

Egeland, E. S., W. Eikrem, J. Throndsen, C. Wilhelm, M. Zapata, and S. Liaaen-Jensen. 1995. Carotenoids from further Prasionophytes. *Biochemical Systematics and Ecology* 23:747–55.

Egeland, E. S., and S. Liaaen-Jensen. 1993. New carotenoids and chemosystematics in the Prasionophyceae. In Proceedings of the 10th International Symposium on Carotenoids, Trondheim, Norway, pp. 4–6.

Falkowski, P. G. 1980. Light-shade adaptation in marine phytoplankton. In *Primary production in the sea*, ed. P. G. Falkowski. New York: Plenum.

Falkowski, P. G., Z. Kolber, and Y. Fujita. 1988. Effect of redox state on the dynamics of photosystem II during steady state photosynthesis in eukaryotic algae. *Biochimica et Biophysica Acta* 933:432–43.

Falkowski, P. G., and T. G. Owens. 1978. Light-shade adaptation: Two strategies in marine phytoplankton. *Plant Physiology* 66:592–95.

Falkowski, P. G., and J. A. Raven. 1997. *Aquatic photosynthesis*. Oxford: Blackwell Publishing.

Fogg, G. E., and J. H. Belcher. 1961. Pigments from the bottom deposits of an English lake. *New Phytology* 60:129–49.

Foy, R. H. 1993. The phycocyanin to chlorophyll a ratio and other cell components as indicators of nutrient limitation in two planktonic cyanobacteria subjected to low light exposure. *Journal of Plankton Research* 15:1263–76.

Garrido, J. L., J. Otero, M. A. Maestro, and M. Zapata. 2000a. The main nonpolar chlorophyll c from *Emiliania huxleyi* (Prymnesiophyceae) is a chlorophyll c2-monogalactosyldiacylglyceride ester: A mass spectrometry study. *Journal of Phycology* 36:497–505.

Garrido J. L., F. Rodríguez, E. Campaña, and M. Zapata. 2000b. Rapid separation of chlorophylls a and b and their demetallated and dehytylated derivatives using a monolithic silica C_{18} column and a pyridine-containing mobile phase. *Journal of Chromatography* 994A:85–92.

Garrido, J. L., and M. Zapata. 2006. Chlorophyll analysis by new HPLC methods. In *Chlorophylls and bacteriochlorophylls: Biochemistry, biophysics, functions and applications*, ed. B. Grimm, R. J. Porra, W. Rüdiger, and U. Scheer. Dordrecht, The Netherlands: Springer.

Gieskes, W. W. C., and G. W. Kraay. 1983a. Dominance of Cryptophyceae during the phytoplankton spring bloom in the central North Sea detected by HPLC analyses of pigments. *Marine Biology* 75: 179–85.

Gieskes, W. W. C., and G. W. Kraay. 1983b. Unknown chlorophyll a derivatives in the North Sea and the tropical Atlantic Ocean as revealed by HPLC analyses. *Limnology and Oceanography* 28:757–66.

Gieskes, W. W. C., G. W. Kraay, A. Nontji, D. Setiapermana, and D. Sutomo. 1988. Monsoonal alternation of a mixed and a layered structure in the phytoplankton of the euphotic zone of the Banda Sea (Indonesia): A mathematical analysis of algal pigment fingerprints. *Netherlands Journal of Sea Research* 22:123–37.

Gilbert, M., C. Wilhelm, and M. Richter. 2000. Bio-optical modelling of oxygen evolution using in vivo fluorescence: Comparison of measured and calculated photosynthesis/irradiance (P-I) curves in four representative phytoplankton species. *Journal of Plant Physiology* 157:307–14.

Goericke, R. 2002. Bacteriochlorophyll *a* in the ocean: Is anoxygenic photosynthesis important? *Limnology and Oceanography* 47:290–95.

Goericke, R., and D. J. Repeta. 1992. The pigments of *Prochlorococcus marinus*: The presence of divinyl chlorophyll a and b in a marine prokaryote. *Limnology and Oceanography* 37:425–33.

Goericke, R., and D. J. Repeta. 1993. Chlorophylls *a* and *b* and divinyl chlorophylls *a* and *b* in the open subtropical North Atlantic Ocean. *Marine Ecology Progress Series* 101:307–13.

Gorbunov, M. Y., P. G. Falkowski, and Z. Kolber. 2000. Measurement of photosynthetic parameters in benthic organisms in situ using a SCUBA-based fast repetition rate fluorometer. *Limnology and Oceanography* 45:242–45.

Gorham, E. 1960. Chlorophyll derivatives in the surface muds of English lakes. *Limnology and Oceanography* 5:29–33.

Haeckel, E. 1890. Plankton-Studien: Vergleichende Untersuchungen über die Bedeutung und Zusammensetzung der pelagischen Fauna und Flora. Jena, Germany: Verlag von Gustav Fischer.

Helfrich, M., A. Ross, G. C. King, A. G. Turner, and A. D. W. Larkum. 1999. Identification of 8-(vinyl) protochlorophyllide *a* in phototrophic prokaryotes and algae: Chemical and spectroscopic properties. *Biochimica et Biophysica Acta* 1410:262–72.

Hooker, S. B., H. Claustre, J. Ras, L. Van Heukelem, J. F. Berthon, C. Targa, D. van der Linde, R. Barlow, and H. Sessions. 2000. The first SeaWiFS HPLC analysis round-robin experiment (SeaHARRE-1). In *NASA Technical Memorandum 2000-206892*, ed. S. B. Hooker and E. R. Firestone. Vol. 14. Greenbelt, MD: NASA Goddard Space Flight Center.

Isailovic, D., H. W. Li, and E. S. Yeung. 2004. Isolation and characterisation of R-phycoerythrin subunits and enzymatic digests. *Journal of Chromatography* 1051A:119–30.

Jacobsen, T. R. 1982. Comparison of chlorophyll a measurements by fluorometric, spectrophotometric and high performance liquid chromatography methods in aquatic environments. *Archiv für Hydrobiology* 16:35–45.

Jeffrey, S. W., R. F. C. Mantoura, and S. W. Wright. 1997. *Phytoplankton pigments in oceanography*. Paris: SCOR-UNESCO.

Jeffrey, S. W., and M. Vesk. 1997. Introduction to marine phytoplankton and their pigment signatures. In *Phytoplankton pigments in oceanography*, ed. S. W. Jeffrey, R. F. C. Mantoura, and S. W. Wright. Paris: SCOR-UNESCO, pp. 37–84.

Jeffrey, S. W., and S. W. Wright. 1987. A new spectrally distinct component in preparations of chlorophyll c from the micro-alga *Emiliania huxleyi* (Prymnesiophyceae). *Biochimica et Biophysica Acta* 894:180–88.

Joint, I. R., J. M. Gee, and R. M. Warwick. 1982. Determination of fine-scale vertical distribution of microbes and meiofauna in an intertidal sediment. *Marine Biology* 72:157–64.

Kiefer, D. A. 1973. Fluorescence properties of natural phytoplankton populations. *Marine Biology* 22:263–69.

Kirk, J. T. O. 1983. *Light and photosynthesis in aquatic ecosystems*. Cambridge, UK: Cambridge University Press.

Knefelkamp, B., K. Carstens, and K. H. Wiltshire. 2007. Comparison of different filter types on chlorophyll-a retention and nutrient measurements. *Journal of Experimental Marine Biology and Ecology* 345:61–70.

Koblížek, M., M. Mašin, J. Ras, A. J. Poulton, and O. Prášil. 2007. Rapid growth rates of aerobic anoxygenic phototrophs in the ocean. *Environmental Microbiology* 9:2401–6.

Kolber Z., K. D. Wyman, and P. G. Falkowski. 1990. Natural variability in photosynthetic energy conversion efficiency: A field study in the Gulf of Maine USA. *Limnology and Oceanography* 35:72–79.

Kolbowski, J., and U. Schreiber. 1995. Computer-controlled phytoplankton analyser based on a 4-wavelengths PAM chlorophyll fluorometer. In *Proceedings of the Xth International Photoshynthesis Congress Photosynthesis: From Light to Biosphere*, ed. P. Mathis. Vol. 5. Dordrecht, The Netherlands: Kluwer Academic Publisher.

Lami, R., T. Cottrell, J. Ras, O. Ulloa, I. Obernosterer, H. Claustre, D. L. Kirchman, and P. Lebaron. 2007. High abundances of aerobic anoxygenic bacteria in the South Pacific Ocean. *Applied and Environmental Microbiology* 73:4198–4205.

Lohr, M., and C. Wilhelm. 1999. Algae displaying the diadinoxanthin cycle also possess the violaxanthin cycle. *Proceedings of the National Academy of Sciences USA* 96:8784–89.

Manning, W. M., and H. H. Strain. 1943. Chlorophyll d: A green pigment in red algae. *Journal of Biological Chemistry* 151:1–19.

Mantoura, R. F. C., and C. A. Llewellyn. 1983. The rapid determination of algal chlorophyll and carotenoid pigments and their breakdown products in natural waters by reversed-phase HPLC. *Analytica Chimica Acta* 151:297–314.

Meeks, J. C. 1974. Chlorophylls. In *Algal physiology and biochemistry*, ed. W. D. P. Stewart. Oxford: Blackwell Publishing.

Millie, D. F., H. W. Paerl, and J. P. Hurley. 1993. Microbial pigment assessments using high performance liquid chromatography—A synopsis of organismal and ecological applications. *Canadian Journal of Fisheries and Aquatic Sciences* 50:2513–27.

Miyashita, H., K. Adachi, N. Kurano, H. Ikemot, M. Chihara, and S. Miyach. 1997. Pigment composition of a novel oxygenic photosynthetic prokaryote containing chlorophyll *d* as the major chlorophyll. *Plant Cell Physiology* 38:274–81.

Miyashita, H., H. Ikemoto, N. Kurano, K. Adachi, M. Chihara, and S. Miyachi. 1996. Chlorophyll *d* as a major pigment. *Nature* 383:402.

Moss, G. P. 1987. Nomenclature of tetrapyrroles. *Pure Applied Chemistry* 57:779–832.

Nichols, B. W. 1973. Lipid composition and metabolism. In *The biology of the blue-green algae*, ed. N. G. Carr and B. A. Whitton. Oxford: Blackwell Scientific Publications.

Nicklisch, A., and P. Woitke. 1999. Pigment content of selected planktonic algae in response to simulated natural light fluctuations and a short photoperiod. *International Review of Hydrobiology* 84:479–95.

Parker, W. R. 1991. Quality control in mud coring. *Geo-Marine Science* 11:132–37.

Pelletier, F., and J. B. Caventou. 1818. Sur la matière verte des feuilles. *Annales des Chimie et des Physique* 9:194–96.

Pinckney, J. L., R. Papa, and R. Zingmark. 1994. Comparison of high performance liquid chromatographic spectrophotometric and fluorometric methods for determining chlorophyll concentrations in estuarine sediments. *Journal of Microbiological Methods* 19:59–66.

Porra, R. J. 1991. Recent advances and re-assessments in chlorophyll extraction and assay procedures for terrestrial, aquatic and marine organisms, including recalcitrant algae. In *Chlorophylls*, ed. H. Scheer. Boca Raton, FL: CRC Press, pp. 31–58.

Post, A. F. 1986. Transient state characteristics of adaptation to changes in light conditions for the cyanobacterium *Oscillatoria agardhii*. I. Pigmentation and photosynthesis. *Archives of Microbiology* 145:353–57.

Repeta, D. J. 1989. Carotenoid diagenesis in recent marine sediments. II. Degradation of fucoxanthin to loliolide. *Geochimica et Cosmochimica Acta* 53:699–707.

Repeta, D. J., and R. B. Gagosian. 1984. Transformation reactions and recycling of carotenoids and chlorins in the Peru upwelling region. *Geochimica et Cosmochimica Acta* 48:1265–77.

Ricketts, T. R. 1970. The pigments of Prasinophyceae and related organisms. *Phytochemistry* 9:1835–42.

Riley, J. P., and T. R. S. Wilson. 1965. The use of thin-layer chromatography for the separation and identification of phytoplankton pigments. *Journal of the Marine Biological Association of the United Kingdom* 45:583–91.

Rutledge, P. A., and J. W. Fleeger. 1988. Laboratory studies on core sampling with application to subtidal meiobenthos collection. *Limnology and Oceanography* 33:274–80.

Sampath-Wiley, P., and C. Neefus. 2007. An improved method for estimating R-phycoerythrin and R-Phycocyanin contents from crude aqueous extracts of *Porphyra* (Bangiales, Rhodophyta). *Journal of Applied Phycology* 19:123–29.

Sanger, J. E. 1988. Fossil pigments in palaeoecology and palaeolimnology. *Palaeogeography Palaeoclimatology Palaeoecology* 62:343–59.

Schanz, F., and H. Rai. 1988. Extract preparation and comparison of fluorometric, chromatographic (HPLC) and spectrophotometric determinations of chlorophyll a. *Archiv für Hydrobiologie* 112:533–39.

Scheer, H. 1991. Chemistry of chlorophylls. In *Chlorophylls*, ed. H. Scheer. Boca Raton, FL: CRC Press.

Schreiber, U. 1986. Detection of rapid induction kinetics with a new type of high frequency modulated chlorophyll fluorometer. *Photosynthesis Research* 9:261–72.

Schreiber, U. 1998. Chlorophyll fluorescence: New instruments for special applications. In *Proceedings of the XIth International Photosynthesis Congress: Photosynthesis: From Light to Biosphere*, ed. G. Garab. Vol. 5. Dordrecht, The Netherlands: Kluwer Academic Publisher.

SCOR-UNESCO. 1966. *Determination of photosynthetic pigments in seawater*. Monographs on Oceanographic Methodology 1. Paris: UNESCO.

SCOR-UNESCO. 1980. *Determination of chlorophyll in seawater*. UNESCO Technical Papers in Marine Science 35. Paris: UNESCO.

Sorby, H. C. 1873. On comparative vegetable chromatology. *Proceedings of the Royal Society of London Series* 21:144–83.

Sorby, H. C. 1877. On the characteristic colouring-matter of red groups of algae. *Botanical Journal of the Linnean Society* 15:34–40.

Sournia, A. 1978. *Phytoplankton manual*. Paris: UNESCO.

Steele, J. H., and I. E. Baird. 1962. Carbon-chlorophyll relations in cultures. *Limnology and Oceanography* 7:101–2.

Steele, J. H., and I. E. Baird. 1965. The chlorophyll a content of particulate organic matter in the North Sea. *Limnology and Oceanography* 10:261–67.

Strain, H., and W. A. Svec. 1966. Extraction, separation estimation and isolation of the chlorophylls. In *The chlorophylls*, ed L. P. Vernon and G. R. Seely. New York: Academic Press, pp. 21–66.

Straub, O. 1976. *Key to the carotenoids: Lists of natural carotenoids*. Basel, Switzerland: Birkhäuser.

Strom, S. L., and N. A. Welschmeyer. 1991. Pigment-specific rates of phytoplankton growth and microzooplankton grazing in the open subarctic Pacific Ocean. *Limnology and Oceanography* 36:50–63.

Stokes, G. G. 1864. On the supposed identity of biliverdin with chlorophyll, with remarks on the constitution of chlorophyll. *Proceedings of the Royal Society of London Series* 13:144–45.

Svec, W. A. 1991. The distribution and extraction of the chlorophylls. In *Chlorophylls*, ed. H. Scheer. Bota Raton, FL: CRC Press, pp. 89–102.

Tett, P., M. G. Kelly, and G. M. Hornberger. 1975. A method for the spectrophotometric measurement of chlorophyll a and phaeophytin a in benthic microalgae. *Limnology and Oceanogaphy* 20:887–96.

Tswett, M. 1906. Adsorptionsanalyse und chromatographische Methode. Anwendung auf die Chemie des Chlorophylls. *Berichte der Deutschen Botanischen Gesellschaft* 24:384–93.

Turner, G. K. 1985. Measurement of light from chemical or biochemical reactions. In *Bioluminnescence and chemiluminescence: Instruments and applications*, ed. K. Van Dyke. Boca Raton, FL: CRC Press, pp. 43–78.

Utermöhl, H. 1958. Zur Vervollkommung der quantitativen Phytoplankton-Methodik. *Mitteilungen der Internationalen Vereinigung für Limnologie* 9:1–38.

Vallentyne, J. R. 1954. Biochemical limnology. *Science* 119:605–6.

Van Den Hoek, C., D. G. Mann, and H. M. Jahns. 1997. *Algae: An introduction to phycology.* Cambridge, UK: Cambridge University Press.

Van Heukelem, L., and C. S. Thomas. 2001. Computer-assisted high-performance liquid chromatography method development with applications to the isolation and analysis of phytoplankton pigments. *Journal of Chromatography* 910A:31–49.

Wilhelm, C., I. Lenartz-Weiler, I. Wiedemann, and A. Wild. 1986. The light-harvesting system of a *Micromonas* species (Prasinophyceae): The combination of three different Chl species in one single Chl-protein complex. *Phycologia* 25:304–12.

Wilhelm, C., I. Rudolph, and W. Renner. 1991. A quantitative method based on HPLC-aided pigment analysis to monitor structure and dynamics of the phytoplankton assemblages—A study from Lake Meerfelder Maar (Eifel, Germany). *Archiv für Hydrobiologie* 123:21–35.

Wiltshire, K. H. 1992. Untersuchungen zum Einfluß des Mikrophytobenthos auf den Nährstoffaustausch zwischen Sediment und Wasser in der Tide-Elbe. PhD thesis, University of Hamburg.

Wiltshire, K. H. 2000. Algae and associated pigments of intertidal sediments, new observations and methods. *Limnologica* 30:205–14.

Wiltshire, K. H., J. Blackburn, and D. Paterson. 1997. The Cryolander: A new method for the in situ sampling of intertidal surface sediments minimizing distortion of sediment structure. *Journal of Sedimentary Research* 67:977–81.

Wiltshire, K. H., S. Harsdorf, B. Smidt, G. Blöcker, R. Reuter, and F. Schroeder. 1998a. The determination of algal biomass (as chlorophyll) in suspended matter from the Elbe estuary and the German Bight: A comparison of HPLC, delayed fluorescence and prompt fluorescence methods. *Journal of Experimental Marine Biology and Ecology* 222:113–31.

Wiltshire, K. H., T. Tolhurst, D. M. Paterson, I. Davidson, and G. Gust. 1998b. Pigment fingerprints as markers of erosion and changes in cohesive surface properties in simulated and natural erosion events. *Geological Society, London, Special Publications* 139:99–114.

Wiltshire K. H., M. Boersma, A. Möller, and H. Buhtz. 2000. The extraction and analyses of pigments and fatty acids from the green alga *Scenedesmus obliquus* (Chlorophyceae). *Aquatic Ecology* 34:119–26.

Withers, N. W., A. Fiksdahl, R. C. Tuttle, and L. S. Jensen. 1981. Carotenoids of the Chrysophyceae. *Comparative Biochemistry and Physiology* 68B:345–49.

Wright, S. W., S. W. Jeffrey, and R. F. C. Mantoura. 1997. Evaluation of methods and solvents for pigment extraction. In *Phytoplankton pigments in oceanography*, ed. S. W. Jeffrey, S. W. Wright, and R. F. C. Mantoura. Paris: UNESCO, pp. 104–116.

Yentsch, C., and D. Menzel. 1963. A method for the determination of phytoplankton chlorophyll and phaeophytin by fluorescence. *Deep Sea Research* 10:221–31.

Yentsch, C. S., and D. A. Phinney. 1985. Spectral fluorescence: A taxonomic tool for studying the structure of phytoplankton populations. *Journal Plankton Research* 7:617–32.

Yentsch, C. S., and R. Vaccaro. 1958. Phytoplankton nitrogen in the oceans. *Limnology and Oceanography* 3:443–48.

Yentsch, C. S., and C. M. Yentsch. 1979. Fluorescence spectral signatures: The characterisation of phytoplankton populations by the use of excitation and emission spectra. *Journal of Marine Research* 37:471–83.

Zapata, M., B. Edvardsen, F. Rodríguez, M. Maestro, and J. L. Garrido. 2001. Chlorophyll c2 monogalactosyldiacylglyceride ester (chl c2-MGDG). A novel marker pigment for *Chrysochromulina* species (Haptophyta). *Marine Ecology Progress Series* 219:85–98.

Zapata, M., and J. L. Garrido. 1991. Influence of injection conditions in reversed-phase high-performance liquid chromatography of chlorophylls and carotenoids. *Chromatographia* 31:589–94.

Zapata, M., F. Rodríguez, and J. L. Garrido. 2000. Separation of chlorophylls and carotenoids from marine phytoplankton: A new HPLC method using a reversed-phase C_8 column and pyridine-containing mobile phases. *Marine Ecology Progress Series* 195:29–45.

11 Determination of DMS, DMSP, and DMSO in Seawater

Jacqueline Stefels

CONTENTS

11.1 INTRODUCTION

Lovelock et al. (1972) was the first to report dimethylsulfide (DMS) measurements from ocean areas and to recognize its importance for the global sulfur cycle. DMS is a semivolatile organic sulfur compound that accounts for 50% to 60% of the total natural reduced sulfur flux to the atmosphere, including emissions from volcanoes and vegetation (Andreae, 1990; Bates et al., 1992; Spiro et al., 1992). Of the total global DMS emissions, about 95% come from the oceans, with estimates of the emission ranging from 15–33 Tg S y^{-1} (Kettle and Andreae, 2000). All surface ocean water is supersaturated with DMS, and typical concentrations for open oceans range between 0.1 and 5 nM, but local increases can be observed in phytoplankton blooms. In the late 1980s, the hypothesis that DMS is involved in the biological regulation of global climate was put forward (Bates et al., 1987; Charlson et al., 1987). After emission to the atmosphere, DMS is oxidized to sulfur dioxide (SO_2) and other products. From SO_2, non-sea-salt (nss) sulfate is produced, which can form sulfate (SO_4^{2-}) particles that act as condensation nuclei for water vapor. These nuclei affect the radiative properties of the atmosphere and clouds, with implications for climate. Higher numbers of condensation nuclei will reflect more incoming solar radiation back into space and thereby reduce the temperature on earth.

Currently, anthropogenic SO_2 production exceeds natural SO_2 production by a factor of 3 (Bates et al., 1992), but the impact of the former on aerosol production is largely confined to industrialized areas of the northern hemisphere. The oceans, on the other hand, cover approximately 70% of the earth's surface, and much of this area is remote from man-made atmospheric contaminants. Consequently, the exchange of marine DMS is of high regional importance and may affect climate globally. For instance, in the southern hemisphere, where anthropogenic sulfate emission is low, DMS plays a major role in the production of atmospheric nss sulfate, with maximum contributions over the Southern Ocean in excess of 80% during summer (Ayers and Gillett, 2000; Gondwe et al., 2003).

Since the publication of a global inventory of DMS data by Kettle and co-authors (Kettle et al., 1999), it has become possible to include DMS in global climate models (e.g., Bopp et al., 2003, 2004; Gabric et al., 2004; Gunson et al., 2006). With these new developments, caveats in our understanding of the DMS-climate coupling have became apparent, showing that an improved coverage of ocean data and greater understanding of the biological processes are necessary to quantify the role of DMS in climate feedback mechanisms. In this respect, the Global Surface Seawater Dimethylsulfide Database (http://saga.pmel.noaa.gov/dms/) is an invaluable tool, to which individual researchers can submit their data and from which data can be retrieved.

The production of DMS is almost exclusively through biogenic processes and shows strong seasonal and latitudinal variation (Kettle et al., 1999). DMS mainly results from the enzymatic cleavage of dimethylsulfoniopropionate (DMSP), a compound that is produced in several groups of marine phytoplankton. A complex network of production and consumption pathways of both DMSP and DMS involves most of the microbial food web and determines the concentration of DMS in surface water, and consequently its flux to the atmosphere. Physical and chemical ecosystem parameters all affect this network, potentially resulting in dramatic shifts in DMS emissions, the magnitude of which we are only beginning to understand. During the past decade, many reviews have been written on aspects of the marine sulfur cycle (see Stefels et al., 2007, and references therein). One of the emerging pictures is not only that this cycle is of interest for global climate, but that DMS and DMSP are compounds that are central to the microbial food web in their own right. In the past 10 years, another important compound, dimethylsulfoxide (DMSO), has been found to be intimately connected to the cycling of DMS and DMSP. DMSO is a photochemical and biological oxidation product of DMS with concentration ranges comparable to those of DMS and DMSP (del Valle et al., 2007; Hatton et al., 2004; Simo and Vila-Costa, 2006). It is thought to be produced both in algae and in the water phase. The major pools and processes involved in the marine sulfur cycle are depicted in Figure 11.1.

For a proper understanding of the marine sulfur cycle it is a necessity to analyze all major pools described in Figure 11.1. Therefore, the method described in this chapter not only concerns the accurate measurement of DMS, but also that of DMSP and DMSO in both the dissolved fraction ($DMSP_d$ and $DMSO_d$), which is taken to be less than the 0.7 µm filter fraction (i.e., the nominal pore size of Whatman GF/F filters), and the fraction associated with particles ($DMSP_p$ and $DMSO_p$). The latter is operationally defined as that captured on the same Whatman GF/F filters. All compounds are gas chromatographically analyzed as DMS: DMSP after base hydrolysis with NaOH and DMSO after reduction with titanium III chloride.

Development of a purge-and-trap method coupled with gas chromatographic analysis of volatile organic sulfur compounds took place between 1980 and 1990 (e.g., Andreae and Barnard, 1983; Turner et al., 1990), and most current methods are derivations: volatile compounds are stripped from a water sample with high-purity purge gas, focused on a cryogenic trap, and subsequently injected onto a gas chromatographic column. With the development of easy-to-use sulfur-specific detectors and the focus on DMS and not other reduced sulfur compounds, this method became increasingly simpler. Also, we now know more about biological conversion processes of DMS and associated compounds, which gives insight into many errors associated with sample preparation that, in the past, have resulted in unexplained variations of the analyses. These new insights will be discussed in the sample preparation section.

11.2   SAMPLING AND SAMPLE PREPARATION

11.2.1   Materials

All materials used should be gas-tight and acid cleaned. Cleaning procedure: soak overnight in 0.1 N HCl, rinse five times with ultrapure water, and dry, if possible at 70°C. Deactivation of glassware in order to prevent compounds adhering to glass surfaces, as prescribed in many of the early methods,

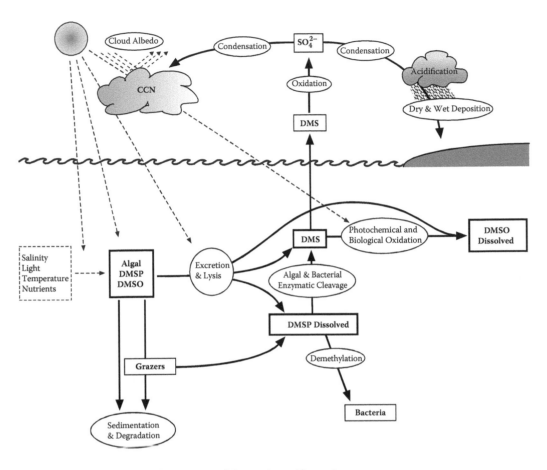

FIGURE 11.1 Major pools and processes of the marine sulfur cycle.

is necessary for many sulfur compounds, but not for DMS. The use of inert material is, however, recommended, such as Teflon tubing and glass vials for sample storage.

11.2.2 SAMPLING

Always treat samples gently in order to minimize gas exchange. When samples are taken with Niskin bottles, or a comparable device, gently flow seawater from the Niskin into dark bottles, allow the sample to overflow in order to remove all bubbles, and seal without headspace; store dark and at the in situ temperature and prepare within 2–3 hours of sampling.

When only surface samples are taken, a bucket can be used. Prevent bubble formation during sampling. When samples can be prepared and analyzed directly upon bulk sampling, one can use a prepared filter unit as described in Figure 11.2 to scoop up the sample from the bucket. Close the air outlets of the bottom reservoir before sampling in order to prevent water from the bucket flowing in. Scoop up water into the top part. Before starting the filtration by opening one of the air outlets in the bottom reservoir, take subsamples for total fractions from the top reservoir as described below. Only use the first milliliters of the filtrate as explained below.

11.2.3 SAMPLE PREPARATION

Since intracellular DMSP and DMSO are solutes in the cytoplasm of algae, samples need to be prepared with extreme care. As with many intracellular solutes, cell breakage during sample preparation

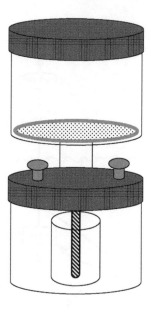

FIGURE 11.2 Sartorius 47 mm polycarbonate filter holder, adapted with a Teflon outflow tube that guides the filtrate onto the bottom of a glass vial (approximate volume of 20 mL).

will result in release of these compounds to the medium. Microflagellates and ciliates are especially vulnerable to breakage upon filtration. Contrasting effects of filtration on DMS concentrations were noticed by Turner et al. (1990), sometimes resulting in higher, but mostly in lower concentrations, especially when the ubiquitous DMS(P)-producing flagellate *Phaeocystis* sp. was present in the sample. In this respect, one should be aware of two potentially damaging processes during sample preparation and analysis, which should be prevented.

First, purging unfiltered samples for analysis of DMS can result in cell breakage and is therefore *not* recommended. Although it may be less of a problem in oligotrophic areas (Dacey et al., 1998), the physical stress of severe bubbling of the sample can result in breakup of fragile cells that will release DMSP to the medium. This is an uncontrolled process that cannot be foreseen. We now know that several algal species contain enzymes that rapidly convert DMSP into DMS (Stefels and Van Boekel, 1993; Steinke et al., 1996; for an overview, Stefels et al., 2007). Breakup of cells and the subsequent release of DMSP and DMSP-lyase enzymes will thus artificially increase the DMS levels. For instance, in blooms of *Phaeocystis* sp. or *Emiliania huxleyi*, species that are known for their DMSP-lyase abilities, particulate DMSP is ten to fifty times higher than DMS. In unfiltered samples from such blooms, agitation of only a small part of the cells will result in a severe increase of DMS, compared with properly filtered samples.

The second potentially damaging process is, ironically, the filtration itself. Filtration can damage fragile cells and result in release of cell solutes, especially the moment cells run dry on the filter. Kiene and Slezak (2006) observed that sample preparation procedures such as in-line filtration, syringe pressure filtration, and gravity filtration of relatively large-volume samples through glass fiber filters caused release of $DMSP_d$ from particulate material. So, although filtration is a necessity for obtaining accurate DMS concentrations, it can severely increase $DMSP_d$ and potentially $DMSO_d$ levels. Another adverse effect of filtering is that when using an open-funnel filtration setup, DMS can escape from the sample during the filtration procedure, thereby artificially reducing DMS levels in the filtrate. This process increases with temperature; hence, samples should be kept as cold as possible without stressing the organisms, that is, slightly under or at the in situ temperature. In checking the effect of filtering on DMS levels in samples from the Sargasso Sea ($T = 28°C$) we observed DMS concentrations in filtrates to be 86% (±3.4%, $n = 8$) of concentrations in unfiltered samples (Stefels and Dacey, unpublished data). Although we do not know whether this was due to either increased production in the unfiltered sample or loss during the filtration, it indicates that if loss occurs, it is a relatively minor and reproducible fraction in these oligotrophic samples.

With the above warnings in mind, the following recommendations for sample preparation can be made. In general, samples should be kept at or just below the in situ temperature as long as possible; work quickly, prevent physical disturbance (do not use pressure or vacuum filtration; let samples flow gently onto the bottom of sample vials), minimize gas exchange with the atmosphere, and work under indoor light conditions. In case large amounts of *Phaeocystis* colonies are present in the samples, gentle prescreening of the sample over a 20 μm mesh is recommended, before filtering the sample onto a Whatman GF/F filter for DMS and dissolved DMSP and DMSO measurements, as described below.

Materials to be used:

- 20 mL glass crimp-cap vials (23 × 75 mm, rim diameter 20 mm). For long storage periods, crimp caps with Teflon-faced septa (20 mm) can be used; for short storage periods and direct DMS analysis with subsequent DMSP or DMSO analysis in the same sample, it is easier to use Teflon-coated grey butyl stoppers (Wheaton) that can be removed by hand and which are reusable.
- 47 mm filtration setup (Sartorius, Polycarbonate) for use with Whatman GF/F filters (nominal retention, 0.7 μm), with the outlet below the filter adapted with a Teflon tube to gently guide the filtrate onto the bottom of a vial inside the receptacle (Figure 11.2).
- 25 mm filtration setup for use with Whatman GF/F filters. Use muffled (5 hours at 350°C) 25 mm filters for $DMSO_p$ measurements.
- 10 mL pipets for sampling of the various fractions.
- Chemicals: NaOH pellets (analytical grade), 15% $TiCl_3$ in 10% HCl (Merck), ultrapure water.

NaOH pellets are used directly in the procedure for base hydrolysis of DMSP to DMS and are preferred over high molar solutions of NaOH. Except for the ease of use, an important reason is that when adding high molar solutions to the sample, this will result in an immediate burst of DMS that can partly escape before closing the vial. When adding pellets to the sample, these will slowly dissolve and raise the pH, giving ample time to close the sample vial without loss of DMS. The weight of one pellet should be approximately 0.1 to 0.2 g. In a 10 mL solution this results in a pH of 13.4 to 13.7, which is well above the commonly used pH of 13, necessary for a fast conversion of DMSP to DMS (Dacey and Blough, 1987).

The analysis of DMSO by reduction with $TiCl_3$ is based on the method by Kiene and Gerard (1994), but with some simplifications. In this method, seawater samples are treated with an acidic solution of $TiCl_3$, sealed with gas-tight stoppers, and either heated (1 hour at 50°C and 1 hour cooling) or allowed to stand for 3 days to allow reduction to proceed. Finally, the DMS produced is quantified with the standard purge-and-trap system as described below. Because Kiene and Gerard (1994) used a basification step with NaOH to remove acid vapors from the $TiCl_3$ reagent, their DMSO sample included DMSP, for which they compensated by subtraction of measured DMSP values in a parallel sample. We found that this may sometimes result in erroneous data, and therefore prefer trapping the acid vapors during the analytical procedure with a trap made of sodium bicarbonate or potassium carbonate. In the strong acidic $TiCl_3$ solution, the conversion of DMSP to DMS is prevented, and therefore the reduction only yields DMS from DMSO. When using the heating procedure, make sure that vial rims are dry, so that stoppers seal well, and that temperature is kept below 55°C.

The sampling scheme given in Figure 11.3 assumes that analysis of DMS subsamples can be done within 2 to 3 hours from sampling. Sample preparation is described in relevant order. Before taking subsamples from the storage vessel, homogenize gently by inverting.

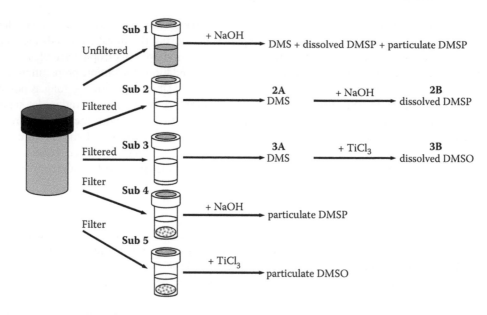

FIGURE 11.3 Sample handling for DMS(P/O) analysis. All compounds are measured gas chromatographically as DMS: DMS → DMS; DMSP + NaOH → DMS; DMSO + TiCl$_3$ → DMS.

Subsample 1: Total DMS + dissolved DMSP + particulate DMSP. Take a subsample of 10 mL into a 20 mL vial, add one pellet of NaOH (~0.2 g), seal quickly, and leave at room temperature to react overnight.

Subsample 2: (A) DMS, (B) dissolved DMSP. Here, sequential DMS and DMSP$_d$ analysis in the same sample is recommended, but only when DMS analysis is done immediately after filtration. This is because some bacteria can pass a Whatman GF/F filter and consume DMSP. This can be fast and might become a problem when DMS analysis takes long. In such cases, it is recommended to take two separate samples, as described in Section 11.2.4.

If possible, put a large subsample (>150 mL) in the filter holder for gravity filtration over a 47 mm Whatman GF/F filter. Use a carefully cleaned and dried filtration unit to prevent carryover from the previous sample and collect only the first 10–11 mL in a small glass vial inside the filtrate collector, as shown in Figure 11.2. The large volume reduces the relative amount of DMS to escape from the sample; collecting only the first 11 mL of filtrate minimizes the filtration artifacts on DMSP due to damaged cells. In fact, Kiene and Slezak (2006) recommend using even smaller filtration volumes for the analysis of DMSP$_d$, but this depends also on analytical potential and is not always needed, depending on the nature of the samples.

In any case, never let the filter run dry before removing the filtrate. From the filtrate a 10 mL sample is transferred to a 20 mL vial. Close with a stopper and analyze for DMS as described below. After analysis, add one pellet of NaOH, seal quickly, and leave at room temperature to react overnight.

Subsample 3: (A) DMS, (B) dissolved DMSO. Repeat the same procedure and take 8 mL and transfer to a 20 mL vial. Close with a stopper and analyze for DMS. This gives you a duplicate DMS measurement. After analysis, add 2 mL TiCl$_3$, seal, and leave to react at room temperature for 3 days or use the heating procedure when samples can be analyzed the same day. In some field samples, like those from oligotrophic waters, or in cultures of robust algal species, cell breakage on filters may be less of a problem and subsamples 2 and 3 can be taken from the same filtrate. It is recommended to check this by taking sequential fractions from one sample and checking for differences in DMSP$_d$ (Kiene and Slezak, 2006).

Subsample 4: Particulate DMSP. Filter a separate 10 mL subsample onto a 25 mm GF/F filter. Put 10 mL of ultrapure water in a 20 mL vial, add the filter, add a NaOH pellet, quickly seal, and leave overnight to react. Although we already discussed that this sample is likely affected by filtration artifacts, and as such is not used for $DMSP_p$ calculations, it can be a valuable backup to evaluate your samples and to compare with particulate DMSO data. Moreover, not all natural communities will suffer from filtration, especially when care has been taken to do gravity filtration only and to minimize the time the filter runs dry.

Subsample 5: Particulate DMSO. Filter a second subsample of 10 mL onto a muffled 25 mm GF/F filter and put the filter into 8 mL of ultrapure water and add 2 mL $TiCl_3$. Close the vial and leave to react at room temperature for 3 days, or use the heating procedure. As for particulate DMSP, this sample may be hampered by filtration artifacts, but seems to be the best we have. Given the fact that $DMSO_p$ levels are usually relatively low, analyzing an unfiltered sample for DMS + total DMSO and calculating the particulate fraction by subtraction will in most cases result in $DMSO_p$ levels that are within or close to the range of procedural and analytical error. Hence, filtration may give a better estimation of the particulate fraction. One should do some trials to see whether bigger subsamples need to be taken for accurate analysis of $DMSO_p$ levels.

With these subsamples analyzed, all individual pools can be calculated by subtraction as follows:

DMS = average of **2A** and **3A** after volume correction
$DMSP_d$ = **2B**
$DMSP_p$ = **1** – average (**2A**, **3A**) – **2B**
$DMSO_d$ = **3B**
$DMSO_p$ = **5** multiplied with a correction factor when $DMSP_p$ from **4** does not yield the same value as $DMSP_p$ derived from **1** and **2**.

11.2.4 SAMPLE STORAGE

It is always best to sample quickly, and prepare and analyze directly. However, this is not always possible, so when this cannot be done within 2 to 3 hours after collection, a preservation procedure is recommended. Of the three compounds, DMS is most vulnerable to changes, first, because of its volatile nature, and second, because its concentration is usually lowest and therefore most susceptible to large relative variations. Therefore, preservation methods are based on either the removal or conversion of DMS and subsequent stabilization of DMSP or DMSO.

For short storage periods (1 to 2 weeks), DMSP samples can be preserved by basification to DMS and subsequent storage at –20°C. The same sample procedure can be used as described above, but **2A** and **2B** should be separate samples, so that DMS can be calculated by subtraction. This implies that purge gas needs to be available to remove DMS from sample **2B**. Then, both can be stored with base added. $TiCl_3$-reduced DMSO samples are also stable for at least several weeks, but do not need to be stored frozen. It should be emphasized that this procedure results in less accurate DMS values when DMS concentrations are low relative to DMSP or DMSO.

For longer storage periods, preservation of DMSP by acidification appears to be better (Kiene and Slezak, 2006; modified from Curran et al., 1998). Per milliliter of sample, 5 µL of 50% H_2SO_4 should be added to obtain a pH < 2. The acid oxidizes all DMS to nonvolatile compounds within 24 hours (Kiene and Slezak, 2006) and stabilizes DMSP. H_2SO_4 appears to work better in oxidizing DMS than HCl, as used by Curran et al. (1998). H_2SO_4-preserved samples are stable for months. Before analysis, DMSP is converted to DMS by adding two pellets of NaOH. One should be careful, however, because in some plankton blooms of species that have DMSP-lyases with an acidic optimum (Stefels et al., 2007), DMSP conversion to DMS can take place during the preservation procedure, as was observed in a *Phaeocystis antarctica* bloom (Kiene, personal communication). Stronger acid to preserve may overcome this problem partly (pH < 1), but one should still be aware of the potential biases using this method.

11.3 ANALYTICAL PROCEDURE

11.3.1 PURGE-AND-TRAP SYSTEM

The purge-and-trap method described here (Figure 11.4) has been developed for its simplicity and can be used as the basis for further sophistication.

Materials to be used:

- All tubing is 1/8-inch Teflon tubing.
- Use Teflon unions and connectors. We have very good experience with Swagelok PFA connectors.
- A thermostatic water bath with a methanol-water solution for –15°C cooling of the gas stream.
- A cold trap made of the same 1/8-inch Teflon tubing, looped once to form a coil of about 5 cm.
- A Dewar with liquid nitrogen, which can be replaced easily with a water boiler.
- A Teflon six-port switching valve.
- For DMSO measurements, an acid trap is needed: Use a 10 cm piece of ½-inch Teflon tubing, fill with 5 cm NaHCO$_3$, and cap both ends with silane-treated glass wool. Fit the trap in the sample line with Teflon connectors (Swagelok PFA) (Figure 11.4).
- Long and short hypodermic needles.

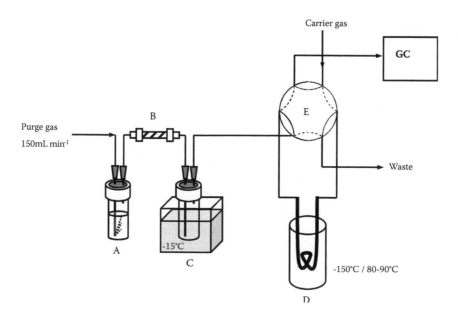

FIGURE 11.4 Purge-and-trap system for DMS analysis. (A) 20 mL sample vial with 10 mL sample, including long and short hypodermic needles for purge gas in- and outlet. (B) Acid scrubber for DMSO analyses only. (C) Drying step: empty glass vial in –15°C thermostatic bath. (D) Teflon cold trap. (E) Six-port switching valve.

The purge gas used to sweep DMS from the sample should be the same as that used for carrier gas in the gas chromatograph. Often nitrogen is used, but helium can be used as well. DMS is purged directly from the sample vials using two hypodermic needles (Figure 11.4A): a long needle reaching the bottom of the vial for gas inflow and a short needle above the sample surface for gas outflow. Needles are reusable, but be aware of clogging with septum remains. In our method we use a flow rate of 150 mL min^{-1} (check regularly at the waste outlet of the six-port valve), but one can experiment with other rates if needed. For DMSO samples, an additional acid trap is needed to prevent corrosion of downstream instrumentation (Figure 11.4B). The gas effluent then needs to be dried in order to prevent clogging of the cold trap. Many researchers use Perma Pure Nafion gas dryers (DM series), but in our experience a thermostatic water bath with a methanol-water mixture set at –15°C works very reliably (Figure 11.4C). Moreover, the empty vial that sits in the water bath is easy to exchange and check. After the drying step, DMS is trapped in a cold trap immersed in a Dewar with liquid nitrogen (Figure 11.4D), via a six-port switching valve (Figure 11.4E). When all DMS is transferred from the sample onto the cold trap, injection onto the analytical column proceeds by switching the valve to the inject position, removing the Dewar and immersing the trap in a water boiler with hot water (>80°C) until the compound elutes from the analytical column. Then the valve can be switched back and a new sample attached.

The amount of time needed to purge a sample depends on the volume of the sample and the concentration of DMS (Figure 11.5). A standard purge time of 20 minutes is normally sufficient. When filters are inside the sample, care should be taken that the filter does not obstruct the gas flow. Maneuvering the filter against the glass wall of the vial, but keeping it under water, should do this.

11.3.2 STANDARDS FOR CALIBRATION

For the analytical procedure, DMS standards can be used. We have very good and reproducible results with stock solutions in methanol and subsequent working solutions in filtered deep-sea water with zero DMS concentrations (if available) or in ultrapure water. Prepare the stock solution as follows: Add 10 mL methanol to a 10 mL crimp-cap vial and seal with a Teflon-faced septum (20 mm). Then add 73 µL pure DMS through the septum, using a glass syringe. To improve accuracy, it is recommended to check both amounts added to the vial gravimetrically. This gives a stock solution of 100 mM DMS that is stable for several months. Working solutions are made in either filtered, DMS-free, deep-sea water, or ultrapure water using a 100 mL volumetric flask to which 10 µL of the stock solution is added. The 10 µL is retrieved from the stock solution through the septum with a glass syringe. The working solution has a concentration of 10 µM DMS, is stored in the fridge,

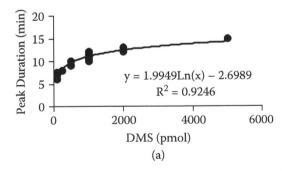
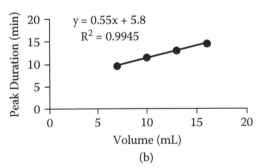

FIGURE 11.5 Peak duration of DMS standards in 20 mL vials at room temperature, purged with a gas flow of 150 mL min^{-1}. Peak development was monitored directly with a mass spectrometer; the end of the peak is defined as the moment the mass spectrometer signal returned to baseline plus 1 minute. (a) DMS concentration range in 10 mL filtered deep-sea water. (b) 1 nmol DMS in various volumes of filtered deep-sea water.

and should be refreshed at least every 4 days, depending on the use and headspace in the volumetric flask. From the working solution, prepare a range of DMS standards directly in the 20 mL vials that are attached to the purge-and-trap system. First put 10 mL DMS-free deep-sea water or ultrapure water in the vial, then add microliter quantities of the working solution and seal with a Teflon-coated stopper. If very low DMS concentrations are needed, an additional dilution step should be made from the working solution before preparing the vials. Always prepare fresh secondary working solutions. Standard curves should be run two to three times a day, depending on the number of samples that are to be analyzed. Alternatively, when comparable samples are analyzed, standards of one concentration can be regularly analyzed in between samples.

For DMSO analysis, background levels can be high and care should be taken during cleaning procedures for glassware (Kiene and Gerard, 1994). Although we have eliminated at least part of the contamination by simplifying the procedure, the $TiCl_3$ solution should be checked for contamination with 8 mL ultrapure water, followed by the same procedure as a sample. Kiene and Gerard (1994) also advised to check the efficiency of the reduction step with DMSO standards, each time a new $TiCl_3$ solution is used.

When working on ships that produce fresh water from seawater, regular checks of the ultrapure water system for contamination with either one of the sulfur compounds is needed, especially when traversing through dense plankton blooms. We observed that at high ambient concentrations, substantial amounts of these compounds pass the purification system, thereby increasing the blank.

11.3.3 INSTRUMENTATION

For the analysis a gas chromatograph equipped with a sulfur-specific detector is used. There are several options, but given the relative ease of analysis, only a basic instrument is needed. We have used a Varian 3600 with pulsed flame photometric detector (P-FPD), which has the advantage over normal FPDs that it uses less detector gas. The configuration of this system is as follows:

- Analytical column: 15 m DB1 wide-bore column with 5 μm film thickness
- Carrier gas: High-purity-grade helium at 4 mL min^{-1}
- Typical detector gas flow rates: 13 mL min^{-1} for H_2 and 30 mL min^{-1} for air
- Column temperature: 60°C isotherm; detector temperature: 200°C

With this configuration, the retention time for DMS is approximately 2 minutes. The square root of the peak area is linear with the total amount of DMS injected, although polynomial fits may give a better description of the response to standard curves when a wide range of concentrations are used.

Another column that is often used in DMS analyses is a Teflon column filled with Chromosil 330 (1/8 inch diameter, 2 m long, Supelco) (Turner et al., 1990). This column needs higher carrier gas flow rates, which has the advantage that the sample is swept from the cold trap within seconds and the valve can be returned to the purge setting while the analysis is still running.

11.4 FINAL REMARKS

In the method presented here we have used a standard volume of 10 mL for subsamples. With open-ocean concentrations of around 0.1 to 1 nM of DMS this gives a total amount of 1 to 10 pmol in the sample. This should be sufficient for most gas chromatographs, but several researchers have lowered the volume to 2 to 4 mL in order to reduce purge time, thereby maximizing sample throughput.

We have also used relatively large filtration volumes (11 to 20 mL), which, in our experience and with the above procedure, worked well. However, Kiene and Slezak (2006) observed filtration artifacts on $DMSP_d$ when volumes larger than 10 mL were filtered. These artifacts are unpredictable and may be due to the species composition of the sample and the physiological condition of the algae. For instance, at the end of a bloom period, most algae are more susceptible to lysis, and thus more easily destroyed due to physical stress. It is therefore advisable to check this procedure when new ecosystems are going to be sampled.

As a final remark it should be mentioned that we have chosen for the $TiCl_3$ reduction method to analyze DMSO. Other methods have been developed to measure DMSO: the cobalt-doped borohydride method (Simo and Vila-Costa, 2006) and the enzyme-linked reduction method (Hatton et al., 1994; Simo and Vila-Costa, 2006). The $TiCl_3$ reduction method is, however, relatively cheap and easy to perform and yields reliable analyses of DMSO.

ACKNOWLEDGMENT

I thank Ron Kiene for valuable comments and suggestions on the manuscript.

REFERENCES

Andreae, M. O. 1990. Ocean-atmosphere interactions in the global biogeochemical sulfur cycle. *Marine Chemistry* 30:1–29.

Andreae, M. O., and W. R. Barnard. 1983. Determination of trace quantities of dimethyl sulfide in aqueous-solutions. *Analytical Chemistry* 55:608–12.

Ayers, G. P., and R. W. Gillett. 2000. DMS and its oxidation products in the remote marine atmosphere: Implications for climate and atmospheric chemistry. *Journal of Sea Research* 43:275–86.

Bates, T. S., R. J. Charlson, and R. H. Gammon. 1987. Evidence for the climatic role of marine biogenic sulphur. *Nature* 329:319–21.

Bates, T. S., B. K. Lamb, A. Guenther, J. Dignon, and R. E. Stoiber. 1992. Sulfur emissions to the atmosphere from natural sources. *Journal of Atmospheric Chemistry* 14:315–37.

Bopp, L., O. Aumont, S. Belviso, and P. Monfray. 2003. Potential impact of climate change on marine dimethyl sulfide emissions. *Tellus Series B: Chemical and Physical Meteorology* 55:11–22.

Bopp, L., O. Boucher, O. Aumont, S. Belviso, J. L. Dufresne, M. Pham, and P. Monfray. 2004. Will marine dimethylsulfide emissions amplify or alleviate global warming? A model study. *Canadian Journal of Fisheries and Aquatic Sciences* 61:826–35.

Charlson, R. J., J. E. Lovelock, M. O. Andreae, and S. G. Warren. 1987. Oceanic phytoplankton, atmospheric sulphur, cloud albedo and climate. *Nature* 326:655–61.

Curran, M. A. J., G. B. Jones, and H. Burton. 1998. Spatial distribution of dimethylsulfide and dimethylsulfoniopropionate in the Australasian sector of the Southern Ocean. *Journal of Geophysical Research: Atmospheres* 103:16677–89.

Dacey, J. W. H., and N. Blough. 1987. Hydroxide decomposition of dimethylsulfoniopropionate to form dimethylsulfide. *Geophysical Research Letters* 14:1246–49.

Dacey, J. W. H., F. A. Howse, A. F. Michaels, and S. G. Wakeham. 1998. Temporal variability of dimethylsulfide and dimethylsulfoniopropionate in the Sargasso Sea. *Deep Sea Research Part I: Oceanographic Research Papers* 45:2085–2104.

del Valle, D. A., D. J. Kieber, and R. P. Kiene. 2007. Depth-dependent fate of biologically-consumed dimethylsulfide in the Sargasso Sea. *Marine Chemistry* 103:197–208.

Gabric, A. J., R. Simo, R. A. Cropp, A. C. Hirst, and J. Dachs. 2004. Modeling estimates of the global emission of dimethylsulfide under enhanced greenhouse conditions. *Global Biogeochemical Cycles* 18, article GB2014.

Gondwe, M., M. Krol, W. Gieskes, W. Klaassen, and H. de Baar. 2003. The contribution of ocean-leaving DMS to the global atmospheric burdens of DMS, MSA, SO2, and NSS SO4=. *Global Biogeochemical Cycles* 17, article 1056.

Gunson, J. R., S. A. Spall, T. R. Anderson, A. Jones, I. J. Totterdell, and M. J. Woodage. 2006. Climate sensitivity to ocean dimethylsulphide emissions. *Geophysical Research Letters* 33, doi: 10.1029/2005GL024982.

Hatton, A. D., L. Darroch, and G. Malin. 2004. The role of dimethylsulphoxide in the marine biogeochemical cycle of dimethylsulphide. *Oceanography and Marine Biology: An Annual Review* 42:29–56.

Hatton, A. D., G. Malin, A. G. McEwan, and P. S. Liss. 1994. Determination of dimethyl-sulfoxide in aqueous-solution by an enzyme-linked method. *Analytical Chemistry* 66:4093–96.

Kettle, A. J., and M. O. Andreae. 2000. Flux of dimethylsulfide from the oceans: A comparison of updated data seas and flux models. *Journal of Geophysical Research: Atmospheres* 105:26793–808.

Kettle, A. J., M. O. Andreae, D. Amouroux, T. W. Andreae, T. S. Bates, H. Berresheim, et al. 1999. A global database of sea surface dimethylsulfide (DMS) measurements and a procedure to predict sea surface DMS as a function of latitude, longitude, and month. *Global Biogeochemical Cycles* 13:399–444.

Kiene, R. P. and G. Gerard. 1994. Determination of trace levels of dimethylsulfoxide (DMSO) in seawater and rainwater. Marine Chemistry 47:1–12.

Kiene, R. P., and D. Slezak. 2006. Low dissolved DMSP concentrations in seawater revealed by small volume gravity filtration and dialysis sampling. *Limnology and Oceanography: Methods* 4:80–95.

Lovelock, J. E., R. J. Maggs, and R. A. Rasmussen. 1972. Atmospheric dimethyl sulphide and the natural sulphur cycle. *Nature* 237:452–53.

Simo, R., and M. Vila-Costa. 2006. Ubiquity of algal dimethylsulfoxide in the surface ocean: Geographic and temporal distribution patterns. *Marine Chemistry* 100:136–46.

Spiro, P. A., D. J. Jacob, and J. A. Logan. 1992. Global inventory of sulfur emissions with $1° \times 1°$ resolution. *Journal of Geophysical Research* 97:6023–36.

Stefels, J., M. Steinke, S. Turner, G. Malin, and S. Belviso. 2007. Environmental constraints on the production and removal of the climatically active gas dimethylsulphide (DMS) and implications for ecosystem modelling. *Biogeochemistry* 83:245–75.

Stefels, J., and W. H. M. Van Boekel. 1993. Production of DMS from dissolved DMSP in axenic cultures of the marine-phytoplankton species *Phaeocystis* sp. *Marine Ecology Progress Series* 97:11–18.

Steinke, M., C. Daniel, and G. O. Kirst. 1996. DMSP lyase in marine macro- and microalgae: Intraspecific differences in cleavage activity. In *Biological and environmental chemistry of DMSP and related sulfonium compounds*, ed. R. P. Kiene, P. T. Visscher, M. D. Keller, and G. O. Kirst. New York: Plenum Press.

Turner, S. M., G. Malin, L. E. Bagander, and C. Leck. 1990. Interlaboratory calibration and sample analysis of dimethyl sulfide in water. *Marine Chemistry* 29:47–62.

Andrew R. Bowie and Maeve C. Lohan

CONTENTS

12.1 INTRODUCTION

Originating from the Anglo-Saxon word *iren*, its chemical symbol taken from the Latin *ferrum*, the element iron (Fe) has been known to humans since ancient times, and no other element has played a more important role in man's material progress. Iron is a vital constituent of plant and animal life, and with a relative abundance of 5.6%, it is the fourth most abundant element in the earth's crust (Taylor, 1964). The element is arguably the most important trace metal in seawater, although it exists at extremely low concentrations (subnanomolar) in open-ocean environments (Martin and Fitzwater,

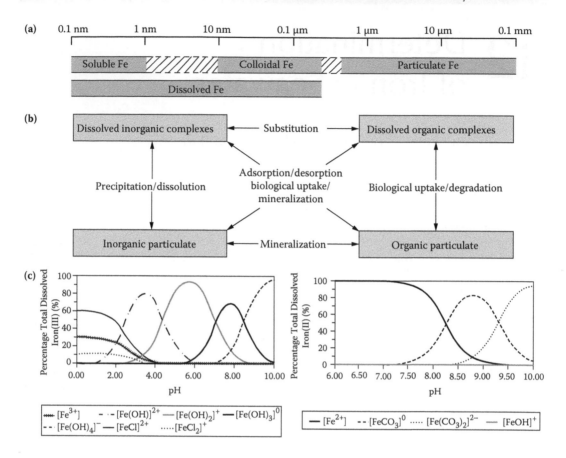

FIGURE 12.1 (a) Chemical forms, (b) phase transfers, and (c) inorganic speciation model of Fe(III) and Fe(II) in seawater (calculated using the hydrolysis stability and chloride stability constants of Liu and Millero (2002) and Millero et al. (1995) measured in seawater; dissolved inorganic carbon modeled as an open system in equilibrium with the atmosphere under ambient conditions). (Adapted from Ussher et al., 2004, with kind permission of CSIRO Publishing, www.publish.csiro.au/nid/190/issue/886.htm.)

1988). Iron plays an important role in ocean biogeochemistry, being a key micronutrient that regulates marine primary productivity in large areas of the world's oceans (Martin and Fitzwater, 1988; Boyd et al., 2000), and thus influencing global carbon cycling and the earth's climate. Conceptual and numerical models of ocean biogeochemistry must therefore include Fe as a limiting component (Moore et al., 2002), and the accuracy of such models will only be improved by routinely including Fe analysis alongside the major nutrients during modern-day oceanographic field campaigns. An interesting paradox in the ocean is that while planktonic microorganisms mediate the chemistry and cycling of biologically significant trace metals such as Fe, the same metals exert an influence on the growth of the organisms and their cycling of the major nutrients (e.g., carbon and nitrogen).

The marine chemistry of Fe in seawater is illustrated in Figure 12.1. The speciation of Fe in natural waters affects its thermodynamic solubility (Byrne and Kester, 1976; Millero, 1998), redox transformations (Millero et al., 1995), and interaction with biota (Anderson and Morel, 1982; Sunda et al., 1991). The chemical form of Fe can be defined by physical size fractionation, usually achieved by using membrane or cartridge filtration. Several different chemical species exist within each different size fraction in natural waters, including free ions (Morel and Hering, 1993), soluble coordination complexes with inorganic (Millero et al., 1995) and organic ligands (Gledhill and van den Berg, 1995), and a variety of colloidal (Wu et al., 2001) and particulate forms (Frew et al., 2006). A large fraction (80%–99%) of dissolved Fe species in seawater are complexed with dissolved organic ligands (Rue and Bruland, 1995; Gledhill and van den Berg, 1995), preventing the formation of insoluble oxyhydroxides and buffering the concentration of dissolved Fe in solution. The origin of these ligands remains largely unknown, although they may be represented by two or more classes (Rue and Bruland, 1995), be produced by phytoplankton or bacteria as an Fe acquisition strategy (Hutchins et al., 1999), and exist in the colloidal phase (Wu et al., 2001). Compounds of Fe are present in two oxidation states in natural waters, the Fe(II) or *ferrous* form and the Fe(III) or *ferric* form. Fe(III) species are dominant in oxygenated seawater and are highly insoluble through the formation of oxyhydroxides (Wells et al., 1995). Fe(II) species are thermodynamically unstable in oxygenated seawater and are rapidly oxidized to Fe(III), although appreciable Fe(II) concentrations may be present in oxygen minimum zones or anoxic environments (Moffett et al., 2007). Potential sources of Fe(II) are photoreduction of Fe(III) in surface waters (Miller et al., 1995), atmospheric deposition (Zhuang et al., 1992), in situ remineralization of sinking organic debris, and diffusion from anoxic sediments (Hong and Kester, 1986).

The distribution of Fe across dissolved and particulate fractions varies strongly between regions. In coastal waters, Fe occurs predominantly in the particulate form due to the supply from rivers or resuspended shelf sediments (Boyle et al., 1977). Dissolved Fe concentrations in surface river plumes vary greatly (e.g., 1 to 60 nM; Lohan and Bruland, 2006). Conversely, in remote open-ocean regions, dissolved Fe concentrations can exceed the particulate Fe concentration in the water column (Bruland et al., 1994). Where Fe limitation of phytoplankton growth occurs in high-nutrient, low-chlorophyll regimes, dissolved Fe concentrations are extremely low (e.g., <0.1 nM; Martin and Gordon, 1988). Deep water (>1,000 m) dissolved Fe concentrations are more uniform (0.3–0.6 nM; Johnson et al., 1997), but may be enriched below dust plumes (Bergquist and Boyle, 2006; Measures et al., 2008a) and depleted in the Southern Ocean.

Historically, operationally defined measurements of dissolved Fe (dFe; <0.2 or 0.45 μm) in seawater have been widely used to gauge the biological availability of Fe to algae and bacteria. Recent studies suggest that a significant portion of the dissolved pool exists in the colloidal size range (nominally >1,000 kDa to <0.2 μm), which are bound to strong Fe-complexing ligands in surface waters (Wu et al., 2001). Colloidal Fe may be less bioavailable than truly soluble (<0.02 μm) forms (Rich and Morel, 1990), although it may undergo the most dynamic changes during phytoplankton blooms (Nishioka et al., 2001). Size fractionation and ultrafiltration studies within the dissolved pool provide further insight into biologically mediated Fe uptake and recycling processes.

The particulate Fe pool is important for remineralization, scavenging, and export processes in the ocean (Frew et al., 2006). Particulate Fe exists as either biotic (e.g., cellular Fe, detritus) or abiotic/lithogenic (e.g., aluminosilicate clays) material, which may be solubilized for uptake by protozoan (Barbeau et al., 1996) or zooplankton (Hutchins and Bruland, 1994) grazers. Little information exists on the coupling between the biogeochemical cycles of Fe and carbon (C) in the ocean, due to the scarcity of data on the Fe content of suspended or exported particles. Iron/carbon ratios are essential to determining the carbon sequestration efficiency and hence the impact of increased Fe supply on atmospheric carbon dioxide (CO_2) drawdown and global climate in the contemporary ocean (Blain et al., 2007) and in the geological past (Boyd et al., 2007).

The first attempts to measure Fe in seawater date back to the 1930s (Cooper, 1935), although open-ocean data prior to the late 1970s was probably inaccurate due to insensitive analytical techniques or contamination during sampling. Modern analytical techniques used for the determination of Fe in seawater can be categorized as either shipboard or land based, depending on size, weight, and fragility of instrumentation. Shipboard methods include both spectrophotometry and chemiluminescence (often coupled to flow injection manifolds) and stripping voltammetry (mainly used for organic speciation studies). Graphite furnace atomic absorption spectrometers and inductively coupled plasma mass spectrometers have been used on land, with the latter providing additional isotopic information.

This chapter presents practical sampling and analytical guidelines for the determination of the concentration and physicochemical forms of Fe in seawater, focusing on each stage of the complete procedure, including sample collection, filtration, storage, analysis, and quality assurance. This task represents one of the most difficult analytical challenges for the modern marine chemist due to its extremely low concentration in seawater, its complex speciation, and its ubiquity as a contaminant. Recent studies have benefited from improved trace metal clean sample acquisition, handling, and processing protocols, undoubtedly key aspects of the overall measurement process. The choice of materials and precleaning of samplers, sample storage devices, and analytical labware are critical. This has resulted in the generation of reliable profiles (Martin and Gordon, 1988) and sections (Measures et al., 2008a) in dissolved, total dissolvable (unfiltered), and particulate operationally defined classes. The advent of new sensitive analytical techniques has enabled the first international intercalibration exercises for Fe to take place at realistic seawater concentrations (Bowie et al., 2006) and led to the generation of an open-ocean reference material (Johnson et al., 2007). The oceanographic trace metal community is now beginning to standardize the sampling, analytical, and quality assurance procedures used in major upcoming international programs such as GEOTRACES (www.geotraces.org). Many of the guidelines described below, which are closely aligned with GEOTRACES protocols and recommendations, will also be suitable for other trace elements and isotopes (see Chapter 14). There is no substitute for experience in obtaining reliable seawater trace metal data, and therefore new researchers interested in this field are encouraged to participate in a research cruise or visit a laboratory of one of the groups routinely publishing good data for Fe in the ocean.

12.2 SAMPLING

Iron is arguably one of the most difficult metals to determine in seawater due to its ubiquity, which increases the risk of inadvertent contamination. Clean room container laboratories, equipped with ISO class 5 (formerly class 100) HEPA filtered air systems, are deemed necessary at sea for all sample handling and processing, including filtration, acidification, and analysis. It is advisable to routinely check sampling systems for Fe contamination using portable shipboard analysis systems (Section 12.4.3), prior to initiating sample collection. Elevated concentrations of other contamination-prone elements (e.g., zinc, lead, or copper) may also indicate possible contamination during sampling. Even after stringent cleaning of sampling equipment, using acid and ultra-high-purity deionized water (UHPW) $\geq 18\ M\Omega\ cm^{-1}$ (e.g., Milli-Q, Millipore), thorough rinsing and conditioning with copious amounts of seawater is often necessary. Sampling systems must not alter the phase (size) or chemical speciation of Fe.

12.2.1 DISCRETE WATER COLUMN PROFILING

The first reliable open-ocean profiles were obtained in the late 1970s using Teflon-coated Go-Flo sampler bottles suspended on nonmetallic hydroline (e.g., Kevlar) (Bruland et al., 1979). Special winches, solid Teflon or PVC messengers, and epoxy-coated end weights were also necessary. Different discrete sampler types have also proved to be successful, including custom-built polycarbonate bottles (Sedwick et al., 1997) and externally closing Niskin-X bottles (Figure 12.2A) (Sedwick et al., 2005). For all samplers, care must be taken to minimize contamination by removing or replacing metallic parts. Specifically, samplers should be Teflon lined and incorporate silicone or Viton rubber O-ring seals, PTFE or PFA stopcocks (including a top fitting for nitrogen gas or compressed air), plastic clips and crimps, nylon monofilament lanyards, and external silicone rubber closure mechanisms or epoxy-coated springs. Where metallic pins, bolts, and springs are necessary, it is advisable to use high-purity titanium (or similar) or to ensure they have a sealed Teflon lining or coating. Citric acid-based cleaners (e.g., CitriSurf 2000) may be used to passify stainless steel parts. Exposed metal areas can be sealed using silicone glue. Prior to deployments, sampler bottles should be cleaned by carefully filling with a mild detergent (3%, 1 day), followed by a weak acid

FIGURE 12.2 Photo of deployment of (a) Niskin-1010X trace metal samplers on an autonomous model 1018 intelligent rosette (General Oceanics) and (b) PPS3/3 free-floating sediment trap (Technicap, France) on the RSV *Aurora Australis* in the Southern Ocean.

solution (~1% HCl, 1 day), with thorough (seven times) rinsing with copious amounts of UHPW between and after each stage. No acid should contact the outside of the bottle or nylon components. At sea, samplers should initially be deployed on a test cast (ideally in a low-Fe, open-ocean region) and thoroughly rinsed with seawater.

Recent advances include the use of the Go-Flo or Niskin-X sampler bottles on rosette frames that have been specially modified for trace metal use (e.g., Measures et al., 2008b; Figure 12.2A). Frames are epoxy powder coated or sheathed in polyurethane, with all U-bolts and shackles made from high-quality marine-grade stainless steel. Sacrificial anodes on the frame and instruments (e.g., conductivity-temperature-depth [CTD] cable) should also be replaced with those made from nontarget trace metals (e.g., magnesium). The block through which the Kevlar hydroline passes should be made from nylon. Rosette frames typically allow for 12 Go-Flo or Niskin-X sampler bottles (ranging from 5 to 12 L in size) to be deployed through the water column at each station. Samplers can be tripped from the deck by either communicating with the pylon via four conductor Kevlar cables or autonomously using a sensor on the pylon (e.g., tripped using pressure, temperature, or time). Bottles may be sent down open, but tops are covered with plastic shower caps and taps covered with Ziploc bags, which are removed just prior to deployment. To prevent contamination from the frame, it is advisable to close sampler bottles while the rosette frame is moving upwards through clean water at 5–10 m min^{-1}. Such systems vastly reduce the deployment times (approximately 1 hour for a 12-bottle cast to 1,000 m), which is important on oceanographic research vessels costing several thousand dollars a day to run, and increase dataset resolution.

An alternative discrete water sampling system suitable for extended deployment on deep-sea moorings has been developed by Bell et al. (2002). The MITESS device collects unfiltered 500 mL samples by opening and closing a bottle originally filled with dilute acid (passively replaced by denser seawater). The device can be suspended below a wire to collect water column samples, and is vaned to allow deployment on conventional hydrowire.

12.2.2 Surface Water Pumping

In addition to discrete sampling devices, a variety of surface (down to 50 m) pumping and tubing systems have been used to deliver seawater to the shipboard laboratory (e.g., towed torpedo fish; Vink et al., 2000; Bowie et al., 2003; see Figure 1.5). These systems pump water up to 10 L min^{-1} and, as such, are particularly effective for bulk clean seawater collection (e.g., for Fe incubation experiments). The location of the pumping system intake is critical in order to avoid contamination from the ship (e.g., the sipper must be outside of any breaking bow waves). One unit developed at the University of California Santa Cruz consists of a PVC depressor vane 1 m above a 20 kg weight enclosed in a PVC towfish (Bruland et al., 2005). Tubing should be made from Teflon PFA or polyethylene, and the pump should be an all-Teflon PTFE, inert diaphragm type (e.g., Osmonics Bruiser). Pneumatically driven double-diaphragm pumps are recommended over electric pumps and require oil-free compressors to drive them. Surface water (0–1 m) samples can also be collected using a pole sampler, with collection bottles secured in a Perspex frame at the end of a 5 m bamboo pole. Rubber workboats have often been operated remotely from the main research vessel to collect surface water where contamination from the ship is suspected. Investigators must ensure that discrete bottle and pumping systems obtain comparable samples.

12.2.3 In Situ Pumping

The discrete water column sampling systems described above allow for the collection of filtered particles from limited volumes (5–30 L depending on sampler size), with later processing in shipboard clean containers using filters (47–142 mm in diameter) held in Teflon or polypropylene holders. These volumes may be suitable for the determination of suspended particulate Fe in certain oceanic regions with high concentrations and may also allow for sequential size fractionation (Cullen and Sherrell, 1999). At present, it is uncertain whether this form of sampling represents the true spectrum of marine particles (e.g., Bishop et al., 1977). In situ pumping and filtration systems (e.g., McLane Research Laboratories, Inc., Challenger Filtration Ltd.) that are capable of sampling 100 L h^{-1} may be required to collect sufficient amounts of particulate material (Sherrell and Boyle, 1992). Large-volume continuous flow centrifugation has also been used (Kuss et al., 2001). In situ pumps need to be vaned or deployed on nonmetallic or sheathed wire. Particular attention needs to be paid to filter blanks. No studies to date have compared bottle samples with in situ pumps, with particular reference to contamination and particle collection efficiency.

12.2.4 Sediment Traps

The determination of the flux of sinking particulate Fe in exported particles can only be quantified using in situ sediment traps, either free-floating (short-term) or moored (seasonal). The ratios of Fe to C and major nutrients (e.g., nitrogen) in exported particles are key constraints on the lifetime and spatial influence of Fe inputs to the surface ocean. Successful recent deployments include the PPS3/3 (Technicap, La Turbie, France; Figure 12.2B) trap in the Southern Ocean (Blain et al., 2007) and the Particle Interceptor Traps (MultiPITs) at the BATS station in the North Atlantic. There remains uncertainty as to the accuracy of estimating upper ocean particle fluxes with sediment traps due to hydrodynamics and swimmers (Buesseler et al., 2007). Hydrodynamic problems occur as sediment traps are moored or drifting and, as a result, rarely move with the flow of surrounding water. The fluid flow over and within the trap can affect how sinking particles move in and out of the funnel- or cylindrical-type trap used (Gardner, 2000). These problems are relevant to particle collection for all elements, and are being addressed using neutrally buoyant sediment traps (Salter et al., 2007).

Although sediment traps may collect relatively large quantities of material (milligrams to grams), care is needed to adapt the equipment to minimize or exclude possible contamination sources. Wires, shackles, and frames need to be modified and metallic parts replaced where possible. Trap bodies need to be thoroughly rinsed with a mild detergent, fresh water, and UHPW, and collection cups need acid cleaning prior to use. For longer-term deployments (>1 week), a preservative (e.g., formaldehyde) may be required to minimize bacterial degradation or zooplankton feeding, and this reagent may require further cleaning to minimize impurities (Section 12.4.1).

12.3   SAMPLE PROCESSING

All sample handling should take place in a clean laboratory area, preferably ISO class 5. To minimize contamination, it is best for sample handling to be a two-person operation: one to open up the outside sample bottle bag, and the other to then open the inside bag and remove the previously labeled bottle and rinse/fill the bottle. Teflon PFA or equivalent tubing should be used for connections among sampler, filtration device, and sample bottles. Sample bottles should be emptied of the weak acid solution (see Section 12.3.2) and rinsed three times with samples by placing sample water in the bottle, screwing the lid back on, shaking, and then pouring the sample out over the lid. The sample should be filled to the bottle's shoulder and all bottles filled to the same amount so that acidification of each sample is identical (i.e., same pH in all bottles). Sample bottles should be capped tightly and stored inside double Ziploc plastic bags or boxes.

12.3.1   FILTRATION

Filtration is a vital first step in the analysis of Fe, especially when the chemical speciation, rates, and mechanisms of transfers between different species are poorly known. Since Fe is strongly associated with particulate and colloidal phases, different operationally defined fractions of Fe can be determined after filtration. The filtration size cutoffs for traditional dissolved (dFe, sample passing through a <0.2–0.45 µm filter) and particulate (pFe, material remaining on a >0.45–0.2 µm filter) fractions have been decided upon based on the biological role of Fe and the need to exclude living cells from the dissolved phase. Filters with a pore size of 0.2 µm are now recommended, as some heterotrophic bacteria exist in the 0.2–0.5 µm size range. Conventional membrane filters with sharp well-defined size cutoffs such as Nuclepore polycarbonate track-etched (PCTE) housed in Teflon sandwiches are recommended (Bruland et al., 1979; Landing et al., 1995). Membrane filters are, however, somewhat fragile, and their decreased capacity only allows for low flow rates. Both filters and sandwiches require rigorous acid cleaning (~20% HCl bath) and rinsing before use. Filters can be stored wet in UHPW or dried in a laminar flow bench and stored unfolded in acid-cleaned petri dishes or plastic bags.

Filter cartridges or capsules (e.g., Pall AcroPak, Sartorious Sartobran) have become popular for larger-volume (>20 L) applications or when the user is not interested in the particulate phase, since the material remaining on the filter cannot be collected for later analysis. Small-volume cartridges with low hold-up volumes can also be used with discrete sampler bottles (<10 L). Cartridges are depth filters consisting of either multiple layers or a single layer of a medium having depth, and capture contaminants within their structure, as opposed to on the surface. They typically have a prefilter to maximize throughput but have less well-defined cutoffs (compared to membranes) and may allow larger particles to penetrate. Combination filters, which combine depth media and a membrane filter to create self-contained serial filter units, are highly recommended. Cartridges are often cleaned using only weak (~2% HCl) acid followed by UHPW (five times), since the housings are often not resistant to harsher treatments. Cartridges should be flushed with copious amounts of seawater (at least 2 L) before use. One cartridge can be used for a complete depth profile, but should be stored in the freezer between uses to minimize growth. Table 4 in Achterberg et al. (2001) lists characteristics of currently used filtration materials.

There is increasing evidence that a significant fraction of the Fe within the dissolved phase consists of colloidal material (Wu et al., 2001). Several researchers have used ultrafiltration (see Chapter 1) (<200 kDa or <0.02 µm) techniques to distinguish between the soluble and colloidal pool within the traditional <0.2 μm dissolved phase (Nishioka et al., 2001). Whatman Anotop sterile syringe filters, which contain a glass microfiber prefilter, are a popular choice.

Several researchers have reported results from the analysis of unfiltered seawater samples to provide additional information on the biogeochemical cycling of Fe (e.g., Sedwick et al., 2005). This eliminates the risk of contamination due to filtration and allows for increased sample through-put. Such an operationally defined fraction is often referred to as total dissolvable (TDFe) or acid-soluble Fe (i.e., the fraction of Fe released by acidifying an unfiltered sample), and will include dissolved, colloidal, and labile particulate phases. The amount of labile Fe that is solubilized from the particulate phases depends on the type of particles (i.e., biogenic, amorphous oxyhydroxides, aluminosilicate clay materials). In general, only acid-resistant forms such as matrix-bound alumino-silicates and unreactive crystalline particles would not be determined. Variations in the acid type, concentration, length of storage, and temperature will lead to variable amounts of particulate Fe that are solubilized. For these reasons, it is advisable that unfiltered samples are stored for at least 6 months before analysis (Bowie et al., 2004) and dissolved Fe measurements be made alongside the analyses of unfiltered samples.

Particulate fractions (suspended and sinking) of Fe need to be collected on a filter type that can then be subjected to later digestion and analysis (Section 12.4.5). Membranes with a small pore size (e.g., 0.2 μm) may not allow sufficient water to pass across them to obtain a high enough sample-to-blank ratio. The choice of polymer material is critical, with polycarbonate and polyester filters showing the lowest blanks (Cullen and Sherrell, 1999). Sequential size fractionation within the par-ticulate phase (e.g., 0.2, 2, 20, and 55 μm screens held within a Teflon PFA stack) will provide addi-tional information on the Fe associated with different biological classes. After filtration, a vacuum must be connected to the stack to draw water from the filter (and minimize salt content). In situ pumps use 142 mm diameter filters and allow for larger volumes (hundreds of liters) to pass across the filter, overcoming problems associated with high blanks for quartz fiber filters (e.g., Whatman QMA). In all cases, the total volume of water passing through the filter must be recorded.

For all filtrations, samples should be drawn from the bottles in a HEPA filtered air environment. Once on deck, plastic bags are placed over the spigots of the Go-Flo or Niskin-X samplers and bot-tles individually transported from the rosette into a HEPA filtered environment. A low overpressure of 10–50 kPa (maximum) of filtered (0.2 μm PTFE) high-quality nitrogen gas or compressed air is required to obtain a sufficient flow across the filters, while minimizing cell rupture or lysis.

12.3.2 Bottle Type and Storage

Sample storage must prevent biological growth and adsorption onto the container walls, and ensure the analyte concentration remains stable over time (months to years). The bottle type, bottle clean-ing, and choice of preservative are critical. Most researchers have now agreed that low-density poly-ethylene (LDPE) is the most suitable material for the measurement of dissolved Fe. For speciation samples, it is recommended that either fluorinated pothethylene (FPE), Teflon FEP, or Teflon PFA bottles are used, which are all suitable for freezing. The recommended GEOTRACES cleaning protocol for LDPE sample bottles is shown in Table 12.1 and that for Teflon in Chapter 14.

12.3.3 Preservation

Most trace metal oceanographers preserve samples for dissolved analysis using acidification, but differ on the type of acid and final pH of the solution. A recent intercalibration study (Sampling and Analysis of Fe [SAFe]) established that accurate shipboard-dissolved Fe analysis required sample acidification to pH < 2 for at least 12 hours (room temperature) before all of the Fe could be detected (i.e., released from organic complexes or colloids) (Johnson et al., 2007). Acidification with high-purity HCl (preferably Seastar Baseline grade or equivalent; Section 12.4.1) is most appropriate because seawater is already high in chloride and results in minimum alteration of the sample matrix.

TABLE 12.1
Recommended GEOTRACES Cleaning Protocols for LDPE Sample Bottles

1. Rinse bottles in reverse osmosis (ROW) or deionized water (DIW).
2. Soak bottles for 1 week in an alkaline detergent such as Decon. This process can be sped up by soaking at 60°C for 1 day.
3. Rinse four times with ROW/DIW.
4. Rinse three times with UHPW.
5. Fill bottles with 6 M HCl (reagent grade) and submerge in a 2 M HCl (reagent grade) bath for 1 month. Again, this can be sped up by heating for 1 week.
6. Rinse four times with UHPW inside a laminar flow hood (ISO class 5).
7. Fill bottles with 0.7 N HNO_3 (trace metal grade) or 1 N HCl (trace metal grade) for at least 1 month (i.e., transport to cruise filled with this). Should be stored double bagged or boxed.

Source: GEOTRACES Standards and Intercalibration Committee, *Draft Sampling and Sample-Handling Protocols for GEOTRACES-Related IPY Cruises, 2007/2008*, 2007, available at www.geotraces.org.

Time-series studies on stored SAFe samples have shown no significant change in Fe concentration under these storage and preservation conditions for a period of 20 months (Johnson et al., 2007). In order to assess blanks associated with sample preservation, a sample of the acid should be archived in a container that is least subjected to degradation or increasing blanks (i.e., Teflon FEP). The acidification step is particularly hazardous for the introduction of contamination since concentrated acids will readily mobilize contaminants. All manipulations should be done under ISO class 5 filtered air conditions.

Samples for Fe speciation should be stored in a refrigerator and analyzed within 24 hours or stored frozen for later analysis (with bottles not completely filled, to prevent cracking). While this has been the typical method for assessing metal-ligand interactions using competitive ligand equilibration–cathodic stripping voltammetry (Section 12.4.3.2), no experiments have assessed the effect of temperature and sample storage conditions for other speciation work (e.g., redox or size fractionation studies). Where possible, shipboard determinations for speciation studies are recommended, ensuring minimal change to the seawater after sampling. In situ analyzers (Chapin et al., 2002) or in situ solid-phase extraction (Okumura et al., 1997) may overcome problems associated with speciation changes in the period between sampling and analysis.

Filters containing material for particulate sample analysis should be stored in acid-cleaned petri dishes or small LDPE vials. Filters may be folded inwards to prevent loss of material, but it is recommended that filters are not cut, punched, or sectioned, to minimize contamination. Filters should be dried in a laminar flow bench after sampling. Some researchers recommend freezing of the filter samples prior to digestion (Cullen et al., 2001).

12.4 ANALYTICAL PROCEDURES

12.4.1 CHEMICALS AND REAGENTS

Systematic contamination is often caused by the use of impure reagents in the analysis of Fe and is indicated by errors that are nonrandom in nature (Howard and Statham, 1993). Therefore, all chemicals and reagents used for Fe analysis must be high-purity items (specified Fe concentrations less than 20 pg g^{-1}) or involve cleaning steps to reduce the Fe impurities. For example, HCl used to acidify samples can be purchased directly from Seastar Chemicals (Baseline grade), where certified Fe concentrations are <20 pg g^{-1}, or trace-metal-grade HCl (Fisher Scientific) may be cleaned using in-house, double subboiling distillation (Q-HCl). All reagents should be made up in UHPW.

Ammonium hydroxide (NH_4OH) is prepared by isopiestic distillation, which involves equilibrating reagent-grade concentrated NH_4OH with UHPW in a clean container (Q-NH_4OH). Ammonium acetate buffer (Q-NH_4Ac) is commonly used for Fe analysis. This is prepared by mixing Q-NH_4OH with quartz-distilled acetic acid and diluting with UHPW. Another common method for cleaning reagents is to pass the reagents over a chelating resin, which may remove any Fe impurities from the solution. Further details are contained in Howard and Statham (1993). For each reagent used, the blank Fe concentration must always be determined (see Section 12.5.1). When primary standards are prepared from solids, the preparation method should be well documented, and standards should be exchanged between researchers to ensure analytical intercalibration.

12.4.2   Analyte Preconcentration and Extraction

Dissolved Fe concentrations in seawater are extremely low (pM to nM), while the concentration of the major ions is 10^6 to 10^{10} times greater (Riley and Chester, 1971). Analysis of dissolved Fe therefore requires a preanalysis step that entails both the separation of dissolved Fe from the bulk seawater matrix and the preconcentration of analyte in order to increase the sensitivity of the overall method. This step has been achieved in a variety of ways: (1) using chelating ion-exchange resins (Landing et al., 1986; Lohan et al., 2005), (2) chelation/solvent extraction methods (Landing and Bruland, 1987; Martin and Gordon, 1988), (3) coprecipitation methods (Wu and Boyle, 1998), and (4) voltammetric techniques (Rue and Bruland, 1995; Gledhill and van den Berg, 1995). Specific examples are described in Sections 12.4.3 and 12.4.4.

12.4.3   Shipboard Methods

The kinetics of marine biogeochemical processes are rapid, and thus the chemical speciation of a sample can be modified by storage. Therefore, many chemical speciation measurements can only be made in the field using reliable shipboard analytical techniques such as flow injection (FI) and cathodic stripping voltammetry (CSV).

12.4.3.1   Flow Injection (FI)

The advantages of FI systems include low sample and reagent consumption, high sample throughput, portability, and near-real-time measurements, thus allowing scientists to examine high-resolution spatial and temporal trends in the oceanic distribution and cycling of Fe, and reducing the chance of sample contamination. A commonality in all FI methods is the preconcentration of Fe onto a chelating resin. The majority of FI techniques reported in the literature utilize 8-hydroxyquinoline (8-HQ) as the functional chelating group immobilized on a chemically resistant vinyl polymer resin (e.g., Toyopearl TSK), as developed by Landing et al. (1986). Recent studies have used commercially available chelating resins such as NTA Superflow (Lohan et al., 2006) and Toyopearl (Hurst and Bruland, 2007), thereby eliminating the synthesis step involved in using 8-HQ.

  Regardless of which chelating resin is chosen, the pH dependency on the recovery of Fe(III) and Fe(II) is an important consideration. For example, Fe(III) is recovered by 8-HQ at pH values 3.0–4.2, while at pH 5.2–6.0 both Fe(III) and Fe(II) are quantitatively recovered (Obata et al., 1993). Therefore, FI analysis may also allow the redox speciation of dissolved Fe to be determined by careful choice of pretreatment pH, reagent conditions, and appreciation of possible interferences (Bowie et al., 2002; Hopkinson and Barbeau, 2007). Samples for dissolved Fe analysis should be acidified to pH < 2 (Johnson et al., 2007) using subboiled distilled Q-HCl (i.e., 2 mL of concentrated HCl per liter of sample, 0.024 M HCl), and stored for at least 12 hours before determination to ensure Fe is released from organic ligands. For total dissolved Fe(II+III) analysis, either an oxidation step (addition of 10 µM hydrogen peroxide (H_2O_2); Lohan et al., 2006) or a reduction step (addition of 100 µM sodium sulfite; Bowie et al., 1998) is required prior to preconcentration.

There are two different types of detection system used in FI analysis: chemiluminescence (CL) and spectrophotometry (SP). FI-CL methods are based on the catalytic effect of either Fe(II) or Fe(III) ions on the oxidation of luminol (5-amino-2,3-dihydro-1,4-phthalazinedione) to generate blue luminescence ($\lambda_{max} \sim 440$ nm), which is detected using a photomultiplier tube (PMT). Two variants exist. FI-CL [FeII] methods (e.g., Bowie et al., 1998) determine Fe(II) and require acidified samples to be reduced off-line using sodium sulfite. Reduced samples are buffered in-line using Q-NH_4Ac to pH 5 prior to preconcentration on an 8-HQ resin for 60–120 seconds. The resin is then rinsed with UHPW for 30 seconds to remove any salts. Fe(II) ions are eluted from the resin using 0.09 N Q-HCl for 60 seconds, and carried to the PMT, where they mix with a luminol/carbonate buffer (pH 10.4) and a signal is generated in the form of a narrow peak. FI-CL [FeIII] methods (e.g., Obata et al., 1993; de Jong et al., 1998) determine Fe(III). The acidified sample should be readjusted to around pH 3 prior to 8-HQ preconcentration, and H_2O_2 should be added to ensure Fe(III) ions are present in the sample. Elution is achieved using 0.3 N Q-HCl. An additional 2 m heated coil is required in the manifold to efficiently mix the eluted Fe(III) with the luminol/triethylenetetramine, ammonia, and H_2O_2 reagent streams before detection at the PMT, and to ensure complete oxidation of the luminol.

FI-SP was first reported for the determination of dissolved Fe by Measures et al. (1995), and involves the catalytic oxidation of DPD (N,N-dimethyl-p-phenylenediamine dihydrochloride) by Fe(III) cycled with H_2O_2. The catalytic nature of the reaction increases the sensitivity of this method, as the amount of oxidized DPD is proportional to the concentration of Fe. H_2O_2 (10 µM) should be added to the sample to ensure oxidation to Fe(III). Depending on the resin used for preconcentration, the samples may need to be buffered in-line. After a load time of 1 to 5 minutes, the resin is rinsed with either UHPW or an ammonium acetate (pH 3.5) solution for 15 seconds. Fe is eluted from the resin and mixes with DPD/buffer (optimum pH 5.5–6) and H_2O_2 (Lohan et al., 2006), producing colored semiquinone derivates, which are detected spectrophotometrically at 514 nm. A typical FI-SP analytical manifold using DPD is shown in Figure 12.3.

12.4.3.2 Cathodic Stripping Voltammetry (CSV)

CSV involves the preconcentration of Fe onto a working electrode (usually a hanging mercury drop, HMDE), followed by reduction back into solution during the stripping step, and then the reductive current response is measured. The following voltametric parameters are typical: deposition time, 60–300 seconds; deposition potential, –0.15 V; linear sweep potential scan at a step rate of 10 s^{-1}; and a potential step of 4 mV. To increase the sensitivity of this technique, a well-characterized Fe-binding ligand (AL) is added to seawater that forms an electro-active adsorptive complex with Fe (AdCSV). Filtered samples are UV irradiated to break down Fe-complexing ligands that occur naturally in seawater and break down any surfactants that could adsorb onto the HMDE (Achterberg and Braungardt, 1999). In high-Fe coastal waters, UV oxidation is not sufficient to release all the Fe from complexing ligands, and samples need to be acidified to pH 1.7, followed by a brief microwave step (Bruland et al., 2005). The sample is adjusted to pH 8 using 1 N Q-NH_4OH, and the ligand and pH buffer (required since the formation of Fe-AdCSV ligand complexes is highly pH dependent) are added to the sample.

Four different ligands have been reported: TAC (2-2 thiazolyazo-p-cresol; Croot and Johansson, 2000), SA (salicylaldoxime; Rue and Bruland, 1995), DHN (2,3 dihydroxynaphthalene; van den Berg, 2006), and NN (1-nitroso-2-naphthol; Gledhill and van den Berg, 1995). For TAC, SA, and DHN the pH is controlled at pH = 8 using borate or EPPS buffer, while for NN the pH is adjusted to 6.9 using PIPES buffer. The concentration of dissolved Fe is then determined by adsorption of the Fe-AL complex onto the HMDE for an appropriate adsorption period, followed by a cathodic stripping step where the Fe(III) in the adsorbed complex is reduced and the resultant reduction current is measured. Standard additions are then made to the sample and concentration determined by linear regression of the standard curve.

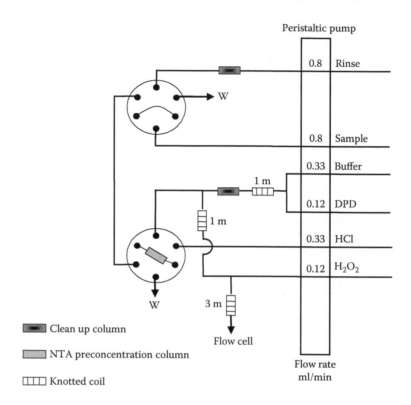

FIGURE 12.3 Manifold used for the preconcentration (on NTA Superflow resin column) and determination of Fe in seawater by flow injection-spectrophotometry using DPD. See text for operating procedure. (Redrawn after Lohan et al., 2006, with kind permission of the American Society of Limnology and Oceanography.)

While this technique can be used to measure total dissolved Fe, it is more time-consuming than FI techniques and is therefore used primarily to determine the complexation of Fe(III) with natural organic ligands present in seawater. At present, this is the only technique available that can determine the concentration of free Fe(III) ions and natural metal-complexing ligands [L] as well as the conditional stability constants of Fe-complexing natural ligands ($K_{Fe_L Fe'}^{Cond}$). Competitive ligand equilibration (CLE) followed by AdCSV is used to determine the electrochemically active Fe-added ligand complex after a competing equilibrium has been established between Fe(III)′, the added ligand, and any naturally occurring Fe(III)-binding ligands in seawater. A titration procedure is employed whereby a buffered seawater sample is divided into twelve 10 mL subsamples in FEP vials, two of which are unmodified and used to measure labile Fe, and to the other ten increasing amounts of Fe are added. Many workers also add the ligand at this stage and equilibrate the samples for ~12 hours. Buck and Bruland (2007) argue that the sample should equilibrate with the natural ligands for 2 hours before adding the well-characterized ligand. After equilibration, the labile Fe concentrations are determined in each of the ten subsamples. Linear (Ružič, 1982) or nonlinear (Gerringa et al., 1995) data transformations allow determination of the stability constants and total concentrations of one or more unknown Fe-binding ligands. (For more details on organic speciation of trace metals, see Chapter 14.) The Fe(III)′ concentration is calculated from [L] and $K_{Fe_L Fe'}^{Cond}$. A recent debate in the literature (see Town and van Leeuwen, 2005, and references therein) has questioned the slow kinetics of Fe with organic ligands, and therefore if thermodynamic considerations are valid for these measurements.

It is important to note that the natural ligands determined using CLE-AdCSV are operationally defined. The detection window of this technique is based on the concentration and the stability constant of the added ligand (Bruland and Rue, 2001). The ligand classes are operationally defined, and they represent the average values for a continuum of possible Fe(III)-binding organic molecules with the assumption that 1:1 Fe(III)-L complexes are formed. The identity, origin, and chemical characteristics of the organic Fe-binding ligands are largely unknown. For a more detailed and theoretical background on AdCSV, see Bruland and Rue (2001) and Achterberg and Braungardt (1999). Many studies on Fe speciation are carried out on samples that have been frozen after collection. Van den Berg (1995) has shown that freezing the samples does not impact on the organic speciation of iron.

12.4.4 LABORATORY METHODS

With the advent of sophisticated analytical techniques in the late 1970s, such as graphite furnace atomic absorption spectrometers (GFAASs) and high-resolution inductively coupled plasma mass spectrometers (ICP-MSs), dissolved and particulate Fe analyses have commonly been carried out in the laboratory. The advantage of these techniques is high sample throughput, high sensitivity, and with high-resolution ICP-MS, the exclusion of isobaric interferences and simultaneous determination of other trace elements of interest. These instruments do not allow for redox or organic speciation determinations. There are three main preconcentration and matrix separation procedures used for Fe analysis by GFAAS and ICP-MS: (1) dithiocarbamate/solvent extraction, (2) chelating resins, and (3) magnesium coprecipitation (Mg(OH)$_2$).

Preconcentration using dithiocarbamate/solvent extraction involves the pH adjustment of acidified samples (pH 1.8) to pH 4.0–4.5 using an ammonium acetate buffer, along with the addition of a large excess of ligands, pyrrolidine dithiocarbamate (PDC) and diethyl dithiocarbamate (DDC). The ligands are added at the same time as the ammonium acetate, resulting in a quantitative complexation of Fe as dithiocarbamate complexes (Bruland et al., 1979). This ensures that Fe will not reestablish equilibrium with low concentrations of any natural Fe-binding ligands or precipitate as insoluble ferric hydroxides. A double extraction of the Fe-dithiocarbamate-neutral complexes into chloroform or Freon is then carried out. The chloroform or Freon fractions are combined in quartz beakers and nitric acid is added. This is slowly evaporated to dryness and subsequently reconstituted in weak nitric acid prior to determination by GFAAS (Bruland et al., 1979) or ICP-MS. This method has a high precision and accuracy but requires large liter-scale sample volumes and is relatively labor-intensive in both sample collection and analysis.

Alternatively, preconcentration can be achieved using a chelating resin such as 8-HQ (Beauchemin and Berman, 1989) or NTA (Lohan et al., 2005). H$_2$O$_2$ (10 µM) should be added to acidified seawater samples to convert any Fe(II) present to Fe(III). Samples are then loaded onto the column for 30 seconds, rinsed with UHPW to remove any salts, and eluted with 1.5 N HNO$_3$. This can be carried out using two pumps controlled by the ICP-MS software and directly injected into the ICP-MS (Lohan et al., 2005), or carried out off-line where the eluent is collected in LDPE vials for subsequent analysis by ICP-MS. Acidified low-Fe seawater (10 µM H$_2$O$_2$ added) is used to prepare the working standards, with blank correction by standard additions. Other chelating resins have also been used, such as Chelex-100 or Toyopearl (Hurst and Bruland, 2007), but in this case the samples must be UV oxidized and buffered online to pH 5.5 before analysis.

The Mg(OH)$_2$ preconcentration technique involves the addition of a base (such as NH$_4$OH) resulting in the precipitation of Fe with magnesium, which is already present in the samples as a major constituent of seawater (Wu and Boyle, 1998). A known volume of seawater sample is spiked with an excess stable isotopic spike (e.g., ^{57}Fe) and allowed to equilibrate at room temperature (2 minutes). A small aliquot of Q-NH$_4$OH is then added to the sample and Mg(OH)$_2$ is precipitated. The sample is then centrifuged, the supernatant discarded, and precipitate dissolved in a small volume of dilute HNO$_3$ for analysis by ICP-MS (Wu and Boyle, 1998). This method has been improved by the addition of a second Mg(OH)$_2$ coprecipitation step (Wu, 2007). The ^{57}Fe spike acts as an internal standard and is used prior to sample preconcentration and ICP-MS measurement, eliminating the need to correct for Fe recovery during preconcentration and also minimizing interferences due to sample matrix and instrument sensitivity variations. The isotope dilution is performed at low pH (<2) to minimize the influence of inorganic/organic Fe speciation on the isotopic equilibration between the enriched Fe isotope spike and naturally abundant Fe isotopes in the sample.

12.4.5 PARTICULATE FE ANALYSES

For details on filtration see Section 12.3.1. The particulate Fe pool is generally characterized as labile and refractory. To characterize the readily exchangeable or labile particulate Fe, an acetic acid (HAc) leach (2 mL of 25% Q-HAc onto the filter for 2 hours at room temperature; Chester and Hughes, 1967) is used. The leachate and UHPW rinses of the filter are transferred into a quartz beaker and heated to dryness. The resulting residue is redissolved in 1 N HNO_3 and spiked with an internal standard (e.g., indium or gallium) prior to analysis by ICP-MS (Section 12.4.4). This leach includes Fe associated with organic phases, carbonates, and Mn/Fe oxides (Landing and Bruland, 1987). To determine the refractory mineral phase, either a heating or pressure digestion step using HNO_3, HCl, or hydrofluoric acid is carried out in Teflon PFA bombs or vials (Cullen et al., 2001). The use of other particulate elemental data and crustal ratios can help differentiate between lithogenic and biogenic particulate Fe.

In culture or shipboard incubation experiments the partitioning of particulate Fe has been classified into the scavenged (onto suspended particles or phytoplankton cell surfaces) and intracellular pool. The intracellular or biological fraction is needed to calculate the cellular Fe quota. Hudson and Morel (1990) used a reductive Ti-citrate EDTA rinse to distinguish between these two pools in radiotracer studies. This has been further developed by Tovar-Sanchez et al. (2003) and Tang and Morel (2006) using an oxalate-citrate EDTA rinse, thereby eliminating the need for radiotracers.

12.5 QUALITY ASSURANCE

12.5.1 BLANK ESTIMATION AND CORRECTION

Since a substantial amount of random error may reside in the blank, good precision on low-concentration Fe samples is a prerequisite to obtaining accurate data. The detection limit of Fe is not restricted by the sensitivity of the instrumental technique but by the blank value. Most laboratories measuring Fe in seawater report their detection limit as "3s of a low Fe sample or blank" (Bowie et al., 2006). Instrument and procedural blanks must be determined with each analysis. Instrumental blanks range from a manifold blank in FI analysis to an ICP-MS instrument blank. The determination of a manifold blank involves running an analytical cycle without the sample (e.g., 1 minute loading of sample buffer only onto an 8-HQ column followed by the rinse phase). The ICP-MS instrumental Fe blank varies with sample acidity and is difficult to correct using isotope dilution (Wu, 2007), and therefore a higher sample preconcentration ratio (500:1) is used to lower its relative contribution (Bowie et al., 2007).

Procedural blanks are those from the chemicals (e.g., sodium sulfite, H_2O_2, Q-NH_4OH) used in analysis and acidification (Q-HCl) of the sample. This is quantified by carrying out multiple additions to UHPW or low-Fe seawater. In general, the blank contribution due to the addition of sodium sulfite has been found to be negligible (<3 pM Fe; Bowie et al., 2007), while hydrogen peroxide contributed 15 pM to the procedural blank (Lohan et al., 2006). For ICP-MS work, the procedural blank can be determined precisely using isotope dilution (Wu, 2007). This is carried out by using 0.4 mL of low-Fe seawater throughout the whole procedure. The lowest blanks reported for Fe analysis are from the $Mg(OH)_2$ precipitation method (0.6 pM; Wu, 2007). These blanks must be determined and subtracted from the measured sample concentrations. Table 12.2 shows the detection limits, blanks, and relative standard deviation of all methods outlined above.

For particulate Fe analysis, the filter blank is the largest contributor to the blank correction (Cullen et al., 2001). Filter blank values are obtained by filtering a small volume of UHPW or

TABLE 12.2
A Comparison of Detection Limits, Blanks, and Precision of Various Methods for Seawater Fe Analysis

Analytical Method	Detection Limit, pM	Blank, pM	Precision, RSD%
Flow injection 8-HQ preconcentration with Fe(III)-chemiluminescence detection (de Jong et al., 1998)	21	22	<5
Flow injection 8-HQ preconcentration with Fe(II)-luminol chemiluminescence detection (Bowie et al., 1998)	40	30	3.2
Flow injection 8-HQ preconcentration with spectrophotometric detection (Measures et al., 1995)	25	Negligible	2.5
Flow injection NTA preconcentration with spectrophotometric detection (Lohan et al., 2006)	24	28	4.2
Cathodic stripping voltammetry using TAC (Croot and Johansson, 2000)	13	40	7
Cathodic stripping voltammetry using SA (Rue and Bruland, 1995)	10	nd	6
Cathodic stripping voltammetry using DHN (van den Berg, 2006)	15	50	4
Cathodic stripping voltammetry using NN (Gledhill and van den Berg, 1995)	30	20	nd
APDC/DDDC solvent extraction, GFAAS (Landing and Bruland, 1987)	30	23	4
Flow injection NTA in-line preconcentration HR-ICPMS (Lohan et al., 2005)	20	28	4.4
Single $Mg(OH)_2$ coprecipitation ID-ICPMS (Wu and Boyle, 1998)	50	40	3
Double $Mg(OH)_2$ coprecipitation ID-ICPMS (Wu, 2007)	2	0.6	4

Note: nd, not determined.

previously filtered seawater through an acid-cleaned filter and digesting this filter using the same protocols as for the samples. Often many blank filter samples are analyzed in order to reduce the uncertainty of the blank. Instrument blanks must also be quantified (see above).

12.5.2 CONTROL OF RECOVERIES

There are several different analytical methods for determining dissolved Fe in seawater (Section 12.4). It is therefore important to quantify the efficiency of the extraction of Fe from the seawater matrix during preconcentration, since different methods can measure different fractions of Fe (e.g., specific for Fe(II) or Fe(III), or preferential extraction of Fe-organic compounds; Ussher et al., 2005). Methods using preconcentration onto a chelating resin quantify the recovery of Fe by loading a known standard of Fe onto the resin and analyzing the concentration eluted from the resin using ICP-MS, where the data are normalized with an internal standard (e.g., scandium). Each chelating resin behaves differently, and quantitative recoveries depend on pH. One hundred percent recovery of Fe(III) is observed at pH 2 using an NTA resin (Lohan et al., 2006), while for 8-HQ Fe(III) is quantitatively recovered at pH 3 (Obata et al., 1993). Fe(II) is quantitatively recovered at pH 5.5 for both NTA and 8-HQ. A major attraction of isotope dilution is that the method quantifies all of the Fe species present in the sample, and using isotopic equilibrium (e.g., examining the ratio of spiked ^{57}Fe to ^{56}Fe present in the sample), it is not essential to quantitatively recover the Fe during each step.

12.5.3 PRECISION AND ACCURACY

Precision is defined as "the degree of mutual agreement characteristic of independent measurements as the result of repeated application of the process under specific conditions" (see Chapter 1). It is concerned with the closeness of results (Taylor, 1987). Precision is therefore a measure of random errors in a method or procedure, in other words, the variability of individual measurements or analytical reproducibility. Precision is quantified by replication and for Fe analysis consists of both field replicates and analytical replicates. Analytical replication is the repeated analysis of a single sample and is a measure of the greatest precision possible for a particular analysis. Field replication is the analysis of two or more samples taken from a single sampling bottle and has an added component of variance due to subsampling, storage, and natural within-sample variability. The variance of field and analytical replicates should be equal when sampling and storage have no effect on the analysis (assuming the analyte is homogeneously distributed within the sampling bottle). Therefore, the difference between field and analytical replicates provides a first-order evaluation of the field sampling procedure.

Accuracy is defined as "the degree of agreement of a measured value with the true or expected value of the quantity of concern" (Taylor, 1987). Accuracy therefore includes random and systematic errors, and is determined by the use of reference standards (Section 1.5.4). The assessment of accuracy must be based upon primary standards that have the highest metrological qualities (and whose value is accepted without reference to other standards) and must be prepared in the same matrix as the analyte (i.e., seawater).

Precision and accuracy are therefore not necessarily coupled, and both should be evaluated independently and reported alongside any oceanographic data. An analysis may be precise yet inaccurate, whereas the mean of a variable result may be quite accurate.

12.5.4 Reference Materials

The accuracy of chemical oceanographic measurements depends on calibration against reference materials to ensure comparability over time and among laboratories. Historically, there has been a lack of availability of appropriate reference materials for subnanomolar concentrations of Fe in seawater. For example, the National Research Council of Canada NASS-5 solution contains 3.71 ± 0.63 nM Fe, and the IRONAGES standard sample collected from the South Atlantic Ocean was not stable until 1 year after collection (Bowie et al., 2006). This has meant that the scientific community has had little ability to correlate oceanographic observations, and distinguish between environmental variability, analytical data quality, and measurement drift, leading to a number of Fe intercomparison exercises over the last decade (Landing et al., 1995; de Jong et al., 2000; Bowie et al., 2003, 2004, 2007; Lohan et al., 2006).

The concentration of a trace element in a reference material is normally assigned by an analytical method that readily establishes traceability in measured results through comparisons to recognized standard quantities (e.g., isotope dilution ICP-MS; Section 12.4.4). However, the extremely low concentration of Fe in seawater, importance of the analytical blank, and risk of contamination make direct assessment of traceability difficult. During the SAFe intercalibration study in the North Pacific Ocean (Johnson et al., 2007), the analytical results were combined to assign consensus concentrations and uncertainties, following recommended procedures for reference materials whose concentrations were determined by interlaboratory study. Results from SAFe suggest that there is now substantial agreement between analytical methods for dissolved Fe in seawater, with systematic differences that appear to be <0.05 nM in surface samples, and depth profiles that are oceanographically and historically consistent (Figure 12.4), although individual operators may produce inconsistent values. Homogeneity and stability were confirmed on the SAFe samples over a 20-month time interval, and therefore it is recommended that all future measurements of dissolved Fe in seawater should be unambiguously tied to these reference materials. Two types of SAFe samples are available (surface and 1,000 m deep), which are contained in individually numbered 500 mL LDPE bottles. Currently there is some consensus that the SAFe samples will become valid reference materials for the following elements: Al, Mn, Fe, Co, Ni, Cu, Zn, Cd, and Pb.

At present, there is no appropriate reference material for Fe concentration in suspended marine particulate samples, although reference materials do exist for trace elements in marine plankton (Institute for Reference Materials and Measurements: BCR-414) and sediments (National Research

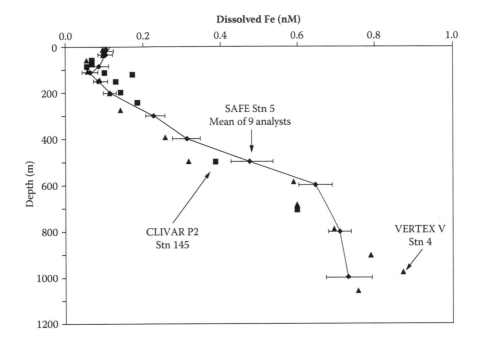

FIGURE 12.4 Dissolved Fe profile at the Sampling and Analysis of Fe (SAFe) station (30°N, 140°W) measured onboard ship during the SAFe cruise (diamonds) (sampled using the trace metal rosette reported in Measures et al., 2008b), during the Vertical Transport and Exchange (VERTEX) program in 1986 by organic solvent extraction GFAAS (triangles) (Martin and Gordon, 1988), and during the Climate Variability and Predictability (CLIVAR) P2 cruise in 2004 by FI-SP using DPD (squares) (C. Measures, unpublished data). (Data first shown in Johnson et al., 2007.)

Council of Canada: HISS, MESS, and PACS). A similar situation exists for Fe speciation measurements (e.g., redox, organic), which will hamper progress in analysis of these chemical forms. The new international GEOTRACES program plans to collect marine particulate samples and undertake speciation intercalibration exercises for Fe.

12.5.5 DATA EVALUATION AND REPORTING

Data evaluation and reporting are the final and essential steps in any quality assurance program. Data should be reviewed in context of the entire sample collection, processing, storage, and analytical process. It is recommended that analysts record the following ancillary information alongside their measured Fe concentrations: standard deviation (1s), quality flag (e.g., 0 = good, 1 = unknown, 4 = questionable, 8 = bad), number of replicates, method and procedural blanks, detection limit (3s), date of analysis, and literature reference for the analytical method.

Discrepancies or anomalous results should be carefully noted at various stages of the analytical process and the final data evaluated for correctness. This evaluation is often termed oceanographic consistency, and is achieved by plotting the analyte profile versus depth and investigating those points that are outside the anticipated data envelope (i.e., possible outliers). These data should not be automatically flagged as "bad," but rather investigated for the source of the problem through the sample documentation. If the problem cannot be identified, the data may be flagged "questionable" if the values are outside the 95% confidence interval (e.g., greater than 2 standard deviations from the historical mean), and "good" if within this error envelope. If a source for the discrepancy is discovered, the data are flagged "bad." Clearly, there are sections of the world's oceans where no historical data currently exist (particularly deep waters), and thus it is not always possible to assign quality flags. This use of quality flags is consistent with the reporting of hydrographic data (e.g., in WOCE exchange format) and is readily amenable to filtering of large datasets by oceanographic graphical display software (e.g., Ocean Data View).

12.6 SUMMARY AND PERSPECTIVES

The biogeochemical cycling of Fe and its role as a micronutrient is pivotal to research addressing the effect of climate on changing ocean chemistry, and fundamental to studies on carbon cycling, marine ecosystem dynamics, and environmental impacts. Over the last two decades, we have made great advances in our knowledge of the distribution and the complex marine biogeochemistry of Fe in the marine environment. This has been achieved through the development of careful sample handling, filtration, and pretreatment steps in combination with rapid, accurate, sensitive, and well-characterized shipboard and laboratory-based analytical techniques. However, there are still uncertainties concerning the sources, sinks, and in particular, the internal cycling and speciation of dissolved Fe in ocean basins. Organic speciation analysis has revealed the ubiquitous presence of strong Fe-binding chelates, but little is known about the molecular structure of these ligands. New method development and optimization of existing methods should focus particularly on the redox and organic speciation of Fe, as well as improved methods for sample collection, filtration, and storage, especially for the particulate phase. Promising new techniques include multiple-collector plasma source mass spectrometers (MC-ICP-MS) for studying Fe isotopic fractionation (de Jong et al., 2007), bioreporters to assess marine microorganism Fe acquisition and bioavailability (Hassler et al., 2006), molecular characterization of Fe-binding ligands such as siderophores (McCormack et al., 2003), and high-temperature fusion methods for the digestion of Fe in oceanic particulate material (Huang et al., 2007).

It is now well established that the marine phytoplankton production and ecosystem structure is limited by the availability of Fe in over 40% of the world's surface oceans. Increased nutrient utilization efficiency of the ocean's biological pump (~50% under modern conditions) could have been responsible for climate-related changes in atmospheric CO_2 during the last ice age. Changes in the delivery and availability of Fe may be the single largest driver of ocean ecosystem productivity and health in the next century, and thus is intrinsically linked to changes in global carbon cycling and climate. As we move forward, the focus on global climate change in ocean systems will become a major area of intensive study. In order to assess and predict the impact of global change on Fe distributions in the ocean, and conversely the effect of Fe supply on ocean productivity and climate, it is imperative that we fully understand the dynamic cycling of Fe in the marine environment. At present, the international CLIVAR/CO_2 Repeat Hydrography (Measures et al., 2008a) and GEOTRACES programs are aiming to determine three-dimensional global ocean sections of key trace elements and isotopes (including Fe), with a focus on evaluating their sources, sinks, and internal cycling. This measurement strategy will provide maximum scientific reward by significantly increasing the spatial resolution of Fe across all ocean basins, providing comprehensive datasets for ocean modelers, and allowing for assessment of lateral transport, particulate fluxes, and speciation of Fe.

One major limitation at present is the expensive nature of ship time and the lack of temporal data, which cannot be realized through the GEOTRACES program. With the expansion of ocean observatory initiatives (e.g., ORION, IMOS), there is a need to develop in situ analytical techniques for the determination of Fe. One potential tool is the voltammetric in situ profiler (VIP), although substantial method development is required for this system to operate at the low concentrations observed in coastal and open-ocean environments. Another potential application is the deployment of chelating resins coupled to in situ pumping systems, which allow Fe to be preconcentrated from seawater samples in the field and the resins returned to onshore laboratories for analysis.

Finally, there is growing interest in ocean iron fertilization as a possible method to increase the ocean's capacity to draw CO_2 out of the atmosphere and act as a climate mitigation strategy (Boyd et al., 2007). While this topic remains controversial, much has been learned from controlled mesoscale iron fertilization experiments, and it is important for marine chemists to continue to develop innovative sampling and analytical techniques to assess the natural mode of Fe supply and its subsequent mobilization and retention by upper ocean processes. Key constraints on the efficiency of Fe enrichments are Fe/C molar ratios of algal communities and exported particles, and future biogeochemical studies should focus on evaluating temporal and spatial controls on this important parameter in both high-Fe and low-Fe ocean regimes. This is an exciting field, and due to the establishment of reliable methods for the analysis of different forms of Fe in seawater, we are now poised to take the next steps in understanding and predicting future global change.

ACKNOWLEDGMENTS

ARB acknowledges the Australian Research Council for grant DP0342826 and the Australian Government Cooperative Research Centres Program through the Antarctic Climate and Ecosystems CRC. We thank Bill Landing, Delphine Lannuzel, Christel Hassler, and Thomas Trull for helpful comments that improved the chapter.

REFERENCES

Achterberg, E. P., and C. Braungardt. 1999. Stripping voltammetry for the determination of trace metal speciation and in-situ measurements of trace metal distributions in marine waters. *Analytica Chimica Acta* 400:381–97.

Achterberg, E. P., T. W. Holland, A. R. Bowie, R. F. C. Mantoura, and P. J. Worsfold. 2001. Determination of iron in seawater. *Analytica Chimica Acta* 442:1–14.

Anderson, M. A., and F. M. M. Morel. 1982. The influence of aqueous iron chemistry on the uptake of iron by the coastal diatom *Thalassiosira weissflogii*. *Limnology and Oceanography* 27:789–813.

Barbeau, K., J. W. Moffett, D. A. Caron, P. L. Croot, and D. L. Erdner. 1996. Role of protozoan grazing in relieving iron limitation of phytoplankton. *Nature* 380:61–64.

Beauchemin, D., and S. S. Berman. 1989. Determination of trace-metals in reference water standards by inductively coupled plasma mass spectrometry with on-line preconcentration. *Analytical Chemistry* 61:1857–62.

Bell, J., J. Betts, and E. Boyle. 2002. MITESS: A moored in situ trace element serial sampler for deep-sea moorings. *Deep Sea Research I* 49:2103–18.

Bergquist, B. A., and E. A. Boyle. 2006. Dissolved iron in the tropical and subtropical Atlantic Ocean. *Global Biogeochemical Cycles* 20:GB1015, doi: 10.1029/2005GB002505.

Bishop, J. K. B., J. M. Edmond, D. R. Ketten, M. P. Bacon, and W. B. Silker. 1977. Chemistry, biology and vertical flux of particulate matter from upper 400 m of equatorial Atlantic Ocean. *Deep Sea Research* 24A:511–48.

Blain, S., B. Queguiner, L. Armand, S. Belviso, B. Bombled, L. Bopp, A. Bowie, C. Brunet, C. Brussaard, and F. Carlotti. 2007. Effect of natural iron fertilization on carbon sequestration in the Southern Ocean. *Nature* 446:1070–74.

Bowie, A. R., E. P. Achterberg, R. F. C. Mantoura, and P. J. Worsfold. 1998. Determination of sub-nanomolar levels of iron in seawater using flow injection with chemiluminescence detection. *Analytica Chimica Acta* 361:189–200.

Bowie, A. R., E. P. Achterberg, S. Blain, M. Boye, P. L. Croot, H. J. W. De Baar, P. Laan, G. Sarthou, and P. J. Worsfold. 2003. Shipboard analytical intercomparison of dissolved iron in surface waters along a north-south transect of the Atlantic Ocean. *Marine Chemistry* 84:19–34.

Bowie, A. R., E. P. Achterberg, P. L. Croot, H. J. W. De Baar, P. Laan, J. W. Moffett, S. Ussher, and P. J. Worsfold. 2006. A community-wide intercomparison exercise for the determination of dissolved iron in seawater. *Marine Chemistry* 98:81–99.

Bowie, A. R., E. P. Achterberg, P. N. Sedwick, S. Ussher, and P. J. Worsfold. 2002. Real-time monitoring of picomolar concentrations of iron(II) in marine waters using automated flow injection-chemilumines-cence instrumentation. *Environmental Science and Technology* 36:4600–7.

Bowie, A. R., P. N. Sedwick, and P. J. Worsfold. 2004. Analytical intercomparison between flow injection-chemiluminescence and flow injection-spectrophotometry for the determination of picomolar concen-trations of iron in seawater. *Limnology and Oceanography: Methods* 2:42–54.

Bowie, A. R., S. J. Ussher, W. M. Landing, and P. J. Worsfold. 2007. Intercomparison between FI-CL and ICP-MS for the determination of dissolved iron in Atlantic seawater. *Environmental Chemistry* 4:1–4.

Boyd, P. W., T. Jickells, C. S. Law, S. Blain, E. A. Boyle, K. O. Buesseler, et al. 2007. Mesoscale iron enrich-ment experiments 1993–2005: Synthesis and future directions. *Science* 315:612–17.

Boyd, P. W., A. J. Watson, C. S. Law, E. R. Abraham, T. Trull, R. Murdoch, D. C. E. Bakker, A. R. Bowie, K. O. Buesseler, and H. Chang. 2000. A mesoscale phytoplankton bloom in the polar Southern Ocean stimulated by iron fertilization. *Nature* 407:695–702.

Boyle, E. A., J. M. Edmond, and E. R. Sholkovitz. 1977. The mechanism of iron removal in estuaries. *Geochimica et Cosmochimica Acta* 41:1313–24.

Bruland, K. W., R. P. Franks, G. A. Knauer, and J. H. Martin. 1979. Sampling and analytical methods for the determination of copper, cadmium, zinc and nickel at the nanogram per litre level in seawater. *Analytica Chimica Acta* 105:233–45.

Bruland, K. W., K. J. Orians, and J. P. Cowen. 1994. Reactive trace-metals in the stratified central North Pacific. *Geochimica et Cosmochimica Acta* 58:3171–82.

Bruland, K. W., and E. L. Rue. 2001. Analytical methods for the determination of concentrations and specia-tion of iron. In *Biogeochemistry of iron in seawater*, ed. D. R. Turner and K. A. Hunter, 250–90. New York: John Wiley & Sons.

Bruland, K. W., E. L. Rue, G. J. Smith, and G. R. Ditullio. 2005. Iron, macronutrients and diatom blooms in the Peru upwelling regime: Brown and blue waters of Peru. *Marine Chemistry* 93:81–103.

Buck, K. N., and K. W. Bruland. 2007. The physicochemical speciation of dissolved iron in the Bering Sea, Alaska. *Limnology and Oceanography* 52:1800–8.

Buesseler, K. O., A. V. Antia, M. Chen, S. W. Fowler, W. D. Gardener, O. Gustafsson, et al. 2007. An assessment of the use of sediment traps for estimating upper ocean particle fluxes. *Journal of Marine Research* 65:345–416.

Byrne, R. H., and D. R. Kester. 1976. Solubility of hydrous ferric oxide and iron speciation in seawater. *Marine Chemistry* 4:255–74.

Chapin, T. P., H. W. Jannasch, and K. S. Johnson. 2002. In situ osmotic analyzer for the year-long continuous determination of Fe in hydrothermal systems. *Analytica Chimica Acta* 463:265–74.

Chester, R., and M. J. Hughes. 1967. A chemical technique for the separation of ferromanganese minerals, carbonate minerals and adsorbed trace elements for pelagic sediments. *Chemical Geology* 2:249–62.

Cooper, L. H. N. 1935. Iron in the sea and in marine plankton. *Proceedings of the Royal Society of London* 118B:419–38.

Croot, P. L., and M. Johansson. 2000. Determination of iron speciation by cathodic stripping voltammetry in seawater using the competing ligand 2-(2- thiazolylazo)-p-cresol (TAC). *Electroanalysis* 12:565–76.

Cullen, J. T., M. P. Field, and R. M. Sherrell. 2001. Determination of trace elements in filtered suspended marine particulate material by sector field HR-ICP-MS. *Journal of Analytical Atomic Spectrometry* 16:1307–12.

Cullen, J. T., and R. M. Sherrell. 1999. Techniques for determination of trace metals in small samples of size-frac-tionated particulate matter: Phytoplankton metals off central California. *Marine Chemistry* 67:233–47.

de Jong, J. T. M., M. Boye, V. F. Schoemann, R. F. Nolting, and H. J. W. de Baar. 2000. Shipboard tech-niques based on flow injection analysis for measuring dissolved Fe, Mn and Al in seawater. *Journal of Environmental Monitoring* 2:496–502.

de Jong, J. T. M., J. den Das, U. Bathmann, M. H. C. Stoll, G. Kattner, R. F. Nolting, and H. J. W. de Baar. 1998. Dissolved iron at subnanomolar levels in the Southern Ocean as determined by shipboard analysis. *Analytica Chimica Acta* 377:113–24.

de Jong, J. T. M., V. Schoemann, J.-L. Tison, S. Becquevort, F. Masson, D. Lannuzel, J. Petit, L. Chou, D. Weis, and N. Mattielli. 2007. Precise measurement of Fe isotopes in marine samples by multi-collector inductively coupled plasma mass spectrometry (MC-ICP-MS). *Analytica Chimica Acta* 589:105–19.

Frew, R. D., D. A. Hutchins, S. Noddler, S. Sanudo-Wilhelmy, A. Tovar-Sanchez, K. Leblanc, C. E. Hare, and P. W. Boyd. 2006. Particulate iron dynamics during Fe cycle in subantarctic waters southeast of New Zealand. *Global Biogeochemical Cycles* 20:GB1S93, doi: 10.1029/2005GB002558.

Gardner, W. D. 2000. Sediment trap technology and surface sampling in surface waters. In *The changing ocean carbon cycle: A midterm synthesis of the joint Global Ocean Flux Study*, ed. R. B. Hanson, H. W. Ducklow, and J. G. Field. Cambridge, UK: Cambridge University Press.

Gerringa, L. J. A., P. M. J. Herman, and T. C. W. Poortvliet. 1995. Comparison of the linear van den Berg Ružič transformation and a non-linear fit of the Langmuir isotherm applied to Cu speciation data in the estuarine environment. *Marine Chemistry* 48:131–42.

Gledhill, M., and C. M. G. van den Berg. 1995. Measurement of the redox speciation of iron in seawater by catalytic cathodic stripping voltammetry. *Marine Chemistry* 50:51–61.

Hassler, C. S., M. R. Twiss, R. M. L. McKay, and G. S. Bullerjahn. 2006. Optimization of iron-dependent cyanobacterial (Synechococcus, Cyanophyceae) bioreporters to measure iron bioavailability. *Journal of Phycology* 42:324–35.

Hong, H. S., and D. R. Kester. 1986. Redox state of iron in the offshore waters of Peru. *Limnology and Oceanography* 31:512–24.

Hopkinson, B. M., and K. A. Barbeau. 2007. Organic and redox speciation of iron in the eastern tropical North Pacific suboxic zone. *Marine Chemistry* 106:2–17.

Howard, A. G., and P. J. Statham. 1993. *Inorganic trace analysis: Philosophy and practice*. Chichester, UK: John Wiley & Sons,

Huang, S., E. R. Sholkovitz, and M. H. Conte. 2007. Application of high-temperature fusion for analysis of major and trace elements in marine sediment trap samples. *Limnology and Oceanography: Methods* 5:13–22.

Hudson, R. J. M., and F. M. M. Morel. 1990. Iron transport in marine phytoplankton: Kinetics of cellular and medium coordination reactions. *Limnology and Oceanography* 35:1002–20.

Hurst, M. P., and K. W. Bruland. 2007. An investigation into the exchange of iron and zinc between soluble, colloidal, and particulate size-fractions in shelf waters using low-abundance isotopes as tracers in shipboard incubation experiments. *Marine Chemistry* 103:211–26.

Hutchins, D. A., and K. W. Bruland. 1994. Grazer-mediated regeneration and assimilation of Fe, Zn and Mn from planktonic prey. *Marine Ecology Progress Series* 110:259–69.

Hutchins, D. A., A. E. Witter, A. Butler, and G. W. Luther. 1999. Competition among marine phytoplankton for different chelated iron species. *Nature* 400:858–61.

Johnson, K. S., E. Boyle, K. Bruland, K. Coale, C. Measures, J. Moffett, et al. 2007. Developing standards for dissolved iron in seawater. *Eos, Transactions of the American Geophysical Union* 88:131–32.

Johnson, K. S., R. M. Gordon, and K. H. Coale. 1997. What controls dissolved iron concentrations in the world ocean? *Marine Chemistry* 57:137–61.

Kuss, J., C. D. Garbe-Schonberg, and K. Kremling. 2001. Rare earth elements in suspended particulate material of North Atlantic surface waters. *Geochimica et Cosmochimica Acta* 65:187–99.

Landing, W. M., and K. W. Bruland. 1987. The contrasting biogeochemistry of iron and manganese in the Pacific Ocean. *Geochimica et Cosmochimica Acta* 51:29–43.

Landing, W. M., G. A. Cutter, J. A. Dalziel, A. R. Flegal, R. T. Powell, D. Schmidt, A. Shiller, P. Statham, S. Westerlund, and J. Resing. 1995. Analytical intercomparison results from the 1990 Intergovernmental Oceanographic Commission open-ocean baseline survey for trace metals—Atlantic Ocean. *Marine Chemistry* 49:253–65.

Landing, W. M., C. Haraldsson, and N. Paxeus. 1986. Vinyl polymer agglomerate based transition metal cation chelating ion-exchange resin containing the 8-hydroxyquinoline functional group. *Analytical Chemistry* 58:3031–35.

Liu, X., and F. J. Millero. 2002. The solubility of iron in seawater. *Marine Chemistry* 77:43–54.

Lohan, M. C., A. M. Aguilar-Islas, and K. W. Bruland. 2006. Direct determination of iron in acidified (pH 1.7) seawater samples by flow injection analysis with catalytic spectrophotometric detection: Application and intercomparison. *Limnology and Oceanography: Methods* 4:164–71.

Lohan, M. C., A. M. Aguilar-Islas, R. P. Franks, and K. W. Bruland. 2005. Determination of iron and copper in seawater at pH 1.7 with a new commercially available chelating resin, NTA Superflow. *Analytica Chimica Acta* 530:121–29.

Lohan, M. C., and K. W. Bruland. 2006. Importance of vertical mixing for additional sources of nitrate and iron to surface waters of the Columbia River plume: Implications for biology. *Marine Chemistry* 98:260–73.

Martin, J. H., and S. E. Fitzwater. 1988. Iron-deficiency limits phytoplankton growth in the northeast Pacific sub-Arctic. *Nature* 331:341–43.

Martin, J. H., and R. M. Gordon. 1988. Northeast Pacific iron distributions in relation to phytoplankton productivity. *Deep Sea Research* 35A:177–96.

McCormack, P., P. J. Worsfold, and M. Gledhill. 2003. Separation and detection of siderophores produced by marine bacterioplankton using high-performance liquid chromatography with electrospray ionization mass spectrometry. *Analytical Chemistry* 75:2647–52.

Measures, C. I., W. M. Landing, M. T. Brown, and C. S. Buck. 2008a. High resolution Al and Fe data from the Atlantic Ocean CLIVAR-CO2 repeat hydrography A16N transect: Extensive linkages between atmospheric dust and upper ocean geochemistry. *Global Biogeochemical Cycles* 22:GB1005, doi: 10.1029/2007/GB003042.

Measures, C. I., W. M. Landing, M. T. Brown, and C. S. Buck. 2008b. A commercially available rosette system for trace metal clean sampling. *Limnology and Oceanography: Methods*, 6: 384–394.

Measures, C. I., J. Yuan, and J. A. Resing. 1995. Determination of iron in seawater by flow injection-analysis using in-line preconcentration and spectrophotometric detection. *Marine Chemistry* 50:3–12.

Miller, W. L., D. W. King, J. Lin, and D. R. Kester. 1995. Photochemical redox cycling of iron in coastal seawater. *Marine Chemistry* 50:63–77.

Millero, F. J. 1998. Solubility of Fe(III) in seawater. *Earth and Planetary Science Letters* 154:323–29.

Millero, F. J., W. S. Yao, and J. Aicher. 1995. The speciation of Fe(II) and Fe(III) in natural waters. *Marine Chemistry* 50:21–39.

Moffett, J. W., T. J. Goeffert, and S. W. A. Naqvi. 2007. Reduced iron associated with secondary nitrite maxima in the Arabian Sea. *Deep Sea Research I* 54:1341–49.

Moore, J. K., S. C. Doney, D. M. Glover, and I. Y. Fung. 2002. Iron cycling and nutrient limitation patterns in surface waters of the world ocean. *Deep Sea Research II* 49:463–508.

Morel, F. M. M., and J. G. Hering. 1993. *Principle of aquatic chemistry*. New York: John Wiley & Sons.

Nishioka, J., S. Takeda, C. S. Wong, and W. K. Johnson. 2001. Size-fractionated iron concentrations in the northeast Pacific Ocean: Distribution of soluble and small colloidal iron. *Marine Chemistry* 74:157–79.

Obata, H., H. Karatani, and E. Nakayama. 1993. Automated determination of iron in seawater by chelating resin concentration and chemiluminescence detection. *Analytical Chemistry* 65:1524–28.

Okumura, M., Y. Seike, K. Fujinaga, and K. Hirao. 1997. In situ preconcentration method for iron(II) in environmental water samples using solid phase extraction followed by spectrophotometric determination. *Analytical Sciences* 13:231–35.

Rich, H. W., and F. M. M. Morel. 1990. Availability of well-defined iron colloids to the marine diatom *Thalassiosira weissflogii*. *Limnology and Oceanography* 35:652–62.

Riley, J. P., and R. Chester. 1971. *Introduction to marine chemistry*. London: Academic Press.

Rue, E. L., and K. W. Bruland. 1995. Complexation of iron(III) by natural organic-ligands in the central North Pacific as determined by a new competitive ligand equilibration adsorptive cathodic stripping voltammetric method. *Marine Chemistry* 50:117–38.

Ružič, I. 1982. Theoretical aspects of the direct titration of natural-waters and its information yield for trace-metal speciation. *Analytica Chimica Acta* 140:99–113.

Salter, I., R. S. Lampitt, R. Sanders, A. Poulton, A. E. S. Kemp, B. Boorman, K. Saw, and R. Pearce. 2007. Estimating carbon, silica and diatom export from a naturally fertilised phytoplankton bloom in the Southern Ocean using PELAGRA: A novel drifting sediment trap. *Deep Sea Research II* 54: 2233–59.

Sedwick, P. N., T. M. Church, A. R. Bowie, C. M. Marsay, S. J. Ussher, K. M. Achilles, P. J. Lethaby, R. J. Johnson, M. M. Sarin, and J. McGillicuddy. 2005. Iron in the Sargasso Sea (Bermuda Atlantic Time-series Study region) during summer: Eolian imprint, spatiotemporal variability, and ecological implications. *Global Biogeochemical Cycles* 19:GB4006, doi: 10.1029/2004GB002445.

Sedwick, P. N., P. R. Edwards, D. J. Mackey, F. B. Griffiths, and J. S. Parslow. 1997. Iron and manganese in surface waters of the Australian sub-Antarctic region. *Deep Sea Research I* 44:1239–53.

Sherrell, R. M., and E. A. Boyle. 1992. The trace-metal composition of suspended particles in the oceanic water column near Bermuda. *Earth and Planetary Science Letters* 111:155–74.

Sunda, W. G., D. G. Swift, and S. A. Huntsman. 1991. Low iron requirement for growth in oceanic phytoplankton. *Nature* 351:55–57.

Tang, D. G., and F. M. M. Morel. 2006. Distinguishing between cellular and Fe-oxide-associated trace elements in phytoplankton. *Marine Chemistry* 98:18–30.

Taylor, J. K. 1987. *Quality assurance of chemical measurements*. Chelsea, MI: Lewis Publishers.

Taylor, S. R. 1964. Abundance of chemical elements in the continental crust: A new table. *Geochimica et Cosmochimica Acta* 28:1273–85.

Tovar-Sanchez, A., S. A. Sañudo-Wilhelmy, M. Garcia-Vargas, R. S. Weaver, L. C. Popels, and D. A. Hutchins. 2003. A trace metal clean reagent to remove surface-bound iron from marine phytoplankton. *Marine Chemistry* 82:91–99.

Town, R. M. and H. P. van Leeuwen. 2005. Measuring marine iron(III) complexes by CLE-AdSV. *Environmental Chemistry* 2: 80–84.

Ussher, S. J., E. P. Achterberg and P. J. Worsfold. 2004. Marine biogeochemistry of iron. *Environmental Chemistry* 1: 67–80.

Ussher, S. J., M. Yaqoob, E. P. Achterberg, A. Nabi, and P. J. Worsfold. 2005. Effect of model ligands on iron redox speciation in natural waters using flow injection with luminol chemiluminescence detection. *Analytical Chemistry* 77:1971–78.

van den Berg, C. M. G. 1995. Evidence of organic complexation of iron. *Marine Chemistry* 50:139–47.

van den Berg, C. M. G. 2006. Chemical speciation of iron in seawater by cathodic stripping voltammetry with dihydroxynaphthalene. *Analytical Chemistry* 78:156–63.

Vink, S., E. A. Boyle, C. I. Measures, and J. Yuan. 2000. Automated high resolution determination of the trace elements iron and aluminium in the surface ocean using a towed fish coupled to flow injection analysis. *Deep Sea Research I* 47:1141–56.

Wells, M. L., N. M. Price, and K. W. Bruland. 1995. Iron chemistry in seawater and its relationship to phytoplankton: A workshop report. *Marine Chemistry* 48:157–82.

Wu, J. F. 2007. Determination of picomolar iron in seawater by double $Mg(OH)_2$ precipitation isotope dilution high-resolution ICPMS. *Marine Chemistry* 103:370–81.

Wu, J. F., and E. A. Boyle. 1998. Determination of iron in seawater by high-resolution isotope dilution inductively coupled plasma mass spectrometry after $Mg(OH)_2$ coprecipitation. *Analytica Chimica Acta* 367:183–91.

Wu, J. F., E. Boyle, W. Sunda, and L. S. Wen. 2001. Soluble and colloidal iron in the olgotrophic North Atlantic and North Pacific. *Science* 293:847–49.

Zhuang, G. S., Z. Yi, R. A. Duce, and P. R. Brown. 1992. Link between iron and sulfur cycles suggested by detection of Fe(II) in remote marine aerosols. *Nature* 355:537–39.

13 Radionuclide Analysis in Seawater

Mark Baskaran, Gi-Hoon Hong, and Peter H. Santschi

CONTENTS

13.1   INTRODUCTION

Investigations of the distribution of radionuclides in the marine environment have provided a wealth of information on their speciation behavior, mobility, fate, and transport. A number of processes control the radionuclides concentration, thus allowing the discernment of their different sources, or a better understanding of the governing biogeochemical and hydrodynamic processes from their partitioning among particulate, colloidal, and dissolved phases. Because of their radioactive decay, they can act as clocks and allow the determination of removal or turnover residence times in marine and freshwater systems. Over the past five to six decades, a large volume of literature has been published on the utility of radionuclides as tracers and chronometers in the marine environment. Reviewing and writing the methodology for all the radionuclides that are used as tracers and chronometers in the marine system would be an overwhelming task. However, several recent papers and book chapters bring out the current methodology for a number of radionuclides, and hence we will refer to those key references wherever it is appropriate (Rutgers van der Loeff and Moore, 1999; Chen et al., 2001; Rutgers van der Loeff et al., 2006; Hou and Roos, 2008) and focus on those nuclides for which the methodology has not been updated for a number of years.

All the radionuclides that are present in the marine environment are grouped into three major categories:

1. Naturally occurring, primarily derived from primordial isotopes, such as ^{238}U-, ^{235}U-, and ^{232}Th-series isotopes that include a total of twelve elements and forty-five isotopes that have half-lives ranging from <1 second to 2.48×10^5 years, and a wide range of geochemical properties, reflecting those of many other elements in the periodic table, and hence can be effectively used as analogs for other elements.
2. Cosmogenic radionuclides (such as ^{7}Be, ^{10}Be, ^{14}C, ^{26}Al, ^{32}Si, ^{36}Cl, ^{129}I, etc.) that are primarily produced in the atmosphere by spallation reactions of protons and neutrons with atmospheric gases, which are subsequently washed out from the atmosphere primarily by precipitation and are deposited to the earth's surface. The depositional fluxes of these can be well characterized.
3. Anthropogenic radionuclides (such as ^{3}H, ^{14}C, ^{36}Cl, ^{90}Sr, ^{129}I, ^{137}Cs, ^{238}Pu, ^{239,240}Pu, ^{241}Am, etc.) mainly derived from weapons testing and emissions from nuclear power reactors and nuclear fuel reprocessing plants.

Earlier analytical methods of most of these nuclides involved collection of relatively large volumes of water, followed by preconcentration and subsequent radiochemical processing and measurements using analysis resulting from radioactive decay (using alpha, beta, and gamma ray spectrometers) and atom (i.e., mass spectrometry) counting. However, developments in instrumentation and technology in sample collection, preconcentration, and analysis have reduced the sample size as well as the time involved in processing and measurement of samples for many radionuclides. For example, for the mean life of ^{234}U, 3.579×10^5 years, for a counting time of 1 week, only one out of 1.86×10^7 ^{234}U atoms will be detected by alpha counting. On the other hand, for a typical ionization efficiency of 10^{-3} with thermal ionization mass spectrometry, 18,600 ^{234}U ions will be detected by mass spectrometric method in less than 2 hours, thus leading to higher precision and faster throughput. This translates to 1/136th of the statistical uncertainty in alpha counting. Although prior to 1973 the mass spectrometric measurements of U isotopes by thermal ionization mass spectrometry (TIMS) and alpha counting techniques were comparable in sample size and precision (summarized in Chen et al., 1992), the higher ionization efficiency obtained in the mid-1980s resulted in smaller volume (e.g., TIMS, 10 mL for ^{238}U concentration to a precision of ±2 per mil, 2σ, ^{232}Th measurements in subliter water sample, Chen et al., 1986; 200–300 mL seawater sample with a precision of 0.3 per mil for ^{238}U with a multicollector sector field inductively coupled plasma mass spectrometer (MC-ICP-MS), Andersen et al., 2007). Higher precision with the mass spectrometric method compared to the alpha spectrometric method can be evaluated as follows: If N is the total number of atoms of a nuclide with a decay constant of λ, η_c is the counting efficiency (e.g., alpha detector assumed to be 30%), T is the counting time (in minutes), and η_i is the ionization efficiency for a given nuclide in the mass spectrometer, then the precision in the alpha spectrometry (P_α) is inversely proportional to the alpha particle detected by the detector.

$$P_\alpha = 1/\sqrt{N\lambda T\eta_c} \tag{13.1}$$

The precision in a mass spectrometer (P_{ms}) is inversely proportional to the number of ions recorded:

$$P_{ms} = 1/\sqrt{N\eta_i t} \tag{13.2}$$

where η_i is the ionization efficiency for a given nuclide and t is the time the mass spectrometer spent in counting a given nuclide (typically 10%; Chen et al., 1992). To have the same precision in the alpha and mass spectrometry techniques, and for a counting of 5 days by alpha spectrometry, the calculated λ value is $\sim 7 \times 10^{-5}$ day^{-1}, which corresponds to a half-life of ~ 28 years. This implies that alpha spectrometric methods will have higher precision in measuring alpha emitting nuclides with a half-life of less than 28 years. This is of the same order of magnitude as the value of 70 years reported by Chen et al. (1992). Thus, for radionuclides that have half-lives of less than 100 years, alpha spectrometry may still provide comparable or better precision than mass spectrometric techniques. Therefore, mass spectrometry and radiometry (counting methods) should be seen as complementary rather than competitive techniques for the measurement of radionuclides in seawater, each with its own advantages and disadvantages.

13.2 SAMPLING PROCEDURES

13.2.1 WATER SAMPLING

Radionuclide analysis in seawater samples involves a proper and clean sampling procedure that does not introduce any artifacts to the natural samples (GEOTRACES Science Plan, 2006; Measures and SCOR Working Group, 2007). The most commonly used procedures to collect water samples are:

1. Surface seawater samples either through seawater supply hose in the ship or submersible pump. Seawater samples collected from supply hose in the ship appear to desorb significant amounts of Ra isotopes from the adsorbed Th in the pipe. Care must be exercised on the blank levels of short-lived Ra when supply hoses are used. The sewage discharged from the ship could be a problem, as this could contaminate the water collected through seawater supply hose or submersible pumping system where the pump is near the point of effluent discharge.
2. Rosette sampling, using Niskin or Go-Flo bottles at predetermined depths. This method will enable collection of $\sim 10^1$–10^2 L of water samples, depending on the number and volume of bottles in the rosette (see Chapter 1). Clean samples from various depths can be obtained by this method.
3. Sampling with large-volume containers (such as Gerard water sampling bottle, although these bottles are not in common use; Broecker et al., 1986). The issue with the collection of water samples with such containers is that the improper closure of the lid could result in exchange of water from other depths.
4. In situ pumps have been extensively utilized to extract radionuclides (in situ pumping system (MIPS), Figure 13.1; Nozaki et al., 1981; Bacon and Anderson, 1982; Buesseler et al., 1992; Baskaran et al., 1993; Guo et al., 1995; Rutgers van der Loeff and Berger, 1993; Cochran et al., 1995; Moran et al., 1997). Many of these pumping systems include

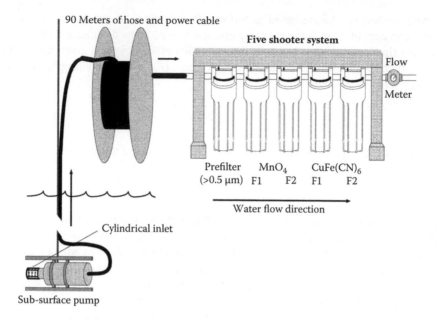

FIGURE 13.1 Schematic diagram of the subsurface pump. The first cartridge is to extract particulate matter; the second and third for extracting Th, Pb, and Ra isotopes; and last two for extracting [137]Cs. This setup can also be used for collecting three different particle sizes using special polypropylene cartridge filters with different pore sizes (nominal) and extracting dissolved Th, Pb, and Ra isotopes using MnO_2-coated filter cartridges.

a prefilter to retain particulate matter (details given below) and two MnO_2 cartridges (or other sorbents; Table 13.1) to extract dissolved radionuclides (Th, ^{210}Pb, [231]Pa, etc.). The advantage of this method is to leave the water behind after extracting the radionuclides in situ. Furthermore, simultaneous collection of particulate and dissolved water samples at multiple depths using in situ pumps improves considerably the efficiency of sampling operations. McLane pumps have been utilized with standard winches, through which water filtration can be done at multiple depths simultaneously, thus reducing the wire time. The disadvantage is that assumption of uniform extraction efficiency for radionuclides will result in under/over estimation of dissolved radionuclide concentration.

13.2.1.1   Recommended Water Sampling Procedure

It is not possible to recommend a single procedure for water sampling for all radionuclides, and the choice depends on the questions that one wishes to answer. In particular, the variables that determine the volume of water and type of sampling include concentrations of the radionuclides, the chemical behavior of the nuclides of interest in the marine system, the analytical methods that will be employed to measure the concentrations of radionuclides, and the precision needed. The general considerations in sampling procedures for [234]Th are summarized in Rutgers van der Loeff et al. (2006). The overall strategy is to optimize the precision, labor, and time needed for collection and analysis and the cost involved for sampling and analysis.

13.2.2   Extraction of Particulate Matter from Seawater

Most studies dealing with particle-reactive radionuclides require that one measures the concentrations of both dissolved and particulate phases (summarized in Cochran and Masque, 2003). It is

TABLE 13.1
List of Sorbents Used for Radionuclide Extraction from Seawater and Fresh Water

Sorbent	Radionuclides	References
MnO_2 imp. fibers	^{234}Th, ^{232}Th, ^{230}Th, ^{231}Pa, ^{228}Ra, ^{226}Ra, ^{224}Ra, ^{223}Ra, ^{210}Pb	Moore et al., 1985; Moore, 1984; Levy and Moore, 1985; Cochran et al., 1987; Baskaran et al., 1992, 1993; Rutgers van der Loeff and Berger, 1993; Moore and Arnold, 1996; Hartman and Buesseler, 1994; Guo et al., 1997; Dulaiova and Burnett, 2004; Trimble et al., 2004; Bourquin et al., 2008; Hung et al., 2008.
$MnO_2/Fe(OH)_3$ imp. fibers	^{227}Ac	Nozaki, 1984
Cu_2 Fe (CN) imp. fibers	^{137}Cs	Buessler et al., 1990
MnO_2 imp. anion exchange resin and MnO_2-coated discs	$^{228,230,232}Th$, $^{223,224,226,228}Ra$, $^{239,240}Pu$	Eikenberg et al., 2001; Moon et al., 2003; Dimova et al., 2007; Varga, 2007
$Fe(OH)_3$ imp. fibers	^7Be, ^{32}Si, ^{238}U	Lee et al., 1991; Kadko and Olsen, 1996
$Fe(OH)_3$ precip.	^{238}U, ^{235}U, ^{234}U; ^{234}Th; ^{210}Po, ^{210}Pb, ^{10}Be, $^{239,240}Pu$, ^{241}Am	Bhat et al., 1969; Southon et al., 1982; Coale and Bruland, 1985; Nozaki, 1991; Nozaki and Nakanishi, 1985; Fleer and Bacon, 1984; Sarin et al., 1990, 1999; Baskaran and Santschi, 1993; Baskaran et al., 1997; Chen et al., 2001
Co-APDC	^{210}Po, ^{210}Pb	Fleer and Bacon, 1984
Eichrom TRU resin	^{231}Pa, actinides (III, IV, VI), Ln(III)	Pickett et al., 1994; Spry et al., 2000; Jakopic et al., 2007
Eichrom TEVA resin	$^{99}Tc(VII)$, Th(IV), Np(IV), Pu(IV)	Horwitz et al., 1995
Eichrom UTEVA resin	$^{238}U(VI)$	Horwitz, 1998; Croudace et al., 2006; Fujiwara et al., 2007
Eichrom actinide resin	Am, Th, U isotopes	Horwitz et al., 1997; Burnett et al., 1997
Anfezh resin	^{137}Cs and others	Bandong et al., 2001
Anion exchange resin	^{129}I—iodide; iodate after reduction, organo-iodine after dehydrohalogenation; ^{99}Tc	Schwehr and Santschi, 2003; Schwehr et al., 2005b; Chen et al., 2001
Acid and vacuum extraction	^{14}C	Schoch et al., 1980
Cool trap	^{222}Rn	Key et al., 1979; Mathieu et al., 1988

widely accepted that the separation of particles from seawater is operationally defined, and hence there is only an approximate pore size or filter material for all the radionuclides of interest near 0.5 μm. The sampling methods for the dissolved and particulate particle-reactive constituents involve filtration of water samples, either through in situ pumps (through a prefilter to retain particulate matter followed by absorber matrix, such as MnO_2-coated fiber material, or prefilter cartridge followed by two MnO_2 cartrridges for Th isotopes and ^{210}Pb; Cochran et al., 1987; Buesseler et al., 1992; Baskaran et al., 1993; Rutgers van der Loeff and Berger, 1993; Guo et al., 1995; Baskaran and Santschi, 2002; Hung et al., 2008) or through a filter paper. A number of pumps have been utilized that include McLane, Challenger, KISP, MULVFS, MIPS, and RAPPID, although there is no consensus on the best type of pump to use.

A number of different filters with different pore sizes and composition have been utilized to separate dissolved and particulate phases of radionuclides. For example, the polycarbonate track-etch filters (Nucleopore and Poretics) are widely accepted as standard filters for trace metal work, but these filters have low porosity and thus significantly reduced flow rates, with a tendency for early clogging by particulate matter present in seawater. Filtering volumes of 20–50 L water samples with polycarbonate track-etch filters in the field will require relatively large amounts of time.

Colloidal material (size: 1 nm–1 μm) plays a major role in metal scavenging, cycling, and transport in the marine environment (Doucet et al., 2007). The commonly used methods to isolate colloidal material include ultracentrifugation (Wells and Goldberg, 1991, 1994), dialysis (de Mora and Harrison, 1983), selective adsorption to XAD™ resin, and cross-flow filtration (also called tangential-flow filtration, and more generically, ultrafiltration; Carlson et al., 1985; Whitehouse et al., 1986, 1990; Baskaran et al., 1992; Moran and Buesseler, 1992; Baskaran and Santschi, 1993; Guo et al., 1997; see also Chapter 1). The volume requirement depends upon the radionuclide as well as the methodology used to measure the concentrations of that nuclide. For example, >100 L seawater samples are required for colloidal ^{234}Th measurements using gamma ray spectroscopy (Baskaran et al., 1992, 2003; Moran and Buesseler, 1992; Guo et al., 1997). For such large volumes, the commonly utilized technique to separate the colloidal material from solutes is to use cross-flow filtration (CFF). The composition of the CFF material seems to affect the extraction efficiency of colloidal material (intercalibration results from marine colloids are summarized in Buesseler et al., 1996). More information on the tangential-flow ultrafiltration can be obtained from a recent in-depth review article (Guo and Santschi, 2007).

13.2.2.1 Recommended Procedure for Suspended Particulate Matter

Generally, the suspended particulate matter is of interest only when one wants to study the particle dynamics (generation, removal, and cycling) using particle-reactive radionuclides. The water-soluble radionuclides are generally used as tracers for water masses, and hence radionuclides measurements in particulate matter are of less interest. When distribution of size-fractionated particulate matter is utilized to investigate particle dynamics, relatively large-volume samples are needed and usually filter cartridges with in situ pumps are recommended. For small-volume filtration, polycarbonate track-etch filters are recommended. The composition of the cross-flow filter does have an effect on the extraction efficiency of colloidal material, and hence we recommend a polysulfone matrix for CFF work.

13.2.3 Storage Artifacts

The accuracy of the nuclides concentrations could be affected by artifacts that may be introduced when water samples are stored without acidification for later analysis. The artifacts may stem from the adsorption of the particle-reactive radionuclides in the sample to the wall of the container or release of container material into the sample, although the latter is rare for the radionuclides; however, this is common for stable elements. Some nuclides such as ^{210}Po (in situ ^{210}Po) and ^{210}Pb (via the in-growth of its granddaughter, ^{210}Po) are assayed using alpha spectrometry, and thus the methodology requires that the in situ ^{210}Po is assayed (or at least separated from ^{210}Pb) as quickly as possible after the sample collection. When a large number of samples are collected in the field, time-consuming methods of preconcentration and equilibration (spikes + Fe(OH)$_3$ precipitation-filtration) may require holding time of the order of several days, and hence the storage artifacts need to be evaluated over a period of about a week for the loss of Po and Pb. Earlier works showed that acidification of water samples in polyethylene bottles to pH < 1.5 with nitric acid was sufficient to prevent the loss of trace elements onto the container walls (Subramanian et al., 1978). However, many aspects of sample storage effects have been studied extensively, and further studies have shown that this is not also true for many trace element isotopes (Landing et al., 1995; GEOTRACES Science Plan, 2006; Reimann et al., 2007). The storage artifact experiments need to be conducted in precleaned cubitainers with controlled natural levels of radioactive spikes (such as ^{210}Po and ^{210}Pb) to quantify the amount of loss of these nuclides.

13.3 ANALYTICAL PROCEDURES

13.3.1 PRECONCENTRATION OF SAMPLES FOR THE DETERMINATION OF DISSOLVED CONCENTRATION

Determination of the concentration of radionuclides in seawater often requires that the radionuclides are either preconcentrated or separated from large quantities of matrix constituents along with other radionuclides. The preconcentration step, commonly followed by analytical procedures of the most short-lived radionuclides (that utilize alpha, beta, and gamma counting systems) as well as many of the long-lived radionuclides (mass spectrometer), includes: (1) coprecipitation with metal oxides and sulfates (of Fe, Mn, Al, Ba, Pb, etc.), (2) solvent extraction, (3) volatilization/evaporation, (4) ion exchange, and (5) sorption onto Fe and Mn oxides coated on fiber matrix (such as polypropylene, acrylic). Various sorbents have been utilized to preconcentrate radionuclides from seawater, and a summary of various chemical procedures is given in Table 13.1. The most recent and commonly employed methods of analysis for all three groups of radionuclides are given in Table 13.2. Specific details on the most currently used radiochemical methods for different radionuclides are outlined below.

13.3.2 U-TH SERIES RADIONUCLIDES

13.3.2.1 ^{238}U, ^{235}U, and ^{234}U

Prior to major developments in mass spectrometry, the measurements of ^{238}U and ^{234}U were conducted using alpha spectrometry, and the details of the chemical and analytical procedures described in Lally (1992) can be followed. In principle, the activities of ^{238}U can be determined by measuring ^{238}U directly by alpha spectrometry or its daughter, ^{234}Th, by either beta counting (Bhat et al., 1969) or gamma ray spectrometry. Details on the alpha spectrometric methods can be found in Ivanovich and Murray (1992). The gamma ray spectrometric method does not offer the accuracy, resolution, or sensitivity required for low-level U measurements. With the advent of mass spectrometry, the precision and accuracy of the U isotopic analysis has improved significantly. Since the first measurements of ^{238}U, ^{234}U, and ^{232}Th by TIMS in seawater (Chen et al., 1986), considerable developments have been made over the past two decades. Although TIMS remains a high-precision analytical tool, it is labor-intensive and time-consuming with relatively low throughput. Inductively coupled plasma-mass spectrometry (ICP-MS; both quadrupole [Q-ICP-MS] and sector MC-ICP-MS) remains advantageous in that the throughput is much higher than for all other methods and the chemical processing is also considerably less (Zheng and Yamada, 2006; Andersen et al., 2007; Stirling et al., 2007; Weyer et al., 2008). The high temperature (~6,000 K) attained in the plasma source produces efficient ionization (>90%) for most of the elements (Gray, 1985; Jarvis and Jarvis, 1992). With the much faster sample throughput and possibly less sample preparation, MC-ICP-MS offers unique advantages compared to other methods. In particular, recent results indicate that the variations in ^{235}U/^{238}U at the 0.4 epsilon level (2σ; 1 ε = 1 part in 10^4) can be resolved on sample sizes comprising 50 ng of uranium (corresponding to ~16 mL of open-ocean seawater) using MC-ICP-MS (Stirling et al., 2007). It has always been considered that the ^{238}U/^{235}U atomic ratio is a constant value of 137.88, but the more recent work of Stirling et al. (2007) suggests that the analytical precision of MC-ICP-MS (~0.006%, 2σ) will permit precise measurements of ^{238}U/^{235}U ratios at sufficient levels that allow the determination of the variations in ^{238}U/^{235}U ratios due to U removal in anoxic oceanic waters. Such high-precision measurements in pore waters and sediments (both biogenic and abiogenic) provide a handle to reconstruct paleo-redox conditions in the ocean. Details on mass spectrometric measurements can be found in Chen et al. (1992).

TABLE 13.2
Chemical Procedures for Natural, Cosmogenic, and Anthropogenic Radionuclides

Isotope	Half-Life	Decay Mode and Product	Spike Used	Volume of Sample	Method of Analysis	References
				Naturally Occurring		
^{238}U	4.468×10^9 y	α, ^{234}Th	$^{236}U/^{233}U$	<1 L	MC-ICP-MS/TIMS	Stirling et al., 2007
^{235}U	7.038×10^8 y	α, ^{231}Th	$^{236}U/^{233}U$	<1 L	MC-ICP-MS/TIMS	Chen et al., 1986
^{234}U	2.455×10^5 y	α, ^{230}Th	$^{236}U/^{233}U$	<1 L	MC-ICP-MS/TIMS	
^{231}Pa	3.276×10^4 y	α, ^{227}Ac	^{233}Pa	10–20 L	MC-ICP-MS/TIMS	Choi et al., 2001
^{234}Th	24.1 d	β, ^{234}Pa	^{230}Th	2–5 L	Beta counting	Buesseler et al., 2001
			No spike	10^2–10^3 L	γ-Spectrometry	Baskaran et al., 1993
^{232}Th	1.405×10^{12} y	α, ^{228}Ra	^{229}Th	1 L	MC-ICP-MS/TIMS	Chen et al., 1986
^{230}Th	7.538×10^4 y	α, ^{226}Ra	^{229}Th	10–20 L	MC-ICP-MS/TIMS	Choi et al., 2001
^{228}Th	1.913 y	α, ^{224}Ra	^{229}Th	10^2–10^3 L	α-Spectrometry	Trimble et al., 2004
^{227}Th	18.72 d	α, ^{223}Ra	^{229}Th	10^2–10^3 L	α-Spectrometry	
^{227}Ac	21.773	β, ^{227}Th	^{225}Ac	50–100 L	α-Spectrometry	Martin et al., 1995
^{228}Ra	5.78 y	β, ^{228}Ac	No spike	10^2–10^3 L	γ-Spectrometry	Moore et al., 1985
^{226}Ra	1603 y	α, ^{222}Rn	No spike	20–100 L	γ-Spectrometry	Moore et al., 1985
^{224}Ra	3.66 d	α, ^{220}Rn	No spike	60–100 L	DCCS	Moore and Arnold, 1996
^{223}Ra	11.44 d	α, ^{219}Rn	No spike	60–100 L	DCCS	Moore and Arnold, 1996
^{222}Rn	3.83 d	α, ^{218}Po	No spike	1–10 L	Scint. counting	Mathieu et al., 1988
^{210}Pb	22.2 y	β, ^{210}Bi	Stable Pb	20 L	α-Spectrometry	Sarin et al., 1992
^{210}Bi	5.01 d	β, ^{210}Po	Stable Bi	20 L	Beta counting	Church et al., 1994
^{210}Po	138.4 d	α, ^{206}Pb	^{209}Po	20 L	α-Spectrometry	Sarin et al., 1992
				Cosmogenic		
^{3}H	12.4 y	β, ^{3}He				
^{7}Be	53.3 d	EC, ^{7}Li	Stable Be	~200 L	γ-Spectrometry	Baskaran et al., 1997
^{10}Be	1.50×10^6 y	β, ^{10}B	Stable Be	100 L	AMS	Kusakabe et al., 1982
^{14}C	5,730 y	β, ^{14}N	No spike	1–10 L	AMS	
^{36}Cl	3.01×10^5 y	β, ^{36}Ar	Stable Cl	5 L	AMS	Bentley et al., 1986
				Anthropogenic		
^{90}Sr	28.9 y	β, ^{90}Y	Stable Sr	50 L	Beta counting	Wong et al., 1994
^{137}Cs	30.2	β, ^{133}Ba	No spike	50–200 L	γ-Spectrometry	Buesseler et al., 1990
^{238}Pu	87.7 y	α, ^{234}U	^{242}Pu	10^2–10^3 L	α-Spectrometry	Baskaran et al., 1996
^{239}Pu	2.41×10^4 y	α, ^{235}U	^{242}Pu	5–10 L	Mass spectrometry	Kim et al., 2002
^{240}Pu	6571 y	α, ^{236}U	^{242}Pu	5–10 L	Mass spectrometry	Kim et al., 2002
^{241}Am	432.2 y	α, ^{237}Np	^{243}Am	100 L	α-Spectrometry	Wong et al., 1994
^{129}I	1.57×10^7 y	β, ^{129}Xe	^{127}I	2–5 L	AMS	Elmore, 1980

13.3.2.1.1   Quality Assurance

The concentrations of ^{238}U and ^{234}U in open-ocean seawaters are well known (Chen et al., 1986). Seawater (constant salinity) whose values are certified should be used as a standard to compare the methodology. In particular, recent studies indicate that there could be fractionation between ^{235}U and ^{238}U in anoxic waters. Thus, the standard should be chosen from oxic open-ocean waters.

13.3.2.1.2   Common Problems and Troubleshooting

With mass spectrometry, the isobaric interferences are issues that need to be addressed and corrected for.

13.3.2.2 ^{234}Th and ^{231}Th

The disequilibria between the insoluble daughters of ^{238}U (^{234}Th), ^{235}U (^{231}Th), ^{234}U (^{230}Th), and ^{228}Ra (^{228}Th) have been widely used as tracers and chronometers in a marine environment. A large number of studies over the past 15 years have focused on the utility of ^{234}Th as a tracer of particulate organic carbon (POC). Due to the short half-life of ^{231}Th (1.1 days) and low activity (^{235}U = 4.54% of ^{238}U activity), there is no data reported on ^{231}Th activity. The standard original procedure for the analysis of ^{234}Th involves coprecipitation of Th with $Fe(OH)_3$ precipitate from 20 L of seawater (Bhat et al., 1969; Coale and Bruland, 1985). Although this procedure is reliable, it is labor-intensive and time-consuming and alternative methods are available. Rutgers van der Loeff et al. (2006) summarized all the methodologies available in the literature to measure the activities of particulate and dissolved ^{234}Th in the marine system. We refer to recent review papers for a complete discussion on the recent methodologies used (Rutgers van der Loeff and Moore, 1999; Rutgers van der Loeff et al., 2006). A brief summary of these methods follows here.

The choices available for the method of collection and analysis of ^{234}Th depend on the environment and the research questions that we are interested in addressing. Since the disequilibrium between ^{238}U and ^{234}Th is what is used, precise determination of ^{238}U is a key issue. In open-ocean oxic waters, the linear relationship between ^{238}U and salinity seems to be valid (Ku et al., 1977; Chen et al., 1986). However, in coastal and suboxic/anoxic waters, the nonconservative (or less conservative) behavior of U would require that we measure the U concentration precisely, rather than estimating from the U-salinity relationship.

^{234}Th measurements are commonly made by (1) gamma spectroscopy (Section 13.3.2.7.1), (2) beta counting, and (3) liquid scintillation methods. The low branching ratios of gamma lines (~4% at 63 keV and ~5.5% at 93 keV) result in low detection efficiencies, and hence larger-volume samples are needed for the nondestructive gamma spectrometry. A typical 20 L sample from surface waters with ~1 dpm L^{-1} (~17 mBq L^{-1}) of ^{234}Th results in ~0.7 counts per minute (cpm) in a HPGe well detector (1 mL geometry). In 1 day of counting, one can get ^{234}Th with a 1σ propagated error of ~5%, if the processing and counting of the samples are completed within 1 or 2 days after sample collection. This methodology requires that ^{234}Th is quantitatively extracted from the water (such as double precipitation with $Fe(OH)_3$ method; Baskaran et al., 1992). Some amount of U will coprecipitate with Th, and hence a correction for the in-growth of ^{234}Th from the ^{238}U coprecipitation with $Fe(OH)_3$ needs to be applied. The gamma ray spectrometer needs to be calibrated with primary standards (such as NIST ^{238}U standard solution or IAEA-RGU-1 standard; a list of standard reference materials is given in Table 13.3). Large-volume water samples (10^2–10^4 L) can be extracted through MnO_2-coated polypropylene cartridges connected in series. If F_1 and F_2 are the dissolved activities of ^{234}Th in the first and second cartridges, respectively, then the ^{234}Th activity ($A(t)$) is calculated as follows:

$$A(t) = F_1/(1 - F_2/F_1) \tag{13.3}$$

This assumes that the absolute extraction efficiencies of the two cartridges are the same. However, the extraction efficiencies calculated from the activities of ^{234}Th in the two cartridges have been shown to vary from 65% to > 95%, and thus this assumption needs to be rigorously tested. Recent results on a number of laboratory experiments with seawater indicate that there is a large variation in the absolute extraction efficiency of the MnO_2-coated cartridge filters (Baskaran et al., 2008), resulting in variations in the ratio of incoming to retained Th in the first two MnO_2-coated cartridges connected in series. The distinct advantage of the cartridge method is that size-fractionated

TABLE 13.3

List of Currently Available Standard Reference Materials (SRMs) for Marine Radioactivity Studies

Code	Matrix	Place of Origin	Radionuclides Recommended Values	Radionuclides Information Values	Year	Ref.
IAEA-MA-B-3/RN	Fish flesh	Baltic Sea	^{40}K, ^{137}Cs	—	1987	Povinec, 2000
IAEA-134	Cockle flesh	Irish Sea	^{40}K, ^{60}Co, ^{137}Cs, $^{239+240}$Pu	^{90}Sr, ^{106}Ru, ^{125}Sb, ^{154}Eu, ^{155}Eu, ^{210}Pb, ^{226}Ra, ^{228}Ra, ^{228}Th, ^{230}Th, ^{232}Th, ^{234}U, ^{235}U, ^{238}U, ^{238}Pu, ^{241}Am	1993	Ballestra et al., 2000a
IAEA-135	Bottom sediment	Irish Sea	^{40}K, ^{60}Co, ^{134}Cs, ^{137}Cs, ^{154}Eu, ^{155}Eu, ^{226}Ra, ^{228}Ra, ^{232}Th, ^{238}Pu, $^{239+240}$Pu	^{57}Co, ^{90}Sr, ^{106}Ru, ^{125}Sb, ^{210}Pb, ^{228}Th, ^{230}Th, ^{234}U, ^{235}U, ^{238}U, ^{241}Am	1993	Ballestra et al., 2000b
IAEA-300	Bottom sediment	Baltic Sea	^{40}K, ^{60}Co, ^{125}Sb, ^{134}Cs, ^{137}Cs, ^{155}Eu, ^{210}Pb, ^{210}Po, ^{228}Ra, ^{234}U, ^{238}U, $^{239+240}$Pu, ^{241}Am	^{54}Mn, ^{90}Sr, ^{226}Ra, ^{228}Th, ^{230}Th, ^{232}Th, ^{235}U, ^{238}Pu	1994	Ballestra et al., 2000c
IAEA-315	Marine bottom sediment	Arabian Sea	^{40}K, ^{210}Pb, ^{226}Ra, ^{228}Ra, ^{228}Th, ^{232}Th, ^{234}U, ^{238}U	^{230}Th, ^{235}U	1996	Ballestra et al., 2000d
IAEA-352	Tuna fish flesh	Mediterranean Sea	^{40}K, ^{137}Cs, ^{210}Pb, ^{210}Po	^{90}Sr	1990	Ballestra et al., 2000e
IAEA-367	Marine bottom sediment	Pacific Ocean	^{60}Co, ^{90}Sr, ^{137}Cs, $^{239+240}$Pu	^{40}K, ^{155}Eu, ^{226}Ra, ^{228}Th, ^{230}Th, ^{234}U, ^{235}U, ^{238}U, ^{238}Pu, ^{241}Pu, ^{241}Am	1991	Ballestra et al., 2000f
IAEA-368	Bottom sediment	Pacific Ocean	^{60}Co, ^{155}Eu, ^{210}Pb, ^{226}Ra, ^{238}U, ^{238}Pu, $^{239+240}$Pu	^{40}K, ^{90}Sr, ^{137}Cs, ^{228}Th, ^{234}U, ^{235}U, ^{241}Am	2000	Ballestra et al., 2000g
IAEA-381	Water	Irish Sea	^{40}K, ^{90}Sr, ^{137}Cs, ^{238}Pu, $^{239+240}$Pu, ^{241}Am	^{3}H, ^{125}Sb	1999	Povinec et al., 1999; Povinec and Pham, 2001
IAEA-384	Marine bottom sediment	Fangataufa Atoll, South Pacific Ocean	^{40}K, ^{60}Co, ^{155}Eu, ^{210}Pb, ^{210}Po, ^{230}Th, ^{234}Th, ^{234}U, ^{238}U, ^{241}Am	^{90}Sr, ^{137}Cs, ^{214}Bi, ^{214}Pb, ^{226}Ra, ^{228}Ac, ^{232}Th, ^{237}Np, ^{239}Pu, ^{240}Pu, ^{241}Pu	2000	Povinec and Pham, 2000; Povinec et al., 2007
IAEA-385	Marine bottom sediment	Irish Sea	^{40}K, ^{137}Cs, ^{226}Ra, ^{232}Th, ^{235}U, ^{238}U, ^{238}Pu, $^{239+240}$Pu, ^{241}Am	^{90}Sr, ^{208}Tl, ^{210}Po, ^{210}Pb, ^{212}Bi, ^{212}Pb, ^{214}Bi, ^{214}Pb, ^{228}Ac, ^{228}Th, ^{230}Th, ^{234}Th, ^{234}U, ^{239}Pu, ^{240}Pu	2005	Pham et al., 2005; Pham et al., 2008
IAEA-414	Fish flesh	Irish Sea and North Sea	^{40}K, ^{137}Cs, ^{232}Th, ^{234}U, ^{235}U, ^{238}U, ^{238}Pu, $^{239+240}$Pu, ^{241}Am	^{90}Sr, ^{210}Pb(^{210}Po), ^{226}Ra, ^{239}Pu, ^{240}Pu, ^{241}Pu	2004	Pham et al., 2004; Pham et al., 2006
NIST SRM 4350B	River sediment	—	^{60}Co, ^{137}Cs, ^{152}Eu, ^{154}Eu, ^{226}Ra, ^{238}Pu, $^{239+240}$Pu, ^{241}Am	^{40}K, ^{55}Fe, ^{90}Sr, ^{228}Th, ^{230}Th, ^{232}Th, ^{234}U, ^{235}U, ^{238}U	1981	NIST, 1981

	Matrix	Source	Radionuclides		Year	Reference
NIST SRM 4357	Bottom sediment	A blend of sediments from Sellafield, UK, and Chesapeake Bay, United States	^{40}K, ^{226}Ra, ^{228}Ra, ^{228}Th, ^{230}Th, ^{232}Th, ^{90}Sr, ^{137}Cs, ^{238}Pu, ^{239}Pu+^{240}Pu	^{129}I, ^{155}Eu, ^{210}Po, ^{210}Pb, ^{214}Bi, ^{234}U, ^{235}U, ^{237}Np, ^{238}U, ^{241}Am	1997	NIST, 1997; Inn et al., 2001b
NIST SRM 4359	Seaweeds	Coast of Ireland and White Sea	^{40}K, ^{137}Cs, ^{210}Pb, ^{210}Po, ^{228}Ra, ^{232}Th, ^{234}U, ^{235}U, ^{238}U, ^{238}Pu, ^{239}Pu, ^{241}Am	^{3}H, ^{14}C, ^{90}Sr, ^{129}I, ^{208}Tl, ^{212}Bi, ^{212}Pb, ^{224}Ra, ^{226}Ra, ^{228}Th, ^{230}Th, ^{234}Th, ^{237}Np, ^{240}Pu	2002	NIST, 2005; Outola et al., 2006
IAEA-RGU-1	Uranium ore BL-5	Canada	U, Th, K		1987	IAEA, 1987
IAEA-RGTh-1	Thorium ore OKA-2	Canada	Th, U, K			
IAEA-RGK-1	Potassium sulfate	Merck Company	K, U, Th			
NIST SRM 4927F	Distilled water	NIST	^{3}H		1998	NIST 2008; Groning et al., 2001
NIST SRM 4222C	Carbon-14-n-hexadecane	NIST	^{14}C		1991	NIST, 1991
NIST SRM 4325	Beryllium chloride	NIST	^{10}Be		1990	NIST 1990
IAEA-375	Soil	Brjansk, Russia	^{40}K, ^{90}Sr, ^{106}Ru, ^{125}Sb, ^{129}I, ^{134}Cs, ^{137}Cs, ^{226}Ra, ^{232}Th	^{228}Th, ^{234}U, ^{238}U, ^{238}Pu, $^{239+240}$Pu, ^{241}Am	1994	IAEA, 2000
NIST SRM 3230	Ammonium iodide	NIST	^{129}I/^{127}I ratio (low level)	—	2003	NIST 2003a
NIST SRM 3231	Ammonium iodide	NIST	^{129}I/^{127}I ratio (high level)	—	2003	NIST 2003b

particulate matter (Figure 13.1) and dissolved ^{234}Th (along with other nuclides, such as ^{210}Pb, activity ratios of ^{226}Ra/^{228}Ra, etc.) can be collected simultaneously (Buesseler et al., 1992; Baskaran et al., 1993; Baskaran and Santschi, 2002; Guo et al., 2002; summarized in Rutgers van der Loeff et al., 2006). Furthermore, the MnO$_2$-coated cartridges have the advantage that they separate ^{234}Th from ^{238}U in the seawater, and hence in-growth corrections for ^{234}Th are negligible. Detailed procedures on how to prepare MnO$_2$-coated cartridges are given in Rutgers van der Loeff and Moore (1999).

Another method that has been recently employed is the coprecipitation of Th with MnO$_2$ from 20 L or smaller volume of seawater (Buesseler et al., 2001; Benitez-Nelson et al., 2001). Thorium from 2–5 L water samples (either total or filtered) is coprecipitated with MnO$_2$, after addition of a yield tracer (e.g., ^{230}Th). The precipitate is collected on a 25 mm diameter Whatman Multigrade GMC filter (1 μm nominal pore size). The filter paper is dried and beta counted. This method can be easily applied at sea, with an overall precision and accuracy of ≤5% (Benitez-Nelson et al., 2001; Buesseler et al., 2001; Cai et al., 2006, 2008).

13.3.2.2.1　Quality Assurance

The best standard for intercalibration of ^{234}Th techniques is pristine deep-ocean water (>1,000 m), where the activities are expected to be in equilibrium. Earlier studies, however, have shown that some occasional disequilibrium does occur (discussed in Baskaran et al., 2003), and it is not clear whether some physical and/or biogeochemical processes that lead to this disequilibrium are operating or this is due to problems with the analytical procedures. To be safe, the best method is to collect a large-volume sample, filter it through a 0.45 μm cartridge filter, acidify the filtrate to pH < 1, and store it for ~6 months, which will ensure the ^{234}Th/^{238}U activity ratio of 1.0.

13.3.2.2.2　Common Problems and Troubleshooting

When the small-volume (2–5 L) beta counting method is used, other beta emitters (as beta counting is not isotope specific) of the U-Th decay series radionuclides (such as betas from the decay of ^{228}Ac, ^{214}Pb, ^{214}Bi, ^{210}Bi, ^{212}Bi, ^{211}Pb, ^{211}Bi, and ^{40}K) could affect the accuracy of the measurements. This is more serious when beta counting is conducted without purification of the sample (such as direct counting of suspended particulate material or MnO$_2$ precipitates; Rutgers van der Loeff et al., 2006). Beta counting of electroplated sources may be ideal for beta counting (uniform thickness), as nonuniform thickness resulting from uneven loading of particles could alter the self-absorption correction factor. Applying appropriate self-absorption corrections could be challenging, as low-energy beta from ^{234}Th (E$_{max}$ = 0.27 MeV) and high-energy beta from its progeny, ^{234m}Pa (E$_{max}$ = 2.19 MeV), will be absorbed to varying extents in different sample thicknesses. More details can be found in Rutgers van der Loeff et al. (2006). Some of the glass fiber filters have been found to contain detectable natural radioactivity within the filter material, and usage of such filters for low-volume water filtration has created relatively high uncertainly on the ^{234}Th data (Buesseler et al., 2001).

13.3.2.3　^{230}Th, ^{232}Th, and ^{231}Pa

All three nuclides have sufficiently high atom abundances and hence can be measured precisely via mass spectrometric techniques. Chen et al. (1986) showed that ^{232}Th in seawater can be measured on a subliter sample with very high precision (±2% with 3×10^{11} ^{232}Th atoms, 10^{-20} g ^{232}Th, 1 L of seawater). A typical precision in the conventional alpha spectrometry varies between 2% and 10%, with 1 to 2 weeks of counting time (Trimble et al., 2004), while a better precision value has been reported with mass spectrometry (0.3%–1.0% using TIMS; <0.1% for ^{232}Th and 0.1%–0.3% for ^{230}Th and ^{234}U using MC-ICP-MS). The precision of the measurements of these nuclides using various instrument methodologies are summarized in Goldstein and Stirling (2003). The sample size requirement with MC-ICP-MS is also significantly lower (alpha spectrometry: 1–100 μg ^{232}Th; MC-ICP-MS: 10–600 ng ^{232}Th; Goldstein and Stirling, 2003).

The detailed procedures for the measurements of ^{232}Th, ^{230}Th, and ^{231}Pa in seawater can be adopted for any type of ICP-MS and are outlined below (from Choi et al., 2001). For the radiochemical work, only high-purity acids (HCl, HNO$_3$ and HF Seastar™) should be used; all Teflon wares and resins should be thoroughly acid-cleaned.

1. Collect 15–20 L of seawater sample and filter it through 0.45 μm filter paper using an in-line filtration system under clean conditions and collect the filtered sample in a pre-cleaned 20 L polyethylene collapsible cubitainer; weigh the sample precisely to <1%.

2. Acidify the sample with 30 mL 6 M HCl and spike it with ^{229}Th, ^{233}Pa (^{233}Pa can be prepared either by milking ^{237}Np or by neutron activation of ^{232}Th); add 1 mL FeCl$_3$ (= 50 mg mL^{-1} Fe; cleaned by extraction in di-isopropyl ether).

3. Allow the spikes and carrier to equilibrate for at least 12 hours; after equilibration, add ~17 mL NH$_4$OH to adjust the pH to ~8.5–9.0 to precipitate Fe(OH)$_3$ along with Th and Pa.

4. About 4–6 hours after the precipitation, decant the supernatant slowly and separate the precipitate and the solution by subsequent decanting followed by centrifugation.

5. Dissolve the precipitate in 9 M HCl by adding a volume (typically ~10–20 mL) of concentrated HCl equivalent to three times the volume of Fe oxyhydroxide recovered by decantation and centrifugation.

6. Pack ~4 mL column of AG1-X8 resin into a 0.6 × 20 cm polyethylene tube with a polyethylene frit; precondition the column with 9 M HCl. Pass the solution through the column and rinse with an additional 12 mL of 9 M HCl. Collect the solution containing Th in a Teflon beaker.

7. Elute the column into a separate beaker with 12 mL of 9 M HCl + 0.14 M HF. This contains Pa. Discard the U and Fe that remain in the column.

8. Evaporate the Th fraction to a small volume and take up in 8 M HNO$_3$ to be further purified by elution through a second anion exchange column of AG1-X8 preconditioned with 8 M HNO$_3$. After elution of Th with 10 mL of 0.1 M HCl, reduce the solution to a drop in a small screw-cap Teflon vial. Add 3 mL of ultrapure water to the fraction just before the analysis. Thorium isotopes (^{232}Th and ^{230}Th) can be measured using MC-ICP-MS.

9. Spike the Pa fraction with ^{236}U (~70 pg), allow it to equilibrate overnight, and pass through another AG1-X8 column preconditioned with 9 M HCl + 0.14 M HF. The purpose of adding ^{236}U is to check for possible ^{233}U "bleeding" in the final fraction. Thus, ^{236}U addition is mainly to quantify the possible U contribution to the 233 atomic mass unit peak.

10. Elute the column with 12 mL of 9 M HCl + 0.14 M HF and collect the total elute in a Teflon beaker.

11. Reduce the Pa fraction to a drop by evaporation in a small screw-cap Teflon vial. Just prior to analysis, add 0.3 mL Milli-Q water to the Pa fraction.

12. Add 0.3 mL of 1 M HNO$_3$ + 0.14 M HF to the Pa fraction and heat the closed vial to 60°C in a drying oven overnight in order to prevent the loss of hydrolyzed Pa complexes on the walls of the vial. Filter the resulting solution through acid-washed Acrodisk filter (0.2 μm presize) to prevent clogging of the micronebulizer. This solution is used to measure Pa in the MC-ICP-MS.

13.3.2.3.1 Quality Assurance

There are several corrections that need to be applied to obtain high-quality data. Those include:

1. Isobaric interference (e.g., for ^{233}Pa and ^{233}U interferences from hydride ^{232}Th^{1}H): The mode and rate of sample introduction into the plasma can significantly affect the hydride generation (discussed in Choi et al., 2001).

2. Analytical errors arising from variable isotopic fractionation (Choi et al., 2001)

3. Possible contamination of ^{233}U into Pa: If there is any contamination of ^{233}U into Pa, the MC-ICP-MS cannot distinguish between ^{233}U and ^{233}Pa (in TIMS, U ionizes at a lower temperature than Pa and thus can be differentiated).

4. Trailing correction on ^{230}Th from ^{232}Th: Although the highest ^{232}Th/^{230}Th atom ratios are found in surface waters, the trailing corrections on the ^{230}Th peaks are estimated to be usually <5% and often <1% (Choi et al., 2001).

13.3.2.3.2 Common Problems and Troubleshooting

The ^{232}Th, ^{230}Th, and ^{231}Pa blanks in small-volume water samples could be quite significant. Choi et al. (2001) estimated the procedural blank to be ~15% and ~8% of the ^{230}Th and ^{231}Pa expected in ~20 L of surface seawater. Using the standard deviation of the procedural blanks, Choi et al. (2001) estimated the detection limits for ^{230}Th and ^{231}Pa to be 3.2 and 0.4 fg, respectively, which are five to fifteen times lower than the amounts found in 20 L of surface seawater. Oceanographic consistency, such as expected linear increase in the concentrations of these nuclides with depth, should always be checked. Preparation of ^{233}Pa by neutron irradiation of target ^{232}Th may result in ^{232}Th contamination, which may preclude its measurement in the same sample.

13.3.2.4 ^{228}Th and ^{227}Th

Of all the Th isotopic measurements, the determination of these two isotopes still requires large-volume samples (>10^2 L). Preconcentration of Th can be accomplished either by filtering through a series of three cartridges—one prefilter to retain particulate matter to measure particulate Th and two MnO_2 cartridge filters (which also retain ^{228}Ra, and hence appropriate in-growth correction will have to be applied)—connected in series or by coprecipitation of Th with the $Fe(OH)_3$ method. The fibers from filter cartridges (both prefilter and MnO_2-coated filter cartridges) are cut off from the core and are ashed at 550°C. The chemical procedures to separate and purify Th from the ash are given in Figure 13.2 (Krishnaswami and Sarin, 1976). The purified Th is electroplated onto a stainless steel planchet (Figure 13.3). For the $Fe(OH)_3$ coprecipitation, the flowchart diagram for the measurements of dissolved Th is shown in Figure 13.4 (Sarin et al., 1992). The volume should be at least 200 L for the ^{228}Th and ^{227}Th measurements.

13.3.2.4.1 Quality Assurance

For the measurements of ^{228}Th and ^{227}Th, the most suitable spike is ^{229}Th or ^{230}Th. Although ^{230}Th occurs naturally, adding a spike significantly higher than the amount (ten to twenty times the natural level) commonly present in seawater would be a better way than adding ^{229}Th, because the alpha energies emanating from ^{229}Th decay have a much broader range. Periodically, a primary standard in which ^{227}Th-^{227}Ac-^{231}Pa-^{235}U as well as ^{228}Th-^{228}Ac-^{228}Ra-^{232}Th are in secular equilibrium must be run, and the activities of ^{227}Th and ^{228}Th obtained must be compared to the certified values.

13.3.2.4.2 Common Problems and Troubleshooting

Since the activity of ^{227}Th is about an order of magnitude less than that of ^{228}Th and its half-life is shorter ($t_{1/2}$ = 18.2 days), relatively faster separation of ^{227}Th from ^{227}Ac is required. Furthermore, its daughter, ^{223}Ra ($t_{1/2}$ = 11.68 days), grows in rapidly during decay counting and has a very similar alpha energy. In addition, interferences from the daughters of ^{228}Th also could compromise the quality of data. In the final data reduction, the time delay between the plating and counting and time elapsed between the cation column separation and plating should be taken into account to make decay and in-growth corrections. Alpha detector background also could be an issue, and the detector must be well monitored in terms of the background counts in energy regions corresponding to the ^{228}Th daughters.

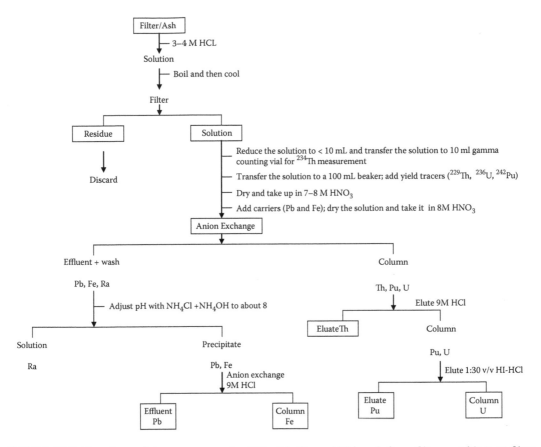

FIGURE 13.2 Flowchart of the measurements of Ra, Pb, Pu, and U in ash from filter cartridges or filter papers.

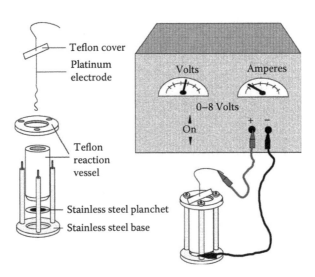

FIGURE 13.3 Electrochemical plating procedure for Pu (and Th for alpha spectrometric method).

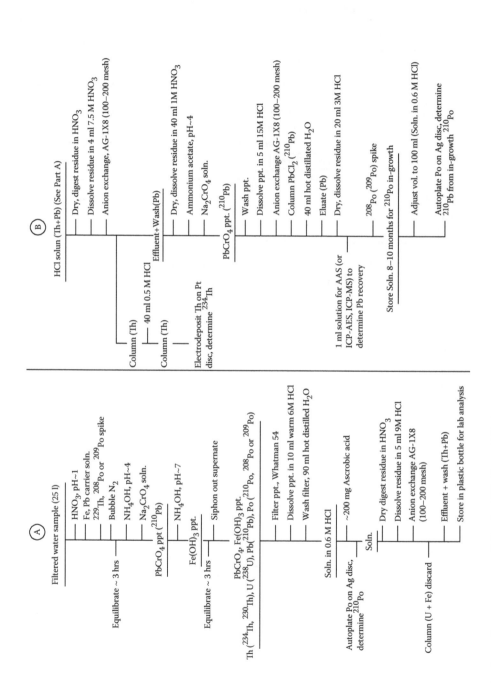

FIGURE 13.4 Flowchart of measurements of dissolved Th, Po, and Pb isotopes.

13.3.2.5 ^{227}Ac

^{227}Ac ($t_{1/2}$ = 21.77 years) is a suitable tracer for the investigation of deep-ocean mixing and upwelling rates. It is a beta emitter, and hence it can be directly assayed using a beta counter. It can also be determined indirectly by measuring its α-emitting daughter nuclide.

Earlier work on ^{227}Ac involved coprecipitation with either Fe(OH)$_3$ or MnO$_2$, followed by purification, and assay is conducted by measuring its daughter/granddaughter nuclides, ^{227}Th or ^{223}Ra (Anderson and Fleer, 1982; Nozaki, 1984; Dickson, 1985; Martin et al., 1995; Geibert and Vöge, 2008).

A recommended procedure is given below (Martin et al., 1995):

1. Collect ~50–100 L of water sample, filter it through a 0.45 μm filter cartridge to remove particulate matter. To the filtrate, add 250 mL concentrated HNO$_3$ (= 5 mL L^{-1}) and a known amount of ^{225}Ac tracer (can be ^{229}Th/^{225}Ra/^{225}Ac solution, in which all three are in secular equilibrium). Add 5 mg L^{-1} Fe (in the form of FeCl$_3$) to the sample. Allow the solution to equilibrate for ~4 hours.

2. Add NH$_4$OH to adjust the pH to ~8 to coprecipitate Ac and Th along with Fe(OH)$_3$. Allow the precipitate to settle, decant the supernatant, and separate the filtrate and precipitate through Whatman 1 filter paper.

3. Dissolve the precipitate in a minimum volume of concentrated HCl and pour into a separating funnel. Separate the Fe using the di-isopropyl ether extraction method. Evaporate the HCl solution to dryness.

4. Dissolve the residue into 100 mL 0.1 M HCl. Add 1 mL 98% H$_2$SO$_4$, 2 g K$_2$SO$_4$ and dissolve. Add 1 mL 0.24 M Pb(NO$_3$)$_2$ solution drop-wise while stirring. Heat, allow the precipitate to settle, and decant the supernatant when cool. Wash with 20 mL 0.1 M K$_2$SO$_4$/0.2 M H$_2$SO$_4$ solution and decant again.

5. Add one drop NH$_4$OH and 5 mL 0.1 M EDTA solution. Adjust the pH to 10 with NH$_4$OH to the PbSO$_4$ precipitate. Warm the beaker to aid dissolution. Add a few drops of an aqueous slurry of anion exchange resin (Bio-Rad AG1-X8) to the solution and allow it to equilibrate for 10 minutes. Add the solution plus resin onto an anion exchange column (Bio-Rad AG1-X8, 100–200 mesh, Cl form, 80 mm height, 7 mm i.d.). Adjust the flow rate to 0.5–0.7 mL min^{-1}. Wash with 13 mL 0.005 M EDTA/0.1 M ammonium acetate at pH 8. Note the time of separation of Ra and Ac.

6. Wash with 5 mL of 0.005 M EDTA/0.1 M ammonium acetate at pH 8 and discard. Wash with 2 mL of 0.005 M EDTA/0.1 M ammonium acetate at pH 4.2, then continue washing with 5 mL aliquots of this solution, monitoring the pH of this solution. When the pH falls to 7, begin collecting the eluate and continue to a volume of 15 mL. Evaporate the eluate to dryness and dissolve the residue in 2–3 mL of 3 M HNO$_3$.

7. Prepare a cation exchange resin column (Bio-Rad AG50W-X8, 200–400 mesh, H form, 80 mm height, 7 mm i.d.), and wash with 15 mL of 3 M HNO$_3$, with a flow rate of ~1 mL min^{-1}. Pour the solution-containing Ac fraction onto the column, and wash with 2 mL aliquots of 3 M HNO$_3$ until a total of 10 mL has been added. Wash the column with 15 mL of 3 M HNO$_3$. Add additional 35 mL of 3 M HNO$_3$ to elute Ac. Collect the eluate, evaporate to ~1 mL, and then bring to dryness at low heat.

8. Dissolve the residue in 1 mL of 0.1 M HNO$_3$, and electroplate on a stainless steel disc at 0.2 A for 30 minutes (Figure 13.3). Add two drops of ammonia 1 minute before stopping. Remove the disc and allow to air-dry.

9. After electrodeposition the disc is stored for at least 15–20 minutes to allow in-growth of the ^{225}Ac daughter ^{221}Fr ($t_{1/2}$ = 4.8 minutes). Determine the recovery of the ^{225}Ac tracer ($t_{1/2}$ = 10.0 day) by counting its granddaughter ^{217}At at 7.06 MeV.

10. A second count is obtained after storing the disc for 2–3 months to allow decay of ^{225}Ac and in-growth of ^{227}Ac and ^{223}Ra. Details on the calculation of specific activity of ^{227}Ac are given in Martin et al. (1995).

13.3.2.5.1 Quality Assurance

Running duplicates will ensure the reproducibility of the measurements. However, if there is a systematic error with the radiochemical separation/purification or counting methods, one may get consistently incorrect data. For the quality assurance, it is therefore recommended to take a known amount (an amount that is comparable to the natural level) of a primary U standard (such as RGU-1 from IAEA; all the daughter products of ^{238}U, ^{235}U, and ^{232}Th in these two standards are in secular equilibrium; these are solid standards and can be easily digested with HF-HNO$_3$-HCl digestion; or UREM-11, Geibert and Vöge, 2008), bring it to solution (if it is solid form), adjust the pH (to ~8.0), and mix with 20 L of seawater sample that was filtered through a set of three cartridges (one prefilter and two MnO$_2$-coated filter cartridges connected in series to remove particulate and dissolved ^{227}Ac and ^{231}Pa). The rest of the procedure is followed as described above, and the activity of the spike obtained should be compared to the amount added (calculated from the certified value). This will ensure the accuracy of the methodology and counting procedures, and that the appropriate corrections are applied.

13.3.2.5.2 Common Problems and Troubleshooting

Although Ra, Po, and Th should have been completely removed during the chemical separation procedures, a check should be made for their presence (Martin et al., 1995). The presence of ^{222}Rn (5.49 MeV) or its daughter nuclide (e.g., ^{218}Po: 6.00 MeV) can interfere with the alpha energy of the nuclide of interest, ^{227}Ac. Other interfering alpha emitting radionuclides, including ^{228}Th and its daughter products (^{224}Ra, ^{212}Bi), have overlapping alpha energies. The presence of these nuclides can be detected from the presence of ^{212}Po peak at 8.78 MeV (Martin et al., 1995).

13.3.2.6 ^{222}Rn

Of all the radon isotopes (^{219}Rn $t_{1/2}$ = 3.92 s, ^{220}Rn $t_{1/2}$ = 55.6 s, ^{222}Rn $t_{1/2}$ = 3.83 days) that are produced in the U-Th decay series, ^{222}Rn is the most important because of its longer half-life. Radon is the only noble gas among twelve elements produced in the U-Th decay chain. It is used for the indirect measurement of its parent, ^{226}Ra, by the emanation method (Broecker, 1965), and also as a recoil flux monitor for daughter products in the U-Th series (Krishnaswami et al., 1982, 1991; summarized in Porcelli and Swarzenski, 2003). The protocol adopted during the GEOSECS cruises involved pumping helium through the water sample to drive all the Rn into an activated charcoal column inside a stainless steel trap that is kept on dry ice (Mathieu et al., 1988). Helium is recirculated through the sample trap using a diaphragm pump that comes with the extraction board (Mathieu et al., 1988). Water vapor and CO$_2$ are usually trapped by serially connected Drierite and Ascarite columns, respectively. Radon from the charcoal column is mobilized by heating, and subsequently quantitatively transferred into a Lucas counting cell. After transferring Rn, the cell is stored for ~2 hours for all the daughter products of ^{222}Rn to reach secular equilibrium (note the long-lived daughter product of ^{222}Rn (before ^{210}Pb) is ^{214}Pb, with a half-life of 26.8 minutes; in 2 hours, ^{214}Pb/^{222}Ra activity ratio = 0.955; in about 3 hours, the activity ratio becomes 0.99). The cell is then counted using an alpha scintillation counter (Mathieu et al., 1988). After the measurement of ^{222}Rn, the water sample can be stored for about 20 days and the concentration of ^{226}Ra can be determined (Mathieu et al., 1988). For in situ ^{222}Rn measurements, appropriate correction for the decay and in-growth should be applied.

More recently, ^{222}Rn has been utilized widely as a tracer to quantify submarine groundwater discharge in coastal areas. A multidetector system that measures ^{222}Rn activities continuously in coastal waters has been developed (Dulaiova et al., 2005). The basic approach of this system is that a constant stream of water is delivered by a submersible pump to an air-water exchange system where radon in the water phase equilibrates with radon in a closed-air loop. The airstream is then fed to three radon-in-air monitors connected in parallel to determine the activity of ^{222}Rn. The precision of this methodology is approximately ±5%–15% for typical coastal seawater concentrations (Dulaiova et al., 2005).

13.3.2.6.1 Quality Assurance

A filtered seawater sample (20 L in a glass container to prevent the loss of Rn through diffusion) should be spiked with a primary ^{226}Ra standard (or RGU-1 standard) and stored for >1 month in a sealed container (to prevent loss of ^{222}Rn). ^{222}Rn should be determined on this spiked sample, and the result obtained by this method should be compared to the certified value.

13.3.2.6.2 Common Problems and Troubleshooting

Collection of proper water samples for ^{222}Rn (or any other gaseous substance) could be challenging. High precision was achieved using large-volume bottles (in particular, during GEOSECS days), but the throughput was poor. However, more recent methodology with a radon monitor yields precision that is less than the previous quantitative extraction methods (Mathieu et al., 1988); however, key factors, including (1) the ease of the analysis in the field, (2) no need to apply decay/growth correction, (3) no in-growth correction, and (4) a much higher throughput, make this method of analysis distinctly advantageous in the investigation of coastal oceanic processes.

13.3.2.7 ^{223}Ra, ^{224}Ra, ^{226}Ra, and ^{228}Ra

^{228}Ra and ^{226}Ra can be determined by a number of methods, first by extraction and beta counting of its immediate daughter, ^{228}Ac ($t_{1/2}$ = 6.13 h; e.g., Krishnaswami et al., 1972; Koide and Bruland, 1975), for ^{228}Ra. This method requires that there is no interference from any other betas, that all ^{228}Ac is quantitatively removed, and the self-absorption corrections are properly applied. The second method is by radiochemical separation, purification, and electroplating of Ra onto stainless steel planchets followed by alpha counting for ^{226}Ra (Koide and Bruland, 1975). For the measurements of ^{228}Ra, this method would require that the in situ ^{228}Th is completely removed and all the in-grown ^{228}Th is assumed to be derived from its parent, ^{228}Ra. An appropriate tracer (^{229}Th or ^{230}Th) should also be added for monitoring the chemical yield. By allowing the planchet to stand for 6–12 months, and recounting the planchet for ^{224}Ra (5.68 MeV, 94.5%, and remaining 5.5% overlaps with ^{228}Th alpha energies), one can determine the activities of ^{228}Th and from this calculate the ^{228}Ra activity. Since some of the alpha energies of ^{222}Rn and its daughter products overlap with ^{228}Th energies, the planchet is flamed to volatilize ^{222}Rn prior to counting in the alpha spectrometer (Koide and Bruland, 1975; Ivanovich and Murray, 1992). The initial activity can be calculated using Bateman's equation (Ivanovich, 1992). The third approach by the ^{222}Rn emanation method (Broecker, 1965; Mathieu et al., 1988; Krishnaswami et al., 1991) is one of the most common utilized to measure ^{226}Ra in seawater. The advantage of this method is that the sample can be utilized several times, after allowing for the in-growth of ^{222}Rn from the decay of ^{226}Ra. Another approach is by gamma ray spectrometry; although ^{226}Ra has a gamma line at 186 keV, there is also some interference from ^{235}U decay. Moreover, the intensity of the 186 keV gamma line is weak. However, gamma ray spectrometry is routinely utilized to assay ^{226}Ra via the three strong gamma lines of the daughters of ^{222}Rn (^{214}Pb: 295 and 352 keV; ^{214}Bi: 609 keV; Table 13.4). Similarly, ^{228}Ra is assayed using the three gamma lines of its daughter product, ^{228}Ac (338, 911, and 969 keV; Table 13.4). For most water samples from shelf, slope, and open-ocean regions, large volumes are required for ^{228}Ra by this

TABLE 13.4
Details of Gamma Rays Measured for Determining Ra Isotopes

Isotope	Measured via	γ Energy (keV)	γ Yield (%)	Interferences
^{224}Ra	^{212}Pb	238.6	43.0	^{214}Pb, ^{224}Ra
^{223}Ra[a]	^{223}Ra	269.6	24.0	^{228}Ac
	^{219}Rn	271.0		
^{228}Ra	^{228}Ac	338	11.3	None reported
	^{228}Ac	911	27.2	None reported
	^{228}Ac	964–969	21.3	None reported
^{226}Ra	^{214}Pb	295	18.9	None reported
	^{214}Pb	353	36.7	None reported
	^{214}Bi	609	46.1	None reported

Source: Data from Firestone et al. (1999).

[a] Sum of gamma yields ^{223}Ra (14%) and ^{219}Rn (10%).

method. Finally, mass spectrometric methods are utilized (TIMS: Volpe et al., 1991; Cohen and O'Nions, 1991; Ghaleb et al., 2004; Yokoyama and Nakamura, 2004; Ollivier et al., 2008; ICP-MS: Lariviere et al., 2005). Although the atom abundance of ^{228}Ra is orders of magnitude lower than that of ^{226}Ra, the ^{226}Ra/^{228}Ra ratios have been measured using TIMS (Yokoyama and Nakamura, 2004; Ollivier et al., 2008).

13.3.2.7.1 Gamma Ray Spectrometry

A number of radionuclides—^{234}Th (Section 13.3.2.2), ^{210}Pb (Section 13.3.2.8), ^{7}Be (Section 13.3.3.1), and ^{137}Cs (Section 13.3.4.1)—can be measured using gamma ray spectrometry. In the case of radium, all Ra isotopes are alpha emitters, except ^{228}Ra, which is a weak beta emitter. However, all four Ra isotopes and their daughter products emit gamma rays (Table 13.4), and hence the activity concentrations of these nuclides can be measured using gamma ray spectrometry. Because of the short half-lives of ^{223}Ra ($t_{1/2}$ = 11.68 days) and ^{224}Ra ($t_{1/2}$ = 3.64 days), it is necessary that the samples are processed as quickly as possible after sample collection. In principle, one can filter large volumes of water samples using MnO_2-coated cartridge filters at relatively high flow rates (7–10 L min⁻¹). Although the quantitative removal of Ra onto MnO_2-coated fiber is unlikely at this flow rate (Baskaran et al., 1993), one can obtain reliable activity ratios of Ra (such as ^{226}Ra/^{228}Ra, ^{223}Ra/^{226}Ra, ^{224}Ra/^{228}Ra) from this method. Using a separate aliquot for ^{226}Ra, one can obtain the specific activities of all four isotopes of Ra (Krishnaswami et al., 1991). This methodology is only efficient if the activities of Ra isotopes are high. The gamma yields of some of the progenies of Ra (Table 13.4) are low, and the absolute detection efficiency of the Ge well detector decreases with increasing energy (Figure 13.5a). For example, one of the most commonly used gamma lines of ^{228}Ac at 338 keV for the measurement of ^{228}Ra has a branching ratio of 11.3%, which implies that 88.7% of the atoms of ^{228}Ac undergoing radioactive decay do not emit a gamma line at 338 keV. Furthermore, the absolute efficiency of the gamma ray detector at 911 keV (^{228}Ac) of 6.49% ± 0.14% (5 mL geometry) is significantly lower than that at 338 keV (17.6% ± 0.4%) (Jweda, 2007).

Quantification of the specific activity requires that the gamma ray detector is well calibrated. In particular, in Ge well detectors, the geometry of the sample is well defined (when sample is put inside the well). When one uses the standard graduated counting vials (such as 10 mL vials obtained from the Milian Company, Columbus, Ohio), the gamma ray spectrometer can be calibrated to

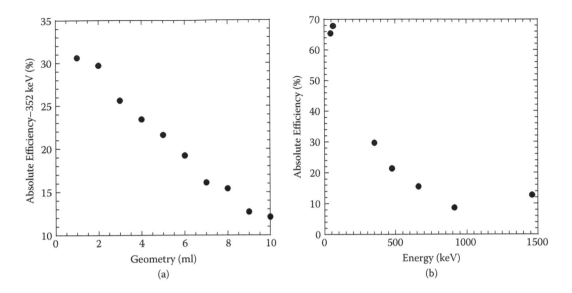

FIGURE 13.5 (a) Absolute efficiency of a Ge well detector (40 mm well depth; Canberra, Inc.) as a function of geometry of the sample in a 10 mL counting vial at 352 keV (^{214}Pb) gamma line. (Data taken from Jweda, 2007.) (b) Absolute efficiency of a Ge well detector (40 mm well depth; Canberra, Inc.) as a function of energy for IAEA-SRM in 2 mL counting vial geometry. (Data taken from Jweda, 2007.)

various geometries (1 to 10 mL), as shown in Figure 13.5b. Using a known set of standards (such as IAEA-RGU-1 and IAEA-RGTh-1), the ratios of disintegrations per minute (dpm) to counts per minute (cpm) can be determined for various geometries. The specific activity on any seawater sample can be calculated as follows:

$$A\ (Bq/m^3) = N \times (dpm/cpm)/60\ T\ V \tag{13.4}$$

where A is the specific activity, N the background subtracted net counts, T the time in minutes, V the volume (m^3), (dpm/cpm) the factor obtained from calibration of standards for that geometry, and 60 the factor used to convert dpm to Bq. This equation does not include corrections for self- and external absorption of gamma rays, as well as coincidence (McCallum and Coote, 1975). When the density of the material is comparable to the standards, then the self- and external absorption corrections for gamma lines > 200 keV seem to be negligible. Even for low-energy gamma lines (such as 46 keV for ^{210}Pb and 63 keV for ^{234}Th), the correction factors seem to be negligible, if the densities of the standard used for calibration and the samples are comparable (Jweda, 2007). The summation corrections are insignificant when the activities are low (less than a few hundred Bq g^{-1}). In the case of ^{226}Ra and ^{228}Ra, there are a number of gamma lines that can be utilized (Table 13.4). When activities are determined for each of the peaks, the weighted mean can be calculated to obtain the final result.

13.3.2.7.2   In Situ Underwater Gamma Counting

In situ underwater radionuclide measurement systems were developed in the late 1950s (Winn, 1995; Jones, 2001) and have been used to screen contaminated areas in emergency situations (Chernayaev et al., 2004), map large seabed areas for natural and anthropogenic radionuclides (Tsabaris and Ballas, 2005; Tsabaris et al., 2005), conduct mineral exploration and other geological purpose mapping (Jones, 2001), and monitor, on a continuous basis, the vicinity of sources or potential sources of release of radionuclides, such as discharged or dumped nuclear wastes, and nuclear weapon test sites (Osvath and Povinec, 2001). The most common detectors used for in situ underwater gamma spectrometry are made of NaI(Tl) and HPGe crystals. The advantage of NaI(Tl)-based systems is primarily related to the high detection efficiency of NaI(Tl) crystals at much lower cost, ruggedness for long-term underwater deployment, shock resistance, and thermal stability of the system, compared to equivalent Ge crystals. The drawbacks of NaI(Tl)-based systems are the high power consumption for the operation of the photomultiplier tube and a relatively poor energy resolution of the detector (~8% resolution for the 662 keV gamma ray from ^{137}Cs for 3 in. [diameter] × 3 in. [long] crystal, as compared to <0.3% for the HP-Ge detector). HPGe-based systems have the advantage of good energy resolution and hence excellent radionuclide identification capability. Development of cadmium zinc telluride (CZT) detectors as gamma radiation detectors offers some promise, but more developmental work needs to be done (Ho et al., 2005).

13.3.2.7.3 Delayed Coincidence Counting

Moore and Arnold (1996) developed a method to measure low levels of ^{223}Ra and ^{224}Ra using a delayed coincidence counter; it has become a widely used method (Moore, 2008). Water samples are passed through MnO_2-coated fiber. The fiber is partially air-dried and placed in an air circulation system, and helium is circulated over the Mn-fiber and through a scintillation cell where the alpha particles from the decay of ^{223}Ra and ^{224}Ra and their daughter products are recorded. Signals from the detector are sent to a delayed coincidence circuit, which discriminates decays of ^{223}Ra (and its daughters) from ^{224}Ra (and its daughters) (Moore and Arnold, 1996). More details on this counting method can be obtained from Rutgers van der Loeff and Moore (1999).

13.3.2.7.3.1 Quality Assurance Periodically, the extraction efficiency of MnO_2-fiber needs to be watched when a single column of fiber is used. Spiking a 20 L of Ra-free seawater (seawater that is passed through three MnO_2-coated cartridges connected in series to remove >95% of the Ra) with ^{238}U, ^{235}U, and ^{232}Th primary standards (or ^{226}Ra or ^{228}Ra primary standards) with all their daughters in secular equilibrium (activities of $^{238}U = {}^{226}Ra$, $^{235}U = {}^{223}Ra$, and $^{232}Th = {}^{228}Ra = {}^{224}Ra$) and measuring the activities of Ra isotopes will ensure the quality of the data.

13.3.2.7.3.2 Common Problems and Troubleshooting Measurements of long-lived Ra isotopes (^{226}Ra and ^{228}Ra) by counting methods involve preconcentration by MnO_2-fiber, MnO_2-coated cartridge filters, $BaSO_4$ precipitation, etc. The fiber is either ashed and directly gamma counted or leached, and subsequently a $Ba(Ra)SO_4$ precipitation is conducted. In this method, quantitative precipitation and transfer to counting vial after filtration is critical, as no yield tracer is added to monitor the potential loss of Ra during the precipitation/filtration/transfer process. Direct ashing of fiber (acrylic) at 820°C for 16 hours needs to be conducted in a covered ceramic crucible to minimize the loss of fiber when it is being ashed. Calibrating the gamma detector with liquid ^{226}Ra is tricky, as some of the ^{222}Rn (and the daughter products from the decay of ^{222}Rn in the air space) tends to stay in the air space, affecting the count rates, and hence it is not recommended to use liquid ^{226}Ra standard for the calibration of the daughter products of ^{226}Ra.

13.3.2.8 ^{210}Po and ^{210}Pb

Although ^{210}Pb concentrations can be measured using a low-energy (46 keV) gamma line of ^{210}Pb, the low branching ratio (4%) requires that a large volume of water sample (10^2–10^3 L) be processed for ^{210}Pb (Baskaran and Santschi, 2002). However, ^{210}Po and ^{210}Po measurements are routinely conducted using alpha spectrometry in the same sample, first by measuring ^{210}Po (called in situ ^{210}Po) and then keeping the sample for a period of 6 months to 2 years for the in-growth of ^{210}Po from ^{210}Pb. The second ^{210}Po (called parent-supported) measurement provides the data on the concentration of ^{210}Pb. The volume required for analysis of dissolved and particulate ^{210}Po and ^{210}Pb ranges from a few liters (Hong et al., 1999) to 20–30 L (Sarin et al., 1992; Church et al., 1994; Radakovitch et al., 1998; Kim et al., 1999; Friedrich and Rutgers van der Loeff, 2002; Masque et al., 2002; Stewart et al., 2007). Concentrations of ^{210}Po and ^{210}Pb in the dissolved, particulate, and total (= particulate + dissolved) phases have been measured. One method for determining the particulate ^{210}Po/^{210}Pb is to measure their respective concentrations in the unfiltered and filtered water (0.7 µm filter cartridge; concentration in particulate phase = concentration in unfiltered phase minus the concentration in dissolved phase). For the total ^{210}Po and ^{210}Pb, the unfiltered water sample is directly processed. The methodology for measuring the dissolved phase is given below:

1. Collect water samples using 30 L Niskin bottles, and filter them through a 0.7 µm GFF filter.
2. Acidify the filtrate immediately after filtration with 25 mL high-purity concentrated HCl. Add 70 mg of Fe (in the form of FeCl$_3$, which is tested for blank levels of ^{210}Po and ^{210}Pb) to the filtrate.
3. Spike the sample with NIST-^{209}Po tracer (SRM 4326—polonium-209 solution; equivalent to 3 dpm) and 5 mg of Pb (high-quality AAS standard, which was tested before for a low blank of ^{210}Pb and ^{210}Po).
4. Allow the sample to equilibrate for ~4 hours.
5. After equilibration of spikes/carriers, add NH$_3$(OH) to a pH ~ 7 to coprecipitate Pb and Po with Fe(OH)$_3$. Allow the precipitate to settle for ~4 hours.
6. Separate the precipitate from the solution by a combination of either decantation-centrifugation or decantation-filtration through a Whatman 41 filter paper.
7. Dissolve the precipitate by adding 10 mL of 6 M HCl, followed by washing of the filter paper with 90 mL of deionized water. To this solution, add 200 mg of ascorbic acid and separate the Po isotopes by spontaneous electroplating onto a silver plate (Flynn, 1968).
8. Dry the electroplated solution completely and dissolve the dried residue in 5 mL of 9 M HCl for the separation of Po and Pb using an anion exchange column (Sarin et al., 1992).
9. After the separation of Pb and Po by an anion ion exchange column, add a known amount of ^{209}Po spike to the purified Pb fraction and store in a clean plastic bottle for about 6 months to 1 year.
10. Measure the activity of ^{210}Pb by the grown-in activity of its granddaughter, ^{210}Po, over the period between the first and second ^{210}Po separations. Calculate the activity of ^{210}Po based on the background corrected peak-area ratios of ^{210}Po/^{210}Po multiplied by the activity of the added ^{209}Po spike. Subtract the mean field blank ^{210}Po activity. Apply appropriate in-growth and decay corrections for ^{210}Po (decay correction for ^{210}Pb). An aliquot of the stored solution (5%, 5 mL out of 100 mL) will be taken in a 25 mL precleaned polyethylene bottle and stored for stable Pb determination (either AAS/ICP-MS or any other suitable instrument) to obtain chemical recovery of Pb.
11. The final activity of ^{210}Pb (A_{Pb}) calculation will involve the in-growth and decay corrections as shown in Equation 13.5.

$$A_{Pb} = A_{Po} \, (\lambda_{Po} - \lambda_{Pb})/(\lambda_{Po} \, (e^{-\lambda_{Pb} t} - e^{-\lambda_{Po} t})) \qquad (13.5)$$

where A_{Po} is the activity of ^{210}Po derived from the in-growth from ^{210}Pb, λ_{Po} and λ_{Pb} are the decay constants for ^{210}Po and ^{210}Pb, respectively, and t is the time elapsed between the last Po/Pb separation (8 M HNO$_3$ column exchange) by the anion exchange column and the Po plating time. A separate correction for the decay of ^{210}Po from the time of plating to mid-counting is also applied ($A_0 = A(t) \, e^{-\lambda_{Po} t_1}$, where t_1 is the time between plating and mid-counting).

13.3.2.8.1   Quality Assurance

A known amount of primary standards of [210]Po and ^{210}Pb should be used to check the precision of the methodology. In particular, [210]Po can be volatile if is in the reduced state, and hence its loss during the chemical procedure (if any) needs to be carefully monitored.

13.3.2.8.2   Common Problems and Troubleshooting

Separation of [210]Po and ^{210}Pb immediately after first plating (in situ [210]Po) is critical, as incomplete plating of [210]Po will result in residual [209]Po and [210]Po remaining in the solution. While about 84% of the residual [210]Po would decay away after 1 year of storage time, only ~0.7% of the residual [209]Po will decay away during this period. Adding secondary silver plates to remove Po does not guarantee the complete removal of Po, and hence it is critical that the Po and Pb are separated through ion-exchange resin immediately after the first plating. The corrections for the in-growth (of the granddaughter, [210]Po) and decay of [210]Po and ^{210}Pb during the time elapsed between collections (first plating, 9 M HCl ion exchange column separation; second plating, counting of the Ag plates) need to be applied. Using the standard Bateman's equations, the explicit correction terms need to be established for the routine procedure.

13.3.3   COSMOGENIC RADIONUCLIDES

13.3.3.1   [7]Be

The determination of [7]Be is routinely conducted using large-volume samples.

13.3.3.1.1   Particulate [7]Be

1. Filter >500 L water through a polypropylene filter cartridge (acrilon fiber cartridge has relatively large ash weight at temperatures of 500°C–600°C).
2. Cut the filter cartridge to remove the inert cartridge core, pack the fiber into a high-temperature-resistant silica crucible, and ash it at 500°C for about 6 hours.
3. The residue is quantitatively transferred (as no yield monitor is added) into a gamma counting vial for gamma ray analysis for the determination of particulate [7]Be (Baskaran et al., 1997; Baskaran and Swarzenski, 2007).

13.3.3.1.2   Dissolved [7]Be

1. Collect ~200 L of the filtrate in a large-volume polyethylene container.
2. Acidify the water sample using 200 mL high-purity concentrated HCl. To the solution, add 5 mg of Be (AAS grade) and 500 mg Fe (in the form of $FeCl_3$).
3. Allow the acidified and spiked sample to equilibrate for about 4 hours.
4. Coprecipitate Be with $Fe(OH)_3$ by adding NH_3OH. Allow the precipitate to settle for about 4–6 hours.
5. Separate the precipitate and the solution by decanting followed by filtering through Whatman 42 paper. The filter paper will be taken to the shore-based laboratory for further analysis.
6. In the laboratory, soak the filter paper containing the precipitate inside a Teflon beaker containing 30 mL of 6 M HCl. After the precipitate has dissolved completely, wash the filter with 6 M HCl three times and combine the solutions. Dry the solution completely and transfer the residue quantitatively into a gamma counting vial and gamma count the residue.

7. After gamma counting, transfer the precipitate quantitatively into a Teflon beaker and dissolve the precipitate in 1 M HNO_3 and make to 100 mL. Take 5 mL out of this solution for stable Be determination (AAS, ICP-MS, ICP, etc.) to determine the chemical yield.

13.3.3.1.2.1 Quality Assurance Getting a certified reference standard for [7]Be measurements is not an easy task, due to its short half-life and the expense of producing it. However, calibration of the gamma ray spectrometer can be commonly carried out by establishing the energy versus absolute efficiency curve. In particular, the two gamma lines with very little interference, 352 keV ([214]Pb) and 609 keV ([214]Bi), can be used to evaluate the absolute efficiency of [7]Be at 476 keV (Jweda et al., 2008). Intercalibration of samples collected from natural waters (surface waters or rainwater) is recommended, as this would help to validate the [7]Be measurements. To our knowledge, so far no intercalibration attempts have been carried out for this nuclide.

13.3.3.1.2.2 Common Problems and Troubleshooting The acrylic cartridge method that involves extracting Be onto Fe-coated cartridges assumes that the efficiency (absolute efficiency in two cartridges connected in series) remains constant for the two cartridges (Kadko and Olson, 1996). However, this might be questionable, as described in Section 13.2.2.2, as rigorous testing has not been done.

13.3.3.2 [10]Be

The conventional method of [10]Be measurements was conducted by the beta counting method. However, development of the AMS technique in the mid-1970s resulted in much better precision, with a considerable decrease in the sample size. More recent advances in tandem accelerators (such as 25 MV tandem accelerators) have the potential to further improve the sensitivity by one to two orders of magnitude (Galindo-Urbarri et al., 2007). For the particulate [10]Be measurements, the water sample needs to be filtered through a cartridge filter and the fiber material needs to be ashed at 550°C for ~4 hours, and the ash is utilized for [10]Be measurements. The filter is brought to solution by digestion. A brief summary of a commonly used method for the dissolved [10]Be is given below (Kusakabe et al., 1982):

1. Collect ~100 L of water samples using 30 L PVC Niskin bottles (for deep waters) or the ship's saltwater pump system for surface waters.
2. Filter the water through a 0.4 μm pore size Nucleopore filtration system directly attached to the ship's pump.
3. Acidify the sample with H_2SO_4 to pH ~ 2, add Fe^{3+} (in the form of $FeCl_3$, 5 mg Fe^{3+} L^{-1}) and stable Be ([9]Be, 4.2 mg $BeSO_4$ or 2.55×10^{19} atoms), and store in plastic containers on board.
4. In the laboratory, add ammonia to pH ~ 8 to form $Fe(OH)_3$. To make the $Fe(OH)_3$ settling more effective, add saturated Na_2CO_3 solution.
5. Separate the precipitate and the filtrate by decanting the supernatant followed by either filtration or centrifugation.
6. Take the precipitate in 3 mL concentrated HCl. Keep adding concentrated HCl until the final solution is in 9 M HCl medium.
7. Centrifuge this solution and the supernatant is now ready for passing through a chromatographic column.
8. Prepare an anion exchange chromatographic column (15 mm ID × 200 mm length, containing 7 mL of anion exchange resin, DOWEX AG 1-X8, 100–200 mesh). Condition the column with 9 M HCl.
9. Load the sample from the centrifuge tubes onto the columns. Wash the column with 30 mL 9 M HCl and collect the effluent.

10. Add 20 mL of ultrapure water to the beaker containing the sample. To the solution add slowly concentrated NH_3OH and adjust the pH to ~7.7 ± 0.3. Centrifuge the sample and decant the supernatant.

11. Redissolve the hydroxide precipitate with 0.5 mL HNO_3. If the precipitate does not dissolve completely, add up to another 2.5 mL of the same HNO_3 solution.

12. Prepare a cation exchange chromatographic column (15 mm ID × 200 mm, containing 10 mL DOWEX AG 50W-X8 cation exchange resin, 100–200 mesh).

13. Condition the column with 100 mL 6 M HNO_3 followed by a wash with 250 mL ultrapure water. After the solution drains off, place the sample collection beakers under the column.

14. Load the sample solution onto the column. After the solution passes through the column, add 50 mL 0.5 M HNO_3 solution to the column.

15. After the solution passes through the column, add 40 mL of 1.0 M HNO_3 into the column and collect it in a separate precleaned beaker. Add another 40 mL 1.0 M HNO_3 onto the column and let it pass through.

16. The solution is evaporated completely. Add 25 mL of 1 M HCl and 35 mL ultrapure water.

17. To the solution slowly add concentrated NH_3OH to adjust the pH to ~7.0 ± 0.3.

18. Centrifuge the samples; decant off the solution. Add a few milliliters of ultrapure water and transfer the slurry into a precleaned (low-boron) quartz crucible. Place the crucible in an oven and dry the hydroxide overnight at 60°C.

19. Cover the crucible with a precleaned quartz lid and transfer it to a furnace. Heat the sample at 350°C for ~30 minutes, and then increase the temperature to 850°C. Keep the crucible for at least 1 hour.

20. Reduce the furnace temperature back to 50°C and allow the furnace to cool down.

21. The BeO oxide is loaded onto the target for the measurements of $^{10}Be/^9Be$ in an accelerated mass spectrometer.

13.3.3.2.1   Quality Assurance

Primary standard for ^{10}Be seems to be out of stock (see Table 13.3). Measurements of ^{10}Be in acidified samples (or sediment samples) that were already measured can serve as a secondary standard.

13.3.3.2.2   Common Problems and Troubleshooting

While the chemical separation and purification is relatively straightforward, isobaric interference from boron isotopes could affect the measurements of ^{10}Be by AMS. Care must be exercised that the chemical separation and purification steps eliminate all boron isotopes.

13.3.3.3 $^{14}CO_2$ Method

The development of AMS in the 1970s led to counting the ^{14}C atoms in the samples, rather than measuring their radioactive decay, which allowed the analysis of much smaller samples (submilligram of carbon as opposed to several grams in the conventional beta counting method). The throughput was also increased significantly, as measurements of ^{14}C required only less than an hour rather than several days or weeks by beta counting. There are a number of AMS facilities available for the measurement of ^{14}C of dissolved inorganic carbon (DIC) in seawater, and they usually provide sampling and handling guidelines. A 500 mL borosilicate glass bottle with a high-quality ground-glass stopper is used to draw water samples from Niskin bottles. Maximum care is needed to avoid contact with air and to stop the bacterial activity (by adding chemicals such as saturated mercuric chloride solution).

13.3.3.4 ^{36}Cl

Chlorine-36 is an ideal oceanic tracer as it does not precipitate or undergo redox changes, and is not removed by particles or organisms. Tandem accelerators with moderate energy (5–10 MeV) have been successfully used to measure ^{36}Cl/Cl ratios in the order of 10^{-12}–10^{-14} (Synal et al., 1990). However, with a 25 MV tandem accelerator, recently it has been shown that very high sensitivity (in ~500 mL seawater) can be obtained (^{36}Cl/Cl ratios 10^{-15}–10^{-16}; Galindo-Uribarri et al., 2007).

Five liters of seawater samples is collected and filtered through a 0.45 μm filter membrane using in-line filters directly into a precleaned container. The chloride is precipitated as AgCl and purified of sulfur interferences (to eliminate the interfering isobar, ^{36}S; Bentley et al., 1986). To the AgCl precipitate, a known amount of Nb powder (325 mesh) is mixed in amounts equal to 50% of the mass of a given AgCl (one part of Nb mixed with three parts of AgCl, in volumetric terms; Galindo-Uribarri et al., 2007). The AgCl-Nb mixture is pressed into a Cu target holder, which is placed in the Cs$^+$ sputter ion source. The ^{36}Cl/Cl ratios are determined by AMS (Elmore et al., 1979; Galindo-Uribarri et al., 2007).

13.3.4   ANTHROPOGENIC RADIONUCLIDES

The preconcentration procedure used for the separation of anthropogenic radionuclides from seawater samples is often based on the coprecipitation of MnO_2 produced from the in situ reaction of $KMnO_4$ and $MnCl_2$ in alkaline medium. In some studies, multiple radionuclides of Pu, Am, Cs, and Sr and other isotopes need to be determined from a single water sample, and the schematic outline of the sequential separation of some key anthropogenic radionuclides is shown in Figure 13.6. The analytical procedures for the key individual radionuclides are described below.

13.3.4.1 ^{137}Cs

^{137}Cs ($t_{1/2}$ = 30.2 years) is a fission product of nuclear reaction and decays to the stable isotope ^{137}Ba through beta decay, but it has a 90.1% probability of emitting a 661.7 keV photon from a transition state ^{137m}Ba($t_{1/2}$ = 2.552 min) to ^{137}Ba. Therefore, both beta (proportional) counter and gamma spectrometry can be used to detect it.

13.3.4.1.1   *AMP Precipitation Method: Batch Process of AMP/Cs Complex*

The mostly widely used ammonium phosphomolybdate (AMP) adsorption methods are described here and largely taken from Wong et al. (1994).

1. Acidify 60 L of seawater with 200 mL of 4 M HCl (pH of ~1).
2. Add 20 mg of Cs carrier (exact amount for calculated yield) or 90 mBq of ^{134}Cs tracer (if available free of ^{137}Cs), and then stir the sample for 1–2 hours by bubbling N_2 gas.
3. Mix 13 g AMP with water to make a slurry. Add AMP slurry and stir for 30 minutes to adsorb the Cs onto the AMP.
4. Let the AMP settle in the container overnight. Siphon off the supernatant and discard it.
5. Collect the AMP quantitatively in a small beaker and allow it to settle. Centrifuge it for 10 minutes and then discard the supernatant.
6. Dissolve the AMP in 10 M NaOH. Centrifuge the solution to remove undissolved solids.
7. Boil the solution to drive off NH_3. Check vapor with wet pH paper until NH_3 emission ceases.
8. Dilute the sample with 500 mL of ultrapure water.
9. Bring the pH to 1–2 by adding 10 M HCl. The solution turns yellow.
10. Add 1 g AMP (as a slurry) to the solution. Stir for 30 minutes and let AMP settle overnight.

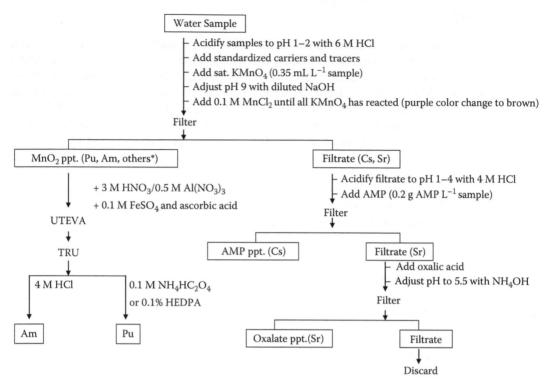

* Isotopes of Sc, V, Cr, Mn, Fe , Co, Ni, Y, Zr, Nb, Ru, Ag, Cd, Sb, La, Ce, Eu, Pb, Bi, Po, Ra, Th, U, NP, Pu, Am, Cm elements have shown to coprecipitate quantitatively with in situ produced MnO_2 in fresh- and seawater samples. (UTEVA, TRU are the extraction chromatographic materials available from Eichrome Technologies, Inc., Darien, IL, USA. HEDPA is hydroxylethylidene diphosphonic acid, and ppt. is precipitate.)

FIGURE 13.6 Flowchart of the chemical procedures for the measurements of dissolved Am, Pu, ^{137}Cs, and [90]Sr. (Modified from Wong et al., 1994.)

11. Transfer the AMP to a 50 mL centrifuge tube, centrifuge, and discard the supernatant.
12. Dissolve the AMP with 5 mL of 10 M NaOH.
13. Transfer the solution to a standardized container for gamma counting using a gamma ray spectrometer. The detector is calibrated by counting a standard solution of ^{137}Cs, using the same counting geometry.
14. The chemical recovery is determined by gamma spectrometry by measuring either the concentration of [134]Cs in the sample or the stable Cs in the final solution using ICP-MS.

If a sample is to be beta counted, further purification has to be done to remove potassium and rubidium activities by ion exchange separation using a strongly acidic cation exchange resin (BIO-REX-40). In particular, [40]K will interfere with the beta counting of [137]Cs. In this case, the purification of Cs from other elements is required.

Recently, combination of AMP precipitation method with an underground (235 m below the sea level, low background) gamma measurement facility (detection limit of ^{137}Cs is 0.18 mBq for a counting time of 10,000 minutes) resulted in the measurement of ^{137}Cs in as little as 250 mL of seawater (Hirose et al., 2005).

13.3.4.1.1.1 Quality Assurance   Adding a known amount of [134]Cs (certified value) to standard seawater (20 L) and following the chemical procedures and comparing the activities to the certified values will ensure the methodology and counting procedures are working satisfactorily.

13.3.4.1.1.2   Common Problems and Troubleshooting   For the small-volume seawater samples, it is important to know background level gamma activity of the reagents, in particular in the ^{137}Cs region. The ^{137}Cs activity in CsCl was found to be less than 0.03 mBq g^{-1} by using extremely low-background gamma spectrometry. The ^{137}Cs activity in AMP was found to be less than 0.008 mBq g^{-1} (Hirose et al., 2005).

13.3.4.1.2   In Situ Sorbent Method

After early research showing that ferrocyanides are useful for the separation of radiocesium from other fission product mixtures either by coprecipitation or by adsorption on preformed salts, several methods have been developed (Watari and Izawa, 1965; Folsom and Sreekumaran, 1970; Wong et al., 1994; Mann and Casso, 1984; Buesseler et al., 1990; Schwantes, 1996). ^{137}Cs measurements based on gamma spectrometry are reviewed in Folsom and Sreekumaran (1970). The conventional procedure as described in Wong et al. (1994) used a methodology that involves batch extraction of Cs onto a microcrystalline cation exchanger, ammonium molybdophosphate (AMP), and subsequent purification from potassium and rubidium activities by ion exchange separation using a strongly acidic cation exchange resin (BIO-REX-40). In particular, ^{40}K will interfere with the beta counting of ^{137}Cs. In this case, the purification of Cs from other elements is required. Development of chemisorption absorbers aimed to preconcentrate it from larger volumes of seawater helped to alleviate this problem. The efficient extraction of ^{137}Cs from large volumes of coastal waters using commercially available absorber material (Anfezh and Fezhel, Eksorb Ltd.) was reported by Bandong et al. (2001). Mann and Casso (1984) developed a method in which seawater was passed through a tandem cartridge consisting of twin beds of ion exchange resin impregnated with cupric ferrocyanide. The inherent assumption in this method is that the collection efficiencies of each bed are the same. In this method, the resin preparation, desorption, and radiochemical purification of the collected Cs involve a considerable amount of laboratory work. Impregnating fiber material with copper ferrocyanide and subsequent ashing of the fiber material, followed by direct gamma counting of the residue, could be much simpler. Since ^{137}Cs is not particle reactive in marine environments, distribution of particulate ^{137}Cs activity is not of much scientific interest. The detailed procedures for the measurement of dissolved ^{137}Cs are given below (Buesseler et al., 1990; Schwantes, 1996).

13.3.4.1.2.1   Impregnation of Copper Ferrocyanide ($Cu_2Fe(CN)$)

1. The cartridge used for impregnation is CUNO Microwynd DCCPY wound cotton fiber filter with a 1 μm nominal pore size (DCCPY1, manufacturer: Eggelhof, Inc., http://www.eggelhof.com/filtration.html). Load the six-shooter housing setup with six filter cartridges, similar to the one shown in Figure 13.1.
2. Dissolve 5 g of $K_4Fe(CN)_6$ in 1 L ultrapure water.
3. Add 5 g of $Cu(NO_3)_2$ in 5 L ultrapure water. Stir this solution on a stir plate until dissolved.
4. Place a new cartridge in the housing (similar to the one shown in Figure 13.1).
5. Pour $K_4Fe(CN)_6$ solution into the middle and allow diffusing across the filter. This will saturate the entire cartridge with the solution.
6. Attach the cartridge housing to a peristaltic pump, with both inlet and outlet hoses from the housing leading into the $Cu(NO_3)_2$ solution.
7. Turn the pump at full speed (flow rate ~12–14 L min^{-1}) for about 2 minutes. Note starting time.
8. Slow down the pump to ~2 L min^{-1} for 30–60 minutes, until the solution is clear, or nearly clear.
9. Remove the inlet hose from the solution and allow the pump to run to partially empty the housing.
10. Remove the cartridge from the housing and allow the fluid to drain.

11. Dry the cartridge in an oven at ~80°C–90°C for ~24 hours. Place the dried cartridge in a sealed plastic bag.

13.3.4.1.2.2 Sample Collection and Processing

1. Place one 0.5 μm pore size (nominal) precleaned polypropylene cartridge (to retain particulate matter) followed by two copper ferrocyanide-coated cartridges in series in a serial housing unit (Figure 13.1). The flow rate meter will read the total volume of water filtered through the fiber cartridge.
2. Pump water through the three-cartridge housing unit at the rate of 7–8 L min^{-1}.
3. To remove the salt in the cartridge, pass distilled water through the used filter cartridges. This will remove the sea salt and thus will decrease the ash content of the filter.
4. After cutting away the plastic core, loosely pack the used cotton filter into two uncovered 1 L high-temperature crucibles.
5. Place the uncovered crucibles in a furnace and combust at 750°C for 24 hours.
6. After the ashing is complete, quantitatively transfer the residue into a bag, weigh it, and then transfer the ash into a 10 mL gamma counting vial. If there is any excess ash, use the ratio of the total ash weight/weight used for gamma counting to calculate the final activity.
7. After determining the activities of ^{137}Cs in both the $Cu_2Fe(CN)_6$-coated cartridges, the extraction efficiency (η) is calculated assuming that both cartridges extract uniformly (i.e., constant extraction efficiency, η), using the formula:

$$\eta\ (\%) = [1 - F_2/F_1] \times 100 \qquad (13.6)$$

where F_1 and F_2 are the activities of ^{137}Cs in the first and second $Cu_2Fe(CN)_6$-coated cartridges, respectively. The specific activity of ^{137}Cs is calculated as follows:

$$\text{Specific activity } (A, \text{ in Bq L}^{-1}) = [(F_1)/\eta]\ (e^{\lambda t})/V \qquad (13.7)$$

where V is the volume of the sample, t is the time elapsed from collection to counting (in years), and λ is the decay constant of ^{137}Cs (2.295×10^{-2} year^{-1}).

13.3.4.1.2.3 Quality Assurance
Adding a known amount of ^{137}Cs (certified value) to a standard seawater (20 L) sample and passing it through a pair of cartridges and comparing the activities to the certified values will ensure the accuracy of the methodology and counting procedures.

13.3.4.1.2.4 Common Problems and Troubleshooting
This method seems to work very well. With every cartridge method, the absolute extraction efficiency needs to be monitored to make sure the values are uniform and remain constant. Sometimes, due to channeling effects, the water may not be in contact with the fiber material long enough and the removal efficiency could be compromised; this needs to be monitored.

13.3.4.2 ^{90}Sr

Most of the ^{90}Sr in surface waters is derived from global fallout. The interest in ^{90}Sr stems from the following:

1. ^{90}Sr is a high-yield product of U and Pu fission.
2. ^{90}Sr is a tracer for Sr, and Sr has similar properties as Ca.
3. ^{90}Sr is the most widely and carefully monitored fallout nuclide in precipitation, aerosols, and soils. Indeed, the bomb test fallout of ^{137}Cs is calculated based on the monitored ^{90}Sr fallout, assuming constancy of the ^{90}Sr/^{137}Cs activity ratio.
4. Its moderate half-life (28.9 years; MacMahon, 2006).

^{90}Sr decays to ^{90}Y ($t_{1/2}$ = 64 hours), which then decays to stable ^{90}Zr. ^{90}Sr measurements are made by separating it from the matrix, and measuring the decay rate of ^{90}Y. The counting of ^{90}Y is commonly done using a variety of gas-filled Geiger-Müller or proportional counters, or with thin plastic film beta scintillation detectors. An excellent review of various methods is given in Bowen (1970). Sr is water soluble, and hence the particulate ^{90}Sr measurements are not of scientific interest. The specific details of the procedure for dissolved ^{90}Sr analysis are given below (Bowen, 1970; Wong et al., 1994; Hong, 2008):

1. Collect a 50 L seawater sample and filter through a 0.5 μm pore size (nominal) filter cartridge and collect the filtrate in a clean plastic container.
2. Add ^{85}Sr tracer (1,000 Bq), 1 g of stable Sr carrier (as a holdback carrier, although we have ~8 mg L^{-1} stable Sr in the open-ocean water), and 2 L saturated oxalic acid solution (40 mL L^{-1} of seawater, 100 g mL^{-1}). Mix the water for ~30 minutes. (Note: The same water can be utilized for Fe or Mn coprecipitation for transuranics or Cs adsorption on AMP or copper ferrocyanide-coated cartridges before strontium oxalate precipitation.)
3. Add NH$_4$OH to pH 1.5–2.0 and precipitate the oxalate. Allow the precipitate to settle overnight.
4. Separate the precipitate from the solution by filtration, or decantation and centrifugation.
5. Dry the precipitate under infrared light and ash the powder at 550°C for about 1 day.
6. Weigh the ash. Add concentrated HNO$_3$; volume of concentrated HNO$_3$ is calculated as follows:

$$\text{Volume of concentrated HNO}_3 \text{ (mL)} = 7 \times \text{weight of ash (g)}$$

7. Allow the precipitate to settle for 5 minutes; wash the solid residue with 40 mL of acetone three times and dry it under infrared light. Make sure the acid is added dropwise to prevent too strong a reaction.
8. Dissolve the precipitate in a minimum volume of 5% HNO$_3$ (3 to 6 mL); add 50 mL concentrated HNO$_3$ to precipitate strontium nitrate. After the solid settles, decant the supernatant solution, wash the residue with acetone, and dry it under infrared light. Steps 7 and 8 are primarily to separate Ca and Sr, from the differences in the solubility of Sr(NO$_3$)$_2$ and Ca(NO$_3$)$_2$.
9. Dissolve the Sr(NO$_3$)$_2$ with water. Add 1 mL of Fe^{3+} (10 mg Fe in the form of FeCl$_3$) and adjust the pH to 9 by adding NH$_4$OH. Centrifuge the solution and discard the precipitate.
10. Transfer the solution into a counting tube and count by gamma spectrometry in a HP-Ge well detector. Count a standard solution of ^{85}Sr with the same geometry and determine recovery of ^{85}Sr. Since recovery is only a relative calculation, the decay constant for ^{85}Sr does not need to be taken into account.
11. Set aside the sample for at least 14 days for the ^{90}Y to reach secular equilibrium with ^{90}Sr.

13.3.4.2.1   First Milking of ^{90}Y

12. Add 10 mg of Y^{3+} and 3 g NH$_4$Cl. Warm the solution and precipitate the hydroxides at pH 9 with concentrated NH$_4$OH. Filter it through a Millipore filter (0.45 μm, 47 mm diameter). Acidify and save the filtrate for future separation of Sr and Y. Record date and time. This is the starting time for ^{90}Y decay correction.
13. Dissolve the precipitate on the filter with concentrated HCl in the same beaker (because yttrium hydroxide sticks very well on any glass wall). Add 10 mg of stable Sr and wash the yttrium hydroxide.

14. Dissolve the precipitate again in the same way, add 1.2 g of oxalic acid, heat until dissolution of the acid, and precipitate yttrium oxalate with NH_4OH at pH 1.5. Maintain the pH at <2 even while rising with ammonium oxalate.

15. Filter the precipitate on preweighed filter paper for oxalate precipitate (Machery-Nagel 640 m, diameter 22 mm). Rinse the beaker then the precipitate with 0.5% ammonium oxalate with pH < 2 and discard the filtrate. Dry the filtered precipitate for 15 minutes under infrared heater and weigh it again. Calculate the Y recovery. For 10 mg Y, 34 mg yttrium oxalate will give 100% chemical yield.

16. Prepare the dry filtered precipitate for beta counting, put it in the counter, and record the time when count acquisition starts. Count for 200 minutes during a 24-hour interval. Calculate the ^{90}Sr concentration for the sample.

13.3.4.2.2   Second Milking of ^{90}Y

17. Two weeks later, the procedures from steps 10 to 15 should be repeated and counting of ^{90}Y is done after equilibrium with ^{90}Sr is reached. The second milking is made in order to check the first result for ^{90}Sr.

18. The appropriate correction for the decay, chemical yield, and self-absorption needs to be applied.

13.3.4.2.3   Quality Assurance

Adding a known amount of ^{90}Sr (certified value) to a standard seawater sample (20 L), and following the chemical procedures and comparing the activities to the certified value will ensure the accuracy of the methodology and counting procedures.

13.3.4.2.4   Common Problems and Troubleshooting

The appropriate corrections for the in-growth of ^{90}Y, decay of ^{90}Sr, self-absorption correction in the beta counter, and monitoring the background and counting efficiency of the beta counter are areas where one has to pay special attention.

13.3.4.3   Pu and Am

In the aquatic environment, the most commonly occurring Pu isotopes are ^{238}Pu (87.7 years), ^{239}Pu (2.41×10^4 years), ^{240}Pu (6571 years), and ^{241}Pu (14.4 years). Of these, the first three are alpha emitters and the last one is a beta emitter. Traditionally, alpha spectrometry was employed extensively to measure ^{238}Pu, ^{239}Pu, and ^{240}Pu. Since the alpha energies of ^{239}Pu (5.106–5.157 MeV) and ^{240}Pu (5.123–5.168 MeV) are close and the alpha spectrometric techniques cannot resolve these two isotopes, their combined concentrations are usually reported as $^{239,240}Pu$. A number of high-precision mass spectrometric techniques have been developed that include TIMS (Perrin et al., 1985), resonance ionization mass spectrometry (RIMS; Ruster et al., 1989), AMS (Fifield et al., 1996), and ICP-MS (Kershaw et al., 1995; Eroglu et al., 1998; Taylor et al., 2001; Kenna, 2002; Ketterer et al., 2002; Kim et al., 2002; reviewed in Ketterer and Szechenyi, 2008). Measurements of ^{238}Pu still require alpha spectrometry, mainly for two reasons: (1) the atom abundance of ^{238}Pu is low (due to low activity and short half-life), and hence its detection is difficult, and (2) isobaric interference from ^{238}U when ICP-MS is used. The methodology described here involves measurements of ^{238}Pu by alpha spectrometry followed by ^{239}Pu and ^{240}Pu measurements by mass spectrometry.

Eighteen isotopes of americium have been characterized, with the most stable being ^{243}Am with a half-life of 7,370 years, and ^{241}Am with a half-life of 432.2 years. All of the remaining radioactive isotopes have half-lives that are less than 51 hours, and the majority of these have half-lives that are less than 100 minutes. This element also has eight meta-states, with the most stable being ^{242m}Am ($t_{1/2}$ = 141 years). ^{241}Am and ^{243}Am emit alpha particles at 5.640 MeV and

5.439 MeV, respectively. ^{241}Am also emits gamma rays at 59.54 KeV (yield 35.7%) and 26.35 keV (yield 2.4%). ^{243}Am emits gamma rays at 43.54 keV (yield 5.1%) and 74.67 keV (yield 60.0%). Therefore, both alpha and gamma spectrometers can be used to detect ^{241}Am and ^{243}Am in environmental samples when the appropriate preconcentration methods are applied. Analysis of seawater for Pu and Am (both particulate and dissolved) requires that large-volume water samples be filtered through either a prefiltration cartridge (such as 0.5 μm pore size [nominal] polypropylene) or a cellulose nitrate membrane filter (0.45 μm). The filtrate is collected in a 100 L polyethylene or glass container and acidified with concentrated HCl (5 mL L^{-1} sample water). The particulate matter retained on the cartridge/membrane filter can be processed (ashing the cartridge filter after taking out the inner core or leaching the membrane filter with concentrated HCl and HNO$_3$) further (e.g., Krishnaswami and Sarin, 1976; Figure 13.2). Once the filter material is brought into solution (after the addition of ^{242}Pu and ^{243}Am as yield tracers), it is converted into 8 M HNO$_3$ medium. The solution is then made basic with concentrated NH$_4$OH to precipitate the hydroxides. Separate the hydroxides by centrifugation and discard the supernatant. The hydroxides are dissolved with a minimum amount of concentrated HNO$_3$, the sample solution transferred to a 50 mL C-tube, and the volume diluted to 15 mL with ultrapure water; 2.5 mL of concentrated HNO$_3$ and 5 mL of 2 M Al(NO$_3$)$_3$ are then added. The Al(NO$_3$)$_3$ is added as a salting agent to improve the column efficiency in the Am purification steps. TRU specification or a combination of Eichrom UTEVA and TRU double-column is widely used for the purification step to separate U (La Rosa et al., 2001; Pilvio and Bickel, 2000; Michel et al., 2008). The step-by-step procedure for the dissolved Pu and Am analysis is given below (from La Rosa et al., 2001).

1. Shipboard, acidify ~100 L of seawater sample (5 mL concentrated HCl L^{-1} of seawater), and add known amount of tracers of ^{242}Pu and ^{243}Am.

2. Add KMnO$_4$ in excess to oxidize organic matter and leave the solution for 1 hour.

3. Add NaOH to make the solution basic to a pH of 9; add 150 mL of 0.4 M MnCl$_2$ solution. A brown hydrated MnO$_2$ coprecipitates Pu and Am with bulky manganese dioxide. Maintain a pH of 9.

4. After ~6 hours of settling, decant the supernatant to a volume of ~20 L. Transfer the residual solution into a 20 L bucket.

5. Separate the residue and solution by subsequent decanting, centrifugation, and filtration. Dissolve the MnO$_2$ (containing Pu, Am) precipitate (ppt) in 2 M HCl with excess NH$_2$OH·HCl (at least 10 mL of 10% solution) to reduce it to Mn(II). Add 50 mg of Fe(III) and 40 mL of 10% hydroxylamine hydrochloride (NH$_2$OH·HCl) solution followed by heating to reduce Fe(III) to Fe(II). The Fe(II) in turn rapidly reduces Pu species to Pu(III).

6. Add NaNO$_2$ excess amount (about 3 g) to oxidize Fe(II) to Fe(III) and Pu(III) to Pu(IV).

7. Add ammonium hydroxide to coprecipitate Pu(IV) and Am (III) along with Fe(OH)$_2$ at pH 8~9. Boil the precipitate to enhance the flocculation of the precipitate, and adjust the pH to 6~7 with 1 M HCl addition to prevent the postprecipitation of MnO$_2$ due to air oxidation of Mn(II) in basic solution. This step enables the bulk of the Mn to be kept in solution. Decant the supernatant and separate the precipitate and the solution by siphoning off the supernatant solution followed by centrifugation.

8. Convert the iron hydroxide to nitrate by adding 2.5 M HNO$_3$ and 0.5 M Al(NO$_3$)$_3$.

9. Add 2 mL of 0.6 M ferrous sulfamate solution per 10 mL of sample load and ascorbic acid (over 500 mg) to reduce Fe(III) to Fe(II).

10. Pass the solution through tandem UTEVA + TRU columns (2 mL, prepacked) with washes to collect Pu(III) and Am (III) together on the TRU column.

11. Elute the column with 4 M HCl first to obtain Am and then with 0.1 M NH$_4$HC$_2$O$_4$ to obtain Pu.

12. Process the Am and Pu fractions for electroplating onto stainless steel discs to be measured by alpha spectrometry with silicon detectors, as given below.

13.3.4.3.1   Electrodeposition of Pu and Am

1. Assemble the plating cell as shown in Figure 13.3.
2. Fill the cell with water to test for leakage.
3. Add 1 mL of concentrated sulfuric acid to the sample from step 11 above.
4. Heat the sample on a hot plate until copious white fumes evolve.
5. Cool the sample to room temperature; carefully rinse the wall of the beaker with 2 mL of 1 N H_2SO_4. Add two drops of 0.1% methyl red indicator.
6. Transfer the sample to the electroplating cell.
7. Rinse the beaker with 2 mL 1 N H_2SO_4. Add the rinse to the plating cell.
8. Repeat previous step 7 three times.
9. Add concentrated NH_4OH dropwise until the color of the sample changes from red (pH 4.4) to yellow (pH 6.2). Mix the solution by swirling the plating cell.
10. Add 0.5 M H_2SO_4 dropwise to the red endpoint (pH 4.4) and then add two drops excess.
11. Complete the assembly of the electroplating cell by attaching the platinum anode with plastic insulation tape as shown in Figure 13.3. Position the platinum wire anode at about 0.5 cm from the stainless steel disc (cathode).
12. Connect the anode and cathode of the electroplating cell to a constant current power supply.
13. Electroplate at 1.0 amp for 60–70 minutes.
14. Before turning off the power supply when electrodeposition is completed, add 1 mL of concentrated NH_4OH to the cell and continue plating for about 1 minute, then turn off the power supply as quickly as possible.
15. Disconnect the cell from the power supply.
16. Discard the solution from the cell.
17. Rinse the cell with diluted NH_4OH from a wash bottle. (Make a diluted NH_4OH solution by adding 0.5 mL concentrated NH_4OH to 500 mL of ultrapure water.)
18. Disassemble the plating cell.
19. Rinse the plated disc with diluted NH_4OH.
20. Rinse the plated disc with acetone and let the disc air-dry on a clean paper tissue.
21. Count the plated disc and determine the activity of the plutonium isotopes by alpha spectrometry.

13.3.4.3.2   Am Measurements by Gamma Spectrometry

Americium measurements can be conducted using a high-resolution gamma ray spectrometer. Radioisotopes present in the suspended material phases are isolated using physical filtration methods, while those in the dissolved phase are extracted using commercially available chemical composite sorbent materials (CUS and ZUS from Eksorb-Chernobyl Company). Seawater is pumped aboard, from hundreds to thousands of liters, through a high-flow industrial filter unit containing a high-capacity, dual-gradient, spun-polypropylene cartridge (Ametek Model DGD-2501). This stage effectively (better than 99%) sequesters suspended material and biomass aggregates down to 1 μm size, which is then measured by gamma spectrometry (Bandong et al., 2001). Extraction and recovery efficiencies are better than 90% for [241]Am.

13.3.4.3.3   Quality Assurance

Adding a known amount of [243]Am or [241]Am and [242]Pu (certified value) to a standard seawater (20 L) sample and following the chemical procedures and comparing the activities to the certified values will ensure the accuracy of the methodology and counting procedures.

13.3.4.3.4   Common Problems and Trouble Shooting

Occasionally there is a large amount of gelatinous silica that accompanies the iron hydroxide. The silica is removed by evaporation with concentrated HF in a PTFE beaker, and residual fluoride is complexed with boric acid after several evaporations with concentrated HNO_3 (La Rosa et al., 2001). Furthermore, self-absorption corrections may be needed for the gamma ray spectrometric method. To reduce environmental background radiation, 4 in. thick high-performance, low-background shielding is used to contain the detector and sample during gamma counting. Detector (counting) efficiency calibration is performed in precisely the same geometrical relationship to the calibrating sources as is used in the measurements on samples of unknown activities. A mixed gamma standard solution (Amersham QCY.48) is used to produce a 12-point energy efficiency calibration of the detector over the 50–3,000 keV regions. In the case of sorbent materials, 150 g of the sorbent is spiked with the standard solution. Shake the mixture well and then quantitatively transfer into a 1 L Marinelli beaker for detector calibration.

13.3.4.4   Determination of ^{129}I

The natural environmental abundances of ^{129}I have been significantly modified by man due to atmospheric nuclear weapon tests and, more recently, emissions from nuclear reprocessing plants (Szidat et al., 2000). ^{129}I has been used as a tracer for water mass mixing in oceans and groundwater, as well as biogeochemical processes (Wong, 1991; Santschi et al., 1996; Moran et al., 1999a, 1999b, 2002; Hou et al., 2001; Santschi and Schwehr, 2004; Schwehr et al., 2005a, 2005b). Radiochemical neutron activation analysis (RNAA) and AMS are used to determine ^{129}I in environmental materials. However, AMS analysis is widely used for seawater samples due to its high sensitivity. Therefore, a method for water sample preparation for AMS measurement is given here. The first high-precision ^{129}I measurement was conducted using tandem-accelerated mass spectrometry (Elmore et al., 1980). Szidt et al. (2000) have recently provided in-depth experience on the sample preparation procedures for AMS measurements of ^{129}I and ^{127}I in environmental materials and some methodological aspects of quality assurance. Dissolved organic iodine (DOI) can make up a major portion of iodine in surface waters, and the ^{129}I/^{127}I ratio can be used as a tracer for the presence of terrestrial organic matter in estuaries (Oktay et al., 2001; Wong and Cheng, 2001; Schwehr and Santschi, 2003; Schwehr et al., 2005a). The determination of this ratio requires the determination of either both isotopes in DOI or the speciated measurement of both isotopes.

13.3.4.4.1   Measurement of the [^{129}I] Concentration in DOI

1. ^{129}I$^-$ and ^{129}IO$_3^-$ are separated by anion chromatography (Schwehr et al., 2005a). The remaining sample portions for both isotopes for IO$_3^-$ and bulk DOI species are frozen in volumes larger than 250 mL, preferably in 1 L containers for ease of processing and storage.
2. Both isotopes are analyzed in individual fractions of I$^-$, IO$_3^-$, and total iodine (TI). Samples to be analyzed are stored refrigerated or in a cool, dark room in containers sealed with parafilm, after the addition of a reducing solution (RS). The RS consists of 2 mL of 1 M HSO$_3$ + 4 mL 1 M NH$_2$OH•HCl, adjusted to pH 6.5, per 1 L sample, to prevent loss of volatile iodine species.
3. The procedure used for measuring the concentration of the different ^{129}I species is similar to that of ^{127}I (Schwehr and Santschi, 2003), except that it is scaled up to allow processing of larger-volume water samples (2–5 L). The low concentrations of ^{129}I in environmental samples (often ~10^7 atoms L^{-1}) require that measurements be made by an AMS on 1 to 2 mg of AgI.
4. Anion chromatography is used to separate and concentrate ^{129}I-I$^-$ and ^{129}I-IO$_3^-$. Both isotopes of I$^-$ are obtained by passing the filtered sample through a column of strong anion exchange resin, AG1-X4 or AG1-X8 (100–200 mesh).
5. I$^-$ is retained on the resin and IO$_3^-$, and most DOI fractions are passed into the effluent and saved. The ^{127}I$^-$ concentration in the eluant is monitored as a yield tracer for breakthrough and recovery calculations using HPLC or ICP-MS.

6. To obtain IO_3^-, RS is added to the sample effluent and the first deionized water (DIW) rinse to reduce IO_3^- species to I^-, and the solution is allowed several hours to react to completion. The rinsing and elution processes are the same as those for the I^- fraction.

7. The concentration of ^{129}I-DOI is obtained as the difference of $[^{129}I$-TI$] - ([^{129}I$-$I^-] + [^{129}I$-$IO_3^-])$. The ^{129}I-TI fraction is determined after dehydrohalogenation, which degrades alkyl and aryl halides. Dehydrohalogenation is accomplished by adding NaOH (4 g per L sample) and ethanol (200 mL per L sample), ultrasonicating at 65°C for 3 hours, then allowing it to react overnight.

8. After repeating these steps, the sample is reduced with RS and acidified, ultrasonicated at 45°C for 1.5 hours, and cooled in a freezer.

9. Next, the sample is gently warmed to room temperature, uncapped, and placed in a boiling water bath for 2 to 3 hours to drive off the ethanol.

10. After cooling, the sample is then passed through Empore C_{18} discs (comparable to the SPE for the small volumes used for ^{127}I) under vacuum to remove any remaining refractory substances, and then processed through a resin column as described for I^-.

11. The Empore C_{18} discs are conditioned prior to use by soaking in methanol overnight and rinsing with ultrapure water.

12. The I^-, IO_3^-, and TI (for DOI by difference) fractions are then processed by liquid-liquid extraction by adding a ^{129}I-free carrier solution of $^{127}I^-$, oxidizing and acidifying the solution with H_2O_2 to transform all iodine species to I_2, then shaking in a separatory funnel with $CHCl_3$ so that all I_2 would partition into the organic $CHCl_3$ phase.

13. This purified I_2 fraction is then back-extracted into a solution of $NaHSO_3$ and H_2SO_4, and precipitated into a pellet of AgI. The AgI pellet is cleaned, dried, and mixed with a pure Ag powder in preparation for measurement of the $^{129}I/^{127}I$ ratio in the iodine species of I^-, IO_3^-, and TI (DOI by difference), by a AMS facility.

13.3.4.4.2   Quality Assurance and Error Calculation for ^{129}I

Analytical accuracy of ^{129}I can be tested against NIST-certified reference materials, SRM 3230 low-level $^{129}I/^{127}I$, and SRM 3231. Sediments can be tested against IAEA SRM 375. Analytical accuracy of stable ^{127}I can be assessed (1) against a certified reference material, SRM 1549, powdered milk (NIST), (3) through the method of standard additions, and (3) against independent measurements made through voltammetry or ICP-MS on 5% of randomly selected samples. The column efficiency of the above method is 99% ± 3% for the recovery of ^{129}I-I^- ($n = 6$) and ^{129}I-IO_3^- ($n = 4$). The dehydrohalogenation efficiencies for ^{129}I-TI are ≥97% ± 3% ($n = 4$). The liquid-liquid extraction recovery is 95% ± 5% ($n = 5$). The counting error for $^{129}I/^{127}I$ by AMS generally averages ±<1% ($n = 18$). Therefore, the final error for ^{129}I-I^- and ^{129}I-IO_3^- after processing through the column, liquid-liquid extraction, and AMS averages ±12% ($n = 4$). The counting error by AMS for ^{129}I-TI averages 6% ($n = 4$). Since ^{129}I-DOI is obtained by difference where $[^{129}I$-DOI$] = [^{129}I$-TI$] - \{[^{129}I$-$I^- + ^{129}I$-$IO_3^-]\}$, the associated propagated error for ^{129}I-DOI ranges from 14% to 25%, and averages 21% ($n = 4$). The highest error is associated with AMS counting of the ^{129}I-I^- fraction. Since this error is associated with the lowest concentration distributions for the measured ^{129}I fractions, the error is lower for higher concentrations or when concentrating a larger volume of water sample.

13.3.5   STANDARD REFERENCE MATERIALS (SRMs) FOR RADIONUCLIDES ANALYSIS

Radionuclide measurement needs in marine studies have become increasingly demanding and are pushing the limits of detection for current practice on the sampling and subsequent handling and instrument performance. A radionuclide measurement methodology for the marine samples must be unbiased to be capable of producing accurate values, with a high precision. A convenient way to address the problems associated with obtaining high precision and accuracy is to utilize standard reference materials (SRMs). These SRMs are utilized not only for the calibration of counting instruments, but also for the method development. A comprehensive discussion on the statistical methods used with the calibration of equipment and the development of SRMs is found in Taylor (1993). Two organizations, the National Institute of Standards and Technology (NIST) in the United States and Marine Environment Laboratory of International Atomic Energy Agency (IAEA-MEL) in Monaco, have been producing commercially available SRMs for marine radioactivity measurements. Both organizations continue to produce the standard and coordinate intercomparison and performance evaluation of SRMs for low-level and environmental mass spectrometry and atom counting (Inn et al., 2001a; Povinec and Pham, 2001). The presently available SRMs are listed in Table 13.3. As the stocks are produced in a limited quantity and new SRMs are constantly produced, it is useful to visit the individual websites of the AQCS Reference Materials division of IAEA-MEL (http://www-naweb.iaea.org/naml/aqcsmaterials.asp) and NIST (http://ts.nist.gov/measurementservices/referencematerials/index.cfm). IAEA-437 (mussel, Mediterranean Sea), IAEA-415 (fish, North Atlantic Ocean), IAEA-418 (water, Mediterranean Sea, ^{129}I), IAEA-412 (sediment, Pacific Ocean), and NIST SRM 4358 (mixed oyster flesh from the southeastern Pacific Ocean blended with mussels from the White and Irish Seas) are under preparation for distribution.

13.4 SUMMARY AND PERSPECTIVES

Modern analytical techniques (such as MC-ICP-MS and other mass spectrometric techniques) have advanced this field of radionuclide analysis by allowing smaller sample size and much higher precision. While the radionuclides that can be measured using mass spectrometric techniques are increasing, still several of the short-lived radionuclides can only be measured by counting methods (such as ^{210}Po, ^{210}Pb, etc.). With significant improvements in some of the analytical methods (such as ^{231}Pa, ^{230}Th, ^{232}Th, ^{238}U, ^{234}U, etc.), the volume requirement for samples has decreased considerably in recent years. In those cases, there is a critical need for blanks and efficient filter materials. The chemical speciation of some of the radionuclides (such as Pu) and the extent of organic binding (such as ^{129}I) require that preservation and storage of water samples is well understood (Choppin, 2006). Temporal variations of the loss of particle-reactive radionuclides due to wall absorption need to be quantified. Although the fraction of radionuclides associated with particles is operationally defined, establishing a standard filter pore size and filter medium still remains a challenge for the chemical oceanographic community. Developing improved fiber or resin matrices on to which radionuclides can be extracted more efficiently and connecting those improved fibers (or resins) to in situ pumps (such as McLane) will help to obtain particulate and dissolved radionuclides simultaneously at multiple depths, thus reducing wire time. Furthermore, establishing a standard in situ pump method also helps to get consistent data across the global oceanographic community.

ACKNOWLEDGMENTS

The writing of this chapter was partially supported by grants from the National Oceanic Atmospheric Administration to MB, PE98120 to GHH, and from the MASINT Consortium (PNNL-NSD-2666) to PHS. We thank the two anonymous reviewers for their constructive comments.

REFERENCES

Andersen, M. B., C. H. Stirling, D. Porcelli, A. N. Halliday, P. S. Andersson and M. Baskaran. 2007. The tracing of riverine U in Arctic seawater with very precise U-234/U-238 measurements. *Earth and Planetary Science Letters* 259:171–85.

Anderson, R. F., and A. P. Fleer. 1982. Determination of natural actinides and plutonium in marine particulate material. *Analytical Chemistry* 54:1142–47.

Bacon, M. P., and R. F. Anderson. 1982. Distribution of thorium isotopes between dissolved and particulate forms in the deep sea. *Journal of Geophysical Research* 87:2045–56.

Ballestra, S., J. Gastaud, J. J. Lopez, P. Parsi, and D. Vas. 2000a. Reference Material IAEA-134. Radionuclides in cockle flesh. Radionuclides in marine fish flesh. Reference sheet, International Atomic Energy Agency, Analytical Quality Control Services, Vienna, Austria.

Ballestra, S., J. Gastaud, J. J. Lopez, P. Parsi, and D. Vas. 2000b. Reference Material IAEA-135. Radionuclides in Irish Sea sediment. Reference sheet, International Atomic Energy Agency, Analytical Quality Control Services, Vienna, Austria.

Ballestra, S., J. Gastaud, J. J. Lopez, P. Parsi, and D. Vas. 2000c. Reference material IAEA-300. Radionuclides in Baltic Sea sediment. Reference sheet, International Atomic Energy Agency, Analytical Quality Control Services, Vienna, Austria.

Ballestra, S., J. Gastaud, J. J. Lopez, P. Parsi, and D. Vas. 2000d. Reference material IAEA-315. Radionuclides in Pacific Ocean sediment. Reference sheet, International Atomic Energy Agency, Analytical Quality Control Services, Vienna, Austria.

Ballestra, S., D. Vas, J. J. Lopez, and V. Noshkin. 2000e. Reference Material IAEA-352. Radionuclides in Pacific Ocean sediment. Reference sheet, International Atomic Energy Agency, Analytical Quality Control Services, Vienna, Austria.

Ballestra, S., J. J. Lopez, J. Gastuad, P. Parsi, D. Vas, and V. Noshkin. 2000f. Reference Material IAEA-367. Radionuclides in Pacific Ocean sediment. Reference sheet, International Atomic Energy Agency, Analytical Quality Control Services, Vienna, Austria.

Ballestra, S., J. J. Lopez, J. Gastuad, P. Parsi, D. Vas, and V. Noshkin. 2000g. Reference Material IAEA-368. Radionuclides in Pacific Ocean sediment. Reference sheet, International Atomic Energy Agency, Analytical Quality Control Services, Vienna, Austria.

Bandong, B. B., A. M. Volpe, B. K. Esser, and G. M. Bianchini. 2001. Pre-concentration and measurement of low levels of gamma-ray emitting radioisotopes in coastal waters. *Applied Radiation and Isotopes* 55:653–65.

Baskaran, M., and P. H. Santschi. 1993. The role of particles and colloids in the transport of radionuclides in coastal environments of Texas. *Marine Chemistry* 43:95–114.

Baskaran, M., and P. H. Santschi. 2002. Particulate and dissolved ^{210}Pb activities in the shelf and slope regions of the Gulf of Mexico waters. *Continental Shelf Research* 22:1493–1510.

Baskaran, M., and P. W. Swarzenski. 2007. Seasonal variations on the residence times and partitioning of short-lived radionuclides (^{234}Th, ^{7}Be and ^{210}Pb) and depositional fluxes of ^{7}Be and ^{210}Pb in Tampa Bay, Florida. *Marine Chemistry* 104:27–42.

Baskaran, M., P. H. Santschi, G. Benoit, and B. D. Honeyman. 1992. Scavenging of Th isotopes by colloids in seawater of the Gulf of Mexico. *Geochimica et Cosmochimica Acta* 56:3375–88.

Baskaran, M., D. J. Murphy, P. H. Santschi, J. C. Orr, and D. R. Schink. 1993. A method for rapid in situ extraction of Th, Pb and Ra isotopes from large volumes of seawater. *Deep Sea Research I* 40:849–65.

Baskaran, M., P. H. Santschi, L. Guo, T. S. Bianchi, and C. Lambert. 1996. ^{234}Th-^{238}U disequilibria in the Gulf of Mexico: The importance of organic matter and particle concentration. *Continental Shelf Research* 16:353–80.

Baskaran, M., M. Ravichandran, and T. S. Bianchi. 1997. Cycling of ^{7}Be and ^{210}Pb in a high DOC, shallow, turbid estuary of southeast Texas. *Estuarine, Coastal and Shelf Science* 45:165–76.

Baskaran, M., P. W. Swarzenski, and D. Porcelli. 2003. Role of colloidal material in the removal of ^{234}Th in the Canada Basin of the Arctic Ocean. *Deep Sea Research I* 50:1353–73.

Baskaran, M., P. W. Swarzenski, and B. S. Biddanda. 2008. Retention of dissolved ^{234}Th on prefilters and MnO_2-coated filter cartridges and variability in the extraction efficiencies. *Geochemistry, Geophysics, Geosystems*, in review.

Benitez-Nelson, C. R., K. O. Buesseler, M. Rutgers van der Loeff, J. Andrews, L. Ball, G. Crossin, and M. A. Charette. 2001. Testing a new small-volume technique for determining Th-234 in seawater. *Journal of Radioanalytical and Nuclear Chemistry* 248:795–99.

Bentley, H. W., E. M. Phillips, S. N. Davis, M. A. Habermehl, P. L. Airey, G. E. Calf, D. Elmore, H. E. Gove, and T. Torgersen. 1986. Chloride-36 dating of very old ground waters. I. The Great Artesian Basin, Australia. *Water Resources Research* 22:1991–2002.

Bhat, S. G., S. Krishnaswami, D. Lal, Rama, and W. S. Moore. 1969. ^{234}Th/^{238}U ratios in the ocean. *Earth and Planetary Science Letters* 5:483–91.

Bourquin, M., P. van Beek, J.-L. Reyss, M. Souhaut, M. A. Charette, and C. Jeandel. 2008. Comparison of techniques for pre-concentrating radium from seawater. *Marine Chemistry* 109:226–37.

Bowen, V. T. 1970. Analyses of sea-water for strontium and strontium-90. In *Reference methods for marine radioactivity studies*. Technical Report Series 118, International Atomic Energy Agency, Vienna, Austria.

Broecker, W. S. 1965. An application of natural radon to problems in oceanic circulations. In *Symposium on diffusion in the oceans and freshwaters*, ed. T. Ichiye. New York: Lamont Geological Observatory.

Broecker, W. S., W. C. Patzert, J. R. Toggweiler, and M. Stuiver. 1986. Hydrography, chemistry and radioisotopes in the Southeast-Asian Basins. *Journal of Geophysical Research* 91:14345–54.

Buesseler, K. O., S. A. Casso, M. C. Hartman, and H. D. Livingston. 1990. Determination of fission-products and actinides in the Black Sea following the Chernobyl accident. *Journal of Radioanalytical and Nuclear Chemistry* 138:33–47.

Buesseler, K. O., J. K. Cochran, M. P. Bacon, H. O. Livingston, S. A. Casso, D. Hirschberg, M. C. Hartman, and A. P. Fleer. 1992. Determination of thorium isotopes in seawater by non-destructive and radiochemical procedures. *Deep Sea Research* 39:1103–14.

Buesseler, K. O., J. E. Bauer, R. F. Chen, T. I. Eglinton, O. Gustafsson, W. Landing, K. Mopper, S. B. Moran, P. H. Santschi, and R. Vernon-Clark. 1996. An intercomparison of cross-filtration techniques used for sampling marine colloids: Overview and organic carbon results. *Marine Chemistry* 55:1–31.

Buesseler, K. O., C. R. Benitez-Nelson, M. Rutgers van der Loeff, J. Andrews, L. Ball, G. Crossin, and M. A. Charette. 2001. An intercomparison of small- and large-volume techniques for thorium-234 in seawater. *Marine Chemistry* 74:15–28.

Burnett, W. C., D. R. Corbett, M. Schultz, E. P. Horwitz, R. Chiarizia, M. Dietz, A. Thakkar, and M. Fern. 1997. Preconcentration of actinide elements from soils and large volume water samples using extraction chromatography. *Journal of Radioanalytical and Nuclear Chemistry* 226:121–27.

Cai, P. H., M. H. Dai, D. W. Lv, and W. F. Chen. 2006. How accurate are ^{234}Th measurements in seawater based on the MnO$_2$-impregnated cartridge technique? *Geochemistry, Geophysics, Geosystems* 7: Q03020, doi: 10.1029/2005GC001104.

Cai, P. H., M. H. Dai, D. W. Lv, and W. F. Chen. 2008. Reply to comment by Cin-Chang Hung et al. on "How accurate are ^{234}Th measurements in seawater based on the MnO$_2$-impregnated cartridge technique?" *Geochemistry, Geophysics, Geosystems* 9: Q02010, doi: 10.1029/2007GC001837.

Carlson, D. J., M. L. Brann, T. H. Mague, and L. M. Mayer. 1985. Molecular weight distribution of dissolved organic materials in seawater determined by ultrafiltration, a re-examination. *Marine Chemistry* 16:155–71.

Chen, J. H., R. L. Edwards, and G. J. Wasserburg. 1986. U-238, U-234 and Th-232 in seawater. *Earth and Planetary Science Letters* 80:241–51.

Chen, J. H., R. L. Edwards, and G. J. Wasserburg. 1992. Mass spectrometry and application to uranium-series disequilibrium. In *Uranium series disequilibrium: Application to environmental problems*, ed. M. Ivanovich and R. S. Harmon. Oxford: Clarendon Press, pp. 174–206.

Chen, Q. J., A. Aarkrog, S. P. Nielsen, H. Dahlgaard, B. Lind, A. K. Kolstad, and Y. X. Yu. 2001. *Procedures for determination of ^{239,240}Pu, ^{241}Am, ^{237}Np, ^{234,238}U, ^{228,230,232}Th, ^{99}Tc, and ^{210}Pb-^{210}Po in environmental materials*. Risø-R-1263(EN) Report by the Norwegian Radiation Protection Authority, Risø National Laboratory, Roskilde, Norway. Pitney Bowes Management Services Denmark A/S.

Chernyaev A., I. Gaponov, and A. Kazennov. 2004. Direct methods for radionuclides measurement in water environment. *Journal of Environmental Radioactivity* 72:187–94.

Choi, M. S., R. Francois, K. Sims, M. P. Bacon, S. Brown-Leger, A. P. Fleer, L. Ball, D. Schneider, and S. Pichat. 2001. Rapid determination of ^{230}Th and ^{231}Pa in seawater by desolvated micronebulization inductively coupled plasma magnetic sector mass spectrometry. *Marine Chemistry* 76:99–112.

Choppin, G. R. 2006. Actinide speciation in aquatic systems. *Marine Chemistry* 99:83–92.

Church, T. M., N. Hussain, T. G. Ferdelman, and S. W. Fowler. 1994. An efficient quantitative technique for the simultaneous analyses of radon daughters Pb-210, Bi-210 and Po-210. *Talanta* 41:243–49.

Coale, K. H., and K. W. Bruland. 1985. ^{234}Th: ^{238}U disequilibria within the California current. *Limnology and Oceanography* 30:22–33.

Cochran, J. K., D. J. Hirschberg, H. D. Livingstone, K. O. Buesseler, and R. M. Key. 1995. Natural and anthropogenic radionuclide distributions in the Nansen Basin, Arctic Ocean: Scavenging rates and circulation time scales. *Deep Sea Research II* 42:1495–517.

Cochran, J. K., H. D. Livingston, D. J. Hirschberg, and L. D. Surprenant. 1987. Natural and anthropogenic radionuclide distributions in the Northwest Atlantic. *Earth and Planetary Science Letters* 84:135–52.

Cochran, J. K., and P. Masque. 2003. Short-lived U/Th series radionuclides in the ocean: Tracers for scavenging rates, export fluxes and particle dynamics. *Reviews in Mineralogy and Geochemistry* 52:461–92.

Cohen, A. S., and R. K. O'Nions. 1991. Precise determination of femtogram quantities by radium by thermal ionization mass-spectrometry. *Analytical Chemistry* 63:2705–8.

Croudace, I. W., P. E. Warwick, and R. C. Greenwood. 2006. A novel approach for the rapid decomposition of Actinide™ resin and its application to measurement of uranium and plutonium in natural waters. *Analytica Chimica Acta* 577:111–18.

de Mora, S. J., and R. M. Harrison. 1983. The use of physical separation techniques in trace metal speciation studies. *Water Research* 17:723–33.

Dickson, B. L. 1985. Radium isotopes in saline seepages, south-western Yilgarn Western Australia. *Geochimica et Cosmochimica Acta* 49:361–68.

Dimova, N., W. C. Burnett, E. P. Horwitz, and D. Lane-Smith. 2007. Automated measurement of Ra-224 and Ra-226 in water. *Applied Radiation and Isotopes* 65:428–34.

Doucet, F. J., J. R. Lead, and P. H. Santschi. 2007. Colloid-trace element interactions in aquatic systems. In *Environmental colloids and particles*, ed. K. J. Wilkonson and J. R. Lead. New York: John Wiley & Sons, pp. 95–117.

Dulaiova, H., and W. C. Burnett. 2004. An efficient method for gamma-spectrometric determination of radium-226,228 via manganese fibers. *Limnology and Oceanography: Methods* 2:256–61.

Dulaiova, H., R. Peterson, W. C. Burnett, and D. Lane-Smith. 2005. A multi-detector continuous monitor for assessment of ^{222}Rn in the coastal ocean. *Journal of Radioanalytical and Nuclear Chemistry* 263:361–65.

Eikenberg, J., A. Tricca, G. Vezzu, S. Bajo, M. Ruethi, and H. Surbeck. 2001. Determination of ^{228}Ra, ^{226}Ra and ^{224}Ra in natural water via adsorption on MnO$_2$-coated discs. *Journal of Environmental Radioactivity* 54:109–31.

Elmore, D., B. R. Fulton, M. R. Clover, J. R. Marsden, H. E. Gove, H. Naylor, K. H. Purser, L. R. Kilius, R. P Beukens, and E. Litherland. 1979. Analysis of chloride-36 in environmental water samples using an electrostatic accelerator. *Nature* 277:22–25.

Elmore, D., H. E. Gove, R. Ferraro, L. R. Kilius, H. W. Lee, K. H. Chang, R. P. Beukens, A. E. Litherland, C. J. Russo, K. H. Purser, M. T. Murrell, and R. C. Finkel. 1980. Determination of I-129 using tandem accelerator mass-spectrometry. *Nature* 286:138–40.

Eroglu, A. E., C. W. McLeod, K. S. Leonard, and D. McCubbin. 1998. Determination of plutonium in seawater using co-precipitation and inductively coupled plasma mass spectrometry with ultrasonic nebulisation. *Spectrochimica Acta* 53B:1221–33.

Fifield, L. K., R. G. Cresswell, M. L. di Tada, T. R. Ophel, J. P. Day, A. P. Clacher, S. J. King, and N. D. Priest. 1996. Accelerator mass spectrometry of plutonium isotopes. *Nuclear Instruments and Methods in Physics Research* 117B:295–303.

Firestone, R. B., C. M. Baglin, and S. Y. F. Chu. 1999. *Table of Isotopes*. 8th ed. New York: John Wiley & Sons.

Fleer, A. P., and M. P. Bacon. 1984. Determination of Pb-210 and Po-210 in seawater and marine particulate matter. *Nuclear Instruments and Methods in Physics Research* 223A:243–49.

Flynn, W. W. 1968. The determination of low levels of ^{210}Po in environmental samples. *Analytica Chimica Acta* 43:221–27.

Folsom, T. R., and C. Sreekumaran. 1970. Some reference methods for determining radioactive and natural cesium for marine studies. In *Reference methods for marine radioactivity studies*, ed. Y. Nishiwaki and R. Fukai. Report 118, International Atomic Energy Agency, pp. 202–210.

Friedrich, J., and M. M. Rutgers van der Loeff. 2002. A two-tracer (^{210}Po-^{234}Th) approach to distinguish organic carbon and biogenic silica export flux in the Antarctic Circumpolar Current. *Deep Sea Research I* 49:101–20.

Fujiwara, A., Y. Kameo, A. Hoshi, T. Haraga, and M. Nakashima. 2007. Application of extraction chromatography to the separation of thorium and uranium dissolved in a solution of high salt concentration. *Journal of Chromatography* 1140A:163–67.

Galindo-Uribarri, A., J. R. Beene, M. Danchev, J. Doupé, B. Fuentes, J. Gomez del Campo, P. A. Hausladen, R. C. Juras, J. F. Liang, A. E. Litherland, Y. Liu, M. J. Meigs, G. D. Mills, P. E. Mueller, E. Padilla-Rodal, J. Pavan, J. W. Sinclair, and D. W. Stracener. 2007. Pushing the limits of accelerator mass spectrometry. *Nuclear Instruments and Methods in Physics Research* 259B:123–30.

Geibert, W., and I. Völge. 2008. Progress in the determination of ^{227}Ac in seawater. *Marine Chemistry* 109:238–49.

GEOTRACES Science Plan. 2006. http://geotraces.org/. Accessed May 12, 2008.

Ghaleb, B., E. Pons-Branchu, and P. Deschamps. 2004. Improved method for radium extraction from environmental samples and its analysis by thermal ionization mass spectrometry. *Journal of Analytical Atomic Spectrometry* 19:906–10.

Goldstein, S. J., and C. H. Stirling. 2003. Techniques for measuring uranium-series nuclides: 1992–2002. Uranium-Series Geochemistry. *Reviews in Mineralogy and Geochemistry* 52:23–57.

Gray, A. 1985. The ICP as an ion source—Origins, achievements and prospects. *Spectrochimica Acta* 40B:1525–37.

Groning, M., C. B. Taylor, G. Winckler, R. Auer, and H. Tatzbar. 2001. Sixth IAEA intercomparison of low-level tritium measurements in water (TRIC2000). Isotope Hydrology Laboratory, International Atomic Energy Agency.

Guo, L., and P. H. Sanstchi. 2007. Ultrafiltration and its applications to sampling and characterization of aquatic colloids. In *Environmental colloids and particles*, ed. K. J. Wilkonson and J. R. Lead. New York: John Wiley & Sons.

Guo, L., P. H. Santschi, M. Baskaran, and A. Zindler. 1995. Distribution of dissolved and particulate ^{230}Th and ^{232}Th in seawater from the Gulf of Mexico and off Cape Hatteras, as measured by SIMS. *Earth and Planetary Science Letters* 133:117–28.

Guo, L., P. H. Santschi, and M. Baskaran. 1997. Interactions of thorium isotopes with colloidal organic matter in oceanic environments. *Colloids and Surfaces* 120A:255–71.

Guo, L., C. C. Hung, P. H. Santschi, and I. D. Walsh. 2002. ^{234}Th scavenging and its relationship to acid polysaccharide abundance in the Gulf of Mexico. *Marine Chemistry* 78:103–19.

Hartman, M. C., and K. O. Buesseler. 1994. Adsorbers for in-situ collection and at-sea gamma analysis of dissolved thorium-234 in seawater. Technical Report WHOI-94-15, Woods Hole Oceanography Institute, Woods Hole, MA.

Hirose, K., M. Aoyama, Y. Igarashi, and K. Komura. 2005. Extremely low background measurements of ^{137}Cs in seawater samples using an underground facility (Ogoya). *Journal of Radioanalytical and Nuclear Chemistry* 263:349–53.

Ho, C. K., A. Robinson, D. R. Miller, and M. J. Davis. 2005. Overview of sensors and needs for environmental monitoring. *Sensors* 5:4–37.

Hong, G.-H. 2008. Radioactivity measurements in KORDI. Marine isotope measurements, March 2008, a compilation of chemical procedures.

Hong, G.-H., S.-K. Park, M. Baskaran, S.-H. Kim, C.-S. Chung, and S.-H. Lee. 1999. Lead-210 and polonium-210 in the winter well-mixed turbid waters in the mouth of the Yellow Sea. *Continental Shelf Research* 19:1049–64.

Horwitz, E. P. 1998. Actinides, Sr in mixed samples using TEVA resin, UTEVA resin, TRU resin, actinide resin, Sr. Paper presented at International Workshop on the Application of Extraction Chromatography in Radionuclide Measurement, IRMM, Geel, Belgium.

Horwitz, E. P., R. Chiarizia, and M. L. Dietz. 1997. DIPEX, a new extraction chromatographic material for the separation and preconcentration of actinides from aqueous solution. *Reactive and Functional Polymers* 33:25–36.

Horwitz, E. P., M. L. Dietz, R. Chiarizia, H. Diamond, S. L. Maxwell III, and M. Nelson. 1995. Separation and preconcentration of actinides by extraction chromatography using a supported liquid anion exchanger: Application to the characterization of high-level nuclear waste solutions. *Analytica Chimica Acta* 310:63–78.

Hou, X., H. Dahlgaard, and S. P. Nielsen. 2001. Chemical speciation analysis of I in seawater and a preliminary investigation to use ^{129}I as a tracer for geochemical cycle study of stable iodine. *Marine Chemistry* 72:145–55.

Hou, X., and P. Roos. 2008. Critical comparison of radiometric and mass spectrometric methods for the determination of radionuclides in environmental, biological and nuclear waste samples. *Analytica Chimica Acta* 608:105–39.

Hung, C.-C., S. B. Moran, J. K. Cochran, L. Guo, and P. H. Santschi. 2008. Comment on "How accurate are ^{234}Th measurements in seawater based on the MnO_2-impregnated cartridge technique?" by Pinghe Cai et al. *Geochemistry, Geophysics, Geosystems* 9: Q02009, doi: 10.1029/2007GC001770.

IAEA. 1987. Preparation and certification of IAEA gamma-ray spectrometry reference materials RGU-1, RGTh-1, and RGK-1. IAEA/RL/148.

IAEA. 2000. Reference Material IAEA-375. Radionuclides and trace elements in soil. Reference sheet.

Inn, K. G. W., D. McCurdy, T. Bell, R. Loesch, J. S. Morton, P. P. Povinec, K. Burns, R. Henry, and N. M. Barss. 2001a. Standards, intercomparisons and performance evaluations for low-level and environmental radionuclides mass spectrometry and atom counting. *Journal of Radioanalytical and Nuclear Chemistry* 249:109–13.

Inn, K. G. W., Z. Lin, Z. Wu, et al. 2001b. The NIST natural-matrix radionuclide standard reference material program for ocean studies. *Journal of Radioanalytical and Nuclear Chemistry* 248:227–31.

Ivanovich, M. 1992. The phenomenon of radioactivity. In *Uranium series disequilibrium: Application to environmental problems*, ed. M. Ivanovich and R. S. Harmon. Oxford: Clarendon Press, 1–33.

Ivanovich, M., and A. Murray. 1992. Spectrometric methods. In *Uranium-series disequilibrium: Applications to earth, marine, and environmental sciences*, ed. M. Ivanovich and R. S. Harmon. Oxford: Clarendon Press, 127–173.

Jakopic, R., P. Tavcar, and L. Benedik. 2007. Sequential determination of Pu and Am radioisotopes in environmental samples: A comparison of two separation procedures. *Applied Radiation and Isotopes* 65:504–11.

Jarvis, I., and K. E. Jarvis. 1992. Plasma spectrometry in the earth sciences: Techniques, applications and future trends. *Chemical Geology* 95:1–33.

Jones, D. G. 2001. Development and application of marine gamma-ray measurements: A review. *Journal of Environmental Radioactivity* 53:313–33.

Jweda, J. 2007. Short-lived radionuclides (^{210}Pb, ^{7}Be and ^{137}Cs) as tracers of particle dynamics and chronometers for sediment accumulation and mixing rates in a river system in Southeast Michigan. M.S. thesis, Department of Geology, Wayne State University, Detroit, Michigan.

Jweda, J., M. Baskaran, E. van Hees, and L. Schweitzer. 2008. Short-lived radionuclides (^{210}Pb and ^{7}Be) as tracers of particle dynamics in a river system in Southeast Michigan. *Limnology and Oceanography*, 53(5): 1934–1944.

Kadko, D., and D. Olson. 1996. Be-7 as a tracer of surface water subduction and mixed layer history. *Deep Sea Research I* 43:89–116.

Kenna, T. C. 2002. Determination of plutonium isotopes and neptunium-237 in environmental samples by inductively coupled plasma mass spectrometry with total sample dissolution. *Journal of Analytical Atomic Spectrometry* 17:1471–79.

Kershaw, P. J., K. E. Sampson, W. McCarthy, and R. D. Scott. 1995. The measurement of the isotopic composition of plutonium in an Irish Sea sediment by mass spectrometry. *Journal of Radioanalytical and Nuclear Chemistry* 198:113–24.

Ketterer, M. E., and S. C. Szechenyi. 2008. Determination of plutonium and other transuranic elements by inductively coupled plasma mass spectrometry: A historical perspective and new frontiers in the environmental sciences. *Spectrochimica Acta* 63B:719–37.

Ketterer, M. E., B. R. Watson, G. Matisoff, and C. G. Wilson. 2002. Rapid dating of recent aquatic sediments using Pu activities and ^{240}Pu/^{239}Pu as determined by quadrupole ICPMS. *Environmental Science and Technology* 36:1307–11.

Key, R. M., R. L. Brewer, J. H. Stockwell, N. L. Guinasso, and D. R. Schink. 1979. Some improved techniques for measuring radon and radium in marine sediments and in sea water. *Marine Chemistry* 7: 254–64.

Kim, G., N. Hussain, T. M. Church, and H. S. Yang. 1999. A practical and accurate method for the determination of ^{234}Th simultaneously with ^{210}Po and ^{210}Pb in seawater. *Talanta* 49:851–58.

Kim, C.-S., C.-K. Kim, and K.-J. Lee. 2002. Determination of Pu isotopes in seawater by an on-line sequential injection technique with sector field ICPMS. *Analytical Chemistry* 74:3824–32.

Koide, M., and K. W. Bruland. 1975. The electrodeposition and determination of radium by isotopic dilution in sea water and in sediments simultaneously with other natural radionuclides. *Analytica Chimica Acta* 75:1–19.

Krishnaswami, S. and M. M. Sarin. 1976. The simultaneous determination of Th, Pu, Ra isotopes, ^{210}Pb, ^{55}Fe, ^{32}Si, and ^{14}C in marine suspended phases. *Analytica Chimica Acta* 83:143–56.

Krishnaswami, S., D. Lal, B. L. K. Somayajulu, F. S. Dixon, S. A. Stonecipher, and H. Craig. 1972. Silicon, radium, thorium and lead in sea water: In situ extraction by synthetic fiber. *Earth and Planetary Science Letters* 16:84–90.

Krishnaswami, S., W. C. Graustein, K. K. Turekian, and J. F. Dowd. 1982. Radium, thorium, and radioactive lead isotopes in groundwaters: Application to the in situ determination of adsorption-desorption rate constants and retardation factors. *Water Resources Research* 18:1633–75.

Krishnaswami, S., R. Bhushan, and M. Baskaran. 1991. Radium isotopes and ^{222}Rn in Shallow Brines, Kharagoda (India). *Chemical Geology* 87:125–36.

Ku, T.-L., K. G. Knauss, and G. G. Mathieu. 1977. Uranium in the open ocean: Concentration and isotopic composition. *Deep Sea Research* 24:1005–17.

Kusakabe, T. L., T. L. Ku, J. Vogel, J. R. Southon, D. E. Nelson, and G. Richards. 1982. ^{10}Be profiles in seawater. *Nature* 299:712–14.

Lally, A. E. 1992. Chemical procedures. In *Uranium series disequilibrium: Application to environmental problems*, ed. M. Ivanovich and R. S. Harmon. Oxford: Clarendon Press, 95–126.

Landing, W. M., G. A. Cutter, J. A. Dalziel, A. R. Flegal, R. T. Powell, D. Schmidt, A. Shiller, P. Statham, S. Westerlund, and J. Resing. 1995. Analytical intercomparison results from the 1990 Intergovernmental Oceanographic Commission open-ocean basin survey for trace metals: Atlantic Ocean. *Marine Chemistry* 49:253–65.

Lariviere, D., V. N. Epov, K. M. Reiber, R. J. Cornett, and R. D. Evans. 2005. Micro-extraction procedures for the determination of Ra-226 in well waters by SF-ICP-MS. *Analytica Chimica Acta* 528:175–82.

La Rosa, J. J., W. Burnett, S. H. Lee, I. Levy, J. Gastaud, and P. P. Povinec. 2001. Separation of actinides, cesium and strontium from marine samples using extraction chromatography and sorbents. *Journal of Radioanalytical and Nuclear Chemistry* 248:765–70.

Lee, T., E. Barg, and D. Lal. 1991. Studies of vertical mixing in the Southern California Bight with cosmogenic radionuclides P-32 and Be-7. *Limnology and Oceanography* 36:1044–53.

Levy, D. M., and W. S. Moore. 1985. ^{224}Ra in continental shelf waters. *Earth and Planetary Science Letters* 73:226–30.

MacMahon, D. 2006. Half-life evaluations for H-3, Sr-90 and Y-90. *Applied Radiation and Isotopes* 64:1417–19.

Mann, D. R., and S. A. Casso. 1984. In-situ chemisorption of radiocesium from seawater. *Marine Chemistry* 14:307–18.

Martin, P., G. J. Hancock, S. Paulka, and R. A. Akber. 1995. Determination of ^{227}Ac by α-particle spectrometry. *Radiation and Isotopes* 46:1065–70.

Masque, P., J. A. Sanchez-Cabeza, J. M. Bruach, E. Palacios, and M. Canals. 2002. Balance and residence times of ^{210}Pb and ^{210}Po in surface waters of the Northwestern Mediterranean Sea. *Continental Shelf Research* 22:2127–46.

Mathieu, G., P. Biscaye, R. Lupton, and D. Hammond. 1988. System for measurements of ^{222}Rn at low levels in natural waters. *Health Physics* 55:989–92.

McCallum, G. J., and G. E. Coote. 1975. Influence of source-detector distance on relative intensity and angular correlation measurements with Ge(Li) spectrometers. *Nuclear Instruments and Methods* 130:189–97.

Measures, C., and SCOR Working Group. 2007. GEOTRACES—An international study of the global marine biogeochemical cycles of trace elements and their isotopes. *Chemie der Erde* 67:85–131.

Michel, H., D. Levent, V. Barci, G. Barci-Funnel, and C. Hurel. 2008. Soil and sediment sample analysis for the sequential determination of natural and anthropogenic radionuclides. *Talanta* 74:1527–33.

Moon, D. S., W. C. Burnett, S. Nour, P. Horwitz, and A. Bond. 2003. Preconcentration of radium isotopes from natural waters using MnO$_2$ resin. *Applied Radiation and Isotopes* 59:255–62.

Moore, W. S. 1984. Radium isotope measurements using germanium detectors. *Nuclear Instruments and Methods in Physics Research* 223A:407–11.

Moore, W. S. 2008. Fifteen years experience in measuring ^{224}Ra and ^{223}Ra by delayed-coincidence counting. *Marine Chemistry* 109:188–97.

Moore, W. S., and R. Arnold. 1996. Measurement of ^{223}Ra and ^{224}Ra in coastal waters using a delayed coincidence counter. *Journal of Geophysical Research* 101:1321–29.

Moore, W. S., R. M. Key, and J. L. Sarmiento. 1985. Techniques for precise mapping of Ra-226 and Ra-228 in the ocean. *Journal of Geophysical Research* 90:6983–94.

Moran, S. B., and K. O. Buesseler. 1992. Short residence time of colloids in the upper ocean estimated from ^{238}U-^{234}Th disequilibria. *Nature* 359:221–23.

Moran, S. B., K. M. Ellis, and J. N. Smith. 1997. ^{234}Th/^{238}U disequilibrium in the central Arctic Ocean: Implications for particulate organic carbon export. *Deep Sea Research II* 44:1593–1606.

Moran, J. E., S. Oktay, P. H. Santschi, and D. R. Schink. 1999a. Atmospheric dispersal of ^{129}iodine from European nuclear fuel reprocessing facilities. *Environmental Science and Technology* 33:2536–42.

Moran, J. E., S. Oktay, P. H. Santschi, D. R. Schink, U. Fehn, and G. Snyder. 1999b. World-wide redistribution of ^{129}iodine from nuclear fuel reprocessing facilities: Results from meteoric, river, and seawater tracer studies. IAEA-SM-354/101. Vienna: International Atomic Energy Agency.

Moran, J. E., S. D. Oktay, and P. H. Santschi. 2002. Sources of iodine and ^{129}iodine in rivers. *Water Resources Research* 38:1–10.

NIST. 1981. Standard reference material 4350B. Environmental radioactivity. National Bureau of Standards Certificate.

NIST. 1990. Standard reference material 4325. Radioactivity standard. National Institute of Standards and Technology Certificate.

NIST. 1991. Standard reference material 4222C. Radioactivity standard for liquid scintillation counting. National Institute of Standards and Technology Certificate.

NIST. 1997. Standard reference material 4357. Ocean sediment environmental radioactivity standard. National Institute of Standards and Technology Certificate.

NIST. 2003a. Standard reference material 3230. Iodine-129 isotopic standard (low level). National Institute of Standards and Technology Certificate.

NIST. 2003b. Standard reference material 3231. Iodine-129 isotopic standard (high level). National Institute of Standards and Technology Certificate.

NIST. 2005. Standard reference material 4359. Seaweed radionuclide standard. National Institute of Standards and Technololgy Certificate.

NIST. 2008. Standard reference material 4927F. Hydrogen-3 radioactivity standard. National Institute of Standards and Technology Certificate.

Nozaki, Y. 1984. Excess ^{227}Ac in deep ocean water. *Nature* 310:486–88.

Nozaki, Y. 1991. The systematics and kinetics of U Th decay series radionuclides in ocean water. *Reviews in Aquatic Sciences* 4:75–105.

Nozaki, Y., and T. Nakanishi. 1985. Pa-231 and Th-230 profiles in the open ocean water column. *Deep Sea Research* 32A:1209–20.

Nozaki, Y., Y. Horibe, and H. Tsubota. 1981. The water column distributions of thorium isotopes in the western North Pacific. *Earth and Planetary Science Letters* 54:203–16.

Oktay, S. D., P. H. Santschi, J. E. Moran, and P. Sharma. 2001. ^{129}I and ^{127}I transport in the Mississippi River. *Environmental Science and Technology* 35:4470–76.

Ollivier, P., C. Claude, O. Radakovitch, and B. Hamelin. 2008. TIMS measurements of ^{226}Ra and ^{228}Ra in the Gulf of Lion, an attempt to quantify submarine groundwater discharge. *Marine Chemistry* 109: 337–54.

Outola, I., J. Filliben, K. G. W. Inn, J. La Rosa, C. A. McMahon, G. A. Peck, et al. 2006. Characterization of the NIST seaweed standard reference material. *Applied Radiation and Isotopes* 64:1242–47.

Osvath, I., and P. P. Povinec. 2001. Seabed γ-ray spectrometry: Applications at IAEA-MEL. *Journal of Environmental Radioactivity* 53:335–49.

Perrin, R. E., G. W. Knobeloch, V. M. Armijo, and D. W. Efurd. 1985. Isotopic analysis of nanogram quantities of plutonium by using a SID ionization source. *International Journal of Mass Spectrometry and Ion Processes* 64:17–24.

Pham, M. K., J. La Rosa, S. H. Lee, and P. P. Povinec. 2004. Report on the worldwide intercomparison exercise IAEA-414 radionuclides in mixed fish from Irish Sea and the North Sea. International Atomic Energy Agency, Marine Environmental Laboratory, Monaco.

Pham, M. K., J. A. Sanchez-Cabeza, and P. P. Povinec. 2005. Report on the worldwide intercomparison exercise IAEA-385 radionuclides in Irish sea sediment. International Atomic Energy Agency, Marine Environmental Laboratory, Monaco.

Pham, M. K., J. A. Sanchez-Cabeza, P. P. Povinec, D. Arnold, M. Benmansour, R. Bojanowski, et al. 2006. Certified reference material for radionuclides in fish flesh sample IAEA-414 (mixed fish from the Irish Sea and North Sea). *Applied Radiation and Isotopes* 64:1253–59.

Pham, M. K., J. A. Sanchez-Cabeza, P. P. Povinec, K. Andor, D. Arnold, M. Benmansour, et al. 2008. A new certified reference material for radionuclides in Irish Sea sediment (IAEA-385). *Applied Radiation and Isotopes*, in press.

Pickett, D. A., M. T. Murrell, and R. W. Williams. 1994. Determination of femtogram quantities of protoactinium in geologic samples by thermal ionization mass-spectrometry. *Analytical Chemistry* 66:1044–49.

Pilvio, T., and M. Bickel. 2000. Actinide separations by extraction chromatography. *Applied Radiation and Isotopes* 53:273–77.

Porcelli, D., and P. W. Swarzenski. 2003. The behavior of U- and Th-series nuclides in groundwater. *Reviews in Mineralogy and Geochemistry* 52:317–61.

Povinec, P. P. 2000. Reference material IAEA-MA-B-3/RN. Radionuclides in marine fish flesh. Reference sheet, International Atomic Energy Agency, Analytical Quality Control Services, Vienna, Austria.

Povinec, P. P., and M. K. Pham. 2000. Report on the intercomparison run IAEA-384 radionuclides in Fangatufa Lagoon sediment. International Atomic Energy Agency, Marine Environmental Laboratory, Monaco.

Povinec, P. P., and M. K. Pham. 2001. IAEA reference materials for quality assurance of marine radioactivity measurements. *Journal of Radioanalytical and Nuclear Chemistry* 248:211–16.

Povinec, P. P., M. K. Pham, and S. Ballestra. 1999. Report on the intercomparison run and certified reference material IAEA-381 radionuclides in Irish sea water. International Atomic Energy Agency, Marine Environmental Laboratory, Monaco.

Povinec, P. P., M. K. Pham, J. A. Sanchez-Cabeza, G. Barci-Funel, R. Bojanowski, T. Boshkova, et al. 2007. Reference material for radionuclides in sediment IAEA-384 (Fangataufa Lagoon sediment). *Journal of Radioanalytical and Nuclear Chemistry* 273:383–93.

Radakovitch, O., R. D. Cherry, M. Heyraud, and M. Heussner. 1998. Unusual Po-210/Pb-210 ratios in the surface water of the Gulf of Lions. *Oceanologica Acta* 21:459–68.

Reimann, C. A., B. Grimstvedt, B. Frengstad, and T. E. Finne. 2007. White HDPE bottles as source of serious contamination of water samples with Ba and Zn. *Science of the Total Environment* 374:292–96.

Rutgers van der Loeff, M., and G. W. Berger. 1993. Scavenging of ^{230}Th and ^{231}Pa near the Antarctic polar front in the South Atlantic. *Deep Sea Research I* 40:339–57.

Rutgers van der Loeff, M., and W. S. Moore. 1999. Determination of natural radioactive tracers. In *Methods of seawater analysis*, ed. K. Grasshoff, M. Ehrardt, and K. Kremling. Weinheim, Germany: Wiley-VCH, pp. 365–397.

Rutgers van der Loeff, M., M. M. Sarin, M. Baskaran, C. Benitez-Nelson, K. O. Buesseler, M. Charette, et al. 2006. A review of present techniques and methodological advances in analyzing ^{234}Th in aquatic systems. *Marine Chemistry* 100:190–212.

Ruster, W., F. Ames, H.-J. Kluge, E.-W., Otten, D. Rehklau, G. Scheerer, C. Hermann, C. Muhleck, J. Riegel, P. Rimke, N. Sattelberger, and A. Trautmann. 1989. A resonance ionization mass spectrometer as an analytical instrument for trace analysis. *Nuclear Instruments and Methods in Physics Research* 281A:547–58.

Santschi, P. H., and K. A. Schwehr. 2004. ^{129}I/^{127}I as a new environmental tracer or geochronometer for biogeochemical or hydrodynamic processes in the hydrosphere and geosphere: The central role of organo-iodine. *Science of the Total Environment* 321:257–71.

Santschi, P. H., D. R. Schink, O. Corapcioglu, S. Oktay-Marshall, P. Sharma, and U. Fehn. 1996. Evidence for elevated levels of iodine-129 in the deep Western Boundary Current in the Middle Atlantic Bight. *Deep Sea Research I* 43:259–65.

Sarin, M. M., S. Krishnaswami, B. L. K. Somayajulu, and W. S. Moore. 1990. Chemistry of uranium, thorium and radium isotopes in the Ganga–Brahmaputra river system: Weathering processes and fluxes to the Bay of Bengal. *Geochimica et Cosmochimica Acta* 54:1387–96.

Sarin, M. M., R. Bhushan, R. Rengarajan, and D. N. Yadav. 1992. The simultaneous determination of ^{238}U series nuclides in seawater: Results from the Arabian Sea and Bay of Bengal. *Indian Journal of Marine Science* 21:121–27.

Sarin, M. M., G. Kim, and T. M. Church. 1999. Po-210 and Pb-210 in the South-equatorial Atlantic: Distribution and disequilibrium in the upper 500 m. *Deep Sea Research II* 46:907–17.

Schoch, H., M. Bruns, K. O. Munnich, and M. Munnich. 1980. A multi-counter system for high-precision C-14 measurements. *Radiocarbon* 22:442–47.

Schwantes, J. 1996. Natural and anthropogenic radionuclides in the marginal seas of Siberia: Implications for the fate and removal of pollutants. M.S. thesis, Texas A&M University, College Station, p. 133.

Schwehr, K. A., and P. H. Santschi. 2003. A sensitive determination of iodine species in fresh and sea water samples, including organo-iodine, using high performance liquid chromatography and spectrophotometric detection. *Analytica Chimica Acta* 482:59–71.

Schwehr, K. A., P. H. Santschi, and J. E. Moran. 2005a. ^{129}Iodine: A new hydrological tracer for aquifer recharge conditions influenced by river flow rate variations and evapotranspiration. *Applied Geochemistry* 20:1461–72.

Schwehr, K. A., P. H. Santschi, and D. Elmore. 2005b. ^{129}I/^{127}I in dissolved organic iodine: A novel tool for tracing terrestrial organic matter in the estuarine surface waters of Galveston Bay, Texas. *Limnology and Oceanography: Methods* 3:326–37.

Southon, J. R., D. E. Nelson, R. Korteling, et al. 1982. Techniques for the direct measurement of natural Be-10 and C-14 with a tandem accelerator. *ACS Symposium Series* 176:75–87.

Spry, N., S. Parry, and S. Jerome. 2000. The development of a sequential method for the determination of actinides and ^{90}Sr in power station effluent using extraction chromatography. *Applied Radiation and Isotopes* 53:163–71.

Stewart, G., J. K. Cochran, J. Xue, C. Lee, S. G. Wakeham, R. A. Armstrong, P. Masque, and J. C. Miquel. 2007. Exploring the connection between [210]Po and organic matter in the northwestern Mediterranean. *Deep Sea Research I* 54:415–27.

Stirling, C. H., M. B. Andersen, E.-K. Potter, and A. N. Halliday. 2007. Low-temperature isotopic fractionation of uranium. *Earth and Planetary Science Letters* 264:208–25.

Subramanian, K. S., C. L. Chakrabarti, J. E. Sueiras, and I. S. Maines. 1978. Preservation of some trace metals in samples of natural waters. *Analytical Chemistry* 50:444–48.

Synal, H.-A., J. Beer, G. Bonani, M. Suter, and W. Wolfli. 1990. Atmospheric transport of bomb-produced Cl-36. *Nuclear Instruments and Methods in Physics Research* 52B:483–88.

Szidat, S., A. Schmidt, J. Handl, D. Jakob, W. Botsch, R. Michel, H.-A. Synal, C. Schnabel, M. Suter, J. M. López-Gutiérrez, and W. Städe. 2000. Iodine-129: Sample preparation, quality control and analyses of pre-nuclear materials and of natural waters from Lower Saxony, Germany. *Nuclear Instruments and Methods in Physics Research* 172B:699–710.

Taylor, J. K. 1993. *Handbook for SRM materials*. NIST Special Publication 260-100, U.S. Department of Commerce, National Institute of Standards and Technology.

Taylor, R. N., T. Warneke, J. A. Milton, I. W. Croudace, P. E. Warwick, and R. W. Nesbitt. 2001. Plutonium isotope ratio analysis at femtogram to nanogram levels by multi-collector ICP-MS. *Journal of Analytical Atomic Spectrometry* 16:279–84.

Trimble, S. M., M. Baskaran, and D. Porcelli. 2004. Scavenging of thorium isotopes in the Canada Basin of the Arctic Ocean. *Earth and Planetary Science Letters* 222:915–32.

Tsabaris, C., and D. Ballas. 2005. On line gamma-ray spectrometry at open sea. *Applied Radiation and Isotopes* 62:83–89.

Tsabaris, C., I. Thanos, and T. Dakladas. 2005. The development and application of an underwater γ-spectrometer in the marine environment. *Radioprotection* 40(Suppl. 1):S677–83.

Varga, Z. 2007. Preparation and characterization of manganese dioxide impregnated resin for radionuclide pre-concentration. *Applied Radiation and Isotopes* 65:1095–1100.

Volpe, A. M., A. M. Olivares, and M. T. Murrell. 1991. Determination of radium isotope ratios and abundances in geological samples by thermal ionization mass spectrometry. *Analytical Chemistry* 63:913–16.

Watari, K., and M. Izawa. 1965. Separation of radiocesium by copper ferrocyanide-anion exchange resin. *Journal of Nuclear Science and Technology* 2:321–22.

Wells, M. L., and E. D. Goldberg. 1991. Occurrence of small colloids in seawater. *Nature* 353:342–44.

Wells, M. L., and E. D. Goldberg. 1994. The distribution of colloids in the North Atlantic and Southern Oceans. *Limnology and Oceanography* 39:286–302.

Weyer, S., A. D. Anbar, A. Gerdes, G. W. Gordon, T. J. Algeo, and E. A. Boyle. 2008. Natural fractionation of [238]U/[234]U. *Geochimica et Cosmochimica Acta* 72:345–59.

Whitehouse, B. G., G. Petrick, and M. Ehrhardt. 1986. Crossflow filtration of colloids from Baltic Sea water. *Water Research* 20:1599–1601.

Whitehouse, B. G., P. A. Yeats, and P. M. Strain. 1990. Cross-flow filtration of colloids from aquatic environments. *Limnology and Oceanography* 35:1368–75.

Winn, W. G. 1995. Environmental measurements at the Savannah River Site with underwater gamma detectors. *Journal of Radioanalytical and Nuclear Chemistry* 194:345–50.

Wong, G. T. F. 1991. The marine geochemistry of iodine. *Reviews in Aquatic Sciences* 4:45–73.

Wong, G. T. F., and X.-H. Cheng. 2001. The formation of iodide in inshore waters from the photochemical decomposition of dissolved organic iodine. *Marine Chemistry* 74:53–64.

Wong, K. M., T. A. Jokela, and V. E. Noshkin. 1994. Radiochemical procedures for analysis of Pu, Am, Cs, and Sr in seawater, soil, sediments, and biota samples. Technical Report UCRL-ID-116497. Lawrence Livermore National Laboratory, Livermore, CA.

Yokoyama, T., and E. Nakamura. 2004. Precise analysis of the Ra-228/Ra-226 isotope ratio for short-lived U-series disequilibrium in natural samples by total evaporation thermal ionization mass spectrometry (TE-TIMS). *Journal of Analytical Atomic Spectrometry* 19:717–27.

Zheng, J., and M. Yamada. 2006. Determination of U isotope ratios in sediments using ICP-QMS after sample cleanup with anion-exchange and extraction chromatography. *Talanta* 68:932–39.

14 Sampling and Measurements of Trace Metals in Seawater

Sylvia G. Sander, Keith Hunter, and Russell Frew

CONTENTS

14.1   INTRODUCTION

Studies of the trace element geochemistry have contributed much to our understanding of the oceans, their circulation, and the complex interactions between chemistry and biology. Lessons learned over the last 30 years coupled with advances in technology mean that, as long as the practitioner pays careful attention to detail, consistent data can readily be produced.

The importance of having reliable techniques for the collection and storage of seawater samples for the analysis of trace metals cannot be overestimated. The extremely low concentrations of many trace metals—often lower than the detection limits of some common analytical techniques—require a rigorous and consistent level of cleanliness in order to avoid the inadvertent introduction of contamination. In the 1980s, both technological advances in analytical techniques and the introduction of clean sampling techniques gave rise to a completely new dataset for the distribution of trace metals in seawater (Patterson and Settle, 1976; Bruland et al., 1979). Before that, most reported analytical results were orders of magnitude too high as a result of contamination throughout the sampling and measurement process (Bruland et al., 1991). In this chapter we describe the crucial procedures that are sufficient for ultratrace metal measurements in the most pristine marine environments.

14.2 TRACE METAL CLEAN HANDLING AND THE WORKPLACE

The foremost requirement for reliable trace element results is compliance with trace metal clean practices throughout the entire process, that is, a clean laboratory atmosphere in which contact of the sample with air and surfaces is minimized. A clean atmosphere is ideally provided by a so-called clean lab, which comprises one or more rooms kept under positive pressure by drawing air through a series of filters, culminating in high-efficiency particle air (HEPA). The clean laboratory may also contain one or more laminar flow benches. The level of air filtration (number and size of particles remaining) is standardized by the U.S. FED Std 209E (old) and ISO 14644-1 (new) and typically class 100, comparable to the new ISO 5 standard clean rooms/flow benches used in marine laboratories doing trace element work. Care should be taken to avoid metallic components in the fan motor, screws, or screens, as they will cause contamination.

However, the most likely source of contamination is the analyst, and to avoid this protective clothing must be worn. This could be a complete clean room uniform with gloves, breathing cover, and specially dedicated shoes.

Onboard ship, when samples have to be prepared for storage or analysis, a full clean room facility will often not be available, even though the shipboard environment is probably more prone to cause contamination than a shore-based facility. Ideally, a dedicated trace metal clean van based on a standard shipping container fitted with HEPA air filtration should be used. However, if this is unavailable, a standard shipboard laboratory can be converted into a temporary clean room by lining all the walls and benches with plastic film and PVC plumbing tubing, and fittings can be used to provide an internal frame to support the plastic sheeting. The use of one or more laminar flow benches to draw air from outside the plastic "cage" enables pressurization. A plastic curtain can be used to provide an air lock space.

In terms of laboratory procedures, the "buddy-up" method, where one person is designated to do all the ultraclean handling, avoiding touching any surface that has not been acid cleaned, and a second person handles all the materials, but never touches the acid-cleaned surfaces, has proven to be very practical and avoids the constant changing of gloves before equipment can be handled again.

14.2.1 CLEANING OF EQUIPMENT AND SAMPLE BOTTLES

Different materials used for sampling, sample handling, and measurement in the lab require different cleaning procedures. Fluorinated plastics known under the brand name Teflon– (polyfluoralcoxy copolymer [PFA], tetrafluoroethylene-hexafluorpropylene copolymer [FEP], and –polytetrafluorethylene [PTFE]) are the cleanest materials, in terms of having the lowest blanks, easiest to prepare, and most reliable in terms of consistency. However, in many cases high costs will limit use of these materials. High-density polyethylene (HDPE) and low-density polyethylene (LDPE) require a more intensive cleaning protocol and are still not suitable for all purposes. Thus, a systematic control of the quality and suitability of materials and procedures used is inevitable to extract authentic data from trace metal analysis. The following cleaning descriptions are practical solutions, approved for subnanomolar levels of most trace metals in seawater (Moody, 1981; Gasparon, 1998; Patterson and Settle, 1976; Ahlers et al., 1990).

14.2.2 CLEANING PROCEDURE FOR TEFLON

The following cleaning procedure for Teflon has proven very reliable in our laboratory:

1. Place new bottles in a laboratory surfactant bath for several days to remove all grease and fat.
2. Rinse bottles and caps thoroughly with ultrapure water several times.
3. Concentrated acid treatment:
   - In a well-ventilated laboratory area, place Teflon bottles into a 50% nitric acid bath (HNO_3 reagent grade 16 M and ultrapure water, ratio 1:1) for 4 weeks.
   - Or, place in a heated bath at 80°C for 1 week.
   - It has also been reported that boiling them in aqua regia (1:3 stoichiometric mixture of HNO_3:HCl) for 12 hours using a 5 L glass beaker covered with a watch glass, followed by rinsing with ultrapure water and subsequently 12 hours boiling in 50% HNO_3, is an efficient way to mobilize and desorb solid phase or adsorbed contaminants from the surface (Gasparon, 1998).
4. Rinse thoroughly with ultrapure water and place into metal-free laminar flow bench or clean laboratory to dry on the outside.
5. For final cleaning, the bottle is filled completely with a 1% (≈ 1.5 M) high-purity HNO_3 prepared by subboiling distillation in quartz (q-HNO_3). The bottles are then double bagged with polyethylene Ziploc bags. The inner bag is ideally acid washed (5% HNO_3 diluted with deionized water), or at least wiped down with a wetted dust-free cloth (e.g., Kimwipes) to remove any particles. The bagged bottles can now be packed into bigger plastic bags and left soaking for at least 2 weeks—ideally longer than the sample will be kept in the bottle afterwards before analysis.

Other Teflon equipment should be cleaned accordingly, but the leaching step in 1% q-HNO_3 should be done in a suitable plastic container covered with a lid.

If Teflon bottles are reused for samples with similar metal levels as the previous sample (e.g., uncontaminated seawater), steps 1–3 can be omitted on reuse. However, if the previous sample had high metal concentrations, the complete acid washing procedure (steps 2–5) should be repeated.

14.2.3 HDPE, LDPE, OR PP BOTTLES

Factory-new HDPE, LDPE, and polypropylene (PP) bottles (and other equipment) have to be cleaned slightly differently, taking into account their lower thermal and chemical resistance.

Cleaning procedure:

1. Surfactant treatment as for Teflon.
2. Rinsing as for Teflon.
3. Concentrated acid treatment:
   - Maximum 15% HNO_3 (≈ 2 M) or 50% HCl (≈ 6 M) at 60°C is recommended to avoid influencing the physical stability of the plastic.
   - At room temperature HNO_3 concentrations should not be higher than 50%. Soak in a 5% acid bath (HNO_3 reagent grade and ultrapure water), or fill with the same solution for at least 5 months (Gasparon, 1998). Make sure the entire surface is in contact with acid at all times (no air bubbles), and where necessary, rotate bottle equipment so all areas are exposed to the acid for a sufficient length of time.
   - It also has been reported that boiling them in aqua regia (1:3 stoichiometric mixture of HNO_3:HCl) for 12 hours using a 5 L glass beaker covered with a watch glass, followed by rinsing with ultrapure water and subsequently 12 hours boiling in 50% HNO_3, is an

efficient way to mobilize and desorb solid phase or adsorbed contaminants from the surface (Gasparon, 1998).

4. Rinse as for Telfon.
5. Dilute high-purity acid treatment as for Teflon.

14.2.4 REAGENT PURIFICATION

Another important source of contamination is chemical reagents of insufficient purity. The chemical blank of a procedure can be evaluated systematically by varying the amount of each chemical used and determining the effect this has on the blank. Some reagents can be analyzed directly. Fe and Zn are probably the elements having the greatest risk of contamination from reagents, but other elements are not free of this problem. In most cases, the use of ultrapure trace metal-grade chemicals (e.g., Supra Pur® grade from Merck, Trace SELECT® from Fluka, SigmaUltra® from Sigma, or Aristar® from BDH) is satisfactory, but these are also very expensive. Thus, in-house purification of chemicals, especially concentrated acids, is common practice in many laboratories. The methods of subboiling distillation of acids and isopiestic distillation of ammonia have already been described in Chapter 12 (see Section 12.4.1). Other purification methods of chemical solutions involve the coprecipitation of the analyte of interest together with, for example, freshly formed iron hydroxide (Boussemart et al., 1992). In some cases a very effective way of removing metal ions from chemical solution is treatment with purified Chelex-100 (Price et al., 1989), especially if the pH of the solutions is in the light acid to neutral range.

14.3   SAMPLING METHODS AND EQUIPMENT

All sampling techniques suitable for trace metals in seawater have already been described in Chapter 12 on iron in seawater. If less suitable equipment is used for trace element sampling, blanks have to be checked carefully. However, for elements such as Fe, Zn, Cu, Ni, Cd, and Pb there is no alternative to the methods described in Chapter 12 if samples are to be free of contamination from the sampling process.

14.3.1 SAMPLE STORAGE

Procedures for sample preservation and storage for total dissolved trace metal concentrations are described in Chapter 12. For speciation analysis it is recommended to analyze the samples as soon as possible, that is, onboard ship, as the storage of the samples is likely to cause changes in the speciation. However, deep-freezing samples immediately after filtration has been shown to be appropriate for the organic speciation of Cu (Sander et al., 2005) in fresh water and also has led to reliable results for organic iron speciation in seawater.

14.4   PARTICULATE TRACE METALS

Ocean waters contain a broad spectrum of particles in both size and substance. Particles in the coastal ocean may be dominated by lithogenic materials and will have considerably different composition and concentration to particles found in the open ocean. The surface waters of the open ocean usually have very low particulate concentrations, a high proportion of those being biogenic particles (cells), and thus the particulates generally contain a small proportion of the trace metal inventory.

Particles play a key role in the cycling of many trace metals. The surfaces of particles are negatively charged, and hence provide adsorption sites for dissolved metals. Fine suspended particles may aggregate or be repackaged by grazers into larger particles that will sink, providing a transport mechanism for adsorbed metals from the surface ocean to the deep. Further processing through remineralization may return labile metals to the water column.

The distinction between particulate and dissolved trace metals has traditionally been an operational one based on what will pass through a fine-pore filter (0.45 or 0.2 µm). It is recognized that there will be very fine colloidal material in this "dissolved" phase, and such colloids may well contain a significant portion of the metals attributed as dissolved. This is particularly important for elements such as Fe (see Chapter 12).

The choice of sampling procedure for particulate material depends greatly on the concentration of particles in the water to be sampled and on the expected concentration of the target analyte in those particles. The higher concentrations found in coastal waters mean samples of 5 to 10 L are usually sufficient, whereas in the open ocean samples of the order of 100 L may be required. The volume of sample to be filtered determines the choice of apparatus. A simple 47 mm filter holder with membrane filter will be suitable for <10 L. Such filter holders are available from Cole Palmer (e.g., cat. KH-06644-62). In the open-ocean situation it is recommended to use 142 mm filters.

Size fractionation may be achieved by stacking filter holders in series (e.g., 20, 5, 2, and 0.2 µm pore sizes; Frew et al., 2006). The sums of these fractions provide an estimate of total particulate metal.

The same stringent procedures are required for collecting particulate sample as for sampling dissolved trace metals. The added complexity arises from the need for much larger volumes of water. Samples from the surface or near surface may be collected using an epoxy-coated towed fish (see Section 1.5.1 and Figure 1.5), which may be deployed at a range of depths in the surface mixed layer about 5 m perpendicular from the vessel using a boom. A simple data logger attached to the fish may be employed to precisely determine the depth at which the sample was collected. Seawater is pumped directly onboard and into a clean laboratory by an all-Teflon pump (e.g., Almatec SL20) or variable-speed high-volume peristaltic pump (Bowie et al., 2006) through Teflon or polyethylene tubing. Teflon fittings (Swagelok) should be used for all joints. All components of the sampling system that will contact the sample must be extensively cleaned with acid (see above for cleaning procedures). Dissolved metal samples may be collected using the same supply system.

Polycarbonate membranes are most commonly used, as they have low blanks and are relatively easy to clean and handle (Cullen et al., 2001; Cullen and Sherrell, 1999). Filters are cleaned by soaking in 10% triple quartz distilled HCl (qHCl) prior to use. If possible, it is preferable to draw a small volume of 10% qHCl through the filter membrane once it is in place in the filter holder. The acid should be followed with an ultrapure water rinse. Particular attention must be given to the determination of the filter blanks, as these are typically the limiting factor in analyzing these samples. Field blanks may be obtained by treating filters exactly as sample filters, except for the filtering of a sample. To filter a sample, flow is adjusted to <1 L min^{-1} and filtrate collected in 20 L carboys to measure total volume collected. Continue filtration until the flow stops, indicating one filter is blocked. Filters should be removed from the holder in a clean bench, folded upon themselves to prevent loss of material, and stored frozen in polypropylene vials.

Many of the issues concerning collecting sufficient sample to ensure the analyte signal is well above the blank can be overcome with in situ pumps (Weinstein and Moran, 2004; Grotti et al., 2003). These pumps are deployed at the depth of interest, and many hundreds of liters of seawater are pumped through their filters.

14.4.1 Ultrafiltration

It has long been recognized that the distinction between dissolved and particulate material based on what passes through a filter membrane is an operational definition of convenience and is an approximation to reality. The determination of trace metals associated with very fine particles such as colloids requires specialized filtration apparatus. Ultrafiltration (UF) is a type of membrane filtration in which hydrostatic pressure forces a liquid against a semipermeable membrane (see Section 1.6.4).

Suspended solids and solutes of high molecular weight are retained, while water and low molecular weight solutes pass through the membrane.

Marine applications of UF were developed for studying colloidal carbon (Buesseler et al., 1996) but can be adapted for trace elements (Chen et al., 2004; Wells, 2003) as the membrane materials are commonly polysulfone (PS), polyethersulfone (PES), or polypropylene (PP) and can be readily cleaned. During UF the sample is continuously circulated across a filter membrane so that, on each pass, a small portion of the solution passes through the membrane. This leaves a slightly more concentrated colloidal fraction behind. The recirculation minimizes concentration polarization effects that arise from the accumulation of high particle and solute concentrations just above the membrane surface.

14.4.2 SEDIMENT TRAPS

Sediment traps provide a means of collecting a large amount of settling particulate material. The principle is based around the interception of sinking material in a tube or device of known area over a known length of time. Traps may be part of a mooring at a particular location and allow seasonal studies of settling particulate material (Nodder and Northcote, 2001). Other studies have utilized drifting sediment traps that are deployed and recovered during a cruise (Waite and Nodder, 2001). Either application requires a large investment in equipment, ship, and personnel time. Additionally, when using samples from sediment traps you must be aware of the limitations of this approach. These have been well documented (Siegel et al., 2008; Stanley et al., 2004; Buesseler et al., 2007). There are three broad issues:

1. Hydrodynamics: The traps are designed to intercept the vertical transport of particles as they settle; however, there is often considerable horizontal transport, and interaction with the trap shape may cause under- or overestimation of vertical flux.
2. Zooplankton swimmers: These animals may bias results by feeding on trapped particles, forming sinking fecal pellets, or disaggregating particles by their actions.
3. Solubilization: Loss of a portion of the trapped material through remineralization and diffusion to the supernatant fluid. This process becomes increasingly important the longer the trap is deployed.

Considerable progress has been made in identifying these issues and developing best practices for correcting for artifacts that may arise (Buesseler et al., 2007).

14.4.3 PARTICULATE ANALYSIS

The digestion procedure employed prior to analysis will depend on the sample type and the rationale for the analysis. If the intention is to determine the biogenic component, then a nitric acid digestion may be appropriate. Researchers have for many years utilized selective leaching schemes to operationally define compartments of the trace metal pool within the particles. As the distinction between components is operationally defined, it is important to analyze certified reference material with your samples to assist in comparisons with other data. A certified extraction scheme and reference material is described in Pueyo et al. (2001). When applying such a scheme for the low levels of metal expected in marine particles, ensure that all labware is polyethylene, polypropylene, or Teflon, and that Milli-Q or equivalent water is used throughout.

It is more common to make a total digestion and estimate the biogenic vs. lithogenic contribution by calculation based on the aluminium content (Frew et al., 2006). Total digestion may be achieved using an approximately 20:1 mix of HNO_3 and HF with heating at 120°C for 4 hours. This results in partial digestion of the filter but total digestion of the particulate material (Cullen et al., 2001). An alternative procedure involves a two-step approach where the filter and particles are digested in a sealed Teflon vessel with HNO_3 or a mixture of HNO_3 and $HClO_4$ before removal of the intact filter material and subsequent total digestion of the particle material with HF (Wang et al., 1996).

A new sample preparation technique for marine particles has recently been reported (Huang et al., 2007). This method involves high-temperature fusion using a lithium metaborate flux for sample digestion and elemental quantification using high-resolution inductively coupled plasma mass spectrometry (HR-ICP-MS). Each analysis consumes only 1 to 2 mg of sample material yet enables simultaneous measurements of eighteen elements (Mg, Al, Si, P, Ca, Sc, Ti, V, Mn, Fe, Co, Ni, Cu, Zn, Sr, Cd, Ba, and Pb) with accuracy of >90% for most elements. The fusion method introduces minimal contamination when appropriate sample handling and procedural precautions are employed. Elemental quantification of samples prepared for ICP-MS analysis using fusion agrees well with those prepared using acid digestion. Additionally, the fusion method has several advantages over acid digestion. No highly toxic reagents like hydrofluoric acid are used. Refractory mineral dissolution is more complete. Si, Ca, and other trace elements can be analyzed without potential losses due to coprecipitation with CaF_2. The simplicity and reproducibility of the procedure makes it especially suitable for routine, ongoing analyses of oceanic particulate materials.

14.5   ANALYSIS OF DISSOLVED TRACE METALS

Commonly used analytical techniques such as atomic absorption spectroscopy (AAS) and ICP-MS generally require the analyte to be removed from the salt matrix. This serves two purposes in that it reduces the possibility of interferences from the salts and provides the opportunity to preconcentrate the analyte. The importance of these steps is highlighted when you consider that the analyte concentration is often five or six orders of magnitude lower than that of the matrix salts. Under conditions where analyte concentrations are high, such as coastal contamination, trace metals can be quantified directly by ICP-MS. However, care must be taken to ensure that the high dissolved solid concentration of such samples does not interfere with the analysis.

14.5.1   PRECONCENTRATION AND SEPARATION TECHNIQUES

Three classes of procedures are commonly employed for this task: liquid-liquid extraction after complexation of the analyte; solid phase extraction using a column with stationary sorbent; and co-precipitation of analyte with a solid phase produced in situ.

A number of liquid-liquid extraction (LLE) schemes are described in the literature. The one that is most commonly used is based on the work of Danielsson et al. (1982) and Bruland et al. (1979) and has proved to be a reliable and widely applicable technique, particularly for the first-row transition metals as well as some heavier elements, such as Cd and Pb. A related technique using liquid membranes has been used to prepare water samples for environmental analysis. The principle of the liquid membrane is that the sample is passed across a membrane that separates it from an acceptor phase. The chemical conditions of the phases are adjusted such that the analyte will form a complex that diffuses across a liquid membrane into an acceptor solution where the complex is decomposed, preventing the analyte ion from reentering the membrane. Liquid membrane systems offer several advantages over LLE, including reduction in the amount of solvent used and the potential of automation (Jönsson and Mathiasson, 1999). Liquid membranes have been applied for the analysis of coastal and contaminated waters where concentrations are relatively high (Irigoyen et al., 2006) but have not been widely used for open-ocean studies where concentrations are low. Here LLE is the method of choice.

The principle of the LLE method is that the analyte is converted to a nonpolar complex through chelation with dithiocarbamate. The organic solvent is immiscible in the seawater, and the distribution of the metal between the solvent and water is given by

$$D = \frac{(M_{org})}{(M_{aq})}$$

(14.1)

The final distribution (D) is a function of the concentration and nature of the chelating agent, the solvent, and the pH of the aqueous phase. It is crucial that the pH be carefully controlled to obtain high and reproducible yields.

Water samples need to be acidified at least the day before analysis with 2 mL of ultrapure HNO_3 per liter of sample to desorb metal ions from the plastic surfaces of the bottle and decompose colloids. The sample aliquot taken for extraction is generally 100 g. Duplicate aliquots are analyzed for every sample. The chelating solution is made up at 0.5% (w/v) each of ammonium pyrrolidine dithiocarbamate (APDC) and sodium diethyl dithiocarbamate (NaDDC). Clean ammonium acetate is prepared by bubbling ammonia gas through quartz distilled acetic acid. A buffer solution is prepared by dissolving 35 g of ammonium acetate in 100 mL of ultrapure water. This solution will last 2 or 3 days before a fresh buffer should be prepared. One milliliter of the buffer solution is added to the chelation solution and vice versa, and then both solutions are purified by extraction with chloroform. The seawater is buffered with a solution of purified ammonium acetate made up so that, on addition of a 1 mL aliquot to 100 g of acidified water sample, a final pH of 4.8 is achieved.

For extraction, an aliquot of the seawater sample is weighed into a 250 mL Teflon separating funnel, and 1 mL each of the ammonium acetate, APDC, and NaDDC solutions are added per 100 g of sample. This is extracted twice using 10 mL aliquots of ultrapure chloroform. Extraction is achieved through gentle agitation for 2 minutes, after which the solution is allowed to sit for 5 minutes for phase separation. Extracts are combined in a 30 mL Teflon screw-cap bottle. The chelated metals are then back extracted by first adding 50 μL of 16 M ultrapure HNO_3 to decompose the complex. Following this the acid is diluted by adding 1.95 mL of ultrapure water to the vial, which is shaken and left for the phases to separate. Note: Metal carbamate complexes are not particularly stable in organic solvents, and so the organic phase should be separated from the water sample as soon as possible after the addition of the complexing agent. The 2 mL final volume obtained here is sufficient for GFAAS but may require dilution to provide sufficient volume for ICP-MS.

There are several parameters of interest that should be monitored with each batch of extractions: blanks, recoveries, and residuals. Reagent blanks can be determined from extractions where ultrapure water is substituted for sample. Residuals may be determined from reextracting a sample that has previously been extracted. Recoveries are determined by spiking a sample that has been previously extracted with known amounts of analyte and then reextracting as a sample. A series of spikes should be determined to span the range of concentrations encountered with the samples. A detailed analysis of the uncertainty sources of this method has been conducted (Pino et al., 2007). The results show that the main contribution to the relative overall uncertainty is the extraction step, emphasizing the need for care in this procedure.

A further technique routinely used for the preconcentration of metals from seawater is solid phase extraction (SPE). SPE has several advantages over LLE in that the volumes of solvent required may be greatly reduced or even eliminated. SPE can also be set up as part of an automated system, enhancing sample throughput and reproducibility.

There are many different resins that have been applied to the separation of trace metals from seawater. Many are quite specific to a particular metal and produce poor recoveries of other metals. The most commonly used general-purpose resin is the iminodiacetic group resin, Chelex-100. Chelex-100 has been used for a great number of trace metal and rare earth species (Rahmi et al., 2007). One issue for multielement applications is that the efficiency of extraction is pH dependent and different elements will have different optimal pH values for extraction. Therefore, a compromise pH has to be selected that will provide high recoveries for the greater number of elements.

14.5.2 Flow Injection Methods

Flow injection is not an analytical technique per se, but rather a means of automation of analysis that could be applied to many detection techniques. Here we group the flow injection techniques utilizing small, portable detectors enabling shipboard analysis of trace metals. The two workhorse

TABLE 14.1
Summary of Key Reagent and Detection Techniques Utilized in FIA Systems

Analyte	Reagent	Detector	Reference
Fe^{2+}	Luminol	Photon counting head	Croot and Laan, 2002
Mn^{2+}	Tiron	Spectrophotometer	Mallini and Shiller, 1993
Al^{3+}	Lumogallion	Fluorimeter	Vink et al., 2000
Zn^{2+}/Cd^{2+}	Anthylazamacrocycle 1	Fluorimeter	Worsfold et al., 2002
Pb^{2+}	8-Hydroxy-7-quinoline sulfonic acid	Phosphorescence	Worsfold et al, 2002
Co^{2+}	Pyrogallol	Photon counting head	Cannizzaro et al., 1999

analytical techniques of GFAAS and ICP-MS are limited in their application in that it is not practical to take such instruments to sea on a ship. In recent years there has been a drive toward designing methods that will provide results while at sea. In some cases near-real-time data can be achieved enabling the science program to be directed by the results. Flow injection techniques provide a means of sample preparation that is almost closed to the environment and associated contamination. Coupled with low-cost and portable detectors, these systems have been applied to measure trace metals in seawater. For example, such techniques have contributed enormously to the success of the mesoscale iron enrichment experiments. The techniques used for the determination of Fe species at sea are presented elsewhere in this book (see Chapter 12).

The primary components of a flow injection system are the multichannel peristaltic pump for careful control of flow rates of sample and reagents, valves for flow switching, a mixing cell, detector, and data acquisition. These instruments are small and relatively cheap, making them ideal for use at sea. They can be built with off-the-shelf components. The prime disadvantage with flow injection systems is that you generally need a separate detector and chemistry for each element. Table 14.1 provides some examples of the primary reagent and detector used for some commonly determined elements.

14.5.3 ISOTOPE DILUTION METHODS

Historically, isotope dilution has been widely used for the analysis of trace elements in seawater, and is often regarded as the gold standard with regard to accuracy. The recent development of ICP-MS technology has brought this technique to the fore again. The principle of isotope dilution is that the seawater sample is first equilibrated with a known spike of a synthetic isotope not normally present in the sample. After this, the isotopes of the element of interest are isolated from the seawater matrix and quantified by some form of mass spectrometry. The elemental concentration in the same can then be calculated from the ratio of naturally occurring isotope to that of the synthetic spike, and the known concentration of synthetic isotope added. For trace elements of medium to high molar mass, there will be no measurable isotopic fractionation during the isolation of the isotopes for mass spectrometric analysis, which means that the yield during isolation does not need to be quantitative or even constant from sample to sample.

In multielement analysis, it is also possible to determine elements for which there is no suitable synthetic isotope spike by comparing the elemental abundance determined by ICP-MS with that of a similar element for which an isotope dilution is possible (Saito and Moffett, 2001). In this case the yield of isolation of both elements needs to be known for each sample.

Recommended methods for isotope dilution ICP-MS (ID ICP-MS) are based on the pioneering work of Wu et al. and Wu and Boyle, in which Cu, Pb, Cd (Wu et al., 1997), and Fe (Wu and Boyle,

1998) were determined by coprecipitating these metals with $Mg(OH)_2$ precipitated by addition of NH_4OH to a small seawater sample (1–15 mL). The $Mg(OH)_2$ precipitate is then isolated by centrifugation and redissolved in dilute acid for ICP-MS analysis. This technique has the advantages that (1) it uses a very small sample volume compared to more traditional solvent extraction or chelating resin methods; (2) it uses a minimum of reagents (NH_4OH, HNO_3, or HCl), all of which are very easily purified; (3) has relatively low blanks; and (4) it is easily capable of multielement determination.

Although it is not strictly necessary to know the yield of isolation, or to ensure that it does not vary sample by sample (at least for those elements for which isotope dilution is possible), it is nonetheless good experimental practice to know about the yield and to control the isolation conditions as closely as possible. Saito and Schneider (Saito and Schneider, 2001) conducted a comprehensive examination of the effect of experimental conditions on the yield of $Mg(OH)_2$ coprecipitation for Fe, Mn, and Co. Their results showed that these elements are incorporated into the initial $Mg(OH)_2$ precipitate as it is first formed, rather than by subsequent adsorption onto the surface of the precipitate afterwards. The yield of extraction is found to vary with the length of time allowed after precipitation and the amount of precipitate, making careful control of this process important. It is also important not to precipitate too much $Mg(OH)_2$ because this leads to issues with respect to clogging of the ICP-MS sample intake system and contamination of the MS cones. Thus, the best procedure represents a fine balance between ensuring high yields and minimizing the amount of Mg in the final analyte solution. Because the concentration of NH_4OH solution varies over time because of volatility, the amount of NH_4OH solution that needs to be added has to be calibrated visually on a daily basis. This calibration will depend on the amount of acid (if any) added to the seawater sample as a preservative.

Wu (2007) has reported an improved double precipitation method for Fe. This achieves higher preconcentration ratios than the previous methods, yet mostly eliminates the uncertainties in yield associated with controlling the amount of NH_4OH added. A much larger 50 mL seawater sample is initially treated with NH_4OH and the resultant $Mg(OH)_2$ precipitate is isolated by centrifugation. The precipitate is then dissolved in a small aliquot of high-purity HCl. After dissolution, the now slightly acidic solution is then evaporated to dryness under an infrared lamp. This step removes any excess HCl. The salt residue, comprising $MgCl_2$ and associated trace elements, is next dissolved in a small volume (1.6 mL) of pH 2 seawater that is low in trace elements. NH_4OH is next added to reprecipitate $Mg(OH)_2$, which is then isolated by centrifugation and dissolved in HNO_3 for ICP-MS analysis in the usual manner. Because the excess HCl has already been removed by evaporation, the second $Mg(OH)_2$ precipitation can be carried out reproducibly. For Fe, this technique has a much lower overall blank and limit of detection than other ID ICP-MS methods. Although its use for other trace metals has not yet been reported, this double precipitation technique warrants further investigation.

The principal drawback of ID ICP-MS is isobaric interferences, a drawback that is shared by the use of ICP-MS with other sample preconcentration methods. For Cu analysis, it is necessary to wash the $Mg(OH)_2$ precipitate with NH_4OH solution to remove residual Na^+, which interferes as $ArNa^+$. The interference of ArO^+ on ^{56}Fe is well known and best overcome by use of reaction cell technology, or a high-resolution MS. Refer to the extensive ICP-MS literature for further information.

14.6 ELECTROCHEMICAL METHODS AND SPECIATION

Many analytical techniques (e.g., AAS, ICP-MS) cannot cope directly with the high salt concentrations present in seawater. Blockages of instrument components such as tubing and spray units are common and often require a pretreatment of the sample. In contrast, voltammetry requires the presence of an electrolyte in the sample to ensure sufficient conductivity, thus making seawater an ideal medium for voltammetry. In addition, voltammetry has high sensitivities and offers the possibility to measure the exact chemical speciation (redox state, biological availability) of trace element. Further advantages of voltammetry are its portability (i.e., onboard analysis) and the suitability for in situ analysis. However, the number of elements that may be analyzed is limited and compared to multielement analyzers such as ICP, the voltammetric determination of total element concentrations is much more time-consuming.

14.6.1 INSTRUMENTATION AND TECHNIQUE

Compared to many other analytical instruments, a typical voltammetric setup is simple and relatively inexpensive. It consists of a set of electrodes and a potentiostat (e.g., 797 Computrace from Metrohm, Switzerland; Autolabs from EcoChemie, Netherlands; Voltalabs- from Radiometer Analytical, France; EG&G, United States; or IVIUM Technologies, Netherlands) controlled by a computer. Although the toxicity of mercury is a disadvantage, the characteristics of the metallic mercury electrode are ideal for the ultratrace determination of many trace metals in seawater. Modern electrode stands (e.g., 663 VA stand from Metrohm; MDE150 Polarographic Stand from Radiometer) use a tiny (typically <1 mm^2) hanging mercury drop (HMDE) as a working electrode, with a platinum or glassy carbon wire/rod counter electrode and Ag/AgCl/3 M KCl reference. The HMDE can be substituted by other electrodes such as static or rotating disc glassy carbon, carbon pastes, iridium, and gold, bare or with a mercury film coating (MFE). However, none of these have a comparable broad span of applications. Interferences from reduction of O_2 are usually eliminated by purging the analyte with N_2 or Ar. Dissolved organic matter (DOM) present in the sample can interfere with the voltammetric determination of total metal concentrations either by forming electroinactive complexes or simply by competing for the available electrode surface. A UV digestion system (e.g., from Metrohm) to destroy organic components is therefore recommended prior to total dissolved metal determinations (Kolb et al., 1992a, 1992b; Achterberg and Van den Berg, 1994; Yokoi et al., 1995). The advantage of the UV digestion is that only very small amounts of ultraclean acid have to be added, as opposed to large amounts of acid required for more conventional oxidative acid digestions.

From all of the electrochemical methods available, stripping voltammetry is most commonly used for the determination of trace elements and their speciation in seawater. The technique consists of a preconcentration step with subsequent redissolution (stripping) and the actual measurement of either oxidative or reductive current during a potential scan. In anodic stripping voltammetry (ASV) the metal ion (e.g., Zn, Cd, Pb, Cu, or Tl) is reduced at the electrode surface by holding at a constant accumulation potential (E_{acc}) for a fixed time (t_{acc}). During the stripping process the analyte, now present in its metallic form or as an amalgam, is reoxidized. In adsorptive cathodic stripping voltammetry (AdCSV) the analyte is converted into an adsorbable metal ion complex through the addition of a known organic complexing reagent. In this case, the preconcentration step is performed at a potential where the metal complex adsorbs at the electrode complex and the stripping step involves a potential scan in the cathodic direction, as a result of which the metal ion (or the ligand) is reduced. In the case of both ASV and AdCSV, the resulting faradic current is proportional to the concentration of the analyte on the electrode surface, which is itself a function of the concentration in the solution and t_{acc}. The quantification of the analyte is commonly done using the standard addition method to account for any variations in sensitivity with differences in the matrix. Modern instruments apply a staircase potential ramp and measure the current differentially, most commonly as a differential pulse (DP) or square wave (SQW) modulations (Thomas and Henze, 2001; Brett and Brett, 1993).

We now describe the main voltammetric methods for speciation analysis of trace metals in seawater (except for iron; see Chapter 12). Total metal concentrations, which are needed for calculating the results of speciation analysis, are often more efficiently measured by GF-AAS and ICP-MS, but can also be measured by voltammetry after UV photooxidation of natural ligands.

14.6.2   REDOX SPECIATION OF METAL IONS: CHROMIUM

There are few published methods for determining the redox speciation of trace metals in seawater, the most important being those for iron (Chapter 12) and chromium, described below. Many other redox-sensitive elements exhibit redox kinetics that are too fast to allow an accurate measurement ex situ. While a number of methods exist for fresh waters, these are not practical in seawater because of interferences caused by the seawater matrix (e.g., high chloride concentration or dissolved organic matter).

Total Cr concentrations in the oceanic water column are usually in the range 3–5 nM, with surface depletion and correlation with nutrients observed (Campbell and Yeats, 1981; Isshiki et al., 1989). Published results on the distributions of Cr(III) and Cr(VI) are still inconsistent and confusing. According to thermodynamic calculations (Elderfield, 1970), Cr(VI) should be the major stable form in oxygenated seawater, which is in agreement with some of the reported data (Isshiki et al., 1989; Sirinawin and Westerlund, 1997). However, Cr(III) concentrations much higher than predicted from the thermodynamic Cr(III)/Cr(VI) ratios have also been observed (Murray et al., 1983; Mugo and Orians, 1993). Hydrogen peroxide (Pettine and Millero, 1990) and manganese oxides (Nakayama et al., 1981a) are potential oxidants of Cr(III). Reduction of Cr(VI) has been shown to be effective with hydrogen sulfide and Fe(II) over a wide range of conditions (Pettine et al., 1994, 1998) and by naturally occurring organic material (Nakayama et al., 1981b). Cr(VI) can cross biological membranes readily, acts as a strongly oxidizing component, and is the more toxic species. Therefore, redox conditions of the environment directly influence Cr bioavailability and toxicity (Sander and Koschinsky, 2000).

The catalytic AdCSV method using diethylenetriaminepentaacetic acid (DTPA; 0.015 M) in the presence of nitrate (1 M $NaNO_3$) and acetate buffer (0.1 M, pH 5.2) is used to measure Cr species in fresh seawater samples. The method for low levels of Cr in seawater has been developed by Boussemart et al. (1992), and the exact electrode reaction was investigated by Sander et al. (2003a).

Analytically detectable chromium species are total dissolved chromium, total chromium (unfiltered) after UV digestion, Cr(VI) after masking of Cr(III), reactive total chromium, and Cr(III). Both redox species of chromium are electroactive after the initial addition of DTPA, but Cr(III) forms an electroinactive complex with DTPA within 30 minutes, which therefore masks its presence, allowing selective detection of Cr(VI). Differentiation between the two redox states is based on utilizing this masking reaction. In Table 14.2, the conditions for the voltammetric determination of all chromium species are summarized (Sander and Koschinsky, 2000).

Figure 14.1 shows a normal distribution for both Cr redox species (A) and a depth profile showing a hydrothermal input at depth (B), found to be in good correlation with hydrothermal parameters such as methane and redox potential (Eh) (Sander et al., 2003b).

14.6.3   ORGANIC SPECIATION OF TRACE METALS

It is now also well established that the physiological availability of most biologically active trace metal ions in the ocean is controlled by the formation of strong complexes with natural organic ligands, many of which are presumed to be produced either to reduce metal ion toxicity (e.g., Cu^{2+}, Moffett and Brand, 1996; Dupont et al., 2004) or to acquire metal ions (e.g., Fe^{3+}, Van den Berg, 1995; see Chapter 12; or Zn^{2+}, Ellwood, 2004). Cd, Co, and Ni have also been found to be dominated in the upper water column by strong organic complexes (Kozelka and Bruland, 1998; Ellwood and Van den Berg, 2001; Saito and Moffett, 2001; Van den Berg and Nimmo, 1987; Donat et al., 1994).

Voltammetric techniques using an equilibrium approach are commonly applied to determine the complexing parameters of seawater toward trace metals such as Fe, Cu, Cd, Zn, Co, and Ni. Both the concentration of metal-binding ligand $[L_M]$ and the conditional stability constant $K_{M'L'}$ of the

TABLE 14.2
Conditions for the Speciation Analysis of Chromium Using Catalytic AdSV with DTPA as Complexing Reagent

Species / Conditions	1. Cr(VI) (as Chromate)	2. Total Dissolved Cr (After UV, Filtrated)	3. Total Cr (Including Particulate, Filtrated After UV)	4. Total Dissolved Cr(III)	5. Total Reactive Cr_a (Cr(VI) + Free and Labile-Bound Cr(III))	6. Reactive Cr(III) (Free and Labile-Bound Cr(III))
Pretreatment of sample	Addition of DTPA 30 min before measurement	Filtration through 0.2 μm cellulose acetate, UV digestion	UV digestion	Calculated from total dissolved Cr–Cr(VI)	Addition of DTPA to sample immediately before measurement	Calculated from total reactive Cr–Cr(VI); because of different calibration slopes, this result must be multiplied with a correction factor[a] to reflect different slopes of Cr(III) and Cr(VI)
Supporting electrolyte	0.015 M DTPA, 1 M NO_3, 0.01 M acetate buffer (pH 5.2); addition of DTPA 30 min before necessary measurement	0.015 M DTPA, 1 M NO_3, 0.01 M acetate buffer (pH 5.2); no waiting time after addition of DTPA	0.015 M DTPA, 1 M NO_3, 0.01 M acetate buffer (pH 5.2); no waiting time necessary after addition of DTPA		0.015 M DTPA, 1 M NO_3, 0.01 M acetate buffer (pH 5.2)	
Voltammetric parameters	DP mode: amplitude = 50 mV, scan rate = 60 mV s^{-1}, potential scan from –1,000 to –1,500 mV; accumulation: t_{acc} = 10 to 30 seconds, U_{acc} = 1,000 mV					

a Correction factor of different calibration slopes for Cr(VI) and Cr(III) has to be determined for the exact conditions used.

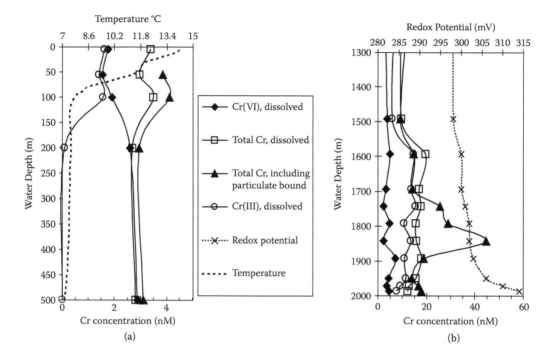

FIGURE 14.1 Chromium redox speciation of (a) undisturbed depth profile from the subantarctic current, New Zealand, with corresponding water temperature and (b) depth profile (including the hydrothermal tracer Eh) from the North Fiji Basin showing hydrothermal Cr(III) input.

complex at ambient seawater pH can be measured using competing-ligand equilibration adsorption cathodic stripping voltammetry (CLE-AdCSV) with a known artificial ligand (AL) used to compete with the naturally occurring ligands of interest. Ligand equilibration anodic stripping voltammetry (LE-ASV) has also been used (Bruland et al., 2000).

The theory behind these so-called metal ligand titrations was described first by Van den Berg (1982) and Ružič (1982) and is based on the assumption of the formation of a 1:1 complex. Briefly, the following equations apply:

$$[M_T] - [M'] = [L_T] - [L'] = [ML] \qquad (14.2)$$

where $[M_T]$ and $[L_T]$ are the total concentrations of trace metal M and unknown ligand L, $[M']$ and $[L']$ are the corresponding concentrations of unreacted species, and $[ML]$ is the concentration of the complex ML.

In the simplest situation, where there is only a single strong ligand L reacting with M, the conditional stability constant $K'_{ML'}$ of ML can be written as

$$K'_{ML'} = \frac{[ML]}{[M'][L']} \qquad (14.3)$$

The free metal M′ unbound to the unknown ligand L is quantified using the sensitivity (*S*) of the titration curve in the concentration range, where the peak current i_p depends linearly on the concentration of metal added to the sample:

$$i_p = S[M'] \qquad (14.4)$$

Rearranging Equation 14.2 using Equation 14.1 results in

$$K'_{ML'} = \frac{[ML]}{[M']([L_T]-[ML])}$$

(14.5)

The total ligand concentration $[L_T]$ and conditional stability constant $K'_{ML'}$ can then be calculated using the Gerringa method (Gerringa et al., 1995), which involves rearrangement of Equation 14.4 to yield a Langmuir isotherm equation:

$$[ML] = \frac{[L_T]K[M']}{1+K[M']}$$

(14.6)

from which $[L_T]$ and K' and the sensitivity S may be obtained by nonlinear least squares fitting (e.g., using the Excel Solver add-in) of the paired $[M']$ and $[ML]$ values calculated from the voltammetric measurements.

Note that the conditional stability constant $K'_{ML'}$ means that it is only valid for M', the combined concentrations of all forms of the metal M that are not bound to strong organic complexes, which corresponds to all inorganic species and complexes with labile, weak organic ligands that might be present. Similarly, L' represents all forms of the unknown ligand L except that bound to the metal ion of interest (e.g., LH, LCa, etc.). Thus, $K'_{ML'}$ does not explicitly take into account the side reactions of M with inorganic or weak organic ligands, nor of those between L and other metals. The latter side reactions of the ligand cannot usually be corrected for, as they are unknown, but those involving M and known inorganic ligands can be accounted for under defined conditions using the ion pairing model. The overall inorganic side reaction coefficient for M is

$$\alpha_{M_{inorg}} = \sum \alpha_i = \sum K_i[L_i]$$

(14.7)

with K_i being the stability constant and $[L_i]$ the concentration of the inorganic ligand L_i (e.g., Cl^-, OH^-, CO_3^{2-} etc.).

Thus, the conditional stability constant $K'_{ML'}$ can be converted to a stability constant $K'_{M^{2+}L'}$ conditional only for the unknown ligand L by dividing $K'_{ML'}$ with α_{Minorg}, resulting in

$$K'_{M^{n+}L'} = \frac{[ML]}{[M^{n+}][L']}$$

(14.8)

Side reaction coefficients can be calculated using the ion pairing model. $K'_{M^{n+}L'}$ is the conditional stability constant relative to the free aquo ion M^{n+}.

The detection window in CLE-AdCSV is determined by the comparative magnitudes of α_{CuL} and that of the artificial ligand α_{CuAL}. It has been estimated that organic complexes with values for α_{CuL} approximately within a decade on either side of α_{CuAL} are measurable (Van den Berg et al., 1990; Nimmo et al., 1989; Apte et al., 1988). This detection window can be shifted to either side by increasing or decreasing the concentration of the artificial ligand (AL), or by using a stronger or weaker AL. The α_{CuAL} is determined by calibration of AL against another ligand with well-known side reaction coefficients with all important seawater constituents (e.g., EDTA).

14.6.4 Cu Speciation Measurements

A number of methods using different competing ligands for AdCLE-CSV and also LE-ASV have been used in the past to determine the Cu-binding ligand concentration ($[L_{Cu}]$) and conditional stability constant ($K_{Cu'L}$) at ambient seawater pH. Town and Filella (2000) have compiled a comprehensive systematic review of complexation parameters reported for Cu in natural seawater and methods applied. However, we recommend the use of AdCLE-CSV with the competing ligand, salicylaldoxime (SA) (Sander et al., 2005; Campos and Van den Berg, 1994), because complexation of Cu with SA in seawater is well characterized and its sensitivity allows the detection of even the lowest ligand and metal concentrations. The voltammetric parameters and concentrations of reagent solutions used are given in Table 14.3.

The actual metal-ligand titrations are performed by measuring the voltammetric peak current i_p of separate aliquots of the sample (e.g., 10–12) in separate Teflon vials, to which a buffer and increasing concentrations of copper standard solution have been added. One hour after the addition of copper, SA is added to a known final concentration well in excess of Cu^{2+} and other metals forming stable complexes with SA (typically between 2 and 20 μM; for side reaction coefficients see Campos and Van den Berg (1994)). This initial 1 hour equilibration time is applied to allow the added Cu to bind to natural ligands, before an excess of SA is added. Without this initial reaction time the achievement of equilibrium is more difficult, as $Cu(SA)_2$ first has to dissociate before the natural ligands can bind to the Cu. A final equilibration minimum time of 12 hours, but typically overnight, has been found to result in stable Cu-SA peaks. Figure 14.2 shows a typical CLE-AdCSV Cu titration curve of Cu (Figure 14.2b), corresponding voltammograms of the Cu-SA peak (Figure 14.2a), and [Cu'] vs. [CuL] and the nonlinear regression (according to Gerringa et al., 1995) used to receive Cu complexation parameters [L] and $K'_{M'L}$ (Figure 14.2c). The sensitivity S was also included in the fitting. Previously S was estimated by visually correcting the slope of a plot of i_p versus [M'], as the final sensitivity is reached asymptotically, a method that is somewhat prone to errors of judgment. The necessity for correction of S by modeling has been discussed by Hudson et al. (2003) and Van den Berg (2005).

The binding of other trace metals by organic ligands present in seawater can be studied in a similar way using CLE-AdCSV with a suitable choice of competing ligands or LE-ASV. Table 14.3 gives an overview of other trace elements where the organic complexation by strong ligands plays an important role in the speciation and bioavailability of the element. In principle, measurements and data treatment are carried as described for the CLE-AdCSV method for Cu with SA as competing ligand.

14.7 SHIPBOARD ANALYSIS

On many cruises it is desirable to get rapid results on trace metal distributions for different reasons: (1) as a control that samples have not been contaminated during sampling, (2) to measure local inputs (e.g., hydrothermal signatures like increased Fe or Mn concentrations), and (3) to measure unstable species, which cannot be preserved without changing the speciation (Sander and Koschinsky, 2000). Voltammetric equipment is very suitable for onboard analysis because of its compact size and robustness. In some cases the vibrations onboard ship may cause the Hg drop to fall off and the electrode stand will have to be damped by means of some foam underlay, or in severe cases, an antivibration electrode support based on a PVC plate suspended from elastic bands (Aldrich et al., 1999) can be used.

14.7.1 IN SITU TRACE METAL ANALYSIS

Although a real-time in situ data acquisition of trace metal concentrations is desirable, the technical complexity to develop such a device is immense. The main hurdles are the extreme low concentrations of trace metals in seawater, pressures up to several hundred bars, and calibration under in situ conditions.

TABLE 14.3

Recommended Methods for Voltammetric Determination of Organic Speciation of Trace Metals in Seawater

Element	Method	Conditions	Voltammetric Parameters	Typical Results	Reference
Cu	CLE-AdSCV with SA	10 mL filtered sample, [Cu] ≈ 0–30 nM, 50 μL HEPES buffer, pH 7.8 (1 M n-(2-hydroxyethyl) piperazine-N'-ethansulfonic acid in ammonia solution, Chelex-100 cleaned), t_{wait} = 1 h, then addition of 1–20 μM SA (10 mM stock solution in 0.1 M HCl), t_{equil} ≥ 12 h	HMDE, t_{acc} = ≥60 s at E_{acc} = –0.05 V, DP scan from –0.05 to –0.5 V, E_{step} = 25 mV, E_{step} = 2.5, E_p = 0.32 V	[L_C] = 2–15 μM; log K'$_{CuL}$ ≥ 15	Campos and Van den Berg, 1994; Sander et al., 2007
Zn	CLE-AdCSV with APDC	10 mL filtered sample, [Zn] ≈ 0–20 nM, 50 μL borate buffer, pH 8.2 (1.5 M borate dissolved in 0.4 M NaOH, Chelex-100 cleaned), t_{wait} = 0 s, addition of 60 μM APDC (0.013 M stock solution, Fisher Scientific used as is), t_{equil} = 12 h	HMDE, t_{acc} = 180 s at E_{acc} = –0.3 V, SQW scan from –0.8 to –1.2 V SQW modulation = 50 Hz, E_{step} = 25 mV	[L_{Zn}] = 2–15 μM; log K'$_{ZnL}$ = 10.0–10.5	Ellwood and Van den Berg, 2000
Co	CLE-AdCSV with nioxime or DMG	8.5 mL filtered seawater, [Co] ≈ 0–510 pM, t_{wait} = 1–2 h, 20 μL of 0.1 M DMG, t_{equil} = 12 h; addition of 50 μL EPPS (0.5 M, pH 8.0) and 1.50 mL NO₂₋ just before measurement	HMDE, t_{acc} = 90 s at E_{acc} = 0.6 V, linear scan from –0.6 to –1.4 V scan rate = 10 V	[L_{Co}] = 9–83 μM; log K'$_{CoL}$ = 16.3 ± 0.9	Ellwood and Van den Berg, 2001; Saito and Moffett, 2001
Ni	CLE-AdCSV with DMG	10 mL thawed sample, 200 μL DMG (0.1 M in 0.25 M NaOH), 100 μL buffer (1 M boric acid, 0.33 M NH₄OH, final pH 8.3) [Ni] ≈ 0–200 nM; equilibrated at room temperature for 36 h	HMDE, no voltammetric parameters given	[L_{Ni}] = 2–4 μM; log K'$_{NiL}$ = 17.3–18.7	Van den Berg and Nimmo, 1987; Martino et al., 2004
Cd	LE-ASV	10 mL filtered sample, [Cd] ≈ 0–10 nM, no buffer used, t_{equil} = 1 h	Hg-film electrode (MFE) E_{acc} = 0.85 V, t_{acc} = 5 min at 3,000 rpm; DP scan from 0.85 to 0.2 V, E_{step} = 50 mV, scan rate = 20 mV/s; E_p = 0.68 V, reconditioning of MFE: E_{cond} = 0.2 V, 1 min	[L_{Cd}] = 1–2.5 nM; log K'$_{CdL}$ = 9.82–10.93	Ellwood, 2004
Pb	LE-ASV	10 mL filtered sample, [Pb] ≈ 0–3 nM, no buffer used	Hg-film electrode (MFE) E_{acc} = 0.75 V, t_{acc} = 15 min at 5,000 rpm; DP scan from 0.75 to 0.15 V, scan rate = 10 mV/s; reconditioning of MFE: E_{cond} = 0.15 V, 5 min	[L_{Pb}] = 0.2–0.5 μM; log K'$_{PbL}$ = 9.7	Capodaglio et al., 1990, 1995

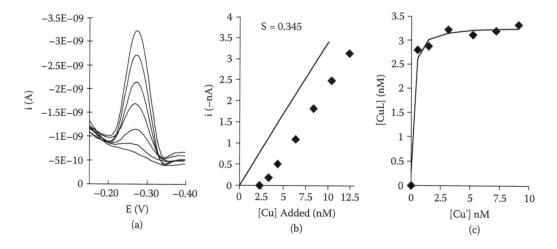

FIGURE 14.2 Cu ligand titration using salicylaldoxime as competing ligand in CLE-AdCSV (unpublished, own data). (a) Voltammograms of increasing concentrations of Cu (0, 1, 2, 4, 6, 8, 10 nM) added to a seawater sample from the Kermadec Arc, New Zealand, from 148 m depth. (b) Respective titration curve. (c) Gerringa plot of data and nonlinear regression (solid line) used to calculate the natural ligand concentration $[L_{Cu}]$ (here 3.29 nM) and the conditional stability constant log $K'_{Cu'L}$ (here 14.92). The $[Cu_T]$ was 2.29 nM and [SA] was 10 µM, resulting in log $\alpha_{CuSA} = 5.044$.

Thus far only a handful of truly *in situ* analyzers have been developed, of which only one, the VIP system, is actually commercially available. However, even this instrument is far from simple routine deployment, such as, e.g., a conductivity-temperature-depth (CTD) profiler is expected to perform. More information on these analyzers can be received from original literature describing these complex systems, like ZAPA using fluorescence spectroscopy to measure Mn at ambient seawater concentrations (Klinkhammer, 1994) and GAMOS successfully using chemiluminescence for Mn in hydrothermal plume observation (Okamura et al., 2001). Also, the voltammetric *in situ* profiling (VIP) system for the simultaneous determinations of Cd, Cu, and Pb based on a gel-integrated microsensor (GIME) is commercially available and was successfully applied for direct *in situ* trace metal analysis in natural waters (Tercier et al., 1998; Tercier-Waeber et al., 2002, 2005; Howell et al., 2003a, 2003b; Buffle and Tercier-Waeber, 2005). Finally, the *in situ* electrochemical analyzer (ISEA) uses gold amalgam electrodes to measure Fe and Mn species in hydrothermal vent systems (Luther et al., 2001, 2008).

14.8 INTERCALIBRATION EXERCISES

The determination of the concentrations of trace elements in seawater is an exacting task. However, with careful attention to detail, precise measurements may be obtained even at the very low levels sometimes encountered. A further and almost as difficult challenge is to ensure that the results obtained are accurate and therefore comparable to data produced elsewhere. Two approaches are required to successfully maintain accuracy: the co-analysis of certified reference materials and participation in interlaboratory comparisons.

Certified reference materials are available from the National Research Council of Canada (NRC; www.nrc-cnrc.gc.ca). These CRMs are intended for method development and can be used to monitor analytical performance. However, the volumes available are small and the concentrations of many of the metals may be quite different for the particular waters you are analyzing. Therefore, each laboratory should collect a volume of seawater to use as a secondary control sample with each analytical run. Instructions for the collection, homogenization, and preservation of such samples may be found at the NRC website.

There have been a small number of interlaboratory comparisons organized by the community (Landing et al., 1995). These are useful in assessing the performance of differing techniques (Bowie et al., 2007; Lysiak-Pastuszak, 2004), providing certainty that results are not skewed by analytical artifacts. The need to be able to determine Fe at very low concentrations drove a major intercomparison program (Bowie et al., 2006). The exercise was conducted as a rigorously blind comparison of seven analytical techniques by twenty-four international laboratories. The comparison was based on a large-volume (700 L), filtered surface seawater sample collected from the South Atlantic Ocean (the IRONAGES sample), which was acidified, mixed, and bottled at sea. The results were interpreted to imply the presence of (1) random variability (inherent to all intercomparison exercises), (2) errors in quantification of the analytical blank, and (3) systematic intermethod variability, perhaps related to secondary sample treatment (e.g., measurement of different physicochemical fractions of iron present in seawater) in the community dataset. Continued participation in such exercises serves to improve analytical performance and enhances the quality of the global dataset.

REFERENCES

Achterberg, E. P., and C. M. G. Van den Berg. 1994. In-line ultraviolet-digestion of natural-water samples for trace-metal determination using an automated voltammetric system. *Analytica Chimica Acta* 291:213–32.

Ahlers, W. W., M. R. Reid, J. P. Kim, and K. A. Hunter. 1990. Contamination-free sample collection and handling protocols for trace-elements in natural fresh waters. *Australian Journal of Marine and Freshwater Research* 41:713–20.

Aldrich, A. P., M. Jahme, and C. M. G. Van den Berg. 1999. Antivibration electrode support for shipboard stripping voltammetry. *Electroanalysis* 11:1155–57.

Apte, S. C., M. J. Gardner, and J. E. Ravenscroft. 1988. An evaluation of voltammetric titration procedures for the determination of trace-metal complexation in natural-waters by use of computer-simulation. *Analytica Chimica Acta* 212:1–21.

Boussemart, M., C. M. G. Van den Berg, and M. Ghaddaf. 1992. The determination of the chromium speciation in sea water using catalytic cathodic stripping voltammetry. *Analytica Chimica Acta* 262:103–15.

Bowie, A. R., E. P. Achterberg, P. L. Croot, H. J. W. de Baar, P. Laan, J. W. Moffett, S. Ussher, and P. J. Worsfold. 2006. A community-wide intercomparison exercise for the determination of dissolved iron in seawater. *Marine Chemistry* 98:81–99.

Bowie, A. R., S. J. Ussher, W. M. Landing, and P. J. Worsfold. 2007. Intercomparison between FI-CL and ICP-MS for the determination of dissolved iron in Atlantic seawater. *Environmental Chemistry* 4:1–4.

Brett, C. M. A., and A. M. O. Brett. 1993. *Electrochemistry principles, methods, and applications.* Oxford: Oxford University Press.

Bruland, K. W., J. R. Donat, and D. A. Hutchins. 1991. Interactive influences of bioactive trace metals on biological production in oceanic waters. *Limnology and Oceanography* 36:1555–77.

Bruland, K. W., R. P. Franks, G. A. Knauer, and J. H. Martin. 1979. Sampling and analytical methods for the determination of copper, cadmium, zinc, and nickel at the nanogram per liter level in sea water. *Analytica Chimica Acta* 105:233–45.

Bruland, K. W., E. L. Rue, J. R. Donat, S. A. Skrabal, and J. W. Moffett. 2000. Intercomparison of voltammetric techniques to determine the chemical speciation of dissolved copper in a coastal seawater sample. *Analytica Chimica Acta* 405:99–113.

Buesseler, K. O., A. N. Antia, M. Chen, S. W. Fowler, W. D. Gardner, O. Gustafsson, K. Harada, A. F. Michaels, M. R. van der Loeffo, M. Sarin, D. K. Steinberg, and T. Trull. 2007. An assessment of the use of sediment traps for estimating upper ocean particle fluxes. *Journal of Marine Research* 65:345–416.

Buesseler, K. O., J. E. Bauer, R. F. Chen, T. I. Eglinton, O. Gustafsson, W. Landing, K. Mopper, S. B. Moran, P. H. Santschi, R. Vernon Clark, and M. L. Wells. 1996. An intercomparison of cross-flow filtration techniques used for sampling marine colloids: Overview and organic carbon results. *Marine Chemistry* 55:1–31.

Buffle, J., and M.-L. Tercier-Waeber. 2005. Voltammetric environmental trace metal analysis and speciation: From laboratory to in situ measurements. *TrAC Trends in Analytical Chemistry* 24:172–91.

Campbell, J. A., and P. A. Yeats. 1981. Dissolved chromium in the northwest Atlantic Ocean. *Earth and Planetary Science* 53:427–33.

Campos, M. L., and C. M. G. Van den Berg. 1994. Determination of copper complexation in sea water by cathodic stripping voltammetry and ligand competition with salicylaldoxime. *Analytica Chimica Acta* 284:481–96.

Cannizzaro, V., A. R. Bowie, A. Sax, E. P. Achterberg, and P. J. Worsfold. 1999. Determination of cobalt and iron in estuarine and coastal waters using flow injection with chemiluminescence detection. *Analyst* 125:51–57.

Capodaglio, G., K. H. Coale, and K. W. Bruland. 1990. Lead speciation in surface waters of the eastern North Pacific. *Marine Chemistry* 29:221–33.

Capodaglio, G., G. Scarpoini, G. Toscano, C. Brabante, and P. Crescon. 1995. Speciation of trace metals in seawater by anodic stripping voltammetry: Critical analytical steps. *Fresenius Journal of Analytical Chemistry* 351:386–92.

Chen, M., W. X. Wang, and L. D. Guo. 2004. Phase partitioning and solubility of iron in natural seawater controlled by dissolved organic matter. *Global Biogeochemical Cycles* 18:GB4013, doi: 10.1029/2003GB002160.

Croot, P. L., and P. Laan. 2002. Continuous shipboard determination of Fe(II) in polar waters using flow injection analysis with chemiluminescence detection. *Analytica Chimica Acta* 466:261–73.

Cullen, J. T., M. P. Field, and R. M. Sherrell. 2001. Determination of trace elements in filtered suspended marine particulate material by sector field HR-ICP-MS. *Journal of Analytical Atomic Spectrometry* 16:1307–12.

Cullen, J. T., and R. M. Sherrell. 1999. Techniques for determination of trace metals in small samples of size-fractionated particulate matter: Phytoplankton metals off central California. *Marine Chemistry* 67:233–47.

Danielsson, L., B. Magnusson, and K. Zhang. 1982. Trace metal determinations in estuarine waters by electrothermal atomic absorption spectroscopy after extraction of dithiocarbamate complexes into freon. *Analytica Chimica Acta* 144:183–88.

Donat, J. R., K. A. Laoand, and K. W. Bruland. 1994. Speciation of dissolved copper and nickel in South San Francisco Bay: A multi-method approach. *Analytica Chimica Acta* 284:547–71.

Dupont, C. L., R. K. Nelson, S. Bashir, J. W. Moffett, and B. A. Ahner. 2004. Novel copper-binding and nitrogen-rich thiols produced and exuded by *Emiliania huxleyi*. *Limnology and Oceanography* 49:1754–62.

Elderfield, H. 1970. Chromium speciation in sea water. *Earth and Planetary Science* 9:10–16.

Ellwood, M. J. 2004. Zinc and cadmium speciation in subantarctic waters east of New Zealand. *Marine Chemistry* 87:37–58.

Ellwood, M. J., and C. M. G. Van den Berg. 2000. Zinc speciation in the northeastern Atlantic Ocean. *Marine Chemistry* 68:295–306.

Ellwood, M. J., and C. M. G. Van den Berg. 2001. Determination of organic complexation of cobalt in seawater by cathodic stripping voltammetry. *Marine Chemistry* 75:33–47.

Frew, R. D., D. A. Hutchins, S. Nodder, S. Sanudo-Wilhelmy, A. Tovar-Sanchez, K. Leblanc, C. E. Hare, and P. W. Boyd. 2006. Particulate iron dynamics during Fe cycle in subantarctic waters southeast of New Zealand. *Global Biogeochemical Cycles* 20: GB1S93, doi: 10.1029/2005GB002558.

Gasparon, M. 1998. Trace metals in water samples: Minimising contamination during sampling and storage. *Environmental Geology* 36:207–14.

Gerringa, L. J. A., P. M. J. Herman, and T. C. W. Poortvliet. 1995. Comparison of the linear Van Den Berg/Ruzic transformation and a non-linear fit of the Langmuir isotherm applied to Cu speciation data in the estuarine environment. *Marine Chemistry* 48:131–42.

Grotti, M., F. Soggia, S. D. Riva, E. Magi, and R. Frache. 2003. An in situ filtration system for trace element determination in suspended particulate matter. *Analytica Chimica Acta* 498:165–73.

Howell, K. A., E. P. Achterberg, C. B. Braungardt, A. D. Tappin, D. R. Turner, and P. J. Worsfold. 2003a. The determination of trace metals in estuarine and coastal waters using a voltammetric in situ profiling system. *The Analyst* 128:734–41.

Howell, K. A., E. P. Achterberg, C. B. Braungardt, A. D. Tappin, P. J. Worsfold, and D. R. Turner. 2003b. Voltammetric in situ measurements of trace metals in coastal waters. *Trends in Analytical Chemistry* 22:828–35.

Huang, S. L., E. R. Sholkovitz, and M. H. Conte. 2007. Application of high-temperature fusion for analysis of major and trace elements in marine sediment trap samples. *Limnology and Oceanography: Methods* 5:13–22.

Hudson, R. J. M., E. L. Rue, and K. W. Bruland. 2003. Modeling complexometric titrations of natural water samples. *Environmental Science and Technology* 37:1553–62.

Irigoyen, L., C. Moreno, C. Mendiguchia, and M. Garcia-Vargas. 2006. Application of liquid membranes to sample preconcentration for the spectrometric determination of cadmium in seawater. *Journal of Membrane Science* 274:169–72.

Isshiki, K., Y. Sohrin, H. Karatani, and E. Nakayama. 1989. Preconcentration of chromium(III) and chromium(VI) in sea-water by complexation with quinolin-8-ol and adsorption on macroporous resin. *Analytica Chimica Acta* 224:55–64.

Jönsson, J. Å., and L. Mathiasson. 1999. Liquid membrane extraction in analytical sample preparation. I. Principles. *TrAC Trends in Analytical Chemistry* 18:318–25.

Klinkhammer, G. P. 1994. Fiber optic spectrometers for in-situ measurements in the oceans: The zaps probe. *Marine Chemistry* 47:13–20.

Kolb, M., P. Rach, J. Schaferand, and A. Wild. 1992a. Investigations of oxidative UV photolysis. I. Sample preparation for the voltammetric determination of Zn, Cd, Pb, Cu, Ni and Co in waters. *Fresenius Journal of Analytical Chemistry* 342:341–49.

Kolb, M., P. Rach, J. Schaferand, and A. Wild. 1992b. Investigations of oxidative UV photolysis. II. Sample preparation for the voltammetric determination of mercury in water samples. *Fresenius Journal of Analytical Chemistry* 344:283–85.

Kozelka, P. B., and K. Bruland. 1998. Chemical speciation of dissolved Cu, Zn, Cd, Pb in Narragansett Bay, Rhode Island. *Marine Chemistry* 60:267–82.

Landing, W. M., G. A. Cutter, J. A. Dalziel, A. R. Flegal, R. T. Powell, D. Schmidt, A. Shiller, P. Statham, S. Westerlund, and J. Resing. 1995. Analytical intercomparison results from the 1990 intergovernmental-oceanographic-commission open-ocean base-line survey for trace-metals—Atlantic-Ocean. *Marine Chemistry* 49:253–65.

Luther, G. W., B. T. Glazer, S. F. Ma, R. E. Trouwborst, T. S. Moore, E. Metzger, C. Kraiya, T. J. Waite, G. Druschel, B. Sundby, M. Taillefert, D. B. Nuzzio, T. M. Shank, B. L. Lewis, and P. J. Brendel. 2008. Use of voltammetric solid-state (micro)electrodes for studying biogeochemical processes: Laboratory measurements to real time measurements with an in situ electrochemical analyzer (ISEA). *Marine Chemistry* 108:221–35.

Luther, G. W. III, T. F. Rozan, M. Taillefert, D. B. Nuzzi, C. Di Meo, T. M. Schank, R. A. Lutz, and S. C. Cary. 2001. Chemical speciation drives hydrothermal vent ecology. *Nature* 410:813–16.

Lysiak-Pastuszak, E. 2004. Interlaboratory analytical performance studies; a way to estimate measurement uncertainty. *Oceanologia* 46:427–38.

Mallini, L. J., and A. M. Shiller. 1993. Determination of dissolved manganese in seawater by flow injection analysis with colorimetric detection. *Limnology and Oceanography* 38:1290–95.

Martino, M., A. Turner, and M. Nimmo. 2004. Distribution, speciation and particle-water interactions of nickel in the Mersey Estuary, UK. *Marine Chemistry* 88:161–77.

Moffett, J. W., and L. Brand. 1996. Production of strong, extracellular Cu chelators by marine cyanobacteria in response to Cu stress. *Limnology and Oceanography* 41:388–95.

Moody, J. R. 1981. Sample handling for trace element analysis. *Analytical Proceedings* 18:337–39.

Mugo, R. K., and K. J. Orians. 1993. Seagoing method for the determination of chromium(III) and total chromium in sea-water by electron-capture detection gas-chromatography. *Analytica Chimica Acta* 271:1–9.

Murray, J. M., B. Spelland, and B. Paul. 1983. The contrasting geochemistry of manganese and chromium in the eastern equatorial tropical Pacific Ocean. In *Trace metals in seawater*, ed. C. S Wong. NATO Symposium Volume, pp. 641–669.

Nakayama, E., T. Kuwamoto, S. Tsurubo, and T. Fujinaga. 1981a. Chemical speciation of chromium in sea-water. II. Effects of manganese oxides and reducible organic materials on the redox processes of chromium. *Analytica Chimica Acta* 130:401–4.

Nakayama, E., T. Kuwamoto, S. Tsurubo, H. Tokoro, and T. Fujinaga. 1981b. Chemical speciation of chromium in sea-water. I. Effect of naturally-occurring organic materials on the complex-formation of chromium(III). *Analytica Chimica Acta* 130:289–94.

Nimmo, M., C. M. G. Van den Berg, and J. Brown. 1989. The chemical speciation of dissolved nickel, copper, vanadium and iron in Liverpool Bay, Irish Sea. *Estuarine, Coastal and Shelf Science* 29:57–74.

Nodder, S. D., and L. C. Northcote. 2001. Episodic particulate fluxes at southern temperate mid-latitudes (42–45 degrees) in the subtropical front region, east of New Zealand. *Deep Sea Research Part I* 48:833–64.

Okamura, K., H. Kimoto, K. Saeki, J. Ishibashi, H. Obata, M. Maruo, T. Gamo, E. Nakayama, and Y. Nozaki. 2001. Development of a deep-sea in situ Mn analyzer and its application for hydrothermal plume observation. *Marine Chemistry* 76:17–26.

Patterson, C. C., and D. M. Settle. 1976. The reduction of orders of magnitude errors in lead analysis of biological materials and natural waters by controlling external sources of industrial Pb contamination introduced during sample collection, handling and analysis. In *Accuracy in trace analysis: Sampling, sample handling, analysis*, ed. D. M. La Fleur. NBS Publication 322, Washington, D.C.

Pettine, M., and F. J. Millero. 1990. Chromium speciation in seawater—The probable role of hydrogen-peroxide. *Limnology and Oceanography* 35:730–36.

Pettine, M., F. J. Millero, and R. Passino. 1994. Reduction of chromium(VI) with hydrogen-sulfide in NACL media. *Marine Chemistry* 46:335–44.

Pettine, M., L. D'Ottone, L. Campanella, F. J. Millero, and R. Passino. 1998. The reduction of chromium (VI) by iron (II) in aqueous solutions. *Geochimica et Cosmochimica Acta* 62:1509–19.

Pino, V., I. Hernandez-Martin, J. H. Ayala, V. Gonzalez, and A. M. Afonso. 2007. Evaluation of the uncertainty associated to the determination of heavy metals in seawater using graphite furnace atomic absorption spectrometry. *Analytical Letters* 40:3322–42.

Price, N. M., G. I. Harrison, J. G. Hering, R. J. Hudson, P. M. V. Nirel, B. Palenik, and F. M. M. Morel. 1989. Preparation and chemistry of the artificial algal culture medium aquil. *Biological Oceanography* 6:443–61.

Pueyo, M., G. Rauret, D. Luck, M. Yli-Halla, H. Muntau, P. Quevauville, and J. F. Lopez-Sanchez. 2001. Certification of the extractable contents of Cd, Cr, Cu, Ni, Pb and Zn in a freshwater sediment following a collaboratively tested and optimised three-step sequential extraction procedure. *Journal of Environmental Monitoring* 3:243–50.

Rahmi, D., Y. Zhu, E. Fujimori, T. Umemura, and H. Haraguchi. 2007. Multielement determination of trace metals in seawater by ICP-MS with aid of down-sized chelating resin-packed minicolumn for preconcentration. *Talanta* 72:600–6.

Ružič, I. 1982. Theoretical aspects of the direct titration of natural waters and its information yield for trace metal speciation. *Analytica Chimica Acta* 140:99–113.

Saito, M. A., and J. W. Moffett. 2001. Complexation of cobalt by natural organic ligands in the Sargasso Sea as determined by a new high-sensitivity electrochemical cobalt speciation method suitable for open ocean work. *Marine Chemistry* 75:49–68.

Saito, M. A. and D. L. Schneider. 2006. Examination of precipitation chemistry and improvements in precision using the $Mg(OH)_2$ preconcentration inductively coupled plasma mass spectrometry (ICP-MS) method for high-throughput analysis of open-ocean Fe and Mn in seawater. *Analytica Chimica Acta* 565:222–233.

Sander, S., J. P. Kim, B. Anderson, and K. A. Hunter. 2005. Effect of UVB irradiation on Cu^{2+}-binding organic ligands and Cu^{2+} speciation in alpine lake waters of New Zealand. *Environmental Chemistry* 2:56–62.

Sander, S., and A. Koschinsky. 2000. Onboard-ship redox speciation of chromium in diffuse hydrothermal fluids from the north Fiji basin. *Marine Chemistry* 71:83–102.

Sander, S., A. Koschinsky, and P. Halbach. 2003b. Redox-speciation of chromium in the oceanic water column of the Lesser Antilles and offshore Otago Peninsula, New Zealand. *Marine and Freshwater Research* 54:745–54.

Sander, S. G., A. Koschinsky, G. J. Massoth, M. Stott, and K. A. Hunter. 2007. Organic complexation of copper in deep-sea hydrothermal vent systems. *Environmental Chemistry* 4:81–89.

Sander, S., T. Navratil, and L. Novotny. 2003a. Study of the complexation, adsorption and electrode reaction mechanisms of chromium(VI) and (III) with DTPA under adsorptive stripping voltammetric conditions. *Electroanalysis* 15:1513–21.

Siegel, D. A., E. Fields, and K. O. Buesseler. 2008. A bottom-up view of the biological pump: Modeling source funnels above ocean sediment traps. *Deep Sea Research Part I* 55:108–27.

Sirinawin, W., and S. Westerlund. 1997. Analysis and storage of samples for chromium determination in seawater. *Analytica Chimica Acta* 356:35–40.

Stanley, R. H. R., K. O. Buesseler, S. J. Manganini, D. K. Steinberg, and J. R. Valdes. 2004. A comparison of major and minor elemental fluxes collected in neutrally buoyant and surface-tethered sediment traps. *Deep Sea Research Part I* 51:1387–95.

Tercier, M.-L., J. Buffle, and F. Graziottin. 1998. A novel voltammetric in-situ profiling system for continuous real-time monitoring of trace elements in natural waters. *Electroanalysis* 10:355–63.

Tercier-Waeber, M.-L., J. Buffle, M. Koudelka-Hep, and F. Graziottin. 2002. Submersible voltammetric probes for in situ real-time trace element monitoring in natural aquatic systems. In *Environmental electrochemistry: Analysis of trace element biogeochemistry*, ed. M. Taillefert. ACS Symposium Series, Vol. 811. Washington, D.C.: American Chemical Society, pp. 16–39.

Tercier-Waeber, M. L., F. Confalonieri, G. Riccardi, A. Sina, S. Noel, J. Buffle, and F. Graziottin. 2005. Multi physical-chemical profiler for real-time in situ monitoring of trace metal speciation and master variables: Development, validation and field applications. *Marine Chemistry* 97:216–35.

Thomas, F. G., and G. Henze. 2001. *Introduction to voltammetric analysis: Theory and practice*. Melbourne, Australia: CSIRO Publishing.

Town, R. M., and M. Filella. 2000. A comprehensive systematic compilation of complexation parameters reported for trace metals in natural waters. *Aquatic Science* 62:252–95.

Van den Berg, C. M. G. 1982. Determination of copper complexation with natural organic ligands in seawater by equilibration with MnO_2. Part 1: Theory. Part 2: Experimental procedures and application to surface seawater. *Marine Chemistry* 11:307–42.

Van den Berg, C. M. G. 1995. Evidence for organic complexation of iron in seawater. *Marine Chemistry* 50:139–57.

Van den Berg, C. M. G. 2005. Chemical speciation of iron in seawater by cathodic stripping voltammetry with dihydroxynaphthalede. *Analytical Chemistry* 78:156–63.

Van den Berg, C. M. G., and M. Nimmo. 1987. Determinations of interactions of nickel with dissolved organic material in seawater using cathodic stripping voltammetry. *Science of the Total Environment* 60:185–95.

Van den Berg, C. M. G., M. Nimmo, P. Daly, and D. R. Turner. 1990. Effects of the detection window on the determination of organic copper speciation in estuarine waters. *Analytica Chimica Acta* 232:149–59.

Vink, S., E. A. Boyle, C. I. Measures, and J. Yuan. 2000. Automated high resolution determination of the trace elements iron and aluminium in the surface ocean using a towed fish coupled to flow injection analysis. *Deep Sea Research Part I* 47:1141–56.

Waite, A. M., and S. D. Nodder. 2001. The effect of in situ iron addition on the sinking rates and export flux of Southern Ocean diatoms. *Deep Sea Research Part II* 48:2635–54.

Wang, C. F., M. F. Huang, E. E. Chang, and P. C. Chiang. 1996. Assessment of closed vessel digestion methods for elemental determination of airborne particulate matter by ICP-AES. *Analytical Sciences* 12:201–7.

Weinstein, S. E., and S. B. Moran. 2004. Distribution of size-fractionated particulate trace metals collected by bottles and in-situ pumps in the Gulf of Maine-Scotian shelf and Labrador Sea. *Marine Chemistry* 87:121–35.

Wells, M. L. 2003. The level of iron enrichment required to initiate diatom blooms in HNLC waters. *Marine Chemistry* 82:101–14.

Worsfold, P. J., E. P. Achterberg, A. R. Bowie, V. Cannizzaro, S. Charles, J. M. Costa, F. Dubois, R. Pereiro, B. San Vicente, A. Sanz-Medel, R. Vandeloise, E. Vander Donckt, P. Wollast, and S. Yunus. 2002. Integrated luminometer for the determination of trace metals in seawater using fluorescence, phosphorescence and chemiluminescence detection. *Journal of Automated Methods and Management in Chemistry* 24:41–47.

Wu, J. 2007. Determination of picomolar iron in seawater by double $Mg(OH)_2$ precipitation isotope dilution high-resolution ICPMS. *Marine Chemistry* 103:370–81.

Wu, J., and E. A. Boyle. 1998. Determination of iron in seawater by high-resolution isotope dilution inductively coupled plasma mass spectrometry after $Mg(OH)_2$ coprecipitation. *Analytica Chimica Acta* 367:183–91.

Wu, J., S. E. Calvertand, and C. S. Wong. 1997. Nitrogen isotope variations in the subarctic Northeast Pacific: Relationships to nitrate utilization and trophic structure. *Deep Sea Research* 44:287–314.

Yokoi, K., T. Tomisaki, T. Koide, and C. M. G. Van der Berg. 1995. Effective UV photolytic decomposition of organic-compounds with a low-pressure mercury lamp as pretreatment for voltammetric analysis of trace-metals. *Fresenius Journal of Analytical Chemistry* 352:547–49.

15 Trace Analysis of Selected Persistent Organic Pollutants in Seawater

Oliver Wurl

CONTENTS

15.1 INTRODUCTION

Persistent organic pollutants (POPs) are globally distributed and readily detected in the environment, even in remote regions of Earth with no historical usage (Iwata et al., 1993; Stern et al., 1997; Kallenborn et al., 1998; Yao et al., 2002). POPs encompass a wide range of xenobiotic chemicals, including polycyclic aromatic hydrocarbons (PAHs), polychlorinated biphenyls (PCBs), polybrominated diphenyl ethers (PBDEs), and organochlorine pesticides (OCPs) such as chlordanes, hexachlorocyclohexane (HCH), dichlorodiphenyltrichloroethane (DDT), and related compounds. Such compounds are of great concern due to their persistence, carcinogenicity, and endocrine-disrupting effects. The tendency of POPs to undergo bioaccumulation in the food chain has also led to concerns regarding their ecotoxicology, and about which residues pose the greatest threat to humans, marine mammals, and other wildlife (Skaare et al., 2002; Tanabe, 2002). POPs have, for example, been reported as having an effect on the endocrine systems of polar bears, thereby indicating the potential of these chemicals to impact mammals at the top of the marine food web (Oskam et al., 2004). The atmosphere is considered to be the predominant pathway for long-range transport of POPs toward polar regions, where condensation and deposition of these compounds occurs on land, sea, and ice. However, less hydrophobic POPs (i.e., HCH isomers) are also carried into the Arctic environment through water currents (Li et al., 2004; Li and Macdonald, 2005), whereas strongly hydrophobic POPs (i.e., highly chlorinated PCBs and brominated PBDEs) are associated with sinking particles and transport pathways in the deep ocean (Gustafsson et al., 1997; Axelman and Broman, 2001). River outflows often carry elevated levels of POPs (i.e., Zhulidov et al., 2000; Zhang et al., 2003; Carroll et al., 2008) and are a major source of such compounds in coastal and estuarine environments. As such, POPs have the potential to threaten fish and bivalve populations (Bayen et al., 2007; Lema et al., 2007) as well as aquaculture operations in coastal waters (Sapkota et al., 2008; Shaw et al., 2008), justifying monitoring programs in these areas.

Data on POPs in seawater are sparse, particularly in oceanic and polar regions (Iwata et al., 1993; Schulz-Bull et al., 1998; Yao et al. 2002; Sobek et al., 2004; Gustafsson et al., 2005), even though quantitative analysis of POPs in such regions is of primary importance in understanding their fate and distribution mechanisms. The collection of such data is challenging due to the need to separate prevailing background levels of POPs from those arising from local sources, and to accurately estimate low background levels with respect to potential sources of contamination incurred during sampling processing and analysis. POP concentrations in oceanic waters can be as low as a few picograms per liter, whereas concentrations in coastal waters can be higher by three orders of magnitude (Wurl and Obbard, 2006). However, significant advances in technology have recently been made, with respect to both sampling large volumes of water and instrumental techniques. One of the newer technologies, presented in this chapter, is gas chromatography-ion trap tandem mass spectrometry (GC-IT-MS/MS), which has the advantage of higher sensitivity and specificity when compared with classical, single-quadrupole MS detection systems. GC-IT-MS/MS instruments are also more affordable than costly high-resolution GC-MS systems, a particularly important consideration with respect to monitoring POPs in developing countries (Muir and Sverko, 2006). Nevertheless, reliable determination of the extremely low POP concentrations (typically pg L^{-1} to ng L^{-1}) found in oceanic waters can be achieved only with rigorous clean sampling and analytical protocols. This chapter describes those protocols.

15.2 SAMPLING, FILTRATION, AND SAMPLE STORAGE

As for other types of trace analysis, sampling is a critical step in the analysis of POPs, and the greatest attention to detail is required of research personnel during sample collection. The conditions and activities normally encountered onboard a research vessel, including engine emissions and the accumulation of exhaust particles on deck, oil and lubricant storage, paints, and waste disposal, are probably the most significant potential sources of contamination. Based on many years of experience, Duinker and Schulz-Bull (1999) showed that conditions on research vessels are often unsuitable for the contamination-free processing of collected samples on board, unless a clean laboratory is available. Gustafsson et al. (2005), for example, constructed an overpressurized clean room with activated carbon- and HEPA-filtered air on the icebreaker ODEN. Only specifically trained research personnel accessed this restricted area through an intermediate air lock, where the researcher changed into clean room clothing. Several hundred liters of seawater were collected via a stainless steel seawater intake system situated at the bow of the ship, and transferred directly into the ultraclean laboratory. Such specialized facilities for the analysis of POPs are rarely available, or affordable. For this reason, collected samples often need to be stored and analyzed in a land-based laboratory. Long-term storage of large volumes of seawater required for POP analysis is often impractical; however, solid adsorbents are well suited to the collection and storage of POPs extracted from seawater samples. The higher concentrations of POPs in estuarine and coastal waters reduce the required water volume to between 5 and 20 L, allowing the more classical method of liquid-liquid extraction to be applied. Both extraction techniques are described in detail in the following sections.

15.2.1 SAMPLING OF OCEANIC WATERS ON ADSORBENT CARTRIDGES

15.2.1.1 Sampling Devices

Self-contained systems for in situ filtration and extraction of organic contaminants are commercially available. Such systems pump water through a filter to remove suspended particulates for analysis, then through a cartridge containing adsorbent material to extract the target analytes from the liquid phase. The manually controlled INFILTREX 300 system (AXYS Technologies, Inc., Sydney, Canada) is a ship-based sampler and therefore restricted to the collection of surface waters. It operates at a flow rate of about 2 L min^{-1}, collecting a total volume of up to 1,000 L. The sampler is made of noncontaminating material, such as stainless steel and fluorinated ethylene propylene (FEP). The Kiel in situ pump (KISP; see Figure 1.6) works on the same principle but is programmable and submersible, collecting about 2,000 L of water at depths of 6,000 m. The KISP is described in detail in Petrick et al. (1996) and is commercially available from Scholz Ingenieur Büro (Fockbek, Germany). The device consists of a pump, filter holder, adsorbent cartridges, battery, and electronic unit to control the operation.

15.2.1.2 Preparation of Adsorbent Columns

Amberlite XAD-2, a hydrophobic cross-linked polystyrene copolymer (typically 20–60 mesh size), is the most widely used adsorbent material for collecting organic contaminants. It allows larger sample volumes to be processed per time unit (e.g., five bed volumes min^{-1}) compared with other adsorbent materials, is available in highly purified form, and has an extraction efficiency of 70%–90%, which is sufficient for oceanic waters. To improve extraction efficiency, two XAD columns can be installed in series.

Purified XAD-2 resins are recommended (i.e., SupelPak 2B, Sigma-Aldrich), although precleaning is still required to achieve the lowest blank values.

XAD-2 batches of poorer quality pro analysis (p.a. quality) contain a considerable amount of smaller particles, which need to be removed to avoid resistance to water flow during sampling. The following procedure for doing this is based on Ehrhardt and Burns (1999). Five hundred grams of XAD-2 is placed in a 3 L Erlenmeyer flask and 2 L of ultrapure water added. The flask is shaken for 2–3 minutes and the larger particles are allowed to settle, while the undesired smaller particles float on top. The separation may take 1–10 hours, depending on the quality of the XAD-2 resin. If initial wetting is insufficient to allow larger particles to settle, acetone can be added until settling is achieved. Ultrapure water is then added through a tube placed below the small, floating particles in order to remove these particles, a procedure that may need to be repeated. The XAD-2 resin is rehydrated in a beaker and covered with methanol for a few hours. Water is then added to replace the methanol while retaining XAD-2 at the bottom of the beaker.

Following the removal of small particles, the XAD-2 adsorbent is cleaned and proofed by Soxhlet preextraction using 1:1 v/v dichloromethane:methanol for about 1 to 5 days, depending on the quality of XAD-2. The solvent is collected, reduced in volume, and injected to a GC with an electron capture detector (ECD) to check for purity. It is well worth purchasing the highest available grade of XAD-2 to minimize the time and solvent costs involved in precleaning the resin. The solvent mixture is exchanged at least three times per batch of XAD-2. Following the last extraction step, exposure of the clean resin to ambient laboratory air should be avoided, since passive adsorption of POPs and other background contaminants from indoor air to XAD-2 may increase crucial blank values above those required for the successful measurement of oceanic POP concentrations. As an alternative to the use of high-maintenance clean rooms, a home-made glove box can be used, as shown in Figure 15.1. The box is filled with purified N_2 through a gas inlet before exposing XAD-2 to the interior environment. All the following steps can then be conducted under a N_2 atmosphere. The XAD-2 is filled as slurry in a polytetrafluorethylene (PTFE) column with an end frit (20 mm i.d. × 300 mm) by gravity, and optimal packing supported by gentle aspiration to remove air bubbles. After the XAD-2 has settled,

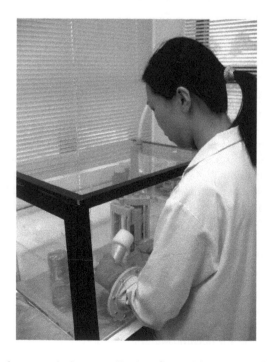

FIGURE 15.1 A glove box home-made from acrylic glass for working under a N_2 atmosphere.

the column is plugged with a preextracted quartz glass wool plug and closed with the frit. The column can then be removed from the N_2 atmosphere and the solvent inside the column replaced with ultrapure water (preferably prewashed by liquid-liquid extraction with hexane). The water should enter the column from below to force out any remaining air bubbles. Sealed columns have been stored at 4°C for 1 year without any significant increase of blank values (Ehrhardt and Burns, 1999).

Alternatively, the column can be prepacked with XAD-2 and inserted in a continuous extractor, as described by Ehrhardt (1987). If such a device is not readily available, a good glass blower should be able to provide the necessary components. Other techniques for cleaning and extracting POPs from samples using XAD-2 include accelerated solvent extraction (ASE; Jang et al., 1999; Dionex Corp., 2002) and microwave-assisted extraction (MAE), although further studies are required to explore the potential of such time- and solvent-saving extraction techniques for the analysis of POPs trapped on XAD-2.

15.2.1.3 Preparation of Filters

Filtration of samples entering the XAD column is necessary to prevent particles from clogging the adsorbent material and reducing the flow rate. This can be achieved using glass fiber (GF) filters of nominal pore size 0.7 µm designed to fit stainless steel or PTFE filter holders with a diameter of 142 or 293 mm (e.g., Whatman GF/C). The filters (which should be free of binders) are baked at a temperature of 400°C to 450°C for 12 hours on individual aluminum sheets, making sure that they are well separated from each other to allow efficient removal of contaminants. After baking, the filters are wrapped individually in the aluminum sheets and stored in an air- and moisture-tight container. The filters should not be treated with any solvents and should be handled with clean tweezers after baking.

15.2.1.4 Storage of Columns and Filters

After passing purified nitrogen through them to remove any residual water, the sealed XAD columns can be stored at 4°C. PAHs have been found to be stable on XAD for a period of up to 3 months (Green and Le Pape, 1987); however, prompt elution and analysis of adsorbed POPs is recommended. Filters should be folded twice, wrapped in aluminum sheeting (prebaked at 450°C), and stored at –20°C in air- and moisture-tight containers.

15.2.2 Sampling of Estuarine and Coastal Waters

Although the aforementioned method for collecting oceanic waters is also applicable to coastal waters, a more classical approach combining liquid-liquid extraction with simple, economical, and readily available sampling devices is described below. Studies in remote coastal areas may still require the use of XAD columns. On the other hand, Yao et al. (2002) have used liquid-liquid extraction to isolate POPs from oceanic waters in the Arctic, although they provide no details regarding the volumes of sample collected.

15.2.2.1 Sampling Devices

Up to 10 L of coastal surface waters can be collected for POP analysis using a 12/24 VDC pump fitted with a Teflon diaphragm (e.g., model PQ-12/24, Greylor, Cape Coral, Florida). Such pumps are effective in collecting samples from depths of at least 15 m (Wurl and Obbard, 2006). Teflon-lined tubing should be used to transfer water through the in-line filtration unit (see Section 15.2.2.2) and precleaned sample container (see Section 15.3.1.1 for cleaning procedure). Prior to sample collection, the pump head and tubing are rinsed with acetone and ultrapure water and stored under clean conditions.

Alternatively, water can be collected using Teflon-lined Niskin or Go-Flo bottles deployed on a Kevlar line (see Section 1.5.2). The water is then transferred through a filtration unit and into the

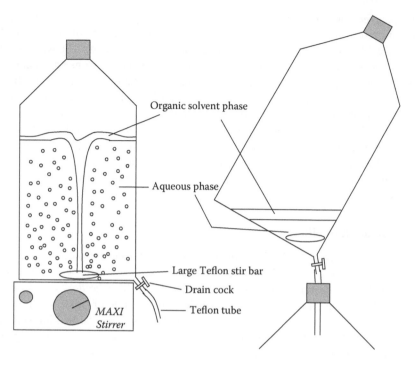

Organic solvent phase

Aqueous phase

Large Teflon stir bar

Drain cock

Teflon tube

MAXI Stirrer

FIGURE 15.2 Setup of liquid-liquid extraction using large-capacity stirrer.

sample container via a pump. Care has to be taken not to introduce potential contamination from ship and deck operations (e.g., dust and engine emissions) during transfer. A custom-made glass bottle with a volume of 10 to 15 L and a drain at the bottom (i.e., aspirator bottle shown in Figure 15.2) makes a very practical sample container. The bottle should contain 200 mL of hexane (purest grade for trace organic analysis) before filling with the filtered sample in order to prevent significant losses of analytes, through adsorption to the bottle wall, before extraction can be performed. The water can be directly extracted in the bottle by liquid-liquid extraction (see Section 15.3.2.2). The sample should be stored at 4°C in the dark and extracted as soon as possible (preferably within 2 to 3 hours of collection).

15.2.2.2 Filtration

GF filters and large stainless steel or PTFE filter holders are recommended, the use and handling of which are described in Section 15.2.1.3.

15.3 SAMPLE PROCESSING

15.3.1 Cleaning Procedures

15.3.1.1 Glassware

All glassware and sample containers should be soaked in detergent solution overnight, rinsed several times with hot tap water and ultrapure water, and then dried at 250°C for at least 12 hours. Acetone is used to rinse glassware several times prior to use. Pasteur pipettes and autosampler vials are cleaned by rinsing with acetone, then heated to 400°C for 12 hours and kept sealed in a glass bottle prior to use. The covered glassware needs to be stored in a dry, clean place.

15.3.1.2 Chemicals

Ultraclean solvents for the trace analysis of organic contaminants are commercially available (e.g., pesticide grade, Tedia; SupraSolv®, Merck; UltraRes Analyzed™, JT Baker) and the distillation of analytical-grade solvents is often no longer necessary (Duinker and Schulz-Bull, 1999). However, after breaking the seal of a solvent bottle the potential to contaminate the batch of solvent is high. A liquid dispenser (e.g., EMCLAB®) made with noncontaminating material and fitted to the solvent bottle enables safe and easy dispensing of the required volumes (i.e., 5 to 100 mL) while preventing airborne contaminants from entering the bottle. It is recommended that the quality of the solvents be checked regularly by concentrating 100 mL of each solvent to 50 µL in a rotary evaporator, using a purified N_2 gas stream (see Section 15.3.3), and injecting 2 µL onto a GC fitted with an ECD detector.

The required reagents, including sodium sulfate (e.g., pesticide grade, Pestanal®, Riedel-de-Haen), silica gel, and aluminum oxide, are baked at 400°C for at least 12 hours and stored in smaller amounts (sufficient to process one batch of sample) in sealed glass bottles. These chemicals have the potential to adsorb water and volatile contaminants, and so exposure to ambient air needs to be restricted as much as possible. Adding 10% by mass of ultrapure water to the baked silica gel and aluminum oxide inhibits adsorption and makes the reagents less likely to undergo activity changes.

15.3.2 Extraction Procedures

Preferably, the extraction and further analytical steps should be done in a clean laboratory fitted with activated carbon and HEPA filters. Such facilities are required for the processing of low-concentration samples from oceanic waters. Coastal waters can be processed in regular laboratories that have clean workspaces designated specifically for organic contaminant analysis. Dust and handling are most likely sources of contamination, and so it is recommended that the work be carried out at a laminar airflow workstation (class 100) by personnel wearing clean laboratory clothing.

15.3.2.1 Elution from XAD-2 Columns

XAD-2 columns are eluted sequentially with 250 mL of methanol followed by 350 mL of dichloromethane delivered at a flow rate of 2 mL min⁻¹ with a reciprocating pump. Each of the two fractions is collected in 500 mL round-bottom flasks and spiked with a known amount of surrogate standard (see Section 15.5.3). The methanol fraction is carefully concentrated to 100 mL by rotary vacuum evaporation. The water bath of the rotary evaporator is kept at room temperature to avoid rapid evaporation, which may lead to the loss of volatile analytes. Prior to use the rotary evaporator should be rinsed by evaporating 200 mL of acetone. For oceanic samples, the rotary evaporator should be purged with purified N_2 gas. Following evaporation to 100 mL, the methanol fraction is transferred to a 250 mL separation funnel and extracted three times with 25 mL of *n*-hexane for 10 minutes. The three *n*-hexane extracts are combined and sodium sulfate is added to remove any remaining water. The *n*-hexane extract is then combined with the dichloromethane extract and reduced in volume to 1–2 mL by rotary evaporation under vacuum.

15.3.2.2 Liquid-Liquid Extraction

The application of liquid-liquid extraction to coastal waters is illustrated in Figure 15.2. The sample is spiked with a known amount of surrogate standard (see Section 15.5.3). The sample is extracted three times with *n*-hexane (about 200 mL for an 8 L sample) by stirring for at least 30 minutes. A heavy-duty stirrer, capable of handling volumes of up to 20 L, and large stir bar (>70 mm) are necessary to ensure that the organic phase is pulled down to the bottom of the bottle. The stirring speed, size of stir bar, and amount of solvent can be adjusted so that the organic phase is just in contact with the stir bar, as illustrated in Figure 15.2. This creates solvent organic bubbles that rise back to the surface, resulting in a continuous cycling of organic solvent that facilitates the transfer of target analytes from the aqueous to the organic phase. After an extraction time of at least 30 minutes, the two phases are allowed to separate (typically for 15 to 20 minutes) in the sample bottle. The aqueous phase is collected in a second sample bottle (Figure 15.2), whereas the organic phase is collected in a 1,000 mL round-bottom flask. The extraction of the aqueous phase is repeated twice. To the three combined n-hexane extracts sodium sulfate is added to remove any remaining water. The n-hexane extract is reduced in volume to 1–2 mL by rotary vacuum evaporation, as described in the previous section.

Alternatively, liquid-liquid extraction can be performed in 10 L separation funnels shaken by hand. However, large separation funnels are difficult to handle and to shake for the required time (i.e., 15–20 minutes for a single extraction step). Extraction efficiencies (typically 75%–95%) are similar for both classical and stirring liquid-liquid extraction methods (Wurl, unpublished).

15.3.2.3 Microwave-Assisted Extraction (MAE) of Suspended Particles

Frozen filters carrying suspended particulate samples are thawed out and transferred to Teflon-lined extraction vessels under a nitrogen atmosphere (Figure 15.1). A known amount of surrogate standard (see Section 15.5.3) is added. After adding the surrogate standard, the extraction vessels are closed, retrieved from the nitrogen atmosphere, and allowed to stand for 2 hours to allow adsorption of the surrogate to the filter. The samples are then extracted with 50 mL of 3:2 v/v acetone:hexane in a microwave oven (e.g., Mars X, CEM Corp.; see Figure 15.3a) for 18 minutes at 115°C. After extraction, the vessels are allowed to cool to room temperature. Under a nitrogen atmosphere, the supernatant is then filtered through glass wool, collected in a 250 mL round-bottom glass flask, and combined with a 2–3 mL acetone:hexane (3:2) wash of the sample residue. The combined extract is reduced in volume to 1–2 mL by rotary evaporation, as described in Section 15.3.2.1. Filter material can be extracted using classical Soxhlet extraction (Sobek and Gustafsson, 2004); however, MAE is generally safer for laboratory personnel (closed system with temperature and pressure control), more economical, faster, and environmentally more friendly, as the amount of solvent is much reduced. Accelerated solvent extraction (ASE) offers similar benefits as MAE and is applicable to the extraction of POPs associated with suspended particles (Bruhn and McLachlan, 2002).

FIGURE 15.3 (a) Microwave-assisted extraction (MAE) instrument for multiple extraction of twelve solid samples. (b) Multineedle device for final evaporation step using a gentle N$_2$ stream.

15.3.3 CLEANUP AND FRACTIONATION OF EXTRACTS

It is essential to remove undesired compounds from XAD-2 and liquid-liquid extracts prior to analysis, as the sample matrix is generally complex (especially for coastal water extracts). Furthermore, different groups of POPs (e.g., PCBs, PBDEs, OCPs, and PAHs) often have similar physicochemical properties, which can lead to coelution of individual compounds from a GC capillary column. Separation of these compounds into different fractions avoids coelution and possible misidentification due to overlapping MS spectra, facilitating sensitive and accurate quantification.

Adsorption chromatography is the most common technique for cleanup and fractionation of solvent extracts for organic contaminant analysis, as described in Section 15.3.3.1. Aluminum oxide, silica gel, and florisil are widely used adsorbents, and are commonly applied to the following types of sample matrix, in order of increasing complexity: water and air > sediment > biological samples. Petrick et al. (1988) used high-performance liquid chromatography (HPLC) for cleanup and fractionation of seawater extracts (see Section 15.3.3.2) in order to overcome some of the drawbacks associated with adsorption chromatography, including relatively poor reproducibility and separation efficiency.

15.3.3.1 Procedure Using Alumina Oxide Chromatography Column

A glass column (200 mm × 5 mm i.d.) with a reservoir at the top and a drain cock at the base is the most practical type for use in adsorption chromatography (Figure 15.4). A small glass wool plug (precleaned at 450°C for > 4 h) is placed above the drain cock to avoid adsorbent material being released. Precleaned aluminum oxide is prepared as a slurry by adding 2 g of the adsorbent to a conical flask containing twice its volume of n-hexane. The glass column is rinsed and filled with n-hexane to just below the reservoir. The slurry is then transferred into the glass column with a precleaned Pasteur pipette, releasing the slurry underneath the surface of the n-hexane. The reservoir is then filled with n-hexane, the drain cock opened, and the solvent drained while tapping the glass column gently to compact the bed of aluminum oxide. The drain cock is closed just before the level of solvent reaches the bed of aluminum oxide, and the glass column capped. The column is allowed to equilibrate for at least 1 hour before flushing it with five bed volumes of n-hexane. The solvent level should be allowed to stand 1 to 2 mm above the adsorbent bed.

The sample extract is transferred to the column using a Pasteur pipette, which is then rinsed with a small amount of n-hexane. A receiving flask is placed under the column and the extract allowed to fall to the level of the adsorbent bed by opening the drain cock. To elute the first and least polar fraction, 6 mL of hexane is added and collected, closing the drain cock just before the solvent level reaches the adsorbent bed. This fraction contains PCBs and PBDEs, which have similar physicochemical properties and there-

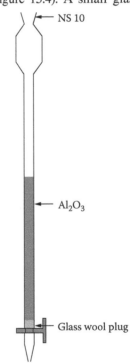

FIGURE 15.4 Al$_2$O$_3$ adsorption column for cleanup of sample extracts.

fore elute at the same time (Oros et al., 2005), although an advanced fractionation procedure using HPLC techniques can be used to separate PCBs and PBDEs into different fractions (see Section 15.3.3.2). After changing the receiving flask, a second fraction, which contains PAHs and OCPs, is eluted using 6 mL of 2:1 v/v dichloromethanehexane (Dachs et al., 1996; Dachs and Bayona, 1997). A constant elution flow rate of about 1 mL min^{-1} is recommended for optimum reproducibility and separation, and can easily be adjusted using nitrogen overpressure. Automatic systems for sample

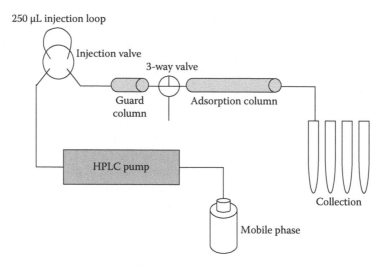

FIGURE 15.5 Schematic illustration of a HPLC system for cleanup of sample extracts.

cleanup and fractionation, including prepacked adsorption columns, are commercially available (e.g., SPE-01, Promochrom Technologies Ltd.; PowerPrep/SPE, Fluid Management System, Inc.). However, such systems need to be carefully checked for potential contamination, particularly the adsorption columns.

15.3.3.2 Procedure Using High-Performance Liquid Chromatography (HPLC)

A typical HPLC system for the cleanup of solvent extracts is illustrated in Figure 15.5. The system consists of a pulse-free HPLC pump, an injection valve with a 250 μL loop, an analytical HPLC column (Nucleosil 100-5; 200 × 4 mm i.d.) and guard column (Nucleosil 50-5; 30 × 4 mm i.d.), a three-way valve, and stoppered conical flasks. The system and procedure are based on Petrick et al. (1988) and Schulz-Bull and Duinker (1999). A new HPLC system should be thoroughly rinsed with 250 mL each of acetone, dichloromethane, 1:1 v/v dichloromethane:pentane, and finally, 500 mL pentane. The system should be calibrated once a week, or every twenty to thirty samples. For calibration, 200 μL of a dilute standard solution containing target analytes (e.g., 1 and 1,000 ng of halogenated compounds and PAHs, respectively) and surrogate standards is injected and eluted with n-pentane (up to 11.0 mL), 4:1 v/v n-pentane:dichloromethane (11.0–15.0 mL), and dichloromethane (>15 mL) at a flow rate of 0.5 mL min^{-1}. The eluates are collected in 0.5 mL fractions and analyzed according to the procedure described in Section 15.4. Table 15.1 summarizes the POPs typically found in four groups of these fractions, although the actual distributions may vary between different systems, and depending on the age of the column; hence, regular calibration is important.

TABLE 15.1
Elution Volumes to Fractionate Various POPs Using HPLC Column

Fraction No.	Volume (ml)	Solvent	Target Analytes
1	0.5–2.0	n-Pentane	Hydrocarbons (alkanes, alkenes)
2	2.0–4.5	n-Pentane	PCBs, PBDEs, few OCPs (DDE, HCB)
3	4.5–11.0	n-Pentane	PAHs
4	11.0–15.0	n-Pentane:DCM (4:1)	OCPs (i.e., DDT, HCHs, Dieldrin)

Source: Data from Schulz-Bull and Duinker, 1999.

HPLC is the only cleanup/fractionation technique currently known to separate PCBs and PBDEs. For example, an isocratic elution from a 2-(1-pyrenyl) ethyldimethylsilylated silica column using 98:2 v/v isooctane:toluene at a flow rate of 1 mL min^{-1} and a column temperature of 45°C has been found to provide optimal fractionation of PCBs (4.0–5.8 minutes) and PBDEs (5.8–9.0 minutes) in food sample extracts (Gómara et al., 2006). Application of this method to seawater extracts has not yet been reported, but its adoption is expected to be straightforward.

15.3.4 FINAL EVAPORATION STEP

The fractionated extracts are reduced slowly in volume to about 1 mL in a rotary evaporator, then transferred into preweighed autosampler vials with Pasteur pipettes. Using a gentle stream of purified nitrogen, the volume is further reduced to 500 µL or less in a clean working environment (e.g., under laminar flow workbench or nitrogen atmosphere). This is a critical step, as significant amounts of analyte may be lost through rapid evaporation, or contaminants introduced to the clean extracts. Typical final volumes of 50 to 500 µL correspond to concentration factors of between 200,000 and 2,000,000 for 100 L of a collected seawater sample. Using a multineedle device, several extracts can be reduced in volume simultaneously (Figure 15.3b), with no additional heating required. It is useful to have a vial containing the desired final volume on hand to enable a quick comparison with, and prompt removal of vials from the gas stream. The vials are closed tightly and the exact volume determined by weight difference using density of the final solvent. For low volumes (50–250 µL), the use of low-volume inserts for the autosampler vials is needed to ensure correct volume injection into the GC instruments. The weight of each vial is noted before storage at –20°C and again just before analysis to check for loss of solvent.

15.4 ANALYTICAL PROCEDURE USING GAS CHROMATOGRAPHY-ION TRAP MASS SPECTROMETRY

A single ion trap can essentially perform the same tandem mass spectrometry (MS/MS) functions as a more expensive triple quadrupole instrument, including multiple reaction monitoring (MRM) of target compounds eluting from an online chromatographic system. However, the selection, fragmentation, and product ion analysis of target compounds are performed sequentially in an ion trap, as opposed to simultaneously in a triple quadrupole instrument, making the latter more sensitive for MRM applications. Nevertheless, ion traps offer significant advantages over single quadrupole instruments, the most important being true MS/MS capability.

The MRM technique described below for POP analysis has three basic steps. In step 1, analyte molecules eluting from the GC system and entering the mass spectrometer at a particular retention time undergo electron ionization (EI), forming both molecular and fragment ions in the process. Selected ions, corresponding to target compounds (i.e., POPs) known to elute at that particular retention time, are then trapped by the application of an oscillating radio frequency (RF) voltage, which causes them to follow stable trajectories within the trap while other ions are ejected. This second step virtually eliminates chemical interference from spurious or unwanted ions. Third, in order to confirm analyte identity, the isolated precursor ions undergo a second fragmentation step achieved by collision-induced dissociation (CID), using a CID voltage applied to the end-cap electrodes of the trap. This further excites the trapped ions, which collide with atoms of an inert gas (helium) introduced into the trap. Detection of the resulting product ions allows for specific, sensitive quantification of individual POPs.

The following procedure was developed on a Varian 4000 GC-IT-MS/MS system equipped with an additional ECD detector (dual-column system) and a CP 8400 autosampler. However, implementation of this procedure on similar instruments should be straightforward for an analytical chemist with experience in gas chromatography and mass spectrometry.

15.4.1 GC INJECTION

Large-volume, splitless injection using 4 μL of sample is recommended to ensure high signal-to-noise ratios during GC-IT-MS/MS analysis. The injection temperature is set to 260°C for the analysis of OCPs, PCBs, and PBDEs, and to 250°C for the analysis of PAHs. After aspirating the sample, the autosampler is programmed to take up an additional 5 μL of air, and to delay injection until 15 seconds after the needle is inserted into the injector. During this delay the stainless steel needle is heated up to induce rapid evaporation of the sample plug within the injector. This preheating of the needle permits maximum vaporization of the sample without any peak distortion. The sampling time is adjusted to 1 minute to allow for quantitative transfer of the sample to the column. An increase in analytical sensitivity by a factor of 1.5 to 2 for various POPs has been observed using this hot needle injection (HNI) technique (Wurl et al., 2007).

15.4.2 GC PARAMETERS, TEMPERATURE PROGRAM, AND COLUMNS

15.4.2.1 OCPs and PCBs

A DB5 fused silica capillary column (60 m × 0.32 mm i.d., film thickness 0.25 μm) provides optimum separation of PCBs and OCPs. Purified helium is used as the GC carrier gas at a flow rate of 1.5 mL min^{-1}. Manifold, ion trap, and transfer line temperatures are set to 60°C, 220°C, and 280°C, respectively. The oven temperature is ramped from 70°C to 140°C at a rate of 25°C min^{-1}; from 140°C to 179°C at a rate of 2°C min^{-1}; from 179°C to 210°C at a rate of 1°C min^{-1}; and from 210°C to 300°C at a rate of 5°C min^{-1}, then held at 300°C for 10 minutes. To increase throughput, PCB and OCP extracts from the same sample can be combined during cleanup (see Sections 15.3.3.1 and 15.3.3.2) and analyzed on the 60 m column with satisfactory separation (Figure 15.6).

15.4.2.2 PBDEs

A CP-Sil8 fused silica capillary column (10 m × 0.53 mm i.d., film thickness 0.25 μm) is used for separation of PBDEs. Purified helium at a flow rate of 1.5 mL min^{-1} is used as the carrier gas. Manifold, ion trap, and transfer line temperatures are set to 60°C, 220°C, and 280°C, respectively. The oven temperature is ramped from 80°C (held for 1.5 minutes) to 250°C at a rate of 12°C min^{-1}, and from 250°C to 300°C at a rate of 25°C min^{-1}, then held for 10 minutes. Typical chromatograms for standards, seawater samples, and blanks are shown in Figure 15.7.

15.4.2.3 PAHs

The following procedure is based on Crozier et al. (2001). Separation of PAHs is performed on a DB5ms fused silica capillary column (30 m × 0.25 mm i.d., film thickness 0.25 μm). Purified helium at a flow rate of 1.5 mL min^{-1} is again used as a carrier gas. Manifold, ion trap, and transfer line temperatures are set to 60°C, 200°C, and 300°C, respectively. The oven temperature is ramped from 100°C (held for 1 minute) to 140°C at a rate of 20°C min^{-1}, and then ramped to 300°C at a rate of 10°C min^{-1}, before holding for 16 minutes. Typical chromatograms for standards and water samples are shown in Figure 3 in Crozier et al. (2001).

15.4.3 DETECTORS

15.4.3.1 Electron Capture Detector (ECD)

ECD detectors are attractive for PCB and OCP analysis because of their low costs and high sensitivity. The sensitivity, a key parameter in the trace analysis of organic contaminants, is about ten times greater than that of GC-MS using a single quadrupole analyzer (Muir and Sverko, 2006) and can be further lowered by a factor of 10 using ECD with micro cell volumes (μ-ECD). ECD detection

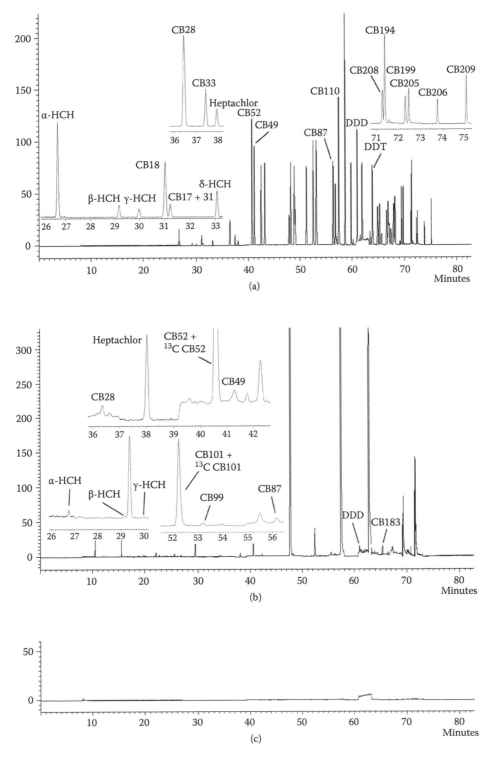

FIGURE 15.6 Typical chromatograms of OCPs and PCBs on a GC-IT-MS/MS: (a) standard solution, (b) extract from seawater sample, and (c) blank extract. The individual cleanup eluates containing OCPs and PCBs have been combined and separated on a 60 m DB5 column to minimize workload. Only selected peaks are labeled with compound name for clarity.

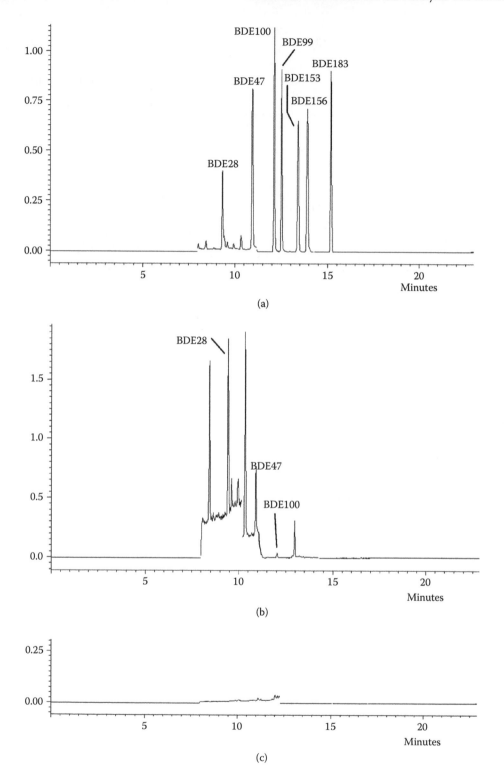

FIGURE 15.7 Typical chromatograms of selected PBDEs on a GC-IT-MS/MS: (a) standard solution, (b) extract from seawater sample, and (c) blank extract.

is straightforward and provides relatively simple and easy-to-interpret chromatograms. At environmentally relevant concentrations, however, ECD detection has a narrow linear range of about one decade. Other drawbacks include poor selectivity and inability to differentiate between target and coeluting compounds. Though lack of specificity limits the application of ECD as a single detection method, its superior sensitivity makes it an excellent tool for trace analysis of POPs when used in combination with mass spectrometry (see Section 15.4.3.2)

Along with OCPs and PCBs, PBDEs have been successfully analyzed using ECD detection, although (as outlined above) ECD is a nonspecific detection technique that responds to all halogenated components (Alaee et al., 2001). A dual-detection system combining the sensitivity of ECD with the specificity of MS/MS is therefore recommended for the analysis of halogenated POPs. PAHs required a flame ionization detector (FID), but the superior sensitivity and selectivity of quadrupole GC-MS analysis obviate the need for dual detection using FID (Jaouen-Madoulet et al., 2000).

15.4.3.2   Ion Trap MS/MS

The ion trap MS/MS system requires careful optimization to achieve both high selectivity and sufficient sensitivity for use in combination with ECD detection. The instrument control software facilitates optimization of parameters such as excitation voltage and storage level to provide the best operating conditions. The Varian instrument software includes an automatic method development (AMD) feature that varies the CID conditions scan by scan, allowing rapid determination of the optimum conditions for each compound. A second, multiple reaction monitoring (MRM) feature is then used to monitor coeluting internal and surrogate standards independently of the analyte.

The ion trap is set up to operate in the internal EI-MS/MS mode (electron ionization) and is typically filled with helium to a pressure of about 1 mtorr. For best sensitivity and selectivity, resonant nonmodulated CID is performed on the precursor ions. The multiplier offset is set to +300 V and the filament emission current to 80 μA. The following ionization parameters are used: electron energy, 70 eV; target total ion counts (TICs), 2,000; maximum ionization time, 25,000 μs; prescan ionization time, 1,500 μs; background mass, 45 m/z; RF dump value, 650 m/z; and ejection amplitude, 20 V.

The first step in the MRM optimization procedure is the selection of an appropriate precursor ion for each compound. Standard solutions are injected into the GC-IT-MS/MS operating in full-scan mode. In general, the most abundant ion observed in the full-scan spectrum is selected as the precursor ion for MRM of that compound. In order to selectively capture the analyte precursor ions, appropriate excitation storage levels must be set. The storage level relates to the strength of the trapping field, and needs to be low enough to capture the target ions, yet high enough to enable strong entrapment. The excitation storage level is selected for each analyte at the minimum value that would allow for dissociation of the precursor (or parent) ion. Higher excitation storage values make the ion more stable, thereby preventing it from dissociating to form the required product (or daughter) ions. Moreover, the excitation storage level sets the lowest mass-to-charge ratio (m/z) that will be observed in the CID spectrum; therefore, the selection of too high a value may cause a loss of important product ions. After the application of the second fragmentation (i.e., CID) step to the selected precursor ions, a product ion MS/MS spectrum is obtained for each compound. Two product ions (generally those of highest relative abundance) are then selected as the characteristic product ions for each compound. The optimum CID voltage required for dissociation of each precursor ion is determined using the AMD feature built into the Varian GC-IT-MS/MS software. Too low a CID voltage may fail to induce fragmentation, whereas too high a voltage can result in annihilation of the precursor ion. During optimization, the CID voltage is increased incrementally until the combined yield of product ions comprises about 80%–90% of the total ion intensity, the intact precursor ion accounting for the remaining 10%–20%. Optimization of CID parameters needs to be performed for all compounds, and the optimum values for selected POPs are listed in Table 15.2.

TABLE 15.2
MS/MS Parameters for Selected POPs

Compound	Precursor Ion (m/z)	Excitation Storage Level (m/z)	CID Voltage (V)	Product Ion (m/z)
α-HCH	219	120	1.10	183
γ-HCH	219	75	0.70	183
p,p-DDT	272	171	1.00	246
p,p-DDD	281	124	1.20	200
p,p-DDE	318	104	0.80	200
CB18	256	75	2.90	186
CB44	292	75	1.10	257
CB49	292	75	1.50	257
CB52	292	75	1.50	257
CB70	292	128	3.60	222
CB74	292	75	3.60	222
CB87	326	90	1.90	291
CB95	326	90	1.60	291
CB101	326	90	2.20	291
CB110/82	326	90	1.50	291
BDE28	246	108	1.50	167
BDE47	326	144	2.60	219
BDE99	566	249	1.00	297
BDE100	566	249	1.00	297

15.4.4 Calibration

The analyst has the choice between external and internal calibration methods. Standard solutions are prepared by gravimetric dilution of a stock solution, or by using micro syringes. Stock solutions containing a wide range of individual compounds or mixtures are commercially available (e.g., from Accustandard, Inc.). A range of fully deuterated surrogate PAHs is available for use as standards in PAH analysis. Isotopically labeled compounds are also available as internal standards for PCBs, OCPs, and PBDEs. Solid standards (e.g., crystalline PAH compounds) should be weighed to a precision of 10^{-5} g. Calibration standards should be stored in the dark, ideally in sealed amber glass ampoules, as some compounds are photosensitive.

In the method of external calibration, at least five standard solutions of differing concentration are injected, and the instrument response (i.e., analyte peak area) plotted against concentration. Alternatively, a plot of (response/injected mass) vs. injected mass may be used as a more sensitive calibration method (Wells et al., 1992). The concentrations of the standard solutions should be chosen so that the expected sample concentrations lie between those of the lowest and highest standard. When the chromatogram is processed using automated integration software the baseline cannot always be defined unambiguously and should therefore always be inspected visually. The calibration coefficient should be better than 0.9900. For external calibration, it is critical that the amount of sample injected be highly reproducible.

Internal standards are spiked (added) into sample extracts just before the analysis and are calibrated in terms of response ratios. This calibration method is independent of injected sample amount and compensates for any instrumental drift. It is the most accurate technique for gas chromatographic quantification of organic contaminants, provided the internal standard is added to each sample in a highly reproducible way. Modern GC autosamplers allow internal standards to be added automatically, thereby avoiding inaccuracies in the spiking procedure. Internal standards should be representative for the range of target compounds; they should, for example, include components with both low and high volatility. As pointed out by Duinker and Schulz-Bull (1999), a further consideration is that added internal standards may not be in the same form as the corresponding target compounds in the origin sample. For routine analysis, it is often sufficient to apply external calibration with regular control standards to monitor instrumental drift.

15.4.5 COMPOUND IDENTIFICATION

The full-scan EI-MS spectra acquired during elution of a specific compound from the GC are combined and averaged. Following background subtraction, the averaged spectrum is compared to a library of reference spectra (e.g., NIST) using the instrument software, which provides a numerical measure of the similarity of spectra of the sample GC-EI-MS peak and that of the pure, reference compound. An index of zero means no similarity at all, whereas an index of 1,000 indicates a perfect match. Three quality assurance criteria are used to ensure correct analyte identification: (1) the signal-to-noise ratio for the analyte peak must be greater than 3; (2) the GC retention time must match that of the corresponding standard compound (± 0.1 minute); and (3) the similarity index calculated using the NIST library software must be greater than a predetermined threshold value. This value can be determined by analyzing the extract of a matrix blank spiked with an amount of target compound equal to the lowest expected sample concentration, and recording the similarity index obtained with reference to the corresponding entry in the NIST database (typically 700 to 800).

Regarding compound identification from ECD chromatograms, this is usually based on retention time or relative retention time (i.e., retention time of the analyte relative to that of a reference compound). Models for identifying PCB congeners based on retention times obtained using different temperature programs on a DB5 column have been reported (Zhang et al., 2004). The calculated retention times have an error of ±4 seconds. Similarly, Öberg (2004) and Korytar et al. (2005) have described extensive PBDE retention time databases for different column materials. In contrast, ECD response factors have limited utility for compound identification, since the ECD response is affected by various parameters, including system temperatures, column dimensions, injection technique, and the amount injected with respect to the linearity of ECD response.

15.5 QUALITY ASSURANCE

The analysis of organic pollutants is a complicated process involving many steps, during which errors can often occur. Concentration factors of between 200,000 and 2,000,000 are common, and the loss of semivolatile POPs can easily occur during the required preconcentration and volume reduction steps. The analysts need to convince scientists and policy makers that the data are precise and accurate. Quality control involves several steps to detect and minimize errors, including the analysis of instrumental and procedural blanks, replicate analyses, analysis of certified reference materials (CRMs), presentation of QA data, and participation in laboratory intercalibration studies (Chapter 1).

15.5.1 FIELD, PROCEDURAL, AND INSTRUMENTAL BLANKS

Procedural blanks are obtained by extraction of fresh XAD-2 cartridges, extraction of blank filters, and rinsing the sample container used for liquid-liquid extraction three times with the same amount of solvent used for sample extraction. Ideally, these media should be transported into the field and exposed briefly to the sampling environment, thus serving as field blanks. The extract is then treated and analyzed in the same way as the samples. A series of procedural blanks is used to determine the limit of detection (LOD), as outlined in Chapter 1. Only concentrations above the LOD are reported. Instrumental blanks are obtained by injection of pure solvent and are usually very low, provided the instrument is well maintained. Instrumental blanks should be assessed regularly between sample batches (e.g., every ten samples).

15.5.2 CERTIFIED REFERENCE MATERIALS (ACCURACY)

Certified reference materials (CRMs) are essential to determine the accuracy of POP analysis (Chapter 1). Unfortunately, no CRMs are available for organic contaminants in aqueous samples, due to the inherent instability of these materials. As a practical alternative, analysts may spike clean extraction media (e.g., XAD-2, GF filter, and preextracted water, preferably deep-ocean water) with known amounts of POP standards (at environmental relevant concentrations) and process their extracts in the same way as samples. The recovery of target compounds needs to meet acceptable criteria (see Section 1.4.1.4). Sediment CRMs are available for a wide range of POPs and provide a suitable substitute for validating the analysis of suspended particulate material. Small amounts of CRM sediment material (equivalent to typical amounts of collected suspended particulate material) should be used for this purpose. Quality control charts for CRMs and spiked samples should be recorded for selected compounds to identify possible errors.

15.5.3 RECOVERY OF SURROGATE STANDARDS

A check on extraction efficiency and cleanup can be performed by spiking samples prior to extraction with a surrogate standard. Surrogates are typically ^{13}C-labeled or (in the case of PAHs) deuterated versions of the target compounds. They have properties identical to those of the corresponding unlabeled target compound, and have become an integral part of the QA protocols for analysis of organic contaminants. A set of several surrogate standards should be used to represent the range of physicochemical properties associated with the target compounds in each analysis; for example, one surrogate for each class of PCB or PBDE homologs. The recovery of each surrogate, determined with reference to the appropriate internal standard (added just prior to analysis; see Section 15.4.4), is essential in assessing the quality of the final data. If low recoveries are observed, solvent samples can be spiked with standards and introduced at different steps in the analytical procedures (e.g., extraction, first or second evaporation step, cleanup, and final volume reduction) to determine at which stage(s) the losses are occurring.

15.5.4 REPEATED SAMPLE ANALYSIS (PRECISION)

Compared with the analysis of sediments or biological samples, the collection of water samples is time-consuming and labor-intensive; hence, the collection of multiple samples with which to assess analytical precision may not be feasible. In this case, the analyst can evaluate the precision of the method by spiking several (at least three) water samples with analyte compounds. Artificial seawater can be prepared with ultrapure and preextracted water; however, it is preferable to use deep-ocean water run through a XAD-2 column to remove target compounds. Precision is usually expressed in terms of relative standard deviation (RSD; Section 1.4.1.4). Alternatively, duplicate samples may be analyzed and the relative percent difference [RPD = (A − B)/(average of A and B)] between duplicate results calculated (ISO/IEC 17025, 2005). However, this is not as good a measure of analytical precision as RSD.

15.6 FUTURE PERSPECTIVES

With an expanding population, growing industrialization, and increasing use of new and diverse chemicals that can enter the environment, the science community will continue to face the challenge of identifying new, emerging POPs. Coastal ecosystems are subject to the discharge of such contaminants via sewage, industrial effluents, storm water runoff, dredged material, and accidental chemical spills. PBDEs are still widely used, and in the absence of strict regulations for disposal, the electronic waste industry in Asia is a major source of these flame retardants in the environment. Martin et al. (2004) reported that in the Chinese province of Guangdong, bordering the Pearl River Estuary, 145 million electronic devices were scrapped in 2002 alone. In Chapter 16, Zhou and Zhang describe the analysis of pharmaceuticals and personal care products (PPCPs), which have recently emerged as chemicals of increasing concern, due to their ecotoxicological effects in coastal ecosystems. Perfluorinated acids (PFAs) and their salts have also appeared as an important class of contaminants in the global environment, including oceanic waters (Yamashita et al., 2004). The impacts of such contaminants on ecosystems often stand in opposition to their potential benefits in many aspects of human life. The science community must continue to identify new emerging chemicals of concern, and to monitor known contaminants, to protect important natural resources and marine ecosystems. This, in turn, will require new developments in analytical chemistry and methodology.

Advanced techniques in mass spectrometry have played a major role in the past for the detection of contaminants, and this trend will continue in the future. Recent developments in this field include desorption electrospray ionization (DESI) and atmospheric pressure chemical ionization (DAPCI), which allow the direct introduction of polar compounds (e.g., pharmaceuticals and fluorinated acids) into a mass spectrometer without chromatographic separation (McEwen et al., 2005; Cotte-Rodriguez et al., 2007). The increasing number of emerging contaminants may promote the use of bioassays in science programs focusing on the detection of unknown chemicals in the environment. Passive samplers for extracting contaminants from the aqueous phase (Chapter 16) in combination with selective pressurized liquid extraction (SPLE; referred to as Accelerated Solvent Extraction, or ASE™, by Dionex; Van Leeuwen and de Boer, 2008) offer attractive benefits for a monitoring program in coastal waters. For example, the rapid, effective, and efficient extraction and cleanup provided by SPLE allows multiple solid samples (e.g., membranes from passive samplers) to be processed per hour. SPLE also has potential to improve the extraction and cleanup procedure for POPs adsorbed onto XAD-2 (Dionex, 2002), thereby simplifying the analysis of POPs in seawater. HPLC fractionation of PBDEs and PCBs appears to be a promising method for optimizing detection of these compounds (Gómara et al., 2006), but may require some development in its application to seawater analysis. The ever-increasing number of contaminants introduced into the environment will also require a broader range of CRMs to be made available, while laboratory intercalibration studies will continue to be important in ensuring that the highest standards in POP analysis are maintained in the future.

REFERENCES

Alaee, M., S. Backus, and C. Cannon. 2001. Potential interference of PBDEs in the determination of PCBs and other organochlorine contaminants using electron capture detection. *Journal of Separation Science* 24:465–69.

Axelman, J., and D. Broman. 2001. Budget calculation for the polychlorinated biphenyls (PCBs) in the northern hemisphere—A single-box approach. *Tellus* 53B:235–59.

Bayen, S., H. K. Lee, and J. P. Obbard. 2007. Exposure and response of aquacultured oysters, *Crassostrea gigas*, to marine contaminants. *Environmental Research* 103:375–82.

Bruhn, R., and M. McLachlan. 2002. Seasonal variation of polychlorinated biphenyl concentrations in the southern part of the Baltic Sea. *Marine Pollution Bulletin* 44:156–63.

Carroll, J., V. Savinov, T. Savinov, S. Dahle, R. Mccrea, and D. C. G Muir. 2008. PCBs, PBDEs and pesticides released to the Arctic Ocean by the Russian rivers Ob and Yenisei. *Environmental Science and Technology* 42:69–74.

Cotte-Rodriguez, I., C. C. Mulligan, and G. Cooks. 2007. Non-proximate detection of small and large molecules by desorption electronspray ionization and desorption atmospheric pressure chemical ionization mass spectrometry: Instrumentation and applications in forensics, chemistry, and biology. *Analytical Chemistry* 79:7069–77.

Crozier, P. W., J. B. Plomley, and L. Matchuk. 2001. Trace level analysis of polycyclic aromatic hydrocarbons in surface waters by solid phase extraction (SPE) and gas chromatography-ion trap mass spectrometry (GC-ITMS). *Analyst* 126:1974–79.

Dachs, J., and J. M. Bayona. 1997. Large volume preconcentration of dissolved hydrocarbons and polychlorinated biphenyls from seawater. Intercomparison between C_{18} disks and XAD-2 column. *Chemosphere* 35:1669–79.

Dachs, J., J. M. Bayona, S. W. Fowler, J.-C. Miquel, and J. Albaigés. 1996. Vertical fluxes of polycyclic aromatic hydrocarbons and organoclorine compounds in the western Alboran Sea (southwestern Mediterreanean). *Marine Chemistry* 52:75–86.

Dionex Corp. 2002. Use of accelerated solvent extraction (ASE) for cleaning and elution of XAD resin. Application Note 347.

Duinker, J. C., and D. E. Schulz-Bull. 1999. Determination of selected organochlorine compounds in seawater. In *Methods of seawater analysis*, ed. K. Grasshoff, K. Kremling, and M. Ehrhardt. Weinheim, Germany: Wiley-VCH.

Ehrhardt, M. 1987. *Lipophilic organic material: An apparatus for extracting solids used for their concentration from sea water*. Techniques in Marine Environmental Sciences 4, ICES, Copenhagen.

Ehrhardt, M., and K. A. Burns. 1999. Preparation of lipophilic organic seawater concentrates. In *Methods of seawater analysis*, ed. K. Grasshoff, K. Kremling, and M. Ehrhardt. Weinheim, Germany: Wiley-VCH.

Gómara, B., C. García-Ruiz, M. J. Gonzáles, and M. L. Marina. 2006. Fractionation of chlorinated and brominated persistent organic pollutants in several food samples by pyrenyl-silica liquid chromatography prior to GC-MS detection. *Analytica Chimica Acta* 565:208–13.

Green, D. R., and D. Le Pape. 1987. Stability of hydrocarbon samples on solid-phase extraction columns. *Analytical Chemistry* 59:699–703.

Gustafsson, O., P. Andersson, J. Axelman, T. Bucheli, P. Komp, M. McLachlan, A. Sobek, and J.-O. Thörngren. 2005. Observations of the PCB distribution within and in-between ice, snow, ice-rafted debris, ice-interstitial water, and seawater in the Barents Sea marginal ice zone and the North Pole area. *Science of the Total Environment* 342:261–79.

Gustafsson, Ö., P. M. Gschwend, and K. O. Buessler. 1997. Settling removal rates of PCBs into the northwestern Atlantic derived from ^{238}U–^{234}Th disequilibria. *Environmental Science and Technology* 31: 3544–50.

ISO/IEC 17025. 2005. *General requirements for the competence of testing and calibration laboratories*. Geneva: Internation Organization for Standardization.

Iwata, H., S. Tanabe, N. Sakai, and R. Tatsukawa. 1993. Distribution of persistent organochlorines in the oceanic air and surface seawater and the role of ocean on their global transport and fate. *Environmental Science and Technology* 27:1080–98.

Jang, J. S., S. K. Lee, Y. H. Park, and D. W. Lee. 1999. Analytical method for dioxin and organo-chlorinated compounds. II. Comparison of extraction methods of dioxins from XAD-2 adsorbent. *Bulletin of the Korean Chemistry Society* 20:689–95.

Jaouen-Madoulet, A, A. Abarnou, A.-M. Le Guellec, V. Loizeau, and F. Leboulenger. 2000. Validation of an analytical procedure for polychlorinated biphenyls, coplanar polychlorinated biphenyls and polycyclic aromatic hydrocarbons in environmental samples. *Journal of Chromatography* 886A:153–73.

Kallenborn, R., M. Oehme, D. D. Wynn-Williams, M. Schlabach, and J. Harris. 1998. Ambient air levels and atmospheric long-range transport of persistent organochlorines to Signy Island, Antarctica. *Science of the Total Environment* 220:167–80.

Korytar, P., A. Covaci, J. de Boer, A. Gelbin, and U. A. Th. Brinkman. 2005. Comprehensive two-dimensional gas chromatography of polybrominated diphenyl ethers. *Journal of Chromatography* 1100A:200–7.

Lema, S. C., I. R. Schultz, N. L. Scholz, J. P. Incardona, and P. Swanson. 2007. Neural defects and cardiac arrhythmia in fish larvae following embryonic exposure to 2,2′,4,4′-tetrabromodiphenyl ether (PBDE 47). *Aquatic Toxicology* 82:296–307.

Li, Y. F., and R. W. Macdonald. 2005. Sources and pathways of selected organochlorine pesticides to the Arctic and the effect of pathway divergence on HCH trends in biota: A review. *Science of the Total Environment* 342:87–106.

Li, Y. F., R. W. Macdonald, J. Ma, H. Hung, and S. Venkatesh. 2004. α-HCH budget in the Arctic Ocean: The arctic mass balance box model (AMBBM). *Science of the Total Environment* 324:115–39.

Martin, M., P. K. S. Lam, and B. J. Richardson. 2004. An Asian quandary: Where have all of the PBDEs gone? *Marine Pollution Bulletin* 49:375–82.

McEwen, C. N., R. G. McKay, and B. S. Larsen. 2005. Analysis of solids, liquids and biological tissues using solids probe introduction at atmospheric pressure on commercial LC/MS instruments. *Analytical Chemistry* 77:7826–31.

Muir, D., and E. Sverko. 2006. Analytical methods for PCBs and organochlorine pesticides in environmental monitoring and surveillance: A critical appraisal. *Analytical and Bioanalytical Chemistry* 386:769–89.

Öberg, T. 2004. Prediction of gas chromatographic separation for PBDE congeners from molecular descriptor. *The ESS Bulletin* 1/2004:74–90.

Oros, D. R., D. Hoover, F. Rodigari, D. Crane, and J. Sericano. 2005. Levels and distribution of polybrominated diphenyl ethers in water, surface sediments and bivalves from the San Francisco Estuary. *Environmental Science and Technology* 39:33–41.

Oskam, I. C., E. Ropstad, E. Lie, A. E. Derocher, Ø. Wiig, E. Dahl, S. Larsen, and J. U. Skaare. 2004. Organochlorines affect the steroid hormone cortisol in free-ranging polar bears (*Ursus maritimus*) at Svalbard, Norway. *Journal of Toxicology Environmental Health* 67:959–77.

Petrick, G., D. E. Schulz-Bull, and D. E. Duinker. 1988. Clean up of environmental samples by high-performance liquid chromatography for analysis of organochlorine compounds by gas chromatography with electron-capature detection. *Journal of Chromatography* 435:241–48.

Petrick, G., D. E. Schulz-Bull, V. Martens, K. Scholz, and J. C. Duinker. 1996. An in-situ filtration/extraction system for the recovery of trace organics in solution and on particles tested in deep ocean water. *Marine Chemistry* 54:97–105.

Sapkota, A., A. R. Sapkota, M. Kucharski, J. Burke, S. McKenzie, P. Walker, and R. Lawrence. 2008. Aquaculture practices and potential human health risks: Current knowledge and future priorities. *Environment International*, 34: 1215–1226.

Schulz-Bull, D. E., and J. C. Duinker. 1999. Clean-up of organic seawater concentrates. In *Methods of seawater analysis*, ed. K. Grasshoff, K. Kremling, and M. Ehrhardt. Weinheim, Germany: Wiley-VCH.

Schulz-Bull, D. E., G. Petrick, R. Bruhn, and J. C. Duinker. 1998. Chlorobiphenyls (PCB) and PAHs in water masses of the northern North Atlantic. *Marine Chemistry* 61:101–14.

Shaw, S. D., M. L. Berger, D. Brenner, D. O. Carpenter, L. Tao, C.-S. Hong, and K. Kannan. 2008. Polybrominated diphenyl ethers (PBDEs) in farmed and wild salmon marketed in the Northeastern United States. *Chemosphere* 71:1422–31.

Skaare, J. U., H. J. Larsen, E. Lie, A. Bernhoft, A. E. Derocher, R. Norstrom, E. Ropstad, N. F. Lunn, and Ø. Wiig. 2002. Ecological risk assessment of persistent organic pollutants in the Arctic. *Toxicology* 181–182:193–97.

Sobek A., and Ö. Gustafsson. 2004. Latitudinal fractionation of polychlorinated biphenyls in surface seawater along a 62 8N–89 8N transect from the southern Norwegian Sea to the North Pole area. *Environmental Science and Technology* 38:2746–51.

Stern, G. A., C. J. Halsall, L. A. Barrie, D. C. G. Muir, P. Fellin, B. Rosenberg, F. Y. Rovinsky, E. Y. Kononov, and B. Pastuhov. 1997. Polychlorinated biphenyls in arctic air. I. Temporal and spatial trends: 1992–1994. *Environmental Science and Technology* 31:3619–28.

Tanabe, S. 2002. Contamination and toxic effects of persistent endocrine disrupter in marine mammals and birds. *Marine Pollution Bulletin* 45:69–77.

Van Leeuwen, S. P. J., and J. de Boer. 2008. Advances in the gas chromatographic determination of persistent organic pollutants in the aquatic environment. *Journal of Chromatography* 1186A:161–82.

Wells, D. E., E. A. Maier, and B. Griepink. 1992. Calibrants and calibration for chlorobiphenyl analysis. *International Journal of Environmental Analytical Chemistry* 46:255–64.

Wurl, O., and J. P. Obbard. 2006. Distribution of organochlorine compounds in the sea-surface microlayer, water column and sediment of Singapore's coastal environment. *Chemosphere* 62:1105–15.

Wurl, O., J. R. Potter, J. P. Obbard, and C. Durville. 2007. New trends in the sampling and analysis of atmospheric persistent organic pollutants over the open ocean. In *Air pollution research advances*, ed. C. G. Bodine. Hauppage, NY: Nova Science Publishers.

Yamashita N., K. Kannan, S. Taniyasu, Y. Horii, T. Okazawa, G. Petrick, and T. Gamo. 2004. Analysis of perfluorinated acids at parts-per-quadrillion levels in seawater using liquid chromatography-tandem mass spectrometry. *Environmental Science and Technology* 38:5522–28.

Yao, Z.-W., G.-B. Jiang, and H.-Z. Xu. 2002. Distribution of organochlorine pesticides in seawater of the Bering and Chukchi Sea. *Environmental Pollution* 116:49–56.

Zhang, Q., X. Liang, J. Chen, P. Lu, A. Yediler, and A. Kettrup. 2004. Correct identification of polychlorinated biphenyls in temperature-programmed GC with ECD detection. *Analytical and Bioanalytical Chemistry* 374:93–102.

Zhang, Z. L., H. S. Hong, J. L. Zhou, J. Huang, and G. Yu. 2003. Fate and assessment of persistent organic pollutants in water and sediment from Minjiang River Estuary, Southeast China. *Chemosphere* 52:1423–30.

Zhulidov, A. V., J. V. Headley, D. F. Pavlov, R. D. Roberts, L. G. Korotova, Y. Y. Vinnikov, and O. V. Zhulidov. 2000. Riverine fluxes of the persistent organochlorine pesticides hexachlorcyclohexane and DDT in the Russian Federation. *Chemosphere* 41:829–41.

16 Pharmaceutical Compounds in Estuarine and Coastal Waters

John L. Zhou and Zulin Zhang

CONTENTS

16.1 INTRODUCTION

Pharmaceuticals and personal care products (PPCPs) are a group of emerging contaminants of environmental concern that have remained largely unrecognized as such until recent advances in trace-level analytical measurements (Cha et al., 2006; Erickson, 2002; Gros et al., 2006; Lindsey et al., 2001). There is growing concern over the occurrence and fate of PPCPs in the environment, with evidence of adverse effects in terrestrial and aquatic organisms, and also the potential of some antibiotics to induce resistance in naturally occurring bacterial strains (Hirsch et al., 1998).

Over three thousand chemical substances are used in human and veterinary medicine (Ternes et al., 2004). Such pharmaceuticals include antiphlogistics/anti-inflammatory drugs, contraceptives, β-blockers, lipid regulators, tranquilizers, antiepileptics, and antibiotics (Ternes et al., 2004; Petrovic et al., 2005). Some typical pharmaceuticals classified by groups according to therapeutic effect and physicochemical properties are listed in Table 16.1. During and after treatment, humans and animals excrete a combination of intact and metabolized pharmaceuticals, many of which are generally soluble in water and have been discharged to the aquatic environment with little evaluation of possible risks or consequences to humans and the environment. In addition, chemicals that are components of personal care products number in the thousands, and are contained in skin care products, dental care products, soaps, sunscreen agents, and hair care products. Annual production exceeds 1×10^6 tonnes worldwide (e.g., >553,000 tonnes was produced in Germany alone in 1993; Daughton and Ternes, 1999). Included in this category are fragrances (e.g., nitro and polycyclic musks), UV blockers (e.g., methylbenzylidene camphor), and preservatives (e.g., parabens). Unlike pharmaceuticals, personal care products enter wastewater and the aquatic environment after regular use during showering or bathing. The environmental fates and effects of many cosmetic ingredients are poorly known, although considerable persistence and bioaccumulation in aquatic organisms have been reported (Daughton and Ternes, 1999; Kallenborn et al., 2001).

In many aquatic environments, particularly in North America and Europe, pharmaceuticals, hormones, metabolites, biocides, musks, and flame retardants have been measured (Ternes, 1998;

TABLE 16.1
Pharmaceuticals and Their Physicochemical Properties

Compound	Therapeutic Class	logK$_{ow}$	pK$_a$	MW	Formula
Ketoprofen	Analgesic/anti-inflammatories	3.12	4.45	254	$C_{16}H_{14}O_3$
Naproxen		3.18	4.15	230	$C_{14}H_{14}O_3$
Ibuprofen		3.97	4.91	206	$C_{13}H_{18}O_2$
Indomethacine		4.27	4.5	358	$C_{19}H_{16}ClNO_2$
Diclofenac		4.51	4.14	296	$C_{14}H_{10}Cl_2NO_2$
Meclofenamic acid		5.12	4.2	241	$C_{15}H_{15}NO_2$
Acetaminophen		0.46	9.38	151	$C_8H_9NO_2$
Propyphenazone	Lipid regulators/cholesterol-	1.94	n/a	230	$C_{14}H_{18}N_2O$
Clofibric acid	lowering statin drugs	n/a	n/a	214	$C_{10}H_{11}O_3Cl$
Gemfibrozil		4.77	n/a	250	$C_{15}H_{22}O_3$
Bezafibrate		4.25	n/a	362	$C_{19}H_{20}ClNO_4$
Pravastatin		3.1	n/a	446	$C_{23}H_{36}O_7$
Mevastatin		3.95	n/a	391	$C_{25}H_{38}O_5$
Carbamazepine	Psychiatric drugs	2.47	7	236	$C_{15}H_{12}NO$
Fluoxetine		3.82	8.7	309	$C_{17}H_{18}F3NO$
Paroxetine		3.95	n/a	329	$C_{19}H_{20}FNO_3$
Lansoprazole	Antiulcer agent	2.58	8.73	369	$C_{16}H_{14}F_3N_3O_2S$
Loratadine	Histamine H$_1$ and H$_2$ receptor	5.20	n/a	383	$C_{22}H_{23}ClN_2O_2$
Famotidine	antagonists	−0.64	n/a	337	$C_8H_{15}N_7O_2S_3$
Ranitidine		0.27	n/a	314	$C_{13}H_{22}N_4O_3S$
Erythromycin	Antibiotics	3.06	8.8	734	$C_{37}H_{67}NO_{13}$
Azythromycin		4.02	8.74	749	$C_{38}H_{72}N_2O_{12}$
Sulfamethoxazole		0.89	6.0	253	$C_{10}H_{11}N_3O_3S$
Trimethoprim		0.91	7.12	290	$C_{14}H_{18}N_4O_3$
Ofloxacin		n/a	n/a	361	$C_{18}H_{20}FN_3O_4$
Atenolol	β-Blockers	0.16	9.6	266	$C_{14}H_{22}N_2O_3$
Sotalol		0.24	n/a	272	$C_{12}H_{20}N_2O_3S$
Metoprolol		1.88	9.68	267	$C_{15}H_{25}NO_3$
Propranolol		1.2–3.48	9.5	260	$C_{16}H_{21}NO_2$
Meberverine	Gastrointestinal	n/a	n/a	429	$C_{25}H_{35}O_5$
Thioridazine	Antidepressant	n/a	n/a	371	$C_{21}H_{26}N_2S_2$
Tamoxifen	Anticancer	n/a	n/a	372	$C_{26}H_{29}NO$
Monensin	Growth promoters	2.75–3.89	6.65	692	$C_{36}H_{61}NaO_{11}$

Note: logK$_{ow}$, log of the octanol-water partition coefficient; pK$_a$, negative log of the dissociation constant; MW, molecular weight; n/a, not available.

Kolpin et al., 2002; Hirsch et al., 1999; Hilton and Thomas, 2003). One of the principal sources is through the release of municipal wastewater. Some pharmaceuticals do not readily biodegrade in a marine environment, and have been detected in seawater (Weigel et al., 2002; Thomas and Hilton, 2004) and sediments (Samuelsen et al., 1992). Weigel et al. (2002) reported a wide distribution of clorfibric acid, caffeine, and *N,N*-diethyl-3-toluamide (DEET, an insect repellent) in concentrations up to 19, 16, and 1.1 ng L^{-1}, respectively, throughout the North Sea, off Scotland, the outer and inner German Bight, as well as the Danish and Norwegian coasts. Samples collected from UK estuaries had clorfibric acid concentrations of approximately 100 ng L^{-1} in two samples (Thomas and Hilton, 2004). Other frequently measured pharmaceutical compounds found in UK estuaries included clotrimazole (a typical antifungal agent, in 59% of samples, up to 22 ng L^{-1}), ibuprofen (an analgestic, in 50% of samples, up to 569 ng L^{-1}), and propranolol (an antihypertensive drug, in 41% of samples, up to 56 ng L^{-1}), with several other drugs appearing in approximately one-third of the samples at lower concentrations (Thomas and Hilton, 2004).

The concentrations of pharmaceuticals present in the aquatic environment are generally in the sub-nanogram per liter range and do not necessarily represent a major threat to drinking water quality. The consequence of a continuous presence of low concentrations of pharmaceuticals for ecosystems is still not fully understood. A discussion of various aspects of ecotoxicology of pharmaceuticals in the environment can be found in recent reviews (Cunningham et al., 2006; Fent et al., 2006; Crane et al., 2006; Hernando et al., 2006). It is quite clear that environmental risk assessment must be based on reliable data about the actual concentrations of pharmaceuticals in aquatic systems. Therefore, efficient analytical methods are of major importance.

Fast progress in the development of analytical procedures for residue analysis of pharmaceutical drugs has been facilitated by the existence of considerable expertise in analysis of other microorganic pollutant residues. Strategies successfully used for routine analysis of traces of some polar organic contaminants have been modified and subsequently applied to residue analysis of pharmaceuticals. In many cases, the common procedures involve sampling, sample treatment (e.g., preconcentration, cleanup step) by solid-phase extraction (SPE) or related techniques, followed by analysis using chromatography in combination with mass spectrometry (MS) as detector. When residue analysis of pharmaceuticals became an important issue in the 1990s, gas chromatography (GC) was the preferred chromatographic technique together with various derivatization procedures for the analytes. Nowadays, GC-MS may still be the perfect technique for certain classes of pharmaceuticals (Togola and Budzinski, 2008), although high-performance liquid chromatography (HPLC) hyphenated with atmospheric pressure ionization-MS has established itself as a better choice for simultaneous determination of pharmaceuticals of widely differing structures (Buchberger, 2007). In this chapter, the analysis of pharmaceuticals in estuarine and coastal waters will be discussed, including the sampling, preconcentration, and instrumental measurement.

16.2 SAMPLING

The first task in any analysis is sampling. The sample being taken should be representative, the composition of which is as close as possible to the whole mass of whatever (e.g., estuarine water) is being analyzed. Obtaining a good sample is a crucial first step in the chemical analysis process. Prior to sampling, a sampling strategy should be drawn concerning the locations of sampling, the number of samples to be taken, where to conduct replicate sampling, size of samples, and storage and transport of samples (see Chapter 1). Preparation should also ensure that all the in situ measurement equipment is calibrated, and necessary sampling tools and containers are cleaned appropriately before sampling.

There are two sampling methods, spot and passive sampling, which are complementary to each other. Currently, the most widely used technique for performing monitoring of organic contaminants is spot sampling followed by laboratory-based extraction and analysis. In general, spot sampling uses a glass sampler or stainless steel sampler such as buckets (Gulkowska et al., 2007), some of which can be opened underwater to prevent the sampling of the surface microlayer (Zhou et al., 1996). The sample volume is typically 1–2 L (Gulkowska et al., 2007; Roberts and Thomas, 2006), although for seawater samples this may be increased to 10–100 L. General biocides such as sodium azide (final concentration of 0.02 M) are added to each sample on site to inactivate bacteria and prevent sample degradation during storage and processing. The samples are stored in a refrigerator below 4°C until filtration and extraction. Prior to use, all glassware is soaked thoroughly with detergents (e.g., Decon-90) and cleaned with ultrapure water, before further treatment (e.g., rinsed with distilled solvents such as dichloromethane and methanol, or baked in a furnace). The procedure should be adjusted for the compounds to be analyzed, based on their physicochemical properties (e.g., solubility, polarity).

Spot sampling is a well-established technique that is easy to perform and inexpensive, and requires limited expertise. However, it only yields an instantaneous measurement of pollutant levels and suffers from the uncertainty of short- and long-term concentration variations that occur in the aquatic environment. An increase in sampling frequency or the use of flow- and time-weighted automatic samplers may reduce such uncertainty; however, the associated increase in costs can be prohibitive. There has been rapid development in the use of passive sampling devices such as the polar organic chemical integrative sampler (POCIS) (Alvarez et al., 2004), Chemcatcher (Mills et al., 2007), and silicon rod (Paschke et al., 2007) that allow continuous monitoring of aqueous pollutants, without the disadvantage of using organisms (a passive sampler could mimic the bioconcentration of pollutants in aquatic organisms but not suffer from adverse effects as organisms). Of the various passive sampling devices, the most widely used is POCIS, which comprises a solid receiving phase (sorbent) sandwiched between two microporous polyethersulfone (PES) membranes (Figure 16.1A; Zhang et al., 2008). POCIS samples from water and thereby enables the chemical concentration to be estimated as follows (Alvarez et al., 2004; Vrana et al., 2005):

$$M_s = C_w R_s t \qquad (16.1)$$

where M_s is the mass of analytes in the receiving phase at time t, and C_w represents time-weighted average concentration in water during the deployment period. R_s is the sampling rate of the system, which may be interpreted as the volume of water cleared of analyte per unit of exposure time by the device (Vrana et al., 2006; Zhang et al., 2008). Although little has been reported on the application of POCIS for pharmaceutical residue measurements in seawater, it has been applied successfully in surface and estuarine water (Jones-Lepp et al., 2004; Petty et al., 2004; Togola and Budzinski, 2007). Figure 16.1B and C shows the operating process for applying POCIS to pharmaceutical monitoring in river and estuarine waters (Zhang et al., 2008).

The POCIS is versatile, and by changing the sequestrating medium, specific chemicals or chemical classes can be targeted. It is common to have POCISs of several different configurations deployed together to maximize the data obtained. There are two configurations of POCIS that are typically used. One is a generic system that is useful for general hydrophilic organic contaminant purposes, and the other is for pharmaceutical sampling. The generic configuration contains the triphasic sorbent admixture of Isolute ENV+ polystyrene divinylbenzene (Argonaut Technologies, Redwood City, California) and Ambersorb 1500 carbon (Rohm and Haas, Philadelphia, Pennsylvania) dispersed on S-X3 Biobeads (200–400 mesh, Bio-Rad, Hercules, California). This mixture exhibits excellent trapping and recovery of many pesticides, natural and synthetic hormones, and other wastewater-related contaminants (Alvarez et al., 2004, 2005). The pharmaceutical configuration uses the Oasis HLB sorbent (Waters, Milford, Massachusetts) for sequestering the chemicals of interest. This configuration is necessary, as many pharmaceuticals, with multiple functional groups, have a tendency to bind strongly to the carbonaceous component of the sorbent admixture. The membrane acts as a semipermeable barrier, allowing chemicals of interest to pass through to the sorbent, while excluding particulate matter, biogenic material, and other large, potentially interfering substances. The polyethersulfone membrane (Pall Gelman Sciences, Ann Arbor, Michigan) contains water-filled pores, 0.1 μm in diameter, to facilitate transport of the hydrophilic chemicals. The POCIS was designed to mimic respiratory exposure of aquatic organisms to dissolved chemicals without the inherent problems of metabolism, depuration of chemicals, avoidance of contaminated areas, and mortalities of test organisms. Also, dietary uptake of polar organic compounds likely represents only a small fraction of residues accumulated in aquatic organism tissues (Huckins et al., 1997).

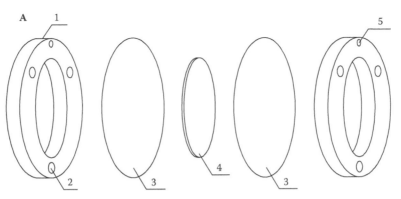

(1) PTFE holder, (2) Screw, (3) Membrane, (4) Sorbe nt, (5) Hole

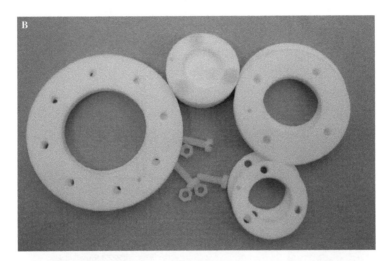

FIGURE 16.1 The POCIS device and its application to pharmaceutical analysis in river and estuarine waters. (A) Component parts of a POCIS, (B) assembled sampling devices, and (C) use of sampler devices in river and estuarine waters.

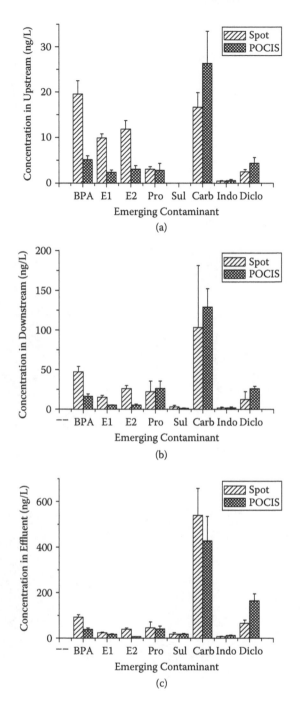

FIGURE 16.2 Comparison of the mean contaminant concentrations between spot water sampling and those predicted by POCIS in (a) upstream, (b) downstream, and (c) effluent outfall, Ouse Estuary, West Sussex, UK. BPA, bisphenol A; E1, estrone; E2, 17β-estradiol; Pro, Propranolol; Sul, sulfamethoxazole; Carb, carbamazepine; Indo, indomethacine; Diclo: diclofenac.

Thus, the POCIS provides a worst-case exposure scenario for aquatic organisms, enables a concentration of sufficient amounts of bioavailable hydrophilic organic chemicals for some biomarker tests, and permits determination of the biologically relevant time-weighted average concentrations in water.

The POCIS devices can be deployed at the sampling site for a duration ranging from 1 week to 2 months (Alvarez et al., 2005; Zhang et al., 2008). Quality control (QC) is achieved using both fabrication and field blanks ($n = 3$) for each analytical technique. Fabrication blanks account for any background contribution due to interferences from POCIS components and for contamination incurred during laboratory storage, processing, and analytical procedures. Field blank POCISs are used as QC samples for transport, deployment, and retrieval procedures (note that these POCIS blanks are sealed again in the same shipping cans and stored frozen during the exposure period). The field blank POCISs are treated identically as the deployed devices, with the exception that they are not exposed to waters at the monitoring sites.

The procedures for the recovery of sequestered chemical residues from the deployed POCISs are as follows: Briefly, the POCISs are disassembled and the sorbent transferred into glass gravity-flow chromatography columns or glass beakers (Alvarez et al., 2005; Zhang et al., 2008). Chemical residues are recovered from the sorbent by organic solvent elution/extraction. Methanol is widely used to recover the pharmaceuticals. The extracts are reduced in volume by rotary evaporation and under a gentle stream of nitrogen, and then are ready for further instrumental analysis.

Zhang et al. (2008) have described the use of POCIS for analysis of emerging contaminants, including pharmaceuticals, and compared the predicted compound concentrations in water with those measured by spot sampling. As shown in Figure 16.2, for the pharmaceuticals propranolol, sulfamethoxazole, carbamazepine, indomethacine, and diclofenac, their mean aqueous concentrations measured by spot sampling varied: 3.0–45.6 ng L^{-1}, <LOD–17.6 ng L^{-1}, 16.6–539 ng L^{-1}, 0.4–7.2 ng L^{-1}, and 2.4–65.2 ng L^{-1}, respectively. Their concentrations predicted by POCIS were 2.8–40.5 ng L^{-1}, <LOD–18.2 ng L^{-1}, 26.3–427 ng L^{-1}, 0.5–11.9 ng L^{-1}, and 4.4–165 ng L^{-1}, respectively. It is apparent that for most samples, the predicted pharmaceutical concentrations by POCIS are similar to those by spot sampling. In addition, POCIS was validated and deployed for monitoring pharmaceuticals in estuarine systems (Togola and Budzinski, 2007), confirming the potential application of passive samplers for routine monitoring of seawater quality.

16.3 SAMPLE PREPARATION

The sample preparation is an important step in analysis, particularly since the concentration levels of pharmaceuticals found in environmental water samples are generally too low to allow a direct injection into a chromatographic system. Therefore, efficient preconcentration steps are necessary that should also result in some sample cleanup. Several techniques have been developed and optimized, with SPE being the most frequent. Also, solid-phase microextraction (SPME), liquid-phase microextraction (LPME), and lyophilization have been applied (Fatta et al., 2007). In a review of thirty-two pharmaceutical studies, Fatta et al. (2007) found that most (twenty-eight studies) used SPE for extraction from water samples. This extraction procedure can be based on multiple equilibria between the liquid phase and the sorbent in SPE cartridges.

Pharmaceuticals of adequate hydrophobicity can easily be preconcentrated using any reversed-phase material such as alkyl-modified silica or polymer-based materials (Buchberger, 2007). Deprotonation of acidic compounds and protonation of basic compounds should be suppressed to ensure sufficient hydrophobicity of the analytes. Therefore, acidic pharmaceuticals should be preconcentrated under acidic conditions, whereas basic analytes should be preconcentrated at an alkaline pH range. Alternatively, mixed-mode SPE materials can be used that exhibit both reversed-phase and cation exchange properties due to the presence of sulfonic acid groups on the hydrophobic surface of the particles. Using acidified sample solutions, acidic and neutral analytes would be extracted by hydrophobic interactions, whereas protonated basic analytes would interact via ion exchange mechanisms. Such an approach has been used by, among others, Stolker et al. (2004) for SPE of a set of thirteen pharmaceuticals of different classes. Mixed-mode materials with reversed-phase and anion exchange properties have been used under slightly basic conditions for antibiotics containing carboxylic acid functionality (Benito-Pena et al., 2006).

From the practical point of view, it might be desirable to extract pharmaceuticals from water samples without any pH adjustment (Buchberger, 2007). Furthermore, various (neutral) pharmaceuticals may exhibit significant hydrophilic properties, which make it difficult to enrich them on conventional alkyl-modified silica materials. SPE procedures for extraction of polar compounds from aqueous samples are still a big challenge in analytical chemistry. A recent review has summarized new SPE materials that can improve the recoveries for polar analytes (Fontanals et al., 2005). These SPE cartridge materials are mainly polymeric sorbents that improve the retention of polar compounds either by novel functional groups in the polymeric structure (resulting in a hydrophilic-hydrophobic balance material) or by considerably increased surface area. Some of these new materials have turned out to be well suited for multiclass analysis of pharmaceuticals in water samples. A number of different SPE stationary phases (Table 16.2) have been evaluated for the extraction of the selected pharmaceutical compounds (Hilton and Thomas, 2003; Weigel et al., 2004; Zhang and Zhou, 2007). Nowadays, one of the most widely used sorbents is a copolymer of divinylbenzene and vinylpyrrolidone, which has been commercialized under the trade name Oasis-HLB by Waters. Weigel et al. (2004) demonstrated that this sorbent can simultaneously extract acidic, neutral, and basic pharmaceuticals at neutral pH. Multiresidue methods for different classes of pharmaceutical using Oasis-HLB at neutral pH have also been reported recently by Barceló and coworkers (Gomez et al., 2006; Gros et al., 2006). Trenholm et al. (2006) developed a comprehensive method for the analysis of fifty-eight potential endocrine-disrupting compounds and pharmaceuticals using a single SPE step based on Oasis-HLB. Various other studies can be found describing the successful use of Oasis-HLB for pharmaceuticals in water (Petrovic et al., 2006; Zhang and Zhou, 2007). A typical multiresidue analysis would include filtration of the seawater, conditioning of the Oasis HLB material (between 60 and 500 mg packed into a suitable cartridge) by several milliliters of methanol and ultrapure water, application of up to 2 L of sample at a flow rate of approximately 10 ml min^{-1},

TABLE 16.2
Different Types of SPE Cartridges Being Used for Pharmaceutical Extraction

Cartridge	Description	Manufacturer
DSC-C$_{18}$	Polymerically bonded, octadecyl	Supelco
DSC-Si	Unbonded acid-washed silica sorbent	Supelco
DSC-SCX	Aliphatic sulfonic acid, Na$^+$ counterion	Supelco
DSC-SAX	Quaternary amine, Cl$^-$ counterion	Supelco
Strata-X-CW	Polymeric weak cation	Phenomenex
Strata SDB-L	Styrene-divinylbenzene polymeric	Phenomenex
Chromabond-Easy	Bifunctionally modified polystyrene-divinylbenzene adsorbent resin	Macherey-Nagel
Chromabond-C$_{18}$ Hydra	Octadecyl-modified silica	Macherey-Nagel
Chromabond-Drug	Modified silica	Macherey-Nagel
Isolute C$_{18}$	Octadecyl	International Sorbent Technology
Isolute C$_{18}$/ENV$^+$	C$_{18}$ hydroxylated polystyrene-divinylbenzene	International Sorbent Technology
Isolute C$_8$	Octadecyl functionalized silica	International Sorbent Technology
Oasis HLB	Poly(divinylbenzene-co-N-vinyl-pyrrolidone)	Waters
Oasis MCX	Poly(divinylbenzene-co-N-vinyl-pyrrolidone, $-SO_3H$)	Waters
Varian Bond Elut C$_{18}$	Irregularly based acid-washed silica	Varian

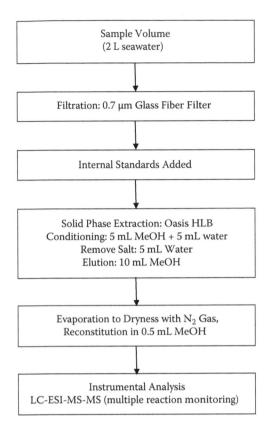

FIGURE 16.3 The procedures involved in the extraction and analysis of pharmaceuticals in estuarine and seawater samples.

rinsing the cartridge with several milliliters of ultrapure water to remove salt, drying the SPE material by applying a vacuum to remove excess water, elution with approximately 10 ml of methanol (it may be necessary to repeat the elution step), evaporation of the extracts under a gentle stream of nitrogen (see Section 15.3.4), and reconstitution in 0.5 ml of methanol or a mixture of methanol and ultrapure water containing internal standards that are then ready for instrumental analysis. The whole analytical procedure is illustrated in Figure 16.3 (Zhang and Zhou, 2007). All the glassware used for the extraction is baked at 400°C for 4 hours to eliminate any organic contaminants. All the solvents are of distilled-in-glass grade.

Hilton and Thomas (2003) have shown that Strata-X is useful for extracting selected pharmaceuticals, after comparing seven types of SPE cartridges. So in a few cases Strata-X (a polydivinylbenzene resin containing piperidone groups manufactured by Phenomenex) has been employed for generic SPE procedures (Hilton and Thomas, 2003; Roberts and Bersuder, 2006). Nebot et al. (2007) also used Strata-X successfully to determine the concentrations of a range of human pharmaceuticals in surface and wastewater, but none of the target compounds were above the limit of detection (LOD) in seawater samples collected offshore Scotland. Strata-X may have properties similar to those of Oasis-HLB, but at present there is not enough data in the literature to allow this comparison to be made.

SPE of pharmaceuticals is often done off-line, and is useful for on-site sampling (performing the preconcentration step in the field, followed by the elution step in the lab). It is a technique that is also well suited for online procedures and automation in the laboratory. The SPE cartridge can be installed in the injection valve instead of the injection loop, and the preconcentrated analytes directly eluted onto the analytical column. An example of this approach is the work of Pozo et al. (2006), who determined sixteen antibiotics in surface and groundwater samples. In such a setup, the SPE cartridge is generally reused for a series of samples. Contrary to this configuration, fully automated SPE procedures with single-use cartridges can be realized by commercially available instrumentation, such as the Symbiosis™ Environ manufactured by Spark Holland. This robotic system includes an automated cartridge exchange module that transfers the cartridge after the preconcentration step into the flow of mobile phase of HPLC (Rodriguez-Mozaz et al., 2007). This approach, which has been used primarily with analysis of pharmaceutical residues, could be applied in seawater monitoring. There are some significant advantages with the approach (Fatta et al., 2007):

1. Direct injection of untreated seawater samples
2. Automatic sample cleanup or analyte enrichment
3. Elimination of conventional manual sample pretreatment steps
4. Faster procedures
5. Methods are less prone to errors, resulting in better reproducibility
6. Reduction of health risks
7. Samples can be run unattended (e.g., overnight or over the weekend)

16.4 INSTRUMENTAL ANALYSIS

The majority of pharmaceuticals lack sufficient volatility and as such are not directly compatible with GC analysis (Buchberger, 2007). Various groups of pharmaceuticals can be derivatized to make them suited for GC analysis. Although such procedures may be time-consuming and can introduce errors due to side reactions during the derivatization, they are still widely in use and well established for routine work. The major advantage of GC-MS is the fact that the usual ionization modes such as electron impact (EI) or chemical ionization (CI) are generally less affected by the matrix of the sample than ionization modes used by, for example, HPLC-MS. Typical derivatization reagents for acidic pharmaceuticals include pentafluorobenzylbromide (Reddersen and Heberer, 2003), methyl chloromethanoate (Weigel et al., 2004), methanol/BF3 (Verenitch et al., 2006), and tetrabutylammonium salts (for derivatization during injection) (Lin et al., 2005). Phenazone-type drugs have been derivatized by silylation using N-tert-butyldimethylsilyl-N-methyltrifluoroacetamide (MTB-STFA) (Zuhlke et al., 2004). Silylation procedures are also commonly used for synthetic estrogens (Quintana et al., 2004; Fernandez et al., 2007), although careful selection of the reagent and the reaction conditions is necessary to avoid side reactions (Shareef et al., 2006). Derivatization reactions that are useful for sorptive extraction combine with thermal desorption GC.

Generally, the use of GC-MS seems to be a well-established approach for residue analysis of pharmaceuticals. Correctly, there is a trend toward tandem MS techniques as the MS component of choice for this type of analysis. The advantages of such instruments will be discussed in more detail in the context of HPLC-MS.

Despite the indisputable merits of GC procedures for residue analysis of certain classes of pharmaceuticals, HPLC shows much more universal applicability (Buchberger, 2007). In some cases, when just a few analytes of a certain class are to be analyzed, even a simple UV absorbance detection may be feasible. This has been demonstrated for residues of oxytetracycline in water, which can be detected at 360 nm (Himmelsbach and Buchberger, 2005). Fluorescence detection may also have some benefits, as shown for the determination of some other compounds, such as anthracycline cytostatics and fluoroquinolones (Mahnik et al., 2006; Golet et al., 2001, 2002). Nevertheless, MS detection involving atmospheric pressure ionization, such as electrospray ionization (ESI), is nowadays state of the art.

Although single-quadrupole instruments were used successfully when HPLC-MS procedures started to be developed for pharmaceutical residue analysis (Ahrer et al., 2001), more sophisticated mass analyzers are nowadays commonly employed that allow an unequivocal confirmation of the identities of the analytes. Triple quadrupole (QqQ) MS instruments have become widely used with HPLC for environmental analysis. When using a QqQ instrument, false positive results can be avoided if the ions of at least two ion-ion transitions are used in combination with at least one ion intensity ratio. Several studies have dealt with HPLC-QqQ/MS for multiclass analysis of pharmaceuticals (Castiglioni et al., 2005; Rodriguez-Mozaz et al., 2004; Miao et al., 2004; Ternes et al., 2005; Gomez et al., 2006; Gros et al., 2006; Zhang and Zhou, 2007). Precursor ions and product ions used for quantification and confirmation purposes have been compiled for a wide range of pharmaceutical compounds (Petrovic et al., 2005).

An example of a typical multipharmaceutical residue analysis by HPLC-QqQ/MS is given in Figure 16.4 (Zhang and Zhou, 2007). The LC separation system is equipped with a Waters Symmetry C_{18} column (4.8 mm × 75 mm, particle size 3.5 μm). The mobile phase is made of eluent A (0.1% formic acid in ultrapure water), eluent B (acetonitrile), and eluent C (methanol). The flow rate of the mobile phase is 0.2 ml min^{-1}. The gradient elution is operated with 10% of eluent B, followed by a 25-minute gradient to 80% of eluent B and a 3-minute gradient to 100% of eluent B, and then changed to 100% of eluent C within 8 minutes, and held there for 10 minutes. The system is reequilibrated for 10 minutes between runs. Typically the injection volumes are 1 to 10 μL.

A general and well-known problem of HPLC-ESI/MS is ionization suppression due to matrix components eluting at the same time as an analyte. On the one hand, ionization suppression effects can reduce the sensitivity of the method considerably; on the other hand, special care must be taken to achieve reliable quantitation. Ideally, isotopically labeled analytes should be used as internal standards but are available only in a few cases for pharmaceuticals. Otherwise, standard addition methods must be applied to obtain correct quantitative data; however, such methods increase the length of the analytical procedure considerably. In all cases it makes sense to optimize the efficiency of the sample cleanup protocols in order to minimize interferences. New-generation MS instruments in some cases allow a dilution of the sample extract before injection into the HPLC system, thereby minimizing any matrix effects. Furthermore, other MS detectors such as time of flight (TOF) are more selective than QqQ, and hence more suited for samples of highly complex matrices, such as wastewater or biological extracts. However, the sensitivity of TOF currently is not as good as QqQ, and requires further development.

16.5 ANALYTICAL QUALITY CONTROLS

To ensure data quality, all the analytical processes should be subject to strict QC procedures to determine systematic and random errors. QC measures in relation to estuarine and coastal water analysis include the collection of blank water samples derived from laboratory-grade or organic-free water to determine if sampling procedures, sampling equipment, field conditions, sample shipment and storage (field blank), or laboratory procedures (laboratory blank) introduced target analytes into environmental samples. The spiked water samples are used to check the precision and recoveries. The blank and spiked water samples are typically made by ultrapure water taken from a Milli-Q system. Typically, several blank and spiked samples are produced with each set of real samples (ten samples for each set). In addition, the random errors involved in sampling are assessed by carrying out replicate sampling of water at the same site and the analysis of sample extracts. Internal standards (usually the target compounds labeled by stable isotopes such as ^{13}C or ^{2}H) are used to compensate for losses involved in the sample extraction and workup, to further characterize the method performance. Prior to use, all glassware is rinsed twice with dichloromethane and methanol, or is baked at 450°C for 4 hours. All the solvents used are of distilled-in-glass grade. All these processes are carried out to minimize the cross-contamination and the loss of analytes through adsorption onto the surface of sampling vessels and the extraction apparatus. As seawater has a very different matrix to freshwater, more appropriate blanks should be used in the future for seawater studies. As an example, synthetic seawater prepared by dissolving appropriate amounts of key ionic substances (e.g., NaCl) in ultrapure water at a concentration identical to that in real

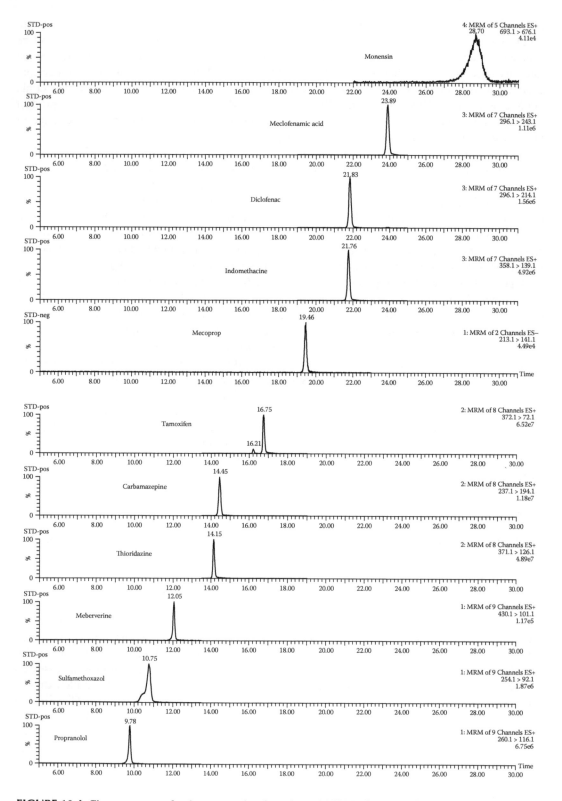

FIGURE 16.4 Chromatograms for the extracted and overlapped MRM of selected pharmaceutical compounds.

seawater of 35‰ salinity should be a more appropriate matrix for making blank samples and spiked recovery samples.

To assess systematic errors, most would use so-called recovery experiments by spiking known amounts of each target compound in seawater, followed by extraction and analysis. This gives a good indication of how reliable your measurement values are (see Chapter 1). In addition, certified reference materials (CRMs) for pharmaceuticals in seawater should be prepared in the future that can identify the closeness between a measured value (from an individual laboratory) and a certified value (from the supplier). As CRMs are vigorously tested under varying environmental conditions by the supplier and independently verified by laboratories worldwide, they become a calibration tool for the international community on pharmaceutical research. Ideally, the relative difference between measured and certified values should be as small as possible (e.g., ±10%). Such materials will ensure that everyone follows the right procedures and generates data of highest quality. In addition, this practice will ensure monitoring data obtained from any marine samples anywhere can be compared against each other, so as to identify hotspots of pharmaceutical pollution, emergence of new pharmaceuticals, and temporal variation of pharmaceutical concentrations in the estuarine and marine environment.

16.6 IMMUNOANALYTICAL TECHNIQUES

Immunoassays show attractive features for organic trace analysis due to the fact that they require little sample pretreatment, exhibit high sensitivity, and are inexpensive in comparison to the instrumental analysis described above. A considerable number of immunoassays have been developed and used for residue analysis of pesticides in water samples, but immunoassays for pharmaceuticals in the aquatic environment are still quite rare. Although test kits for pharmaceuticals are commercially available, these kits are in most cases optimized for biological samples like blood, urine, or food. The applicability to environmental samples has not been investigated in the majority of cases. Deng et al. (2003) developed a highly sensitive and specific indirect competitive enzyme-linked immunosorbent assay (ELISA) for the determination of diclofenac in water samples. When they applied the assay to analysis of diclofenac in tap and surface water as well as wastewater in Austria and Germany, they showed that ELISA-derived diclofenac concentrations in wastewater samples were about 25% higher than those using GC-MS. The technique should be equally applicable to seawater matrices, although the sensitivity and selectivity should be further improved.

16.7 SUMMARY AND PERSPECTIVES

Analysis of emerging pollutants such as pharmaceutically active compounds in the aquatic environment, including seawater, was reviewed in this chapter. As pharmaceutical compounds are usually present at trace levels (e.g., pg L^{-1} to ng L^{-1}) in complex matrices (e.g., seawater), it is common practice to develop extraction and analytical methods that can concentrate the target compounds while minimizing matrix interference. The analytical procedure involves many interrelated steps, including sample pretreatment (e.g., filtration), preconcentration (e.g., SPE), and analysis by advanced techniques (e.g., HPLC-MS). Residues of pharmaceuticals have most probably been present in our aquatic environment since their application, but only recently have advances in analytical chemistry and instrument performance allowed analysis of compounds at low nanogram per liter concentrations. The development of advanced mass spectrometric detectors for chromatography has made a significant contribution to these achievements. The limits of detection of analytical methods may be improved even further during the next few years. Residues of pharmaceuticals in aquatic systems are not yet included in regular monitoring programs. The high costs of instrumental analysis may be prohibitive to more extended studies. A focus on a limited set of pharmaceuticals that are representative in regard to toxic effects may be advantageous (but a final selection of such a set has not yet been done). The importance of reliable and inexpensive biosensors may increase in the future, provided that they meet the criteria of analytical QC in the same way as traditional techniques do.

In addition, general QC procedures must be followed, including appropriate replicate sampling, sample preservation at 4°C, application of isotopically labeled internal standards, suitable blanks, satisfactory recovery of the target compounds, and eventually use of CRMs (either in-house or commercial ones). Although pharmaceutical residues in the environment are a major concern and have been widely studied in freshwater systems, the study of their occurrence, behavior, and impacts in coastal waters has been quite limited. In addition, the residues of pharmaceuticals in aquatic systems are not yet included in the regular monitoring programs of regulatory bodies. Further research on emerging pollutants such as pharmaceuticals should be extended to coastal and marine ecosystems, including polar regions, where our understanding of pharmaceutical occurrence and behavior is very limited. The low cost and robustness of passive sampling such as POCIS is recommended for routine monitoring of pharmaceuticals and other similar pollutants by governmental agencies.

REFERENCES

Ahrer, W., E. Scherwenk, and W. Buchberger. 2001. Determination of drug residues in water by the combination of liquid chromatography or capillary electrophoresis with electrospray mass spectrometry. *Journal of Chromatography* A910:69–78.

Alvarez, D. A., J. D. Petty, J. N. Huckins, and T. L. Jones-Lepp. 2004. Development of a passive, in situ, integrative sampler for hydrophilic organic contaminants in aquatic environments. *Environmental Toxicology and Chemistry* 23:1640–48.

Alvarez, D. A., P. E. Stackelberg, J. D. Petty, J. N. Huckins, E. T. Furlong, S. D. Zaugg, and M. T. Meyer. 2005. Comparison of a novel passive sampler to standard water-column sampling for organic contaminants associated with wastewater effluents entering a New Jersey stream. *Chemosphere* 61:610–22.

Benito-Pena, E., A. I. Partal-Rodera, M. E. Leon-Gonzalez, and M. C. Moreno-Bondi. 2006. Evaluation of mixed mode solid phase extraction cartridges for the preconcentration of beta-lactam antibiotics in wastewater using liquid chromatography with UV-DAD detection. *Analytica Chimica Acta* 556:415–22.

Buchberger, W. W. 2007. Novel analytical procedures for screening of drug residues in water, waste water, sediment and sludge. *Analytica Chimica Acta* 593:129–39.

Castiglioni, S., R. Bagnati, D. Calamari, R. Fanelli, and E. Zuccato. 2005. A multiresidue analytical method using solid-phase extraction and high-pressure liquid chromatography tandem mass spectrometry to measure pharmaceuticals of different therapeutic classes in urban wastewaters. *Journal of Chromatography* A1092:206–15.

Cha, J. M., S. Yang, and K. H. Carlson. 2006. Trace determination of β-lactam antibiotics in surface water and urban wastewater using liquid chromatography combined with electrospray tandem mass spectrometry. *Journal of Chromatography* A1115:46–57.

Crane, M., C. Watts, and T. Boucard. 2006. Chronic aquatic environmental risks from exposure to human pharmaceuticals. *Science of the Total Environment* 367:23–41.

Cunningham, V. L., M. Buzby, T. Hutchinson, F. Mastrocco, N. Parke, and N. Roden. 2006. Effects of human pharmaceuticals on aquatic life: Next steps. *Environmental Science and Technology* 40:3456–62.

Daughton, G. C., and T. Ternes. 1999. Pharmaceuticals and personal care products in the environment: Agents of subtle changes? *Environmental Science and Technology* 107:907–38.

Deng, A. P., M. Himmelsbach, Q. Z. Zhu, S. Frey, M. Sengl, W. Buchberger, R. Niessner, and D. Knopp. 2003. Residue analysis of the pharmaceutical diclofenac in different water types using ELISA and GC-MS. *Environmental Science and Technology* 37:3422–29.

Erickson, B. E. 2002. Analyzing the ignored environmental contaminants. *Environmental Science and Technology* 36:140A–45A.

Fatta, D., A. Nikolaou, A. Achilleos, and S. Meric. 2007. Analytical methods for tracing pharmaceutical residues in water and wastewater. *TrAC Trends in Analytical Chemistry* 26:515–33.

Fent, K., A. A. Weston, and D. Caminada. 2006. Ecotoxicology of human pharmaceuticals. *Aquatic Toxicology* 76:122–59.

Fernandez, M. P., M. G. Ikonomou, and I. Buchanan. 2007. An assessment of estrogenic organic contaminants in Canadian wastewaters. *Science of the Total Environment* 373:250–69.

Fontanals, N., R. M. Marce, and F. Borrull. 2005. New hydrophilic materials for solid-phase extraction. *TrAC Trends in Analytical Chemistry* 24:394–406.

Golet, E. M., A. C. Alder, A. Hartmann, T. A. Ternes, and W. Giger. 2001. Trace determination of fluoroquinolone antibacterial agents in solid-phase extraction urban wastewater by liquid chromatography with fluorescence detection. *Analytical Chemistry* 73:3632–38.

Golet, E. M., A. Strehler, A. C. Alder, and W. Giger. 2002. Determination of fluoroquinolone antibacterial agents in sewage sludge and sludge-treated soil using accelerated solvent extraction followed by solid-phase extraction. *Analytical Chemistry* 74:5455–62.

Gomez, M. J., M. Petrovic, A. R. Fernandez-Alba, and D. Barcelo. 2006. Determination of pharmaceuticals of various therapeutic classes by solid-phase extraction and liquid chromatography-tandem mass spectrometry analysis in hospital effluent wastewaters. *Journal of Chromatography* A1114:224–33.

Gros, M., M. Petrovic, and D. Barcelo. 2006. Development of a multi-residue analytical methodology based on liquid chromatography-tandem mass spectrometry (LC-MS/MS) for screening and trace level determination of pharmaceuticals in surface and wastewaters. *Talanta* 70:678–90.

Gulkowska, A., Y. H. He, M. K. So, L. W. Y. Yeung, H. W. Leung, J. P. Giesy, P. K. S. Lam, M. Martin, and B. J. Richardson. 2007. The occurrence of selected antibiotics in Hong Kong coastal waters. *Marine Pollution Bulletin* 54:1287–1306.

Hernando, M. D., M. Mezcua, A. R. Fernandez-Alba, and D. Barcelo. 2006. Environmental risk assessment of pharmaceutical residues in wastewater effluents, surface waters and sediments. *Talanta* 69:334–42.

Hilton, M. J., and K. V. Thomas. 2003. Determination of selected human pharmaceutical compounds in effluent and surface water samples by high-performance liquid chromatography-electrospray tandem mass spectrometry. *Journal of Chromatography* A1015:129–41.

Himmelsbach, M., and W. Buchberger. 2005. Residue analysis of oxytetracycline in water and sediment samples by high-performance liquid chromatography and immunochemical techniques. *Microchimica Acta* 151:67–72.

Hirsch, R., T. A. Ternes, K. Haberer, and K. L. Kratz. 1999. Occurrence of antibiotics in the aquatic environment. *Science of the Total Environment* 225:109–18.

Hirsch, R., T. A. Ternes, K. Haberer, A. Mehlich, F. Ballwanz, and K. Kartz. 1998. Determination of antibiotics in different water compartments via liquid chromatography-electrospray tandem mass spectrometry. *Journal of Chromatography* A815:213–23.

Huckins, J. N., J. D. Petty, and J. Thomas. 1997. *Bioaccumulation: How chemicals move from the water into fish and other aquatic organisms.* American Petroleum Institute Publication 4656, Washington, D.C.

Jones-Lepp, T. L., D. A. Alvarez, J. D. Petty, and J. N. Huckins. 2004. Polar organic chemical integrative sampling and liquid chromatography-electrospray/ion-trap mass spectrometry for assessing selected prescription and illicit drugs in treated sewage effluents. *Archives of Environmental Contamination and Toxicology* 47:427–39.

Kallenborn, R., R. Gatermann, T. Nygard, J. Knutzen, and M. Schlabach. 2001. Synthetic musks in Norwegian marine fish samples collected in the vicinity of densely populated areas. *Fresenius Environmental Bulletin* 10:832–42.

Kolpin, D. W., E. T. Furlong, M. T. Meyer, E. M. Thurman, S. D. Zaugg, L. B. Barber, and H. T. Buxton. 2002. Pharmaceuticals, hormones, and other organic wastewater contaminants in U.S. streams, 1999–2000: A national reconnaissance. *Environmental Science and Technology* 36:1202–11.

Lin, W. C., H. C. Chen, and W. H. Ding. 2005. Determination of pharmaceutical residues in waters by solid-phase extraction and large-volume on-line derivatisation with gas chromatography-mass spectrometry. *Journal of Chromatography* A1065:279–85.

Lindsey, M. E., M. Meyer, and E. M. Thurman. 2001. Analysis of trace levels of sulfonamide and tetracycline antimicrobials in groundwater and surface water using solid-phase extraction and liquid chromatography/mass spectrometry. *Analytical Chemistry* 73:4640–46.

Mahnik, S. N., B. Rizovski, M. Fuerhacker, and R. M. Mader. 2006. Development of an analytical method for the determination of anthracyclines in hospital effluents. *Chemosphere* 65:1419–25.

Miao, X. S., F. Bishay, M. Chen, and C. D. Metcalfe. 2004. Occurrence of antimicrobials in the final effluents of wastewater treatment plants in Canada. *Environmental Science and Technology* 38:3533–41.

Mills, G. A., B. Vrana, I. Allan, D. A. Alvarez, J. N. Huckins, and R. Greenwood. 2007. Trends in monitoring pharmaceuticals and personal-care products in the aquatic environment by use of passive sampling devices. *Analytical and Bioanalytical Chemistry* 387:1153–57.

Nebot, C., S. W. Gibb, and K. G. Boyd. 2007. Quantification of human pharmaceuticals in water samples by high performance liquid chromatography-tandem mass spectrometry. *Analytica Chimica Acta* 598:87–94.

Paschke, A., J. Brummer, and G. Schuurmann. 2007. Silicone rod extraction of pharmaceuticals from water. *Analytical and Bioanalytical Chemistry* 387:1417–21.

Petrovic, M., M. Gros, and D. Barcelo. 2006. Multi-residue analysis of pharmaceuticals in wastewater by ultra-performance liquid chromatography-quadrupole-time-of-flight mass spectrometry. *Journal of Chromatography* A1124:68–81.

Petrovic, M., M. D. Hernando, M. S. Diaz-Cruz, and D. Barcelo. 2005. Liquid chromatography-tandem mass spectrometry for the analysis of pharmaceutical residues in environmental samples: A review. *Journal of Chromatography* A1067:1–14.

Petty, J. D., J. N. Huckins, D. A. Alvarez, W. G. Brumbaugh, W. L. Granor, R. W. Gale, A. C. Rastall, T. L. Jones-Lepp, T. J. Leiker, C. E. Rostad, and E. T. Furlong. 2004. A holistic passive integrative sampling approach for assessing the presence and potential impacts of waterborne environmental contaminants. *Chemosphere* 54:695–705.

Pozo, O. J., C. Guerrero, J. V. Sancho, M. Ibanez, E. Pitarch, E. Hogendoorn, and F. Hernandez. 2006. Efficient approach for the reliable quantification and confirmation of antibiotics in water using on-line solid-phase extraction liquid chromatography/tandem mass spectrometry. *Journal of Chromatography* A1103:83–93.

Quintana, J. B., J. Carpinteiro, I. Rodriguez, R. A. Lorenzo, A. M. Carro, and R. Cela. 2004. Determination of natural and synthetic estrogens in water by gas chromatography with mass spectrometric detection. *Journal of Chromatography* A1024:177–85.

Reddersen, K., and T. Heberer. 2003. Multi-compound methods for the detection of pharmaceutical residues in various waters applying solid phase extraction (SPE) and gas chromatography with mass spectrometric (GC-MS) detection. *Journal of Separation Science* 26:1443–50.

Roberts, P. H., and P. Bersuder. 2006. Analysis of OSPAR priority pharmaceuticals using high-performance liquid chromatography-electrospray ionisation tandem mass spectrometry. *Journal of Chromatography* A1134:143–50.

Roberts, P. H., and K. V. Thomas. 2006. The occurrence of selected pharmaceuticals in wastewater effluent and surface waters of the lower Tyne catchment. *Science of the Total Environment* 356:143–53.

Rodriguez-Mozaz, S., M. J. Lopez de Alda, and D. Barcelo. 2004. Picogram per liter level determination of estrogens in natural waters and waterworks by a fully automated on-line solid-phase extraction-liquid chromatography-electrospray tandem mass spectrometry method. *Analytical Chemistry* 76:6998–7006.

Rodriguez-Mozaz, S., M. J. Lopez de Alda, and D. Barcelo. 2007. Advantages and limitations of on-line solid phase extraction coupled to liquid chromatography-mass spectrometry technologies versus biosensors for monitoring of emerging contaminants in water. *Journal of Chromatography* A1152:97–115.

Samuelsen, O. B., V. Torsvik, and A. Ervik. 1992. Long-range changes in oxytetracycline concentration and bacterial resistance towards oxytetracycline in a fish farm sediment after medication. *Science of the Total Environment* 114:25–36.

Shareef, A., M. J. Angove, and J. D. Wells. 2006. Optimization of silylation using *N*-methyl-*N*-(trimethylsilyl)-trifluoroacetamide, *N*,*O*-*bis*-(trimethylsilyl)-trifluoroacetamide and *N*-(*tert*-butyldimethylsilyl)-*N*-methyltrifluoroacetamide for the determination of the estrogens estrone and 17 alpha-ethinylestradiol by gas chromatography-mass spectrometry. *Journal of Chromatography* A1108:121–28.

Stolker, A. A. M., W. Niesing, E. A. Hogendoorn, J. F. M. Versteegh, R. Fuchs, and U. A. T. Brinkman. 2004. Liquid chromatography with triple-quadrupole or quadrupole-time of flight mass spectrometry for screening and confirmation of residues of pharmaceuticals in water. *Analytical and Bioanalytical Chemistry* 378:955–63.

Ternes, T. A. 1998. Occurrence of drugs in German sewage treatment plants and rivers. *Water Research* 32:3245–60.

Ternes, T. A., M. Bonerz, N. Herrmann, D. Loffler, E. Keller, B. Bago Lacida, and A. C. Alder. 2005. Determination of pharmaceuticals, iodinated contrast media and musk fragrances in sludge by LC/tandem MS and GC/MS. *Journal of Chromatography* A1067:213–23.

Ternes, T. A., A. Joss, and H. Siegrist. 2004. Scrutinizing pharmaceuticals and personal care products in waste-water treatment. *Environmental Science and Technology* 38:392A–99A.

Togola, A., and H. Budzinski. 2007. Development of polar organic integrative samplers for analysis of pharmaceuticals in aquatic systems. *Analytical Chemistry* 79:6734–41.

Togola, A., and H. Budzinski. 2008. Multi-residue analysis of pharmaceutical compounds in aqueous samples. *Journal of Chromatography* A1177:150–58.

Thomas, K. V., and M. J. Hilton. 2004. The occurrence of selected human pharmaceutical compounds in UK estuaries. *Marine Pollution Bulletin* 49:436–44.

Trenholm, R. A., B. J. Vanderford, J. C. Holady, D. J. Rexing, and S. A. Snyder. 2006. Broad range analysis of endocrine disruptors and pharmaceuticals using gas chromatography and liquid chromatography tandem mass spectrometry. *Chemosphere* 65:1990–98.

Verenitch, S. S., C. J. Lowe, and A. Mazumder. 2006. Determination of acidic drugs and caffeine in municipal wastewaters and receiving waters by gas chromatography-ion trap tandem mass spectrometry. *Journal of Chromatography* A1116:193–203.

Vrana, B., G. A. Mills, I. J. Allan, E. Dominiak, K. Svensson, J. Knutsson, G. Morrison, and R. Greenwood. 2005. Passive sampling techniques for monitoring pollutants in water. *TrAC Trends in Analytical Chemistry* 24:845–68.

Vrana, B., G. A. Mills, E. Dominiak, and R. Greenwood. 2006. Calibration of the Chemcatcher passive sampler for the monitoring of priority organic pollutants in water. *Environmental Pollution* 142:333–43.

Weigel, S., R. Kallenborn, and H. Hühnerfuss. 2004. Simultaneous solid-phase extraction of acidic, neutral and basic pharmaceuticals from aqueous samples at ambient (neutral) pH and their determination by gas chromatography-mass spectrometry. *Journal of Chromatography* A1023:183–95.

Weigel, S., J. Kuhlmann, and H. Hühnerfuss. 2002. Drugs and personal care products as ubiquitous pollutants: Occurrence and distribution of clofibric acid, caffeine and DEET in the North Sea. *Science of the Total Environment* 295:121–41.

Zuhlke, S., U. Dunnbier, and T. Heberer. 2004. Detection and identification of phenazone-type drugs and their microbial metabolites in ground and drinking water applying solid-phase extraction and gas chromatography with mass spectrometric detection. *Journal of Chromatography* A1050:201–9.

Zhang, Z. L., A. Hibberd, and J. L. Zhou. 2008. Analysis of emerging contaminants in sewage effluent and river water: Comparison from spot and passive sampling. *Analytica Chimica Acta* 607:37–44.

Zhang, Z. L., and J. L. Zhou. 2007. Simultaneous determination of various pharmaceutical compounds in water by solid-phase extraction-liquid chromatography-tandem mass spectrometry. *Journal of Chromatography* A1154:205–13.

Zhou, J. L., T. W. Fileman, S. Evans, P. Donkin, R. F. C. Mantoura, and S. J. Rowland. 1996. Seasonal distribution of dissolved pesticides and polynuclear aromatic hydrocarbons in the Humber estuary and Humber coastal zone. *Marine Pollution Bulletin* 32:599–608.

Appendix A: First Aid for Common Problems with Typical Analytical Instruments

The following tables aim to identify the most common causes of problems with analytical instruments described in this book, as well as how to recognize and correct them. Due to the complexity of instruments and different designs among manufacturers, the tables should be considered as a "first aid," and intractable problems may require technical service from the manufacturer.

TABLE A.1
First Aid for Spectrophotometer

Problem	Cause	Solution
Excessive noise over whole wavelength range	Light path is blocked	Check that light path is free.
	Cuvette not installed correctly	Check that cuvette sits correctly in cell holder.
	Air bubbles stick on cuvette	Tap cuvette gently to remove bubbles. Clean cuvette with suitable cleaning solution.
	Source lens dirty or fogged	Clean lenses according to manufacturer's manual.
Excessive noise in part of spectrum	One of the lamps may be turned off during measurements. Measurements taken with deuterium lamp off show noise in the UV wavelength range, and similar measurements taken with tungsten lamp off show noise in the visible wavelength range.	Turn on proper lamp.
	Noise in the UV wavelength range caused by weak or defective deuterium lamp	Replace lamp.
	Incorrect cuvette used, causing low lamp intensity	Change cuvette to standard type. Use cuvette made of quartz for measurements in the UV wavelength range. Check cuvette for scratches and replace if necessary.
	Dirty windows of cuvette, including fingerprints	Clean windows with lint-free tissue.
	Matrix of sample (solvent, buffer, etc.) blocks light in certain wavelength range	Change matrix that is transparent for that wavelength range. Matrix needs to be free of particles.
	Source lens dirty or fogged	Clean lenses according to manufacturer's manual.
Negative values/ no absorbance	No sample	Add sample to solution.
	Wrong wavelength setting	Check and use correct wavelength.
	Instrument zeroed with sample instead of blank solution	Zero instrument with blank solution.
	Cuvette height incompatible with cuvette holder	Use compatible cuvette and holder.

TABLE A.2
First Aid for Flow Injection Analyzer (FIA)

Problem	Cause	Solution
Poor reproducibility for repeated analysis of a single sample	Carryover	Decrease sample frequency. Increase pump rate of carrier stream. Check valve for leakage. Check level of sample in autosampler cups.
Peak response is sluggish in reaching baseline	System has dead volume (i.e., bad connectors, large volume of flow cell, large volume of terminals of flow cell)	Check system for bad connectors. Use low dead-volume connectors. Use microvolume flow cell.
Baseline drift	Deposition of materials in system	Check flow cell, tubing, valve, and connectors for any deposition. Pump suitable washing solution (often diluted acid) through system. Switch valve several times during washing. Replace components if necessary.
Air bubbles	Reagents not deaerated Gases formed as result of chemical reaction Venturi effect in flow cell caused by sudden pressure changes	Deaerate all reagents in an ultrasonic bath. Check reaction media (e.g., pH values of reagents). Insert connecting tubes to flow cell, so that orifices are as close as possible to cell chamber; use 20 cm outlet tubing with 0.5 mm ID before stream is led to remaining waste tube.
Spikes on peaks	Small air bubbles pass flow cell	See above and check for complete filling of sample loop.
Noisy baseline/signals	Bad pulsation of pump Worn pumping tubes	Check movement of pump and correct sitting of pumping tubes (need to be pressed hard against roller). Replace pumping tubes.
Poor reproducibility at high concentrations	Insufficient reagent volume	Increase reagent concentration. Dilute samples of high concentrations.
Negative peaks	Colorless sample is injected in colored carrier stream Matrix effect (different matrix composition between carrier and sample)	Change carrier stream with similar chemical composition and physical properties (color, viscosity, etc.) as sample.

TABLE A.3
First Aid for Continuous Flow Analyzer (CFA)

Problem	Cause	Solution
Bad bubble pattern	Insufficient wetting agent in reagents (air bubbles in plastic tubing flat at back instead of round)	Check application notes for correct concentration of wetting agent. Increase concentration of wetting agent if necessary.
	Reagent pump line sucks air	Refill reagent.
	Back pressure in system	Check system for blockages. Waste line immersed in liquid too long or wrong diameter.
	Bad tubing	Check for correct tubing material, size, and cleanliness.
	Bad glass to tubing connection	Renew connection without having a gap.
	Bad position of air injection tubing under air valves	Check position of tubing.
	Air injection tubing worn	Replace air injection tubing.
	Flow system with dialyser: broken membrane	Replace membrane.
Baseline noise	Particles or small bubbles in reagent	Filter and degas reagents.
	Reaction conditions not optimal	Check reaction conditions (pH, temperature, concentrations).
	Contaminated flow cell	Clean the flow cell and flush the flow cell by forcing cleaning solution (solvent, acid) through with a syringe.
	Pump tubes worn	Replace pump tubes.
	Bubble pattern not regular	See above.
	Regular noise: dirty or worn pump rollers or platen	Clean rollers and platen.
Baseline drift	Contaminated or unstable reagents	Use fresh reagents. Mix reagents well before start of analysis.
	Contaminated flow cell	Clean the flow cell and flush the flow cell by forcing cleaning solution (solvent, acid) through with a syringe.
	Old wavelength filter (appears dark at the outer edge)	Replace the filter.
	Old lamp in photometer module	Replace the lamp.
	Downward drift: old contamination being washed out	Wait until baseline acceptable. Flush system with wash solution.
Elevated baseline after analysis of samples	Contaminated flow cell	Clean the flow cell and flush the flow cell by forcing cleaning solution (solvent, acid) through with a syringe.
	Unstable system operation. System has not reached working temperature (heating bath) or pump tubes are not run in yet.	Warm up the system for a period of 30–60 minutes.
	Reagents not pure	Use chemical of highest purity, especially wetting agents.
Spikes on baseline or peaks	Insufficient wetting agent in reagents	Check application notes for correct concentration of wetting agent. Increase concentration of wetting agent if necessary.

(Continued)

TABLE A.3 (CONTINUED)
First Aid for Continuous Flow Analyzer (CFA)

Problem	Cause	Solution
	Back pressure in system	Check system for blockages. Waste line immersed in liquid too long or wrong diameter.
	Total flow rate too low	Check all pump tubes for correct size.
	Small bubbles in reagent	Degas reagents.
	Flow cell with debubbler: flow rate through flow cell too high	Check pump tube removing air bubbles from the flow stream. Some liquid should be removed with air bubbles.
Sensitivity drift	Pump tubes worn	Replace pump tubes.
	Unstable room temperature	Avoid working space with direct sunlight. Cover system if necessary.
Downward sensitivity drift	Unstable reagents	Use fresh and well-mixed reagents.
	Unstable system operation. System has not reached working temperature (heating bath) or pump tubes are not run in yet.	Warm up the system for a period of 30–60 minutes.
	System with dialyser: different conditions in lower and upper stream (pH, ionic strength, etc.)	Use similar conditions in lower and upper stream.
	System with Cd column: bad reduction efficiency of Cd column	Reactivate Cd column; check pH value of stream. Use fresh wetting agents.
Elevated, but slowly decreasing baseline at end of analysis	Sampler wash solution contaminated	Check purity and flow rate of sampler wash solution.
	Contaminated flow cell	Clean the flow cell and flush the flow cell by forcing cleaning solution (solvent, acid) through with a syringe.
	Interfering substances in samples	Use dialyzer to remove substances.
	Regular noise: dirty or worn pump rollers or platen	Clean rollers and platen.
	System with dialyser: unstable room temperature	Avoid working space with direct sunlight. Cover system if necessary.
Low sensitivity	Bad reagents or standards	Use fresh reagents and standards.
	Bad reaction conditions	Check reaction conditions (pH, temperature, etc.).
	Wrong wavelength	Check for correct filter.
	Ineffective UV digestion	Replace UV lamp.
	Bad dialyzer membrane	Change membrane.
	Old lamp in photometer module	Replace lamp.
	Wrong photometer settings	Check settings.
	NO_3 method: poor reduction efficiency	Check by running NO_2 standard. Recondition Cd column. Check reaction condition (pH, reagent concentration, and temperature).
	NH_3 method: wrong chlorine concentrations	Check composition of reagents.

TABLE A.3 (CONTINUED)
First Aid for Continuous Flow Analyzer (CFA)

Problem	Cause	Solution
High carryover	Loss of intersample air bubble	Verify intersample bubble travels throughout the system and its size (see Chapter 9 for details). Check for correct internal diameter of probe, transmission tubing, T-pieces, and nipples.
	Tubing between sampler and flow module too big or long	Use tubing with equivalent diameter as pump tube. Keep tubing as short as possible.
	Old wetting agent	Use new batch of wetting agent.
	Too much wetting agent	Check and correct concentration of wetting agent in reagents.
	Precipitation formed in system	Check reagents and chemistry. Wash system with washing solutions.
	Wrong reagent composition (especially PO_4 method)	Replace reagents one at a time.
No peak separation	Intersample air bubble does not contact tubing during its travel through system (air bubble too small)	Check for correct ID of transmission tubing, T-pieces, nipples, e.g., should not be greater than sample pump tube.
No steady state	Sample probe clogged	Flush the probe and filter samples prior to analysis.
	Insufficient volume of sample	Fill sample cup.
	Sample:wash ratio too low	4:1 is a typical value.
	High dispersion	Make unsegmented distances as short as possible.
	Stainless steel probe used with samples of acidic matrix	Use PEEK probe.
	System with dialyser: Bad membrane	Replace membrane.

TABLE A.4
First Aid for High-Pressure Liquid Chromatography (HPLC)

Problem	Cause	Solution
Baseline noise	System not equilibrated	Equilibrate with ten volumes of mobile phase.
	Bubbles in flow cell	Degas the mobile phase and pass degassed solvent through the flow cell. Do not exceed the cell's pressure limit.
	Contaminated guard	Replace the guard cartridge.
	Contaminated column	Wash the column using an appropriate solvent. If this does not resolve the problem, replace the column.
	Contaminated detector	Clean the detector according to the manufacturer's instructions.
	Contaminated solvents	Use freshly prepared solvents of HPLC grade.
Baseline drift	Slow elution of previously adsorbed substances	Wash the column using an appropriate solvent. If this does not resolve the problem, replace the column.
	Detector temperature not constant	Allow sufficient time to level off detector temperature.
	Contaminated detector	Clean the detector according to the manufacturer's instructions.

(Continued)

TABLE A.4 (CONTINUED)
First Aid for High-Pressure Liquid Chromatography (HPLC)

Problem	Cause	Solution
Unusually low pressure	Leak in the system	Leak can occur within the pump, injector, and detector. Check for and replace any leaking tubing, fittings, sealings, and gaskets. Check correct syringe size.
	Air in solvent lines or pump	Ensure that the reservoirs and solvent lines are fully primed and the purge valve is fully closed.
	Low-viscosity mobile phase	Confirm expected pressure using the Lozeny-Carmen or similar equation.
Unusually high pressure	Plugged in-line filters	Back flush filter with solvent and clean in ultrasonic bath.
	Sample precipitation	Filter samples before injection.
	Tubing blocked	Working backwards from detector outlet, check source of blockage and replace tubing as necessary.
	Guard blocked	Replace guard cartridge.
	Detector flow cell blocked	Clean the flow cell according to the manufacturer's instructions.
	High-viscosity mobile phase	Confirm expected pressure using the Lozeny-Carmen equation.
Broad peaks	Injected solvent too strong	Ensure that the injected solvent has the same or weaker strength than the mobile phase.
	Injected volume too high	Reduce the injected volumes to avoid overload.
	Injected mass too high	Reduce the sample concentration to avoid mass overload.
	Column volume too high	Reduce diameter and length of connecting tubing. Reduce the volume of the flow cell where possible.
	Old guard cartridge or column	Replace the guard cartridge or column.
	Contaminated column	Wash the column using an appropriate solvent. Replace the column if necessary.
Double peaks	Contaminated column	Wash the column using an appropriate solvent. If this does not resolve the problem, replace the column.
	Old guard cartridge or column	Replace the guard cartridge or column.
Negative peaks	Contaminated solvents	Use freshly prepared solvents.
	Mobile phase more absorptive than sample components	Use mobile phase that is transparent at the wavelength used.
	Vacancy peaks	Originate from great difference in composition between sample solvent and mobile phase. Dissolve sample in mobile phase.
	Refractive index of mobile phase higher than that of solute	Change mobile phase.
Flat-top peaks	Detector overload	Reduce the sample concentration.
	Detector setup	Check the detector attenuation and rezero.
Tailing peaks	Contaminated column	Wash the column using an appropriate solvent. If necessary replace the column.
	Old guard cartridge or column	Replace the guard cartridge or column.
	Injected solvent too strong	Ensure that the injected solvent is the same or weaker strength than the mobile phase.
	Injection volume/concentration too high	Reduce the injection volumes or dilute samples.

TABLE A.4 (CONTINUED)
First Aid for High-Pressure Liquid Chromatography (HPLC)

Problem	Cause	Solution
Fronting peak	Old column	Replace column.
	Sample solvent incompatible with mobile phase	Choose compatible solvent.
Small or no peaks	No sample	Be sure sample is not deteriorated. Check for bubbles in the vials.
	Detector setup	Check the detector attenuation and rezero.
	Damaged or blocked syringe	Replace the syringe.
	Detector off or malfunction	Check detector. Service detector if required.
Missing peak	Degraded sample	Inject a fresh sample.
	Fluctuations in pH	Buffer the mobile phase so that retention of ionizable compounds is controlled.
	Immiscible mobile phase	Check that any solvent already in the column is miscible with the mobile phase.
Ghost peak	Degraded sample	Inject a fresh sample.
	Contaminated column	Wash the column using an appropriate solvent. If this does not resolve the problem, replace the column.
	Old guard cartridge or column	Replace the guard cartridge or column.
	Immiscible mobile phase	Check that any solvent already in the column is miscible with the mobile phase.
	Fluctuations in pH	Buffer the mobile phase so that retention of ionizable compounds is controlled.
Varying retention times	Leak in the system	Leak can occur within the pump, injector, and detector. Check for and replace any leaking tubing, fittings, sealings, and gaskets. Check correct syringe size.
	Change in mobile phase composition	Use freshly prepared mobile phase.
	Air trapped in pump	Prime pump. Check and change seals. Be sure the mobile phase is degassed.
	Sample solvent incompatible with mobile phase	Choose compatible solvent.
	Temperature fluctuations	Stabilize column temperature. Use column oven.
	Detector overloaded	Dilute sample.

TABLE A.5
First Aid for Gas Chromatography (GC)

Problem	Cause	Solution
Baseline noise	Column contaminated	Condition or replace the column.
	Contaminated gas supply	Check the gas purity and installed gas filters.
	Contaminated detector	Clean the detector according to the manufacturer's instructions.
	Detector gas flow too low or too high	Check the detector gas flow.
	Detector temperature higher than column maximum temperature	Reduce the detector temperature accordingly.
	Loose column fittings	Tighten fittings.

(Continued)

TABLE A.5 (CONTINUED)
First Aid for Gas Chromatography (GC)

Problem	Cause	Solution
	Excessive column bleeding	Check connections on the carrier gas line.
	Oxygen contamination is decomposing the stationary phase	Use gas purity with lowest oxygen content and install oxygen filters in gas line. Check system for leaks.
Baseline spiking	Column too close to flame (when using an FID)	Lower the column to the correct position according to the manufacturer's instructions.
	Contaminated detector	Clean the detector according to the manufacturer's instructions.
	FID temperature too low (when using an FID)	Increase the FID temperature to at least 150°C.
Baseline drift	Accumulation of stationary phase	Remove the end (2–5 cm) of the column.
	Contaminated column	Condition column. Check for correct gas purity. Check gas filters and replace if required.
	Carrier gas cylinder pressure too low to allow control	Replace the carrier gas cylinder. Increase the pressure.
	Drifting carrier gas or combustion gas flows	Check the gas controllers.
Baseline falling	Carrier gas leak in the system	Perform leak test. Check the tightness of the connections on the carrier gas line.
	Column temperature not stable	Allow column temperature to stabilize.
	Inadequate purge flow rate	Increase the purge flow rate.
	Purge valve left closed for too long	Shorten the purge time.
	Solvent tail peak	Increase the solvent delay. Shorten the purge time.
Baseline rising	Contaminated column	Condition column. Check for correct gas purity. Check gas filters and replace if required.
	Contaminated detector	Clean the detector according to the manufacturer's instructions.
	Bleeding column	Condition or replace the column.
	Air leaking into system	Perform leak test. Check the tightness of the connections on the carrier gas line.
Baseline high-standing current	Flow rate of carrier gas too high	Reduce flow rate.
	Contaminated column	Condition column. Check for correct gas purity. Check gas filters and replace if required.
	Excessive column bleeding	Ensure oven temperature does not exceed maximum column temperature. Condition or replace column.
	Loose column fittings	Tighten fittings.
Broad peaks	Column flow too high or low	Reduce or increase column flow slightly above optimum.
	Split flow too low in split injection	Increase the flow to 40–50 ml min^{-1}.
	The sample is overloading the column.	Reduce the amount or concentration of the sample.
	Dirty injector	Clean or replace liner.
	Bad column performance	Remove the last two coils from the column. Condition or replace column.

TABLE A.5 (CONTINUED)
First Aid for Gas Chromatography (GC)

Problem	Cause	Solution
Double peaks	Injection too slow	Increase injection speed.
Peak tailing	Column/oven temperature too low	Increase the column/oven temperature. Do not exceed the recommended maximum temperature for the stationary phase.
	Inlet temperature too low	Increase the inlet temperature.
	Dirty injector	Clean or replace liner.
	Bad column performance	Condition or replace column.
	Poor or obstructed column connections	Remake the column inlet connection.
	Wrong stationary phase	Replace the column.
Peak fronting	Column or detector overloaded	Decrease the injected amount. Increase the split ratio.
	Column/oven temperature too low	Increase the column/oven temperature. Do not exceed the recommended maximum temperature for the stationary phase.
	Stationary phase too thin	Use a thicker-film column.
	Poor injection technique	Optimize injection parameter.
No peaks	No sample	Be sure sample is not deteriorated. Check for bubbles in the vials.
	Detector off or malfunction/contaminated	Check detector. Clean and service detector if required.
	Clogged syringe needle	Replace the syringe.
	Column broken or disconnected	Check column and replace if required.
	Carrier gas flow too high	Reduce the carrier gas flow rate.
	Combustion gas flow incorrectly set (FID)	Check and increase/decrease combustion gas flow.
	Incorrect column position in injector	Check the column position.
Ghost peaks	Degraded or contaminated sample	Inject a fresh sample.
	Contaminated column	Wash the column using an appropriate solvent. If this does not resolve the problem, replace the column.
	Contaminated carrier gas	Replace gas cylinder and gas filter.
	Contaminated injector	Remove column and take injector out. Replace liner and septum.
	Incomplete elution of previous sample	Increase the final oven program temperature or total runtime. Increase the column flow rate.
Unresolved peaks	Carrier gas flow rate too high	Reduce the carrier gas flow rate.
	Bad column performance	Condition or replace column.
	Column temperature too high	Lower the column oven temperature.
	Column too short	Use a longer column.
	Incorrect column choice	Install suitable column.
	Poor injection technique	Optimize injection parameter.

TABLE A.6
First Aid for High-Temperature Catalytic Oxidation (HTCO) Analyzer

Problem	Cause	Solution
Poor repeatability	Injection volume too high	Decrease injection volume.
	Carrier gas flow rate too low	Adjust flow rate.
	Bubbles in injection syringe	Remove bubbles and prime syringe.
	Injection syringe leaking	Replace syringe.
	Port valve leaking	Check and tighten connections if required. Check for clogged pathways and clean valve if required.
	Port valve not aligned with correct flow line	Realign.
	Deteriorated IC solution	Use fresh IC solution.
	Inefficient/contaminated catalyst (repeatability poor for TC measurements)	Clean/condition catalyst according to manufacturer. Replace catalyst if required.
Baseline noisy	Incorrect carrier gas flow	Adjust carrier gas to correct flow rate, typically 150 ml min^{-1}.
	Zero point of detector deviated	Set zero point.
	No solution in IC reaction vessel	Check level in IC reagent vessel. Use fresh IC reagent solution.
	No bubbles in IC reaction vessel	Check for disconnected flow line within system. Check for clogged valves and tubing. Tighten connections and caps.
	No liquid in drain pot	Fill drain pot.
	Bubbles in drain pot	Check for clogs in dehumidifier, CO_2 absorber, halogen scrubber, and membrane filter.
	Contaminated catalyst	Clean/condition catalyst according to manufacturer. Replace catalyst if required.
	Malfunction of dehumidifier	Replace dehumidifier.
No or irregular peaks	Incorrect carrier gas flow	Adjust carrier gas to correct flow rate, typically 150 ml min^{-1}.
	No solution in IC reaction vessel	Check level in IC reagent vessel. Use fresh IC reagent solution.
	No bubbles in IC reaction vessel	Check for disconnected flow line within system. Check for clogged valves and tubing. Tighten connections and caps.
	No liquid in drain pot	Fill drain pot.
	Bubbles in drain pot	Check for clogs in dehumidifier, CO_2 absorber, halogen scrubber, and membrane filter.
	Injection syringe leaking	Replace syringe.
	Port valve leaking	Check and tighten connections if required. Check for clogged pathways and clean valve if required.
	Port valve not aligned	Realign.
	Leak or clog in the TC/IC lines or needles	Replace/tighten lines or needles.
	Incorrect filling height of catalyst	Refill and condition catalyst.
	Inefficient/contaminated catalyst	Clean/condition catalyst according to manufacturer. Replace catalyst if required.
	Particles in NDIR flow cell or gold coating dull or eroded	Replace flow cell.

TABLE A.7
First Aid for Inductively Coupled Plasma-Mass Spectrometer (ICP-MS)

Problem	Cause	Solution
Poor repeatability	Worn peristaltic pump tube or improper tension	Check tension of tubing. Replace pump tubing.
	Nebulizer partially clogged	Clean nebulizer according to the manufacturer's instructions.
	Read time too short	Increase read time.
	Contaminated system	Clean individual components according to the manufacturer's instructions.
Noisy background signal	High power setting	Reduce power setting.
	High vacuum pressure	Check vacuum for leaks in system.
	Bad detector performance	Adjust detector parameters according to the manufacturer's instructions.
	Shiny lenses	Clean lenses according to the manufacturer's instructions.
High oxides	High nebulizer argon flow	Decrease argon flow.
	Dirty cones	Clean cones according to the manufacturer's instructions.
High doubled charged ions	Contaminated solutions	Use fresh solution.
	Leaky or worn interface cones, o-rings, and seating of cones	Check interface cones and seating. Replace o-rings if required.
	High power setting	Reduce power setting.
Drift	Contaminated sample introduction system	Clean the following components according to the manufacturer's instructions: nebulizer, injector, cones, lenses. Replace worn pump tubing.
Plasma goes out	Torch out of alignment	Check position of torch and realign if necessary.
	Spray chamber contains liquid	Drain spray chamber completely.
Plasma does not ignite	Torch out of alignment	Check position of torch and realign if required.
	Bad quality of argon gas	Use new gas container of argon.
	Connection of argon gas line loose, causing leak	Check and tighten connector of gas line.
	Spray chamber contains liquid	Drain spray chamber completely.
	Dirty or cracked torch	Replace torch.
Vacuum shutdown after plasma ignition	Problems at interface	Check cones and o-ring. Replace if necessary.
	Torch out of alignment	Check position of torch and realign if necessary.
Vacuum lowers while running	Orifices of cones clogged	Check and clean cones.
Partial or total loss of ions	Wrong tuning parameter	Reset to last good tuning parameters
	Failure of nebulizer	Check and clean nebulizer. Confirm nebulization visually by disconnecting from spray chamber.
	Failure of injection	Check for blocked or melted nebulizer tip.
	Status failure of guard electrode	Check status of guard electrode.
	Cones blocked	Check and clean cones.
	Status failure of magnets	Check status of magnets.
	Accelerating voltage failed	Verify voltage reading and compare with last good one.
	Lens voltage failed	Verify voltage reading and compare with last good one.
	Failure of mass calibration	Perform a scan to identify shift in peak positions.
	Failure of slits	Check and replace slits if necessary.
	Inappropriate SEM voltage	Verify voltage reading and compare with last good one.

TABLE A.8
First Aid for Graphite Furnace Atomic Absorption Spectroscopy (GFAAS)

Problem	Cause	Solution
Furnace does not heat	Flow rate of cooling water too low	Adjust the flow rate to manufacturer'srecommendation. Check for deposition/obstruction in water line.
	Flow rate of purge gas (typically argon) below minimum level	Check flow rate and pressure of argon tank.
	Bad physical contact between graphite tube and electrodes	Check seating of tube. Check for cracked tube. Clean contact area.
Low or no intensity of lamp source	Faulty hollow cathode lamp	Replace lamp.
	Dirty furnace windows	Clean furnace windows.
Temperature is higher than setting	Dirty quartz window protecting temperature sensor	Clean quartz window.
Graphite tubes wearing out quickly	Oxygen present in furnace chamber	Check proper closing of chamber door. Use argon with lowest oxygen content.
No signal	Faulty autosampler	Check operation of autosampler and proper sample delivery.
	Faulty hollow cathode lamp	Replace lamp.
	Graphite tube damaged	Check tube and replace if necessary.
	Pyrolysis temperature too high	Decrease pyrolysis temperature.
	Atomization temperature too low	Increase atomization temperature.
Double peaks	Atomization occurs from more than one surface, e.g., autosampler delivers sample to platform and wall	Check autosampler operation and proper sample delivery.
Memory effects	Retention of analyte between two furnace heatings	Include cleaning step or dry atomization cycle subsequent to sample atomization.
	Contaminated graphite electrodes	Replace electrodes.
Low signal or poor precision	Faulty autosampler	Check operation of autosampler and proper sample delivery.
	Faulty hollow cathode lamp	Replace lamp.
	Graphite tube damaged	Check tube and replace if necessary.
	Pyrolysis temperature too high	Decrease pyrolysis temperature.
	Atomization temperature too low	Increase atomization temperature.
	Improper drying of sample— temperature too high	Decrease temperature and reduce heating rate for drying of sample.
	Internal flow of purge gas during atomization step	Ensure purge gas is off during atomization step.

TABLE A.9
First Aid for Polarography/Voltammetry

Problem	Cause	Solution
No current response	Electrodes not immersed in solution	Ensure that all electrodes are fully immersed.
	Broken wire or connector	Check for broken wires and good connectors.
	Internal thin wire of microelectrode loose or broken	Change electrode.
	No obvious cause	Replace each electrode in turn to identify faulty electrode. Replace each lead, connection, and equipment in turn to identify problem.
Potential shift	Faulty reference electrode.	See below.
Faulty reference electrode (RE)	Loose connector/wire on RE	Check proper connection to RE.
	RE stored in wrong solution	Use freshly made RE. Alternatively, replace solution and allow to equilibrate for a few hours.
	Diaphragm is blocked.	Use freshly made RE.
	Air bubble in the RE	Shake RE to remove air bubble.
Faulty working electrode (WE) (WE is hanging mercury drop electrode [HMDE])	WE does not drop	Check if WE is connected.
		Check inert gas pressure.
		Check mercury level in reservoir.
	WE drops irregularly or continually	Check mercury level in reservoir (do not overfill).
		Rub WE dry, clean with ultrapure water or dilute acid, dry with lint-free cloth.
		Check inert gas pressure.
		Use mercury of highest purity. Change capillary.
Faulty auxiliary electrode (AE)	Loose connector/wire on AE	Check for proper connection.
Low background current or unstable baseline	Drop at HMDE does not remain hanging	See faulty working electrode.
	Interference from oxygen in sample	Degas sample with inert gas.
	Ion concentration in sample too high	Dilute sample.
	Wrong electrolyte and pH of sample	Adjust electrolyte concentration and pH.
	Old electrolyte solution	Use fresh electrolyte solution.
	For dropping mercury electrode (DME): electrode drops irregularly	Clean or change capillary.
		Use high-purity mercury.
		Check level of mercury.
	For rotating disk electrode (RDE): bad electrode surface	Repolish surface with aluminum oxide.
Peak displacement	Wrong electrolyte and pH of sample	Adjust electrolyte composition/concentration and pH.
	Interference from organic components in sample	Carry out UV digestion of sample prior to analysis.
	Old electrolyte solution	Use fresh electrolyte solution.
	Faulty RE	Check RE (see above).

(Continued)

TABLE A.9 (CONTINUED)
First Aid for Polarography/Voltammetry

Problem	Cause	Solution
No peak	Peak is displaced	Adjust half-wave potential and recalculate.
	Incorrect initial and final potential	Adjust potential.
	Sample concentration too small	Increase sample volume.
	Deposition time too short	Increase deposition time.
	No mercury drops	Check WE for fault (see above).
	Interference from organic components in sample	Carry out UV digestion of sample prior to analysis.
	Old electrolyte solution	Use fresh electrolyte solution.
	For rotating disk electrode (RDE): bad electrode surface	Repolish surface with aluminum oxide.
Double peaks	Interference from organic components in sample	Carry out UV digestion of sample prior to analysis.
	Old electrolyte solution	Use fresh electrolyte solution.
	Second element present at same potential	Confirm by adding second element and observing increase in peak height. Second element may be masked with a suitable complexing agent.
	For RDE: bad electrode surface	Repolish surface with aluminum oxide.
Peak at highest μA range	Sample volume too large	Reduce sample volume.
	For HMDE/RDE: deposition time too long	Reduce deposition time.
	For RDE: bad electrode surface	Repolish surface with aluminum oxide.

FURTHER READING

Dean, R. 1999. *A practical guide to the care, maintenance and troubleshooting of capillary gas chromatographic systems*. Weinheim, Germany: Wiley-VCH.

Dolan, J. W., and L. R. Synder. 1989. *Troubleshooting LC systems: A comprehensive approach to troubleshooting LC equipment and separations*. Totowa, NJ: Humana Press.

Furman, W. 1976. *Continuous flow analysis: Theory and practice*. New York: Marcel Dekker.

Gaines, P. 2005. Sample introduction for ICP-MS and ICP-OES. *Spectroscopy* 20, January.

Karlberg, B., and G. E. Pacey. 1989. *Flow injection analysis (techniques and instrumentation in analytical chemistry)*. Amsterdam, The Netherlands: Elsevier.

Patton, C. J., and A. P. Wade. 1990. Continuous-flow analyzer. In *Analytical instrumentation handbook*, ed. G. W. Ewing, 2nd ed. New York: Marcel Dekker.

Ružicka, J., and E. Hansen. 1988. *Flow injection analysis*. New York: John Wiley & Sons.

Appendix B: Chemical Compatibilities and Physical Properties of Various Materials

TABLE B.1
Chemical Compatibilities of Various Materials to Common Cleaning Solutions

Chemical	Acetal (Delrin)	Buna N (Nitrile)	Epoxy	FEP (Teflon®)	HDPE	LDPE	Neoprene	Nylon	PCTFE (Kel-F)	Polycarbonate	PPS (Ryton)	PTFE (Teflon®)	PVC	PVDF (Kynar®)	Silicone	Stainless Steel	Tygon	Viton
Acids, weak or dilute	L	R	R	R	R	R	R	L	R	R	L	R	R	R	L	N	R	R
Acids, strong or concentrated	N	N	N	R	R	R	N	N	R	N	L	R	R	R	N	N	L	R
Bases	R	R	R	R	R	R	R	L	R	N	R	R	R	R	L	R	L	R
Oxidizing agents	N	N	R	R	L	L	L	N	R	N	R	R	R	R	L	R	R	R
Alcohols	R	L	R	R	R	R	R	N	R	R	R	R	R	R	R	R	L	L
Aldehydes	R	N	L	R	R	R	N	L	R	L	R	R	N	R	N	R	N	L
Esters	L	N	R	R	R	R	N	R	R	N	R	R	N	R	N	R	N	N
Ketones	N	N	L	R	R	R	N	R	R	N	R	R	N	N	N	R	N	N
Hydrocarbons, aliphatic	L	L	R	R	R	L	N	R	R	L	R	R	R	R	N	R	N	R
Hydrocarbons, aromatic	R	N	R	R	R	L	N	R	R	N	R	R	N	R	N	L	N	R
Hydrocarbons, halogenated	L	N	L	R	L	N	N	R	R	N	R	R	N	N	N	R	N	R

Note: R, recommended (no or minor effects after 48 hours of exposure); L, limited resistance (moderate effect after 48 hours of exposure); N, not recommended (severe effects within short period of exposure).

This table is to be used only as a guideline for the chemical resistance of various materials. Variations in chemical behavior and effects may occur due to different conditions, such as temperature, pressure, concentration, and composition of materials. It is strongly recommended to test the equipment/material with the chemicals with caution before permanent installation.

TABLE B.2

Chemical Compatibilities of Various Filter Materials to Common Cleaning Solutions

Chemical	Glass Fiber	Polycarbonate	Polypropylene	Polyethersulfone	Nylon	Mixed-Cellulose Esters	Cellulose Acetate	Cellulose Nitrate	Teflon, Hydrophilic	Teflon, Hydrophobic	Silver	Alumina
Hydrochloric acid, 6 mol L^{-1}	R	R	R	R	N	N	L	L	R	R	R	R
Hydrochloric acid, concentrated	R	R	R	R	N	N	N	N	R	R	R	—
Nitric acid, 6 mol L^{-1}	R	R	R	N	N	R	L	L	R	R	N	R
Nitric acid, concentrated	R	R	R	N	N	N	N	N	R	R	N	—
Sulfuric acid, 3 mol L^{-1}	—	R	R	N	N	R	L	L	R	R	N	—
Sulfuric acid, concentrated	—	N	R	N	N	N	N	N	R	R	N	—
Ammonium hydroxide, 6 mol L^{-1}	—	N	R	R	N	N	N	N	R	R	R	R
Potassium hydroxide, 6 mol L^{-1}	R	N	—	—	L	N	N	N	R	R	R	R
Sodium hydroxide, 6 mol L^{-1}	L	N	R	R	L	N	N	N	R	R	R	R
Isopropanol	R	R	R	R	R	L	R	R	R	R	R	R
Propanol	R	R	—	—	R	R	R	R	R	R	R	R
Isobutyl alcohol	R	R	—	—	R	R	R	—	R	R	R	R
Ethyl alcohol	R	R	—	R	R	R	R	—	R	R	R	R
Acetone	R	L	R	N	R	N	N	N	R	R	R	R
Hexane	R	R	L	—	R	R	R	R	R	R	R	R
Pentane	R	R	—	N	R	R	R	—	R	R	R	R
Methylene chloride	R	N	L	N	L	N	N	N	R	R	R	R
Toluene	R	L	L	N	R	R	L	N	R	R	R	R

Note: R, recommended; L, limited resistance; N, not recommended.

This table is to be used only as a guideline for the chemical resistance of various filter materials. Variations in chemical behavior and effects may occur due to different conditions, such as temperature, pressure, concentration, and composition of filter materials. It is strongly recommended to test the equipment/filter material with the chemicals with caution before regular application.

TABLE B.3
Physical Properties of Various Materials

	Maximum Use Temperature (°C)	Minimum Use Temperature (°C)	Autoclavable	Flexibility	Transparency	Permeability (cm^3 mm/m^2 24 h bar)		
						N_2	O_2	CO_2
Acetal (Delrin)	90	−50	Yes	Rigid	Opaque	—	—	—
Buna N (Nitrile)	120	−40	Yes	Moderate	Opaque	—	—	—
Epoxy	205	−50	Yes	Rigid	Opaque/transparent	—	—	—
FEP	205	−270	Yes	Flexible	Translucent	4,960	11,625	34,100
HDPE	120	−100	No	Rigid	Translucent	651	2,868	8,990
LDPE	80	−100	No	Flexible	Translucent	2,790	7,750	41,850
Neoprene	110	−40	No	Flexible	Opaque	—	—	—
Nylon	90	0	No	Rigid	Translucent	—	—	—
PCTFE (Kel-F)	200	−240	Yes	Flexible	Translucent	—	—	—
Polycarbonate	135	−135	Yes	Rigid	Translucent	755	4,650	16,663
Polypropylene	135	0	Yes	Rigid	Translucent	744	3,720	12,400
PPS (Ryton)	230	—	Yes	Flexible/rigid	Opaque/translucent	—	—	—
PTFE	260	−267	Yes	Rigid	Translucent	—	—	—
PVC	70	−30	No	Rigid	Transparent	30–310	62	62
PVDF	110	−62	No	Rigid	Translucent	140	217	7,828
Silicone	170	−65	Yes	Flexible	Opaque/transparent	—	—	—
Tygon	70	−50	No	Flexible	Transparent	—	—	—
Viton	200	−30	Yes	Flexible	Opaque/transparent	—	—	—
Soda-lime glass	400	−50	Yes	Rigid	Transparent	—	—	—
Borosilicate glass	450	−70	Yes	Rigid	Transparent	—	—	—
Stainless steel	1,200	−200	Yes	Excel	Opaque	—	—	—
Ceramic	>1,150	—	Yes	Rigid	Opaque	—	—	—

This table is to be used only as a guideline for the properties of various materials. Variations may occur due to different conditions, such as temperature, pressure, and composition of the materials. It is strongly recommended to test the equipment/filter material under conditions with caution before regular application.

Appendix C: Water Purification Technologies

TABLE C.1
Water Purification Technologies and Their Applications to Remove Various Water Constituents

	Ions	Dissolved Organic Matter	Particulates	Bacteria	Pyrogens
Deionization	Very good	Poor	Poor	Poor	Poor
Distillation	Good	Poor	Very good	Very good	Very good
Filtration	Poor	Poor	Very good	Good	Poor
Ultrafiltration (<0.0005 mm)	Poor	Good	Very good	Very good	Very good
Reverse osmosis	Good	Good	Very good	Very good	Very good
Adsorption	Poor	Very good	Poor	Poor	Poor
UV oxidation (<280 mm)	Poor	Very good	Poor	Good	Poor

Index

17β-estradiol, f356

A

Accuracy, 5, 10
  documentation of, 12
Accelerated solvent extraction, 333, 336,
    347
Acid fuming method, 39, 111
Actinium isotope (^{227}Ac)
  analytical procedure, 275
  electrochemical plating procedure, 275
  common problems, 276
  quality assurance, 276
Adsorbent cartridges, *see* also Water sampler
  chelating resins, 28
  polyurethane foam, 29
  XAD-2, 29
Adsorption chromatography, 337
Alcian Blue, 128
  staining solution, preparation, 130
Americium isotopes,
  analytical methodologies, 290
  analytical procedure, 286f, 291
  common problems, 293
  digestion procedure, 291
  electrodeposition, procedure, 292
  filtration procedure, 291
  gamma spectrometry, measurements with,
    292
  quality assurance, 292
Amino acids, determination of, 67, *see also* Fractionation
    of organic compounds
  analysis as amines, 68
  blanks, 68, 69, 72
  calibration, 74, 75t
  contamination sources, 68
  cromatographic analysis,
    gas chromatography, 68
    high-performance liquid cromatography,
      68
    isotope-ratio mass spectrometer, 68, 113
  derivatization with isopropyl-TFA, 113
  dissolved combined amino acids, 68
  dissolved free amino acids, 68, 70
  enantiomeric amino acid, analysis of, see Enantiomeric
    amino acid analysis
  enantiomeric amino acid, hydrolysis of, 70
  filtration procedure, 69
    materials and reagents, 70t
  Fourier transform ion cyclotron resonance mass
    spectrometry, 75
  high resolution multidimensional nuclear magnetic
    resonance spectrometry, 75

hydrolysis,
  acidic hydrolysis, 70, 113
  alkaline hydrolysis, 70
  liquid-phase hydrolysis,
    procedure of, 70, 71
  materials and reagents, 71t
  vapor-phase hydrolysis, 70
  hydrolyzable amino acids, 70, 180
  isotopic analysis, 113
  particular amino acids, 69
  preservation procedure, 69
    materials and reagents, 70t
  sampling procedure, 68, 69
    materials and reagents, 70t
  simplified analysis (non-enantiomeric approach),
    chromatographic conditions, 75
    derivatization reagents, 75
    limit of detection, 75
    precision, 75
  ultra-high performance liquid chromatography, 76
Ammonium,
  adsorption effects, 168
  baseline stability, improved, 167
  calibration, 169, 169f
  contamination, airborne, 167
  description, method, 166
  detection limit, 166, 169
  detector, settings of, 167
  hydraulic optimization, 166
  interferences, 166
  linear range, 166
  manifold, 166, 167f
  operating protocol, 168
  performance, method, 169
  preparation of calibration standards, 168
  preparation of reagents, 167, 168
  preparation of working standards, 168
  repeatability, 169
  reproducibility, 169
  salt effect, 168
  sample dilution, automated, 167
  sampling rate, 167
Atomic force microscopy (AFM), 129
AutoAnalyzer, 145, 146, 181, 186, *see also* Segmented
    flow analysis

B

Beryllium isotope (^{7}Be),
  analytical procedure,
    dissolved ^{7}Be, 282
    particulate ^{7}Be, 282
  common problems, 283
  quality assurance, 283

Milton Keynes UK
Ingram Content Group UK Ltd.
UKHW050455071024
449327UK00015B/386